Lecture Notes in Computer Science 8325

Commenced Publication in 1973
Founding and Former Series Editors:
Gerhard Goos, Juris Hartmanis, and Jan van Leeuwen

T0190053

Cathal Gurrin   Frank Hopfgartner
Wolfgang Hurst   Håvard Johansen
Hyowon Lee   Noel O'Connor (Eds.)

MultiMedia Modeling

20th Anniversary International Conference, MMM 2014
Dublin, Ireland, January 6-10, 2014
Proceedings, Part I

 Springer

Volume Editors

Cathal Gurrin
Dublin City University, Ireland
E-mail: cgurrin@computing.dcu.ie

Frank Hopfgartner
Technische Universität Berlin / DAI-Labor, Germany
E-mail: frank.hopfgartner@dai-labor.de

Wolfgang Hurst
Universiteit Utrecht, The Netherlands
E-mail: huerst@uu.nl

Håvard Johansen
UiT The Arctic University of Norway
E-mail: haavardj@cs.uit.no

Hyowon Lee
Singapore University of Technology and Design, Singapore
E-mail: hlee@sutd.edu.sg

Noel O'Connor
Dublin City University, Ireland
E-mail: oconnor2n@eeng.dcu.ie

ISSN 0302-9743   e-ISSN 1611-3349
ISBN 978-3-319-04113-1   e-ISBN 978-3-319-04114-8
DOI 10.1007/978-3-319-04114-8
Springer Cham Heidelberg New York Dordrecht London

Library of Congress Control Number: 2013955783
CR Subject Classification (1998): H.3, H.5, I.5, H.2.8, H.4, I.4, I.2
LNCS Sublibrary: SL 3 – Information Systems and Application,
incl. Internet/Web and HCI

Typesetting: Camera-ready by author, data conversion by Scientific Publishing Services, Chennai, India

Printed on acid-free paper

Springer is part of Springer Science+Business Media (www.springer.com)

Preface

These proceedings contain the papers presented at MMM 2014, the 20th Anniversary International Conference on MultiMedia Modeling. The conference was organized by Inisght Centre for Data Analytics, Dublin City University, and was held during January 6-10, 2014, at the wonderful venue of the Guinness Storehouse in Dublin, Ireland. We greeted the attendees at MMM 2014 with the following address: "Táimid an-bhrodúil fáilte a chur romhaibh chuig Baile Átha Cliath agus chuig an fichiú Comhdháil Idirnáisiúnta bliain ar Samhaltú Ilmheán. Tá súil againn go mbeidh am iontach agaibh anseo in Éirinn agus go mbeidh bhur gcuairt taitneamhnach agus sásúil. Táimid an-bhrodúil go háirithe fáilte a chur roimh na daoine ón oiread sin tíortha difriúla agus na daoine a tháinig as i bhfad i gcéin. Tá an oiread sin páipéar curtha isteach chuigh an chomhdháil seo go bhfuil caighdeán na bpáipéar, na bpóstaer agus na léiriń an-ard ar fad agus táimid ag súil go mór le hócaid iontach. We are delighted to welcome you to Dublin for the 20th Anniversary International Conference on Multimedia Modeling. We hope that the attendees have a wonderful stay in Ireland and that their visits are both enjoyable and rewarding. We are very proud to welcome visitors from both Ireland and abroad and we are delighted to be able to include in the proceedings such high-quality papers, posters, and demonstrations."

MMM 2014 received a total 176 submissions across four categories; 103 full-paper submissions, 24 short paper submissions, and 12 demonstration submissions. Of these submissions, 55% were from Europe, 41% from Asia, 3% from the Americas, and 1% from the Middle East. All full paper submissions were reviewed by at least three members of the 110-person Program Committee, for whom we owe a debt of gratitude for providing their valuable time to MMM 2014. Of the 103 full papers submitted, 30 were selected for oral presentation, which equates to a 29% acceptance rate. A further 16 papers were chosen for poster presentation. For short papers, a total of 11 were accepted for poster presentation, representing a 45% acceptance rate. In addition, nine demonstrations from a total of 12 submissions were accepted for MMM 2014. We accepted 28 special session submissions across the five special sessions and six Video Browser Showdown (VBS 2014) submissions. The accepted contributions represent the state of the art in multimedia modeling research and cover a diverse range of topics including: applications of multimedia modelling, interactive retrieval, image and video collections, 3D and augmented reality, temporal analysis of multimedia content, compression and streaming.

As in recent years, MMM 2014 included VBS 2014. This year we made the VBS a half-day workshop, which took place on January 7, 2014. For the first time in 2014, we also co-located the WinterSchool on Multimedia Processing and Applications (WMPA 2014), which ran during January 6-7, 2014.

As is usual for MMM, there were a number of special sessions accepted for inclusion in MMM 2014. Each special session paper was also reviewed by at least

International Liaisons

USA: Alex Hauptmann   Carnegie Mellon University, USA
Europe: Susanne Boll   University of Oldenburg, Germany
Asia: Jialie Shen   Singapore Management University, Singapore

Local Organizing Co-chairs

Rami Albatal   Dublin City University, Ireland
Lijuan Zhou   Dublin City University, Ireland

Website

Yang Yang   Dublin City University, Ireland
David Scott   Dublin City University, Ireland

Program Committee

Amin Ahmadi   Dublin City University, Ireland
Rami Albatal   Dublin City University, Ireland
Laurent Amsaleg   CNRS-IRISA, France
Noboru Babaguchi   Osaka University, Japan
Jenny Benois-Pineau   LABRI/University of Bordeaux, France
Laszlo Boeszoermenyi   Klagenfurt University, Austria
Susanne Boll   University of Oldenburg, Germany
Vincent Charvillat   University of Toulouse, France
Gene Cheung   National Institute of Informatics, Japan
Liang-Tien Chia   Nanyang Technological University, Singapore
Insook Choi   Columbia College Chicago, USA
Konstantinos Chorianopoulos Ionian University, Greece
Wei-Ta Chu   National Chung Cheng University, Taiwan
Tat-Seng Chua   National University of Singapore, Singapore
Kathy M. Clawson   University of Ulster, UK
Matthew Cooper   FX Palo Alto Laboratory, USA
W. Bas de Haas   Utrecht University, The Netherlands
Francois Destelle   Dublin City University, Ireland
Cem Direkoglu   Dublin City University, Ireland
Ajay Divakaran   SRI International, USA
Lingyu Duan   Peking University, China
Stéphane Dupont   University of Mons, Belgium
Thierry Dutoit   University of Mons, Belgium
Maria Eskevich   Dublin City University, Ireland
Jianping Fan   University of North Carolina, USA
Gerald Friedland   ICSI Berkeley, USA
Yue Gao   National University of Singapore, Singapore

Lyndon Nixon	STI International GmbH, Austria
Noel O'Connor	Dublin City University, Ireland
Neil O'Hare	Yahoo Labs, Spain
Vincent Oria	New Jersey Institute of Technology, USA
Marco Paleari	Italian Institute of Technology, Italy
Fernando Pereira	Instituto Superior Técnico - Instituto de Telecomunicações, Portugal
Miriam Redi	Eurecom, France
Mukesh Kumar Saini	University of Ottawa, Canada
Jitao Sang	Institute of Automation, Chinese Academy of Sciences, China
Shin'ichi Satoh	National Institute of Informatics, Japan
Klaus Schoeffmann	Klagenfurt University, Austria
David Scott	Dublin City University, Ireland
Caifeng Shan	Philips, The Netherlands
Jialie Shen	Singapore Management University, Singapore
Koichi Shinoda	Tokyo Institute of Technology, Japan
Mei-Ling Shyu	University of Miami, USA
Alan Smeaton	Dublin City University, Ireland
Jia Su	Nippon Telegraph and Telephone Corporation, Japan
Yongqing Sun	NTT Media Intelligence Laboratories, Japan
Robby Tan	Utrecht University, The Netherlands
Shuhei Tarashima	NTT Media Intelligence Laboratories, Japan
Xinmei Tian	University of Science and Technology of China, China
Dian Tjondronegoro	Queensland University of Technology, Australia
Shingo Uchihashi	Fuji Xerox Co., Ltd., Japan
Egon L. van den Broek	Utrecht University, The Netherlands
Nico van der Aa	Utrecht University, The Netherlands
Coert van Gemeren	Utrecht University, The Netherlands
Jingdong Wang	Microsoft Research Asia, China
Xin-Jing Wang	Microsoft Research Asia, China
Lai Kuan Wong	Multimedia University, Malaysia
Marcel Worring	University of Amsterdam, The Netherlands
Feng Wu	Microsoft Research Asia, China
Peng Wu	Hewlett-Packard, USA
Qiang Wu	University of Technology Sydney, Australia
Xiao Wu	Southwest Jiaotong University, China
Changsheng Xu	Institute of Automation, Chinese Academy of Sciences, China
Keiji Yanai	University of Electro-Communications, Japan
You Yang	Huazhong University of Science and Technology, China
Zheng-Jun Zha	Hefei Institute of Intelligent Machines, Chinese Academy of Sciences, China

Cha Zhang   Microsoft Research, USA
Lijuan Zhou   Dublin City University, Ireland
Roger Zimmermann   National University of Singapore, Singapore

Additional Reviewers

Zheng Song   National University of Singapore
Hanwang Zhang   National University of Singapore

Sponsoring Institutions

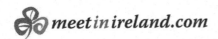

In Cooperation with

Bernauer-Budiman Inc., Reading, Mass.
The Hofmann-International Company, San Louis Obispo, Cal.
Kramer Industries, Heidelberg, Germany

Table of Contents – Part I

Interactive Indexing and Retrieval

Multimedia Collections

Applications

Temporal Analysis

3D and Augmented Reality

Compression, Transcoding and Streaming

Table of Contents – Part II

Special Session: 3D Multimedia Computing and Modeling

Special Session: Social Geo-Media Analytics and Retrieval

Special Session: Multimedia Hyperlinking and Retrieval

Short Papers

Demonstrations

Video Browser Showdown

A Comparative Study on the Use of Multi-label Classification Techniques for Concept-Based Video Indexing and Annotation

Fotini Markatopoulou, Vasileios Mezaris, and Ioannis Kompatsiaris

Information Technologies Institute(ITI), Centre for Research and Technology Hellas
(CERTH), Thermi 57001, Greece
{markatopoulou,bmezaris,ikom}@iti.gr

Abstract. Exploiting concept correlations is a promising way for boosting the performance of concept detection systems, aiming at concept-based video indexing or annotation. Stacking approaches, which can model the correlation information, appear to be the most commonly used techniques to this end. This paper performs a comparative study and proposes an improved way of employing stacked models, by using multi-label classification methods in the last level of the stack. The experimental results on the TRECVID 2011 and 2012 semantic indexing task datasets show the effectiveness of the proposed framework compared to existing works. In addition to this, as part of our comparative study, we investigate whether the evaluation of concept detection results at the level of individual concepts, as is typically the case in the literature, is appropriate for assessing the usefulness of concept detection results in both video indexing applications and in the somewhat different problem of video annotation.

Keywords: Concept detection, concept correlation, stacking, multi-label classification.

1 Introduction

Semantic concept detection in videos, often also referred to as semantic indexing or high-level feature extraction, is the task of assigning one or more labels (semantic concepts) to video sequences, based on a predefined concept list [1]. This process is important for several applications such as video search and retrieval, concept-based annotation and video summarization.

The majority of concept detection systems are based on variations of the following process: Ground-truth annotated videos are segmented into shots, visual features are extracted from each shot, and supervised classifiers are trained separately for each concept. Then, a new, non-annotated video shot can be associated with concept labels by applying the trained concept detectors, to get a set of confidence scores. These scores indicate the belief of each detector that the corresponding concept appears in the shot. Assigning concepts to video shots is

C. Gurrin et al. (Eds.): MMM 2014, Part I, LNCS 8325, pp. 1–12, 2014.
© Springer International Publishing Switzerland 2014

by definition a multi-label classification problem, since multiple concepts may match a single video shot. The process of training each concept detector independently, as described earlier, is known as Binary Relevance (BR) transformation and is the simplest way of solving multi-label learning problems.

In this baseline BR system, any existing semantic relations among concepts are not taken into account (e.g., the fact that *sun* and *sky* will often appear together in the same video shot). Thus, one way of improving the performance of concept detection is to also consider such concept correlations. A group of methods in this category follow a stacking architecture (e.g. [2], [3]). The predictions of multiple BR-trained concept detectors form model vectors that are used as a meta-learning training set for a second learning round. While there is no strict rule for the selection of the meta-learning algorithm, researchers mainly adopt a second round of BR models. In this work we examine the use of elaborate multi-label classification algorithms instead of BR models for the second-layer learning.

In addition to this, a closer look to the way that concept detection is evaluated shows that researchers focus on evaluating it in a concept-based indexing and retrieval setting, i.e. given a concept, measure how well the top retrieved video shots for this concept truly relate to it. However, besides the retrieval problem, another important problem related to video concept detection is the annotation problem, i.e. the problem of estimating which concepts best describe a given video shot. We argue that the retrieval-based evaluation of concept detection results is not sufficient for assessing the goodness of concept detectors in the context of the annotation problem, and we experimentally underline the importance of reporting evaluation results in both these directions.

2   Related Work

Concept correlation refers to the relations among concepts within a video shot. By using this information we can refine the predictions derived from multiple concept detectors in order to improve their accuracy, a process known as Context Based Concept Fusion (CBCF) [3]. Two main types of methods have been adopted in the literature for this: a) Stacking-based approaches that collect the scores produced by a baseline set of concept detectors and introduce a second learning step in order to refine them, b) Inner-learning approaches that follow a single-step learning process, which jointly considers low-level visual features and concept correlation information [1].

In this work we mainly focus on the first category. Stacking approaches aim to detect dependencies among concepts in the last layer of the stack. One popular group is the BR-based stacking approaches. For example, Discriminative Model Fusion (DMF) [2] obtains concept score predictions from the individual (BR-trained) concept detectors in the first layer, in order to create a *model vector* for each shot. These vectors form a meta-level training set, which is used to train a second layer of BR models. Correlation-Based Pruning of Stacked Binary Relevance Models (BSBRM) [4] extends the previous approach by pruning the predictions of non-correlated concept detectors before the training of each individual classifier of the second-layer BR models. Similarly to DMF, the Baseline CBCF (BCBCF) [3] forms model vectors, in this case using the ground truth annotation, in order to train second-layer BR models. Furthermore, the authors of [3] note that not all concepts can take advantage of CBCF, so their method refines only a subset of them. Another group of stacking approaches are the graph-based ones, which model label correlations explicitly [1]. Multi-Cue Fusion (MCF) method [5] uses the ground truth annotation to build decision trees that describe the relations among concepts, separately for each concept. Initial scores are refined by approximating these graphs.

Inner-learning approaches make use of contextual information from the beginning of the concept learning process. For example, the authors of [6] and [7] propose methods that simultaneously learn the relation between visual features and concepts and also the correlations among concepts. In [8] a probabilistic Multi-Label Multi-Instance learning approach is proposed, where the multi-label part models correlations among multiple concepts and the multi-instance part models relations among different image regions. These two parts are combined into a single step in order to develop a complete system that detects multiple concepts in an image. In [9] a combination of a weighted version of kNN and multiple SVM classifiers are used for jointly assessing the semantic similarity between concepts and the visual similarity between images annotated with them. Although inner-learning approaches are out of the scope of this work, they were briefly discussed in this paragraph for the sake of completeness.

In TRECVID 2012 several teams explicitly study label correlations. For example, in [10] the Concept Association Network is used, which is a rule-based system searching for frequent item sets of concepts and extracting association rules. Other systems aim to take advantage of "imply" and "exclude" relations between concepts [11], [12]. However, we did not consider such methods in the present comparative study, because in the TRECVID experiments reported in the aforementioned publications these methods did not exhibit a significant improvement in the goodness of concept detection, compared to the BR baseline.

Label correlation has also been investigated in the broader multi-label learning domain. In [13], multi-label classification methods, including methods that consider contextual relations, are compared on multimedia data. In [9] and [14] such methods are adapted for concept detection. Nevertheless, none of these approaches considers the use of multi-label classification methods as part of a stacking architecture. The latter is the focus of this work, and section 3 describes the way we adapt such methods in order to build models in the second-layer of the stacking architecture that learn the correlations among labels.

3   The Proposed Stacking Architecture

Let $D_1, ..., D_N$ denote a set of N trained concept detectors on N different concepts. Let T denote a validation set of video shots, which will be used for training the second layer of the stacking architecture, and m denote the model vector of a new unlabeled video shot. Figure 1 summarizes the full pipeline from training the second-layer classifiers to using them for classifying an unlabeled sample when using: (1) the BR stacking architecture (Fig. 1(b),(d)), and (2) the proposed stacking architecture (Fig. 1(c),(e)). Both architectures use exactly the same strategy to create the meta-level training set; the trained BR models $(D_1, ..., D_N)$ of the first layer are applied to the validation dataset T and in this way a model vector set M is created, consisting of the scores that each of $D_1, ..., D_N$ has assigned to each video shot of T for every concept (Fig. 1(a)). What distinguishes the two architectures is the way that this meta-learning information is used and therefore the way that the second-layer learning is performed.

During the training phase, the BR stacking architecture builds a new set of BR models $(D'_1, ..., D'_N)$. To train each model, a different subset of M that is ground-truth annotated for the corresponding concept C_n that the meta-concept detector D'_n will be trained for, is used (Fig. 1(b)). In contrast, the proposed architecture uses the whole model vector set and the ground truth annotation at once in order to train a single multi-label classification model D', instead of separate models $D'_1, ..., D'_N$ (Fig. 1(c)).

During the classification phase, a new unlabeled video shot is firstly given to the first layer BR models $(D_1, ..., D_N)$ and a model vector m is returned. On the one hand, the BR stacking architecture will let each of the $D'_1, ..., D'_N$ models to classify m and one score will be returned separately from each model (Fig. 1(d)). On the other hand, the proposed architecture uses the single trained model D' in order to return a final score vector (Fig. 1(e)).

With respect to learning concept correlations, the BR-based stacking methods learn them only by using the meta-level feature space. However, the learning of each concept is still independent of the learning of the rest of the concepts. The rationale behind us proposing the use of multi-label learning algorithms in replacement of the BR models at the second layer of the stacking architecture is based on the assumption that if we choose algorithms that explicitly consider label relationships as part of the second-layer training, improved detection can be achieved. Our stacking architecture learns concept correlations in the last layer of the stack both from the outputs of first-layer concept detectors and by modelling correlations directly from the ground-truth annotation of the meta-level training set. This is achieved by instantiating our architecture in our experiments with different second-layer algorithms that model:

- Correlations between pairs of concepts;
- Correlations among sets of more than two concepts;
- Multiple correlations in the neighbourhood of each testing instance.

To model the correlation information described above we exploit methods from the multi-label learning field [15]. Pairwise methods can consider pairwise relations among labels; similar to the multi-class problem, one versus one models are trained and a voting strategy is adopted in order to decide for the final classification. In this category we choose the Calibrated Label Ranking (CLR) algorithm [16] that combines pairwise and BR learning. Label power set (LP) methods search for subsets of labels that appear together in the training set

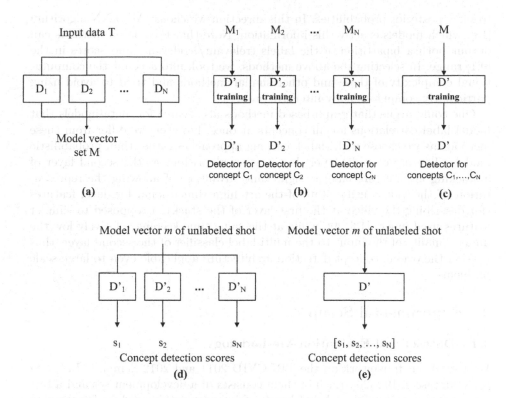

Fig. 1. Comparing BR and the proposed stacking architecture. (a) First layer of a stacking architecture. Video shot set T is given to trained concept detectors $D_1, ..., D_N$, and a model vector set M consisting of the responses of the detectors for each video shot of T is returned. (b) Training of the second layer of a BR-stacking architecture. During the training phase, BR-stacking builds a second set of concept detectors $(D'_1, ..., D'_N)$ separately for each concept, using for training each of $D'_1, ..., D'_N$ a different subset of M according to the availability of ground-truth annotations for each concept. (c) Training of the second layer of the proposed stacking architecture. The proposed architecture uses both the complete model vector set M and the ground truth annotations in order to build a single multi-label model D'. (d)&(e) During the classification phase, a new unlabeled video shot is firstly given to the first layer BR models $(D_1, ..., D_N)$ and a model vector m is returned, to be used as input to the second layer classifiers. (d) The BR stacking architecture will let each of the $D'_1, ..., D'_N$ models to classify m and one score will be returned separately from each model. (e) The proposed architecture uses m as input to the single trained multi-label classification model D'. In both cases, a set of final scores $s_1, ..., s_N$ are produced, corresponding to concepts $C_1, ..., C_N$.

and consider each set as a separate class in order to solve a multi-class problem. We choose the original LP tranformation [15], as well as the Pruned Problem Transformation algorithm (PPT) [17] that reduces the class imbalance problem by pruning label sets that occur less than l times. Finally, lazy style methods most often use label correlations in the neighbourhood of the tested instance, to infer posterior probabilities. In this direction we choose ML-kNN algorithm [18], which models exactly this information. Note that the chosen methods can output both a bipartition of the labels (relevant/irrelevant) and scores in the [0,1] range. In selecting the above methods, we took into account the computational complexity of these and other similar methods and tried to avoid using particularly computationally intensive ones.

One could argue that graph-based methods also search for meta-models that model label correlations for all concepts at once; however, we differ from these methods as we choose multi-label learning approaches rather than probabilistic models. The use of multi-label classification algorithms as the second layer of a stacking architecture has the significant advantage of allowing the representation of the videos using state-of-the-art high dimensional low-level features (for describing the video at the first layer of the stack), as opposed to simpler features used in e.g. [13], [9], while at the same time keeping relatively low the dimensionality of the input to the multi-label classifier of the second layer, thus making the overall concept detection architecture applicable even to large-scale problems.

4   Experimental Setup

4.1   Dataset and Evaluation Methodology

We tested our framework on the TRECVID 2011 and 2012 Semantic Indexing (SIN) datasets [19], [20]. Each of them consists of a development set and a test set (approximately 400 and 200 hours of internet archive videos for training and testing, respectively, for TRECVID 2011, and another 600 and 200 hours for TRECVID 2012). We further partitioned the original test set into 2 sets (validation and test set, 50% each) by using the Iterative Stratification algorithm [21], suitable for multi-label data, and evaluated all techniques on the latter set using the 50 and 46 concepts that were evaluated as part of the TRECVID 2011 and 2012 SIN Task, respectively.

Regarding the annotations for these datasets, we augmented those used by TRECVID in 2011 with the results of collaborative annotation [22], [23] that was carried out for the same dataset (among other datasets) as part of the 2012 edition of the SIN Task. We solve disagreements between the two annotations by using the max operator (where in each collection of annotations, numbers 1, 0, -1 for a given shot-concept pair denote the following: 1=concept appears, -1=does not appear, 0=ambiguous). We further augmented the ground truth by using the concept "imply" relations provided by TRECVID. Finally, ambiguous and missing annotations are ignored during evaluation. A similar process was performed in order to augment the original annotations for the TRECVID 2012 dataset exploiting the results of the 2013 collaborative annotation [22], [23].

As discussed in the introduction, we also want to investigate if the typical way of evaluating concept detection results [19] is suitable for assessing their goodness for different applications. Based on this, we adopt two evaluation strategies:

i) Considering the video indexing problem, given a concept, we measure how well the top retrieved video shots for this concept truly relate to it. ii) Considering the video annotation problem, given a video shot, we measure how well the top retrieved concepts describe it. For each such strategy we calculate Mean Average Precision (MAP) and Mean Precision at depth k (MP@k).

4.2 Baseline Detectors and Comparisons

For the first layer of the stacking architecture (which also serves as the baseline for comparisons) we use one concept detection score per concept, extracted by combing the output of 25 linear SVM classifiers trained for the same concept, following the methodology of [24].

We instantiate the second layer of the proposed architecture with four different multi-label learning algorithms as described in section 3, and will refer to our framework as P-CLR , P-LP, P-PPT and P-MLkNN when instantiated with CLR [16], LP [15], PPT [17] and ML-kNN [18] respectively. The value of l for P-PPT was set to 30.

We compare the proposed framework against BCBCF [3], DMF [2], BSBRM [4] and MCF [5], which were reviewed in section 2. For BCBCF we use the concept predictions instead of the ground truth in order to form the meta-learning dataset, as this was shown to improve its performance in our experiments; we refer to this method as CBCFpred in the sequel. Regarding the concept selection step we use these parameters: $\lambda = 0.5$, $\theta = 0.6$, $\eta = 0.2$, $\gamma =$ the mean of Mutual Information values. For MCF we only use the spatial cue, so temporal weights have been set to zero. Finally, the ϕ coefficient threshold, used by BSBRM, was set to 0.09.

For the purpose of implementing the above techniques the Logistic Regression learning algorithm [25] is used for the classification tasks considered by some of the methods. The WEKA [26] and MULAN [27] machine learning libraries were used as the source of single-class and multi-label learning algorithms, respectively.

5   Results and Discussion

We performed two sets of experiments for each of the two TRECVID datasets[1]. For the TRECVID 2011 dataset, in the first set, the meta-level training set is composed of predictions from 50 concept detectors (the 50 concepts for which ground-truth annotations exist not only in the training set but also in the test set). In the second set of experiments, we include information from 296 more first-layer concept detectors (346 in total). For the TRECVID 2012 dataset the two experiment sets were similarly instantiated with 46 and 346 concepts respectively. Table 1 summarizes the results for the two datasets.

We start the analysis, based on the MAP and MP@k results, separately for each evaluation strategy. Results regarding the indexing problem

---

[1] The experiments were conducted on a PC with 3.5 GHz CPU and 16GB of RAM.

Table 1. Performance, in terms of MAP, MP@k and CPU time, for the different methods that are compared on the TRECVID 2011 and 2012 datasets. The number of concepts that are evaluated on these datasets are 50 and 46, respectively. The meta-learning feature space for the second layer of the stacking architecture is constructed using detection scores for (I) the same 50 and 46 concepts and (II) an extended set of 346 concepts. The $\sim$ symbol indicates that the difference in MP@k between the denoted method and the best-performing method in the same column of the table is not statistically significant (thus, the absence of $\sim$ suggests statistical significance). CPU times refer to mean training (in minutes) for all 50 or 46 concepts on datasets of 67874 and 72818 shots, respectively, and application of the trained second-layer detectors on one shot of the test set (in milliseconds). Evaluation was performed only on shots that are annotated with at least one concept.

	(I) Using the output of 50 (TRECVID 2011) and 46 (TRECVID 2012) detectors for meta-learning								
	TRECVID 2011				TRECVID 2012				
	(a)	(b)	(c)	(d)	(e)	(f)	(g)	(h)	(i)
Method	MAP	MP@100	MAP	MP@3	MAP	MP@100	MAP	MP@3	mean Exec. Time
	(Indexing)		(Annotation)		(Indexing)		(Annotation)		Training /Testing
Baseline	0.3391	0.6608	0.6150	0.3697	0.2052	0.3711	0.6006	0.3251	N/A
DMF[2]	0.4068	0.7448	**0.6878**	**0.4226**	0.2614	0.4350	**0.7610**	**0.4102**	1.33/0.30
BSBRM[4]	0.3744	0.7038	0.6785	0.4174	0.2260	0.3848	0.7499	0.4046	0.39/0.09
CBCFpred[3]	0.3321	0.6146	0.6586	0.4081	0.1675	0.2700	0.6529	0.3582	0.96/0.35
MCF[5]	0.3388	0.6630	0.6122	0.3762	0.2039	0.3654	0.6661	0.3581	24.86/0.32
P-CLR	0.3876	0.7238	0.6876	0.4183	0.1997	0.3335	0.7530	0.4041	3.63/3.85
P-LP	0.3925	0.7404	0.6852	0.4174	0.2667	0.4430	0.7603	0.4074$\sim$	74.27/63.08
P-PPT	0.3614	0.6334	0.6797	0.4162	0.2443	0.4213	0.7536	0.4048	29.88/0.20
P-MLkNN	**0.4727**	**0.7998**	0.6667	0.4073	**0.2760**	**0.4893**	0.7487	0.4021	21.41/17.72

	(II) Using the output of 346 detectors for meta-learning								
	TRECVID 2011				TRECVID 2012				
	(j)	(k)	(l)	(m)	(n)	(o)	(p)	(q)	(r)
Method	MAP	MP@100	MAP	MP@3	MAP	MP@100	MAP	MP@3	mean Exec. Time
	(Indexing)		(Annotation)		(Indexing)		(Annotation)		Training /Testing
Baseline	0.3391	0.6608	0.6150	0.3697	0.2052	0.3711	0.6006	0.3251	N/A
DMF[2]	0.4095	0.7480	0.6833	0.4177	0.2611	0.4383	0.7538	0.4075	10.03/0.48
BSBRM[4]	0.4114	0.7472	0.6905	0.4231$\sim$	0.2778	0.4517	0.7645	0.4111$\sim$	1.66/0.08
CBCFpred[3]	0.3643	0.6782	0.6713	0.4109	0.2294	0.3854	0.7218	0.3824	12.68/0.28
MCF[5]	0.3440	0.6702	0.5979	0.3667	0.2030	0.3628	0.6717	0.3656	131.68/0.81
P-CLR	0.3310	0.6320	0.6731	0.4111	0.2071	0.3578	0.7508	0.4030	28.76/7.47
P-LP	0.4281	0.7684	**0.7001**	**0.4242**	0.2940	0.4761	**0.7733**	**0.4125**	390.99/68.26
P-PPT	0.3710	0.6268	0.6879	0.4176	0.2848	0.4663	0.7622	0.4100$\sim$	144.34/0.23
P-MLkNN	**0.4959**	**0.8078**	0.6810	0.4145	**0.3182**	**0.5278**	0.7704	0.4111$\sim$	135.30/115.82

(Table 1:(a),(b),(e),(f),(j),(k),(n),(o)) clearly show the effectiveness of the proposed stacking architecture when combined with ML-kNN. ML-kNN assumes that similarity among predictions means semantic similarity and also that the same errors that are observed in the first layer will be performed to images with similar concepts. While ML-kNN can model any possible correlation in the neighbourhood of a testing instance, LP and PPT can model only those that have appeared in the training set. Modelling pairwise correlations can not been considered as robust, because CLR exhibits moderate to low performance.

When assessing the performance of detectors in relation to the annotation problem, P-LP appears more suitable: In the first round of experiments (Table 1:(c),(d),(g),(h)), it performs slightly worse than DMF; in the second round (Table 1:(l),(m),(p),(q)) it exhibits the best performance. In general, though, with respect to the annotation problem, there is no clear winner: the performance differences between methods that model label correlations and BR methods are often limited (although in most cases statistically significant).

In order to investigate the statistical significance of the differences in MP@k observed in Table 1, the chi-square test [28] is used together with the following null hypothesis: "there is no significant difference in the total number of correct shots/concepts that appear in the first k positions between the results obtained after the application of the best performing method and the results obtained after the application of another competing approach". This test is performed separately for each of the MP@k columns of Table 1 (columns (b),(d),(f),(h),(k),(m),(o),(q)). Methods that do not have statistically significant difference ($p \geq 0.05$) from the best performing method are indicated with the $\sim$ symbol.

Regarding the second fold of this work, we observe in Table 1 that good results in the indexing-based evaluation do not guarantee the same when the system is assessed with respect to the annotation problem, and vice versa. There is not any method that reaches top performance for both of these problems. The differences in the ordering of the tested methods according to their goodness in the different experiments are striking, thus highlighting the importance of following both evaluation strategies and reporting results in both these directions when evaluating general-purpose concept detection methods. In addition to this, researchers should bear in mind that every top-performing method is shown in our experiments to be most appropriate for addressing only one of these two problems. The results presented in this work could be used as a guide in order for researchers to choose the appropriate method based on the specific task that they are interested in.

Finally, we take a look at the execution times that each method requires (Table 1:(i),(r)). One could argue that the proposed architecture that uses multi-label learning methods requires considerably more time than the typical BR-stacking one. However, we should note here that extracting one model vector from one video shot, using the first-layer detectors for 346 concepts requires approximately 1.33 minutes in our experiments, which is about three orders of magnitude slower than the slowest of the second-layer methods. As a result of the high computational complexity of the first layer of the stack, the execution time differences between all the second-layer methods that are reported in Table 1 can be considered negligible. At this point it would be reasonable to compare the stacking-based multi-label architecture to the one-layer alternative, i.e., building a multi-label classifier directly from the low-level visual features of video shots. However, the high requirements for memory space and computation time that the latter methods exhibit do not make this comparison practically feasible for our datasets on typical PCs, as we explain in the following.

The computational complexity of BR, CLR, LP and PPT when used in a single-layer architecture depends on the complexity of the base classifier, in our case the Logistic Regression, and on the parameters of the learning problem. Let us assume that N concepts need to been detected and m training examples are available for learning to detect them. In this learning problem the BR algorithm, which builds N models (one for each concept), is the simplest one. CLR is the next least complex algorithm, requiring the building of N BR-models and additionally $N * (N - 1)/2$ one-against-one models. LP is the most complex algorithm, since it trains a multi-class model, with the number of classes being equal to the number of distinct label sets in the training set. PPT works in the same fashion as the LP with the difference that only a pruned set of distinct label sets will be used to train the multi-class model. Finally, the training of ML-kNN is linear with respect to the size of the training set and the length of the training vectors, but the algorithm needs to make many calculations that involve the consideration of all k-neighbours of all m training examples. Given that the training datasets used in this work consist of more than 200.000 training examples, and each training example (video shot) is represented by a 4000-element low-level feature vector and is associated with a few tens of concepts (e.g. 46 for TRECVID 2012), according to the above, for the TRECVID 2012 dataset the BR algorithm would build 46 models, CLR would build 46 BR-models and 1035 one-against-one models, LP and PPT would build a multi-class classifier of 1544 and 152 (for pruning threshold equal to 30 as reported in section 4.2) classes, respectively, and finally ML-kNN would compare each training example with all other (200.000) available examples; in all these cases, the 4000-element low-level feature vectors would be employed. Taking into consideration the dimensionality of these feature vectors all above actions require considerably more time compared to the BR alternative that we employ as the first layer in our proposed stacking architecture. In addition to this, the software (MULAN [27]) used in our experiments requires the full training set to be loaded on memory at once, which again is practically unfeasible without extending the MULAN code, which is out of the scope of this work. We conclude that the two major obstacles of using multi-label classification algorithms in a one-layer architecture are the high memory space and computation time requirements, and this finding further stresses the merit of our proposed multi-label stacking architecture.

6   Conclusion and Future Work

This paper proposed an alternative way of employing the stacking architecture, used for concept detection score refinement. Multi-label classification algorithms that consider label correlations appear to be more suitable for a meta-learning training, instead of the commonly used Binary Relevance models. This conclusion is supported by a comparative study on two challenging datasets involving a multitude of diverse concepts. Furthermore, this paper compared concept detection approaches on two different problems, video indexing and annotation. In relation to this comparison, the message that this work aims to pass is that there is not a method able to deal with both these problems in the best possible way; good performance of video indexing according to each concept separately is not a good indicator of the suitability of the method for addressing different problems such as concept-based video annotation. Future directions of work include improving the speed of some of the second-layer learning methods and also experimenting with modifications of methods that gave promising results, such as MLkNN and LP.

Acknowledgements. This work was supported by the EC under contracts FP7-287911 LinkedTV and FP7-318101 MediaMixer.

References

1. Snoek, C.G.M., Worring, M.: Concept-Based Video Retrieval. Foundations and Trends in Information Retrieval 2(4), 215–322 (2009)
2. Smith, J., Naphade, M., Natsev, A.: Multimedia semantic indexing using model vectors. In: 2003 Int. Conf. on Multimedia and Expo, ICME 2003, pp. 445–448. IEEE Press, New York (2003)
3. Jiang, W., Chang, S.F., Loui, A.C.: Active context-based concept fusion with partial user labels. In: IEEE Int. Conf. on Image Processing. IEEE Press, New York (2006)
4. Tsoumakas, G., Dimou, A., Spyromitros-xioufis, E., Mezaris, V., Kompatsiaris, I., Vlahavas, I.: Correlation-Based Pruning of Stacked Binary Relevance Models for Multi-Label learning. In: ECML/PKDD 2009 Workshop on Learning from Multi-Label Data (MLD 2009), pp. 101–116. Springer, Heidelberg (2009)
5. Weng, M.F., Chuang, Y.Y.: Cross-Domain Multicue Fusion for Concept-Based Video Indexing. IEEE Transactions on Pattern Analysis and Machine Intelligence 34(10), 1927–1941 (2012)
6. Qi, G.J., Hua, X.S., Rui, Y., Tang, J., Mei, T., Zhang, H.J.: Correlative multi-label video annotation. In: 15th International Conference on Multimedia, MULTIMEDIA 2007, pp. 17–26. ACM, New York (2007)
7. Zha, Z.J., Mei, T., Wang, J., Wang, Z., Hua, X.S.: Graph-based semi-supervised learning with multiple labels. Journal of Visual Communication and Image Representation 20(2), 97–103 (2009)
8. Zha, Z.J., Hua, X.S., Mei, T., Wang, J., Qi, G.J., Wang, Z.: Joint multi-label multi-instance learning for image classification. In: Computer Vision and Pattern Recognition (CVRP 2008), pp. 1–8. IEEE, New York (2008)
9. Wang, M., Zhou, X., Chua, T.S.: Automatic image annotation via local multi-label classification. In: Int. Conf. on Content-based image and video retrieval, CIVR 2008, pp. 17–26. ACM Press, New York (2008)
10. Zhu, Q., Liu, D., Meng, T., Chen, C., Shyu, M., Yang, Y., Ha, H.Y., Fleites, F., Chen, S.C.: Florida International University and University of Miami TRECVID 2012. In: TRECVID 2012 Workshop, Gaithersburg, MD, USA (2012)
11. Yu, S.I., Xu, Z., Ding, D., Sze, W., Vicente, F., Lan, Z., Cai, Y., Rawat, S., Schulam, P., Markandaiah, N., Bahmani, S., Juarez, A., Tong, W., Yang, Y., Burger, S., Metze, F., Singh, R., Raj, B., Stern, R., Mitamura, T., Nyberg, E., Jiang, L., Chen, Q., Brown, L., Datta, A., Fan, Q., Feris, R., Yan, S., Pankanti, S., Hauptmann, A.: Informedia @TRECVID 2012. In: TRECVID 2012 Workshop, Gaithersburg, MD, USA (2012)

12. Wang, F., Sun, Z., Zhang, D., Ngo, C.: Semantic Indexing and Multimedia Event Detection: ECNU at TRECVID 2012. In: TRECVID 2012 Workshop, Gaithersburg, MD, USA (2012)
13. Nasierding, G., Kouzani, A.Z.: Empirical Study of Multi-label Classification Methods for Image Annotation and Retrieval. In: 2010 Int. Conf. on Digital Image Computing: Techniques and Applications, pp. 617–622. IEEE, China (2010)
14. Kang, F., Jin, R., Sukthankar, R.: Correlated Label Propagation with Application to Multi-label Learning. In: IEEE Computer Society Conf. on Computer Vision and Pattern Recognition, CVPR 2006, pp. 1719–1726. IEEE Press, New York (2006)
15. Tsoumakas, G., Katakis, I., Vlahavas, I.: Mining multi-label data. In: Data Mining and Knowledge Discovery Handbook, pp. 667–686. Springer, Berlin (2010)
16. Fürnkranz, J., Hüllermeier, E., Loza Mencía, E., Brinker, K.: Multilabel classification via calibrated label ranking. Machine Learning 73(2), 133–153 (2008)
17. Read, J.: A pruned problem transformation method for multi-label classification. In: 2008 New Zealand Computer Science Research Student Conference (NZCSRS 2008), New Zealand (2008)
18. Zhang, M.L., Zhou, Z.H.: ML-KNN: A lazy learning approach to multi-label learning. Pattern Recognition 40(7), 2038–2048 (2007)
19. Over, P., Awad, G., Fiscus, J., Antonishek, B., Michel, M., Smeaton, A.F., Kraaij, W., Queenot, G.: Trecvid 2011 – an overview of the goals, tasks, data, evaluation mechanisms and metrics. In: TRECVID 2011, NIST, USA (2011)
20. Over, P., Fiscus, J., Sanders, G., Shaw, B., Awad, G., Qu, G.: Trecvid 2012 an overview of the goals, tasks, data, evaluation mechanisms, and metrics. In: TRECVID 2012, NIST, USA (2012)
21. Sechidis, K., Tsoumakas, G., Vlahavas, I.: On the Stratification of Multi-Label Data. In: Gunopulos, D., Hofmann, T., Malerba, D., Vazirgiannis, M. (eds.) ECML PKDD 2011, Part III. LNCS (LNAI), vol. 6913, pp. 145–158. Springer, Heidelberg (2011)
22. Ayache, S., Quénot, G.: Video corpus annotation using active learning. In: Macdonald, C., Ounis, I., Plachouras, V., Ruthven, I., White, R.W. (eds.) ECIR 2008. LNCS, vol. 4956, pp. 187–198. Springer, Heidelberg (2008)
23. Hradiš, M., Kolář, M., Láník, A., Král, J., Zemčík, P., Smrž, P.: Annotating images with suggestions user study of a tagging system. In: Blanc-Talon, J., Philips, W., Popescu, D., Scheunders, P., Zemčík, P. (eds.) ACIVS 2012. LNCS, vol. 7517, pp. 155–166. Springer, Heidelberg (2012)
24. Moumtzidou, A., Gkalelis, N., Sidiropoulos, P., Dimopoulos, M., Nikolopoulos, S., Vrochidis, S., Mezaris, V., Kompatsiaris, I.: ITI-CERTH participation to TRECVID 2012. In: TRECVID 2012 Workshop, Gaithersburg, MD, USA (2012)
25. Le Cessie, S., Van Houwelingen, J.: Ridge estimators in logistic regression. Journal of the Royal Statistical Society. Series C (Applied Statistics) 41(1), 191–201 (1992)
26. Witten, I., Frank, E.: Data Mining Practical Machine Learning Tools and Techniques, 2nd edn. Morgan Kaufmann, San Francisco (2005)
27. Tsoumakas, G., Spyromitros-xioufis, E., Vilcek, J., Vlahavas, I.: MULAN: A Java Library for Multi-Label Learning. Journal of Machine Learning Research 12, 2411–2414 (2011)
28. Greenwood, P., Nikulin, M.: A guide to chi-squared testing. Wiley-Interscience, Canada (1996)

Coherence Analysis of Metrics in LBP Space for Interactive Face Retrieval

Yuchun Fang, Ying Tan, and Chanjuan Yu

School of Computer Engineering and Science,
Shanghai Univeristy 200444 Shanghai, China
ycfang@shu.edu.cn

Abstract. Interactive retrieval model is a useful solution for the multimedia retrieval applications in case of targets unavailable. The goodness of such model relies on a high coherence between human and machine cognition about the regarded retrieval task. In this paper, we specially perform coherence analysis for interactive face retrieval and explore the influence of metrics to human and machine face recognition in Local Binary Pattern (LBP) feature space. With the collected real user feedback, we discover several new conclusions about unbalanced coherence distribution model and propose an improved correntropy metrics that leads to improved coherence and fast retrieval.

Keywords: Interactive face retrieval, Correntropy, Metrics, Semantic gap, Local Binary Pattern.

1   Introduction

As an important aspect of multimedia retrieval, human face retrieval aims at searching target human face from image databases fast and accurately. It has extensive applications in criminal detection, personal identification, credit card verification and pedestrian surveillance. There have been a lot research advances concerning face image retrieval in recent years. Kim [1] adopted MPEG-7(Moving Picture Experts Group) that decompose face images into multiple components to partly avoid pose and lighting variations, and then utilized Linear Discriminant Analysis (LDA) feature for retrieval. Park [2] believed that there exist certain semantic auxiliary feature points such as moles or scars for retrieval. Brandon proposed to build up shape models of contour, nose, mouth and eyes by users for searching target [3]. While in some scenarios, a physical image of target may not be available, in such case, example based face retrieval will become invalid.

Interactive retrieval is a solution in the case of targets unavailable. Through iterative interaction between human and machine, the relevance feedback driven retrieval model can boost the knowledge about target, and finally fix the target with several iterations of question and answer interaction. In previous research, Zhou [4] summarized many image retrieval strategies and proposed to use multiple interactions to alleviate burdens of user in a single iteration. In face retrieval, Navarrete [5]

C. Gurrin et al. (Eds.): MMM 2014, Part I, LNCS 8325, pp. 13–24, 2014.

proposed a self-organizing mapping in interactive searching. Fang and Geman [6] proposed a Bayesian model for target searching of human face. Ferecatu and Geman [7] also used a probabilistic model for mental image retrieval. A bi-objective model was proposed that took into consideration of both Maximum a Posterior and mutual information in [8]. Since human feedback has important influence to retrieval precision, He et al [9] proposed to train Support Vector Machine (SVM) to adjust human feedback for interactive retrieval.

In interactive retrieval of the mental target, since there is no physical form of target appearance, the semantic gap between human feedback and low level machine perception serves to decide the feasibility of retrieval task. In previous work [6][8], the user feedback for interactive face retrieval is set as a label corresponding to the image regarded by user as the closest one among each display to the target. Based on the user feedback, the interactive retrieval model updates the posterior conditioned to retrieval history. Hence, the convergence of the searching process is influenced by the coherence between human and machine face recognition. High coherence will lead to fast searching of target. Usually, it is irrational to control the subjective user feedback. However, the machine cognition is decided by the low level features and metrics, which can be adjusted and designed to fit the coherence requirement of interactive face retrieval. Moreover, different metrics not only means different coherence for interactive retrieval system, it also affects the similarity among images in low-level feature space and also the topological distribution of images in the high dimensional space.

This paper specially contributes to analyze the interactive face retrieval under unbalanced coherence distribution and explore the control of metrics to promote coherence between human and machine face recognition. With the feedback data sets of real users collected in interactive face retrieval, we compare the coherence of human and machine face recognition with several metrics in the Local Binary Pattern (LBP) feature space and perform simulation experiments to test the performance of interactive retrieval among various metrics.

2   Interactive Face Retrieval

The interactive face retrieval system mentioned in [8] is adopted as the experimental platform in this paper. The system serves to collect real user feedback and perform simulation experiments as shown in Fig 1. Defining a retrieval test as finding one target, the searching is composed of iterations of question and answer. In each iteration, the user is asked to choose a 'closest' X_D to the target image Y from the set of displayed images D $(D = 1, 2, \cdots, n)$. Under each type of parameter setting used by the feedback model, such as the feature space or the metrics, a certain number of tests are performed by real user or machine simulation to obtained statistics to evaluate the retrieval performance. In the simulation tests, machine simulates the real user to provide feedback according to the coherence distribution learned from collected real user feedback. Concretely speaking, the feedback is picked according to a learned probability distribution in simulation tests. For M tests, we record the number of retrieval iterations as $T = \{T_1, T_2, \ldots, T_M\}$. Two measurements to evaluate the retrieval performance are the average number of retrieval iteration in Equation (1) and the cumulative probability of T in Equation (2).

$$E(T) = \frac{1}{M} \sum_{i=1}^{M} T_i \qquad (1)$$

$$P(T \leq t) = \frac{\left|\{T_i \,\middle|\, T_i \leq t, T_i \in T\}\right|}{M} \qquad (2)$$

where $\{T_i \,\middle|\, T_i \leq t, T_i \in T\}$ is the set of T less than t and $\left|T_i \,\middle|\, T_i \leq t, T_i \in T\}\right|$ denotes the size of this set. The meaning of $P(T \leq t)$ is the probability of finding target in less than t retrieval iterations.

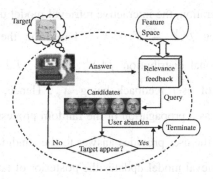

Fig. 1. Interactive face retrieval system

2.1   Coherence Analysis for Interactive Face Retrieval

Coherence is defined in previous work [6][8] to measure the semantic gap. To compare the coherence of metrics in feature spaces, we adopt the distance between target Y and displayed candidate images X to evaluate the relationship of neighborhood among them. The coherence distribution $P(r)$, i.e. the frequency of the r-th closest, is used as the statics of user select X_D among X as the r-th closest neighbor to target Y. $P(r)$ can be calculated according to Equation (3).

$$P(r) = \frac{\sum_{t=1}^{N} F(M_t, r)}{N} \qquad (3)$$

$$F(M_t, r) = \begin{cases} 1 & C(Y^t, X^t, r) = x_D^t \\ 0 & other \end{cases}$$

where N is the number of records of collected real user feedback, $M_t = (Y^t, X^t, X_D^t)$ is the t-th record of M, and $C(Y^t, X^t, r)$ denotes the r-th closest image of target Y^t in displayed candidate set X^t. The coherence distribution at $r = 1$ is the most essential, since it directly denotes the relationship between the closeness defined by machine in low level feature space and the semantic feedback of human. The cumulative distribution $C(r) = \int P(r)dr$ is also adopted to measure coherence [6], which denotes the probability of the user feedback the r-th closest image, $J = \int C(r)dr$, the integration of $C(r)$ is used as another coherence measure [9]. The larger the value of J is, the higher the coherence is and thus the faster the search will be conducted.

2.2   Coherence Distribution vs. Feedback Distribution

At the t-th retrieval iteration, the interactive retrieval model updates the posterior of image s_j^t being target $p(Y = s_j^t \mid x_D^t)$ based on the user feedback x_D^t. Meanwhile, the conditional distribution $p(x_D^t = x_i \mid Y)$ means under a given target, the probability of user feedback being x_i. Hence, each iteration of the retrieval can be viewed as a propagation of the random process $Y \rightarrow x_D \rightarrow s_j$. In the part $Y \rightarrow x_D$, the user provides answer to the candidate display. In the part of $x_D \rightarrow s_j$, the retrieval model updates the posterior of target and provides new display. Among this process, the conditional distribution $p(x_D^t = x_i \mid Y)$ serves as the machine learning from the coherence distribution, which is decided by the metrics defined in feature space. Normally, the higher the coherence, the fast the retrieval will be [6]. In this paper, we perform further test to evaluate the interactive retrieval model in the case that the coherence distribution mismatches real user feedback.

To serve this purpose, under the retrieval setting in [6] and [8], we set three groups of coherence distribution with increasing coherence in each row in Table 1. The value is a permutation around the real user feedback statistic. We also define labels when the three distributions are used as simulated user feedback distribution as A1, B1, C1 and coherence distribution as A2, B2, C2 respectively in the first and last column of Table 1. The test is denoted in a short form as A1 VS A2, meaning the test of user feedback distribution A1 versus coherence distribution A2. Two kinds of tests are performed. One is to fix the coherence of simulated user feedback and update posterior with different coherence distribution, another is to fix the different coherence distribution and accept feedback according to simulated users with different coherence. The comparison is summarized in In Fig. 2 (a) and (b).

Table 1. Three Groups of Coherence Distribution

Label of User feedback	1	2	3	4	5	6	7	8	Label of coherence distribution
A1	0.143	0.134	0.125	0.131	0.12	0.136	0.113	0.098	A2
B1	0.232	0.143	0.196	0.143	0.072	0.071	0.089	0.054	B2
C1	0.333	0.25	0.167	0.083	0.057	0.05	0.03	0.03	C2

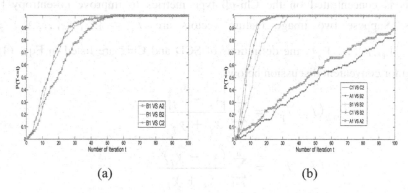

(a)   (b)

Fig. 2. Retrieval precision: (a) Group B1 versus various coherence distributions; (b) Fixed coherence distribution versus various feedback distributions

In Fig. 2(b), by comparing the tree curves of A1 VS A2, B1 VS B2 and C1 VS C2, it can be toll that when the coherence distribution is the same as the user feedback distribution, the more coherent the distribution is, the better the performance of the retrieval system is.

By the curve of B1 VS A2 in Fig. 2 (a), fixing the user feedback distribution B1, the performance of retrieval degrades when the coherence distribution is more coherent than of the feedback distribution. While in the inverse case as for B1 VS C2, the retrieval speed keeps at the same level as the case when the two distributions are in the same coherence level (B1 VS B2). The similar results can be observed in Fig. 2 (b). By comparing the pair of curves of C1 VS C2 and C1 VS B2, the former is much better than the latter. While comparing the pair of curves of A1 VS A2 and A1 VS B2, the latter is much better. These observations lead to some new conclusions about coherence analysis. When the coherence distribution is more coherent than the user feedback distribution, the higher the coherence of the distribution, the better the performance is. While in the opposite case, the searching tends to be slowed down. The coherence distribution serves to adjust the posterior and thus helps to decide the searching direction in retrieval. When it is far from the real user feedback distribution, the learning of posterior tends to be slow. When designing the interactive retrieval model, it is very important to use optimistic coherence distribution to update posterior.

3   Coherence Analysis of Metrics in LBP Feature Space

Based on the above observations of interactive retrieval, we analyze the coherence of user feedback with respect to metrics in feature space. In this paper, we conduct comparison with the general LBP feature and Uniform LBP (ULBP), which are already very popular in face recognition [11]. We explore their usage in interactive face retrieval problem with some usual metrics such as L1, L2, SCD [12], Chi2 [13], Jensen–Shannon divergence based metrics (JS) [14]. And more specifically, the analysis is concentrated on the Chi-αβ type metrics to improve co-entropy [15] metrics. Suppose two image feature vectors are $I_x = \{x_1, x_2, \ldots x_n\}$ and $I_y = \{y_1, y_2, \ldots y_n\}$, the definitions of SCD and Chi2 are listed in Eqn. (4) to Eqn.(5) for convenient discussion below.

$$D_{SCD}(I_X, I_Y) = \sum_{i=1}^{n} \frac{|x_i - y_i|}{x_i + y_i} \tag{4}$$

$$D_{Chi2}(I_X, I_Y) = \sum_{i=1}^{n} \frac{(x_i - y_i)^2}{x_i + y_i} \tag{5}$$

3.1   Analysis of Chi-αβ Metrics

When comparing the definition between SCD and L1, also between Chi2 and L2, the only difference is the item $x_i + y_i$ in denominator as a weight. For two single dimension x_i and y_i, if $(x_i - y_i)$ is fixed, the larger the sum $x_i + y_i$ is, the more similar the two components are. Hence, such weight is especially suitable for features combined of histograms, such as the LBP.

To explore the influence of each single dimension to coherence, we defined the Chi-αβ metrics in Eqn. (6)

$$D(X, Y) = \sum_{i=1}^{n} \frac{(x_i - y_i)^\alpha}{(x_i + y_i)^\beta} \tag{6}$$

If $\alpha \in [0, 1)$, the effect of $(x_i - y_i)$ is weakened, and vice versa, if $\alpha \in (1, \infty]$, the effect of $(x_i - y_i)$ will be strengthened. Similarly, if $\beta \in (0, 1)$, the weight of $x_i + y_i$ is weakened, and if $\beta \in (1, \infty)$, the weight effect is enlarged.

To further explore the influence of parameter α and β to the coherence of Chi-αβ metrics, we vary their values from 0.125 to 3 with a uniform step 0.125 and form 576 parameter pairs. For each parameter pair, the coherence distribution at $r = 1$

is analyzed since it controls the trends of coherence. The MANOVA analysis of factors α and β for coherence is summarized in Table 3. Normally, if P value is smaller than 0.05, the corresponding factor is regarded as significant. The F value denotes the degree of such significance, the larger the F value is, the more significant the factor is. Since in Table 2, both α and β with P values smaller than 0.05, it can be regarded that both factors has significance. Since both P values are very small, their interaction can be neglected.

Table 2. MANOVA analysis to factor α and β

	Sum Sq	Df	Mean Sq	F value	P value
α	0.35483	23	0.01543	25.14	<0.0001
β	0.25137	23	0.01093	17.81	<0.0001
Residuals	0.32464	529	0.00061		
Total	0.93083				

* 'Sum Sq' denotes the sum of square, 'Mean Sq' denotes average sum of square and 'Df' denotes the degree of freedom.

3.2   Improved Correntropy Metrics

Stimulated by the above analysis about α and β for Chi-$\alpha\beta$ metrics, we propose an improved correntropy metrics for promoting coherence as defined in Eqn. (7).

$$D(I_X, I_Y) = \frac{1}{n} \sum_{i=0}^{n} \frac{1}{\sqrt{2\pi\sigma}} \exp\left(-\frac{(x_i - y_i)^2}{2\sigma^2(x_i + y_i)}\right) \qquad (7)$$

The improved co-entropy metrics can be regarded as weighing each dimension of Chi2 metrics with a Gaussian function. We select the parameter σ by analyzing the distribution of metrics variation at dimension level.

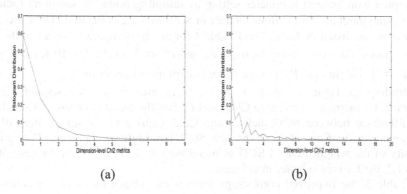

(a)   (b)

Fig. 3. Histogram of dimension-level Chi2 metrics variation: (a) in LBP Feature Space; (b) in ULBP Feature Space

In Fig.3 (a) and (b), we plot the distribution histogram of dimension level variation of Chi2 metrics for both LBP and ULBP features. We find that high-dimensional LBP feature tends to have more small distance at dimension level with Chi2 compared with the ULBP feature. Hence, for the improved correntropy metrics, the value of σ in LBP feature space should be larger than that in ULBP feature space to avoid the influence of small distances at dimension level.

4   Experimental Analysis

With the above mentioned interactive face retrieval model, coherence analysis formetrics, the comparison experiments are performed to validate the performance of the improve correntropy metrics.

4.1   Data Sets

With the interactive face retrieval system in Fig.1, real user feedbacks are collected for our experiments. Two face databases are involved, one is the ID images of college students (4056 Asians aged from 18-21, 1 image per subject), the other is the FERET database (1199 subjects from multiple races, 1 image per subject are randomly picked) [16]. All images are preprocessed into resolution 160*140 facial area after adjusting lighting and pose. As in [6] and [8], the interactive retrieval system randomly picks 1 subject as target, for each target, several groups of display with 8 randomly picked face images will be shown for real user to provide feedback. Each user is required to provide feedback for several targets. In this paper, we collected 2079 records of feedback from 59 persons.

4.2   Comparison of Coherence among Metrics

In experiments, 3304-dimensional ULBP feature and 14336-dimensional LBP feature are adopted with general parameter setting as sampling points 8, sampling radius 2 and blocking number 7*8. Various metrics in Section 3 are compared. The coherence distributions are listed in Table 3 and Table 4 for all the compared metrics in the two feature spaces. For correntropy metrics, we set $\sigma = 1$ in the ULBP feature space and $\sigma = 1.32$ in the LBP feature space as explained in Section 3.2.

According to Table 3, when $r = 1$, SCD metrics has the second highest coherence, L1 metrics is similar to Chi2, and L2 has the worst coherence. Though the only difference between SCD metrics and Chi2 metrics is the square item of the denominator as in Eqn. (4) and (5), the SCD has higher coherence. This proves necessity of the weight item of SCD as mentioned in Section 3.1. In the case of L1 versus L2, the L1 is also better in coherence.

In Table 3, the improved correntropy metrics can obtain much higher coherence compared with other metrics in ULBP feature space. Similar results can be observed for the LBP feature space in Table 4. If $r = 1$, the coherence for correntropy has been promoted to 0.7% compared with the SCD metrics.

When comparing the results in Table 3 and Table 4, if $r = 1$, the coherence of LBP feature is better than ULBP feature for most metrics. Especially for L1 and Chi2 metrics, the coherence has been promoted for up to 1%. The only difference happens for L2 metrics, which is better for the ULBP than for the LBP feature.

Table 3. Coherence distribution of various metrics of ULBP feature on FERET

Metrics \ r	1	2	3	4	5	6	7	8
L1	0.2215	0.1569	0.1314	0.1082	0.1130	0.0938	0.0897	0.0855
L2	0.2108	0.1362	0.1192	0.1200	0.1037	0.0978	0.0986	0.1137
Chi2	0.2200	0.1532	0.1347	0.1104	0.1059	0.0993	0.0912	0.0853
JS	0.2137	0.1510	0.1307	0.1189	0.1041	0.0982	0.0960	0.0874
SCD	0.2311	0.1628	0.1270	0.1178	0.0971	0.1063	0.0849	0.073
Correntropy	0.2396	0.1496	0.1373	0.1178	0.1015	0.0982	0.0904	0.0683

Table 4. Coherence distribution of various metrics of LBP feature on FERET

metrics \ r	1	2	3	4	5	6	7	8
L1	0.2318	0.1576	0.134	0.1163	0.0956	0.0986	0.0882	0.0779
L2	0.2008	0.151	0.1155	0.1178	0.0978	0.1023	0.1004	0.1144
Chi2	0.2333	0.1539	0.1292	0.1203	0.1	0.0982	0.0915	0.0736
JS	0.2303	0.151	0.127	0.1203	0.1059	0.0971	0.093	0.0754
SCD	0.2355	0.1561	0.124	0.1214	0.0993	0.1041	0.0882	0.0714
Correntropy	0.2425	0.1473	0.134	0.1144	0.1059	0.0945	0.0864	0.075

Except the coherence distribution at $r = 1$, the J value is also calculated for comparison as shown in Fig.4 (a) and (b). By J value, it also can be found that the coherence is the highest for the improved correntropy metrics, SCD ranks in the second place. Similar to the result obtained by coherence distribution, the J value prominently increases for L1 and Chi2 metrics in LBP feature space than in the ULBP feature space.

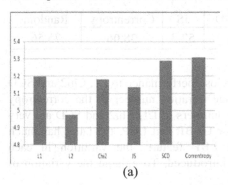
(a)

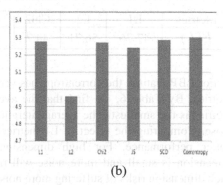
(b)

Fig. 4. Comparison of J value among various metrics (a) in the ULBP feature space; (b) in the LBP feature space

4.3   Simulation Test of Interactive Face Retrieval

The simulation of real user for interactive face retrieval is performed in the FERET database. Each test result is the statistics for 100 simulations. The results are summarized in Fig.5 and Table 5. The case of random search is also plotted for comparison.

By Fig.5, it can be observed that the performance with the improved correntropy is the best among all 6 compared metrics. This is in accordance with their coherence, the higher the coherence, the faster the searching will be. Though the coherence distribution at $r = 1$ of L2 is only 1% less than Chi2, in interactive retrieval, its performance is much worse.

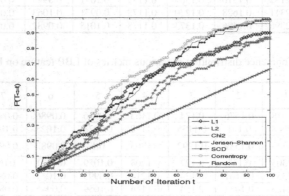

Fig. 5. Cumulative Distribution of retrieval iteration numbers among various metrics in the ULBP feature space

By Table 5, the average number of retrieval iteration of the improved correntropy metrics drops 3 compared with that of the SCD metrics, and 6.6 compared with that of the Chi2 metrics.

Table 5. Average retrieval iteration numbers of various metrics of ULBP feature

Metrics	L1	L2	Chi2	SCD	JS	Correntropy	Random
$E(T)$	47.76	56.71	44.7	41.7	52.3	**38.06**	75.56

For LBP feature, the correntropy has very close performance to L1, Chi2 and SCD metrics. By Table 6, we find that the average iteration number of the correntropy metrics is the smallest. The average iteration numbers of L1, Chi2 and SCD are very close. Comparing the effect of L1 metrics for ULBP and LBP, the latter has much better performance. For high dimensional LBP feature, the variation in each dimension is small and more noise will be introduced. To enlarge the variation of each dimension risks of suffering more noise.

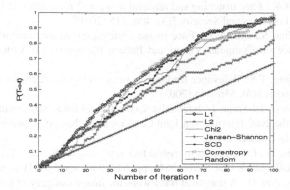

Fig. 6. Cumulative Distribution of retrieval iteration numbers among various metrics in the LBP feature space

Table 6. Average retrieval iteration numbers of various metrics of LBP feature

Metrics	L1	L2	Chi2	SCD	JS	Correntropy	Random
$E(T)$	43.21	59.33	44.6	44.21	49.5	**39.55**	75.56

5   Conclusions

In this paper, based on the collected real user feedback, we mainly concentrate on the coherence analysis of human and machine face recognition in LBP feature space in the sense of metrics. We find several new and interesting conclusions about the importance of coherence in interactive face retrieval. First, the coherence distribution and user feedback distribution are all important for retrieval. Second, the user feedback distribution decides the up-limit of the retrieval performance. Lastly, the posterior should be updated with the coherence distribution no less coherent than user feedback distribution. From the perspective of metrics, we perform a thorough coherence analysis and detailed comparison in the LBP feature space. By exploring the parameter α and β of Chi-αβ type metrics, an improved correntropy metrics is proposed that not only proves to have better coherence, but also can lead to faster interactive face retrieval.

Acknowledgments. The work is funded by the National Natural Science Foundation of China (No.61170155) and the Shanghai Leading Academic Discipline Project (No.J50103).

References

1. Kim, T.K., Kima, H., Hwang, W., Kittler, J.: Component-based LDA face description for image retrieval and MPEG-7 standardisation. Image and Vision Computing 23, 631–642 (2005)

2. Park, U., Jain, A.K.: Face matching and retrieval using soft biometrics. IEEE Transactions on Information Forensics and Security 5(3), 406–415 (2010)
3. Smith, B.M., Zhu, S., Zhang, L.: Face image retrieval by shape manipulation. In: 2011 IEEE Conference on Computer Vision and Pattern Recognition (CVPR), pp. 769–776. IEEE (2011)
4. Zhou, X.S., Huang, T.S.: Relevance feedback in image retrieval: A comprehensive review. Multimedia Systems 8(6), 536–544 (2003)
5. Navarrete, P., Ruiz-del-Solar, J.: Interactive face retrieval using self-organizing maps. In: Proceedings of the 2002 International Joint Conference on Neural Networks, IJCNN 2002, vol. 1, pp. 687–691. IEEE (2002)
6. Fang, Y., Geman, D.: Experiments in mental face retrieval. In: Kanade, T., Jain, A., Ratha, N.K. (eds.) AVBPA 2005. LNCS, vol. 3546, pp. 637–646. Springer, Heidelberg (2005)
7. Ferecatu, M., Geman, D.: A statistical framework for image category search from a mental picture. IEEE Transactions on Pattern Analysis and Machine Intelligence 31(6), 1087–1101 (2009)
8. Fang, Y., Cai, Q., Luo, J., Dai, W., Lou, C.: A bi-objective optimization model for interactive face retrieval. In: Lee, K.-T., Tsai, W.-H., Liao, H.-Y.M., Chen, T., Hsieh, J.-W., Tseng, C.-C. (eds.) MMM 2011 Part II. LNCS, vol. 6524, pp. 393–400. Springer, Heidelberg (2011)
9. He, R., Zheng, W.S., Ao, M., Li, S.Z.: Reducing Impact of Inaccurate User Feedback in Face Retrieval. In: Chinese Conference on Pattern Recognition, CCPR 2008, pp. 1–6. IEEE (2008)
10. Dai, W., Fang, Y., Hu, B.: Feature selection in interactive face retrieval. In: 2011 4th International Congress on Image and Signal Processing (CISP), vol. 3, pp. 1358–1362. IEEE (2011)
11. Ahonen, T., Hadid, A., Pietikainen, M.: Face description with local binary patterns: Application to face recognition. IEEE Transactions on Pattern Analysis and Machine Intelligence 28(12), 2037–2041 (2006)
12. Fan, Z.G., Li, J., Wu, B., Wu, Y.: Local patterns constrained image histograms for image retrieval. In: 15th IEEE International Conference on Image Processing, ICIP 2008, pp. 941–944. IEEE (2008)
13. Gorisse, D., Cord, M., Precioso, F.: Locality-sensitive hashing for chi2 distance. IEEE Transactions on Pattern Analysis and Machine Intelligence 34(2), 402–409 (2012)
14. Cai, Q., Fang, Y., Luo, J., Dai, W.: A novel metrics based on information bottleneck principle for face retrieval. In: Qiu, G., Lam, K.M., Kiya, H., Xue, X.-Y., Kuo, C.-C.J., Lew, M.S. (eds.) PCM 2010, Part I. LNCS, vol. 6297, pp. 404–413. Springer, Heidelberg (2010)
15. Tan, Y., Fang, Y., Li, Y., Dai, W.: Adaptive kernel size selection for correntropy based metric. In: Park, J.-I., Kim, J. (eds.) ACCV Workshops 2012, Part I. LNCS, vol. 7728, pp. 50–60. Springer, Heidelberg (2013)
16. Phillips, P.J., Moon, H., Rizvi, S.A., Rauss, P.J.: The FERET evaluation methodology for face-recognition algorithms. IEEE Transactions on Pattern Analysis and Machine Intelligence 22(10), 1090–1104 (2000)

A Hybrid Machine-Crowd Approach to Photo Retrieval Result Diversification

Anca-Livia Radu[1,2], Bogdan Ionescu[2], María Menéndez[1], Julian Stöttinger[1],
Fausto Giunchiglia[1], and Antonella De Angeli[1]

[1] DISI, University of Trento, 38123 Povo, Italy
{ancalivia.radu,menendez,julian,fausto,
antonella.deangeli}@unitn.it
[2] LAPI, University "Politehnica" of Bucharest, 061071 Bucharest, Romania
bionescu@alpha.imag.pub.ro

Abstract. In this paper we address the issue of optimizing the actual social photo retrieval technology in terms of users' requirements. Typical users are interested in taking possession of accurately relevant-to-the-query and non-redundant images so they can build a correct exhaustive perception over the query. We propose to tackle this issue by combining two approaches previously considered non-overlapping: machine image analysis for a pre-filtering of the initial query results followed by crowd-sourcing for a final refinement. In this mechanism, the machine part plays the role of reducing the time and resource consumption allowing better crowd-sourcing results. The machine technique ensures representativeness in images by performing a re-ranking of all images according to the most common image in the initial noisy set; additionally, diversity is ensured by clustering the images and selecting the best ranked images among the most representative in each cluster. Further, the crowd-sourcing part enforces both representativeness and diversity in images, objectives that are, to a certain extent, out of reach by solely the automated machine technique. The mechanism was validated on more than 25,000 photos retrieved from several common social media platforms, proving the efficiency of this approach.

Keywords: image retrieval, results diversification, crowd-sourcing, image content descriptors, social media.

1 Introduction

The continuously growing number of online personal image collections requires building efficient retrieval systems, e.g., social platforms such as *Panoramio*, *Picasa* or *Flickr* are visited daily by hundreds of millions of users sharing various multimedia resources. Existing social photo search technology is relying mainly on text, image, or more recently on GPS coordinates to provide the user with accurate results for a given query. Retrieval capabilities are however still far from the actual needs of the user. For instance, textual tags tend to be noisy or inaccurate, automatic content descriptors fail to provide high-level understanding of the scene while GPS coordinates capture the position of the photographer and not necessarily the position of the query object.

Social photo retrieval engines focus almost exclusively on the accuracy of the results and may provide the user with near replicas of the query (best matches provide more

C. Gurrin et al. (Eds.): MMM 2014, Part I, LNCS 8325, pp. 25–36, 2014.

or less redundant information). However, most of the social media users would expect to retrieve not only representative photos, but also diverse results that can depict the query in a comprehensive and complete manner, covering different aspects of the query. Equally important, focus should be put also on summarizing the query with a small set of images since most of the users commonly browse only the top results.

In this paper we address these particular aspects of photo retrieval and specifically the issue of result diversification. Research on automatic media analysis techniques reached the point where further improvement of retrieval performance requires the use of user expertise. We propose a hybrid machine-human approach that acts as a top layer in the retrieval chain of current social media platforms. Photo search results are firstly filtered using an automated machine analysis step. We designed a new approach that uses a re-ranking scheme for improving representativeness followed by a clustering mechanism that is specifically designed to ensure diversity. Secondly, we employ crowd-sourcing by designing an adapted study with the objective of using human expertise as a final refinement of the results. In this chain, the automated media part has also the role of a pre-filtering step that diminishes the time, pay and cognitive load and implicitly people's work assuring better crowd-sourcing.

The remainder of the paper is organized as follows: Section 2 presents an overview of the literature and Section 3 situates our approach accordingly. Section 4 deals with the automated media analysis while Section 5 deals with the design of the crowd-sourcing studies. Experimental validation is presented in Section 6 and conclusions in Section 7.

2   Previous Work

Various approaches have been studied in the context of social media to improve search capabilities and specifically the representativeness and diversification of the results. In the following, we detail some of the most popular approaches that are related to our work, namely: re-ranking, automatic geo-tagging, relevance feedback and crowd-sourcing techniques.

Re-ranking techniques are the closest to our machine analysis part. They attempt to re-order the initial retrieval results by taking advantage of additional information, e.g., the initial query is performed using text while the refinement uses visual information. An example is the approach in [6] that aims to populate a database with high precision and diverse photos of different entities by revaluating relational facts about the entities. Authors use a model parameter that is estimated from a small set of training entities. Visual similarity is exploited using classic Scale-Invariant Feature Transform (SIFT). Another example is the approach in [7]. It defines a retrieved image as representative and diverse based on the following properties: should be representative for a local group in the set, should cover as many distinct groups as possible and should incorporate an arbitrary pre-specified ranking as prior knowledge. To determine these properties, authors propose a unified framework of absorbing Markov chain random walks. A different work [8] defines a criterion to measure the diversity of results in image retrieval and propose three approaches to optimize directly this criterion. The proposed methods have been quantitatively evaluated for 39 queries on 20,000 images from the public ImageCLEF 2008 photo retrieval task which incorporates visual and textual information and also qualitatively on a novel product image data set. In the context of video data, the approach in [9] addresses representativeness and diversity in Internet video retrieval. A video near-duplicate graph representing the visual similarity relationship among videos is built. Then, near-duplicate clusters are identified and ranked based on cluster properties and inter-cluster links. The final results are obtained by selecting a representative video from each ranked cluster.

Re-ranking techniques prove to be an efficient solution to the diversification issue as long as there are enough positive examples among the first returned results of the system. Another limitation is in the fact that a re-ranking system has to learn "blindly" which examples are relevant based only on the automatic analysis of data contents.

Automatic geo-tagging techniques are not that close related to our approach but provide an interesting alternative to the diversification part. Automatic geo-tagging deals with automatically determining the geographical position of a picture based on its content description, e.g., textual tags, visual descriptors [1]. This allows for identifying similar images, e.g., from similar locations, without employing any GPS information.

Another perspective to our research problem is to take advantage directly of the human expertise. *Relevance Feedback* (RF) techniques attempt to introduce the user in the loop by harvesting feedback about the relevance of the results. This information is used as ground truth for recomputing a better representation of the data needed. Starting from early approaches, such as Rocchio's algorithm that formulates new queries as a weighted sum of the initial results [2], to current machine learning-based techniques (e.g., Support Vector Machines, boosting techniques) that formulate the RF problem as a two-class classification of the negative and positive examples [3]; relevance feedback proved itself efficient in improving the relevance of the results but more limited in improving the diversification. The main limitation of this approach is the need for the end user to be part of the retrieval system, task that cannot be performed automatically.

A much appealing alternative to relevance feedback is to take advantage of the "crowd" users. *Crowd-sourcing* is defined as "the act of taking a job traditionally performed by a designated agent (e.g., an employee) and outsourcing it to an undefined, generally large group of people in the form of an open call" [4]. Recently, researchers started to explore the potential of crowd-sourcing in tasks which can be subject of individual variations such as annotation of affective videos and perceived similarity between multimedia files [5]. The main benefit of this human-oriented approach consists of humans acting like a computational machine that can be accessed via a computer interface. Although crowd-sourcing shows great potential, issues such as validity, reliability, and quality control are still open to further investigation especially for high complexity tasks.

3   Proposed Approach

To ensure representativeness and diversity of photo retrieval results, we propose a hybrid approach that takes advantage of both machine computational power and human expertise in a unified approach that acts as a top layer in the retrieval chain of current social media platforms (this paper continues our preliminary work presented in [10]). It involves the following steps:

1. machine: photos are retrieved using best current retrieval technology, e.g., using text and GPS tags on current social media platforms. These results are numerous (hundreds) and typically contain noisy and redundant information;

2. machine media analysis: automated machine analysis is used to filter the results and reduce the time and resource consumption to allow better crowd-sourcing. We designed a new approach that re-ranks the results according to the similarity to the "most common" image in the set for improving representativeness followed by a clustering mechanism that is specifically designed to ensure the diversification by retaining only the best ranked representative images in each cluster (see Section 4);

3. crowd-sourcing analysis: to bridge further inherent machine semantic gap, crowd-sourcing is used as a final refinement step for selecting high quality diverse and representative images. To cope with the inherent crowd reliability problem, we designed two adapted studies for both representativeness and diversification. The final objective is to summarize the query with very few results which corresponds to the typical user scenario that browses only the top results.

We identify the following contributions to the current state-of-the-art:

— *re-ranking with diversification of the results:* although the issue of results diversification was already studied in the literature, we introduce a new re-ranking scheme that allows for better selection of both relevant and diverse results and provide an in-depth study of the influence of various content descriptors to this task;

— *better understanding of crowd-sourcing capabilities for result diversification:* we provide a crowd-study of the diversification task which has not been yet addressed in the literature and experimentally assesses the reliability of the crowd-sourcing studies for this particular task;

— *hybrid machine-crowd approach:* we study the perspective and the efficiency of including humans in the computational chain by proposing a unified machine-crowd diversification approach where machine plays the role of reducing the time and resource consumption and crowd-sourcing filters high quality diverse and representative images.

The validation of our approach is carried out on a landmark retrieval scenario using a data set of more than 25,000 photos retrieved from several social media platforms.

4   Machine Media Analysis

We designed a new approach as presented in the sequel as well as in Figure 1 that involves a first step of image ranking in terms of representativeness using the similarity to the rest of the images. Then, all images are clustered and a small number of diverse images coming from different clusters are selected. Finally, a diversity rank is given by means of the dissimilarity to the rest of the selected images. A mediation between the two ranks guarantees the representativeness and diversity in images.

Step 1: for each of the N images in the initial noisy image set, S, we describe the underlying information using visual content descriptors (see Section 6.1). Then, we determine using the features of all images a Synthetic Representative Image Feature (SRI) by taking the average of the Euclidean distances between all image features;

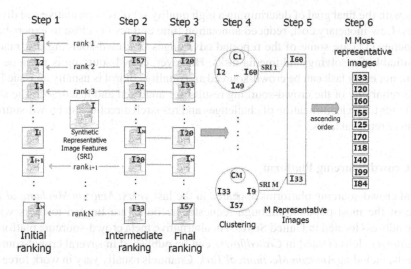

Fig. 1. Proposed re-ranking approach (I represents an image and C denotes the clusters)

Step 2: for each image, I_i, $i = 1, ..., N$, we compute the average of the Euclidean distances to the rest of the images in S which leads to a N-dimensional array. The value of SRI is subtracted from the array which is sorted in ascending order. The new position of each value will account for the intermediate new rank of its corresponding image;

Step 3: supposing that most representative images are among the first returned, for the final re-ranking we average the two ranks (initial rank and the intermediate one determined in Step 2) which yields N average values. Values are again sorted in ascending order and images are re-arranged accordingly;

Step 4: all re-ranked images are clustered in M clusters using a k-means approach based on their visual content. Preliminary tests returns good performance for M around 30;

Step 5: a SRI_j, $j = 1, ..., M$, is computed as presented in Step 1 and a new re-ranking is performed by re-iterating Steps 1, 2 and 3 over the set of images inside each cluster, C_j. Each cluster's first ranked image is considered to be representative (denoted RI_j);

Step 6: from all RI_j images, we select only a small set of P ($P << M$) highest ranked images (ranking according to the final rank computed in Step 3). This will ensure also the diversification of the results.

Experimental validation shows the efficiency of this approach that significantly improves the initial results as well as outperforms another relevant approach from the literature (see Section 6.1).

5   Crowd-Sourcing Analysis

To bridge further the inherent semantic gap of automatic machine analysis techniques, we use a crowd-sourcing approach. We designed two studies that are adapted to our diversification task, where crowd is involved to refine and improve machine-analysis results with the final goal of determining a high quality set of representative and diverse images. Low monetary cost, reduced annotation time and results close to expert-based approaches [11] are some of the reported advantages of crowd-sourcing that made it very suitable for solving multimedia tasks. However, crowd-sourcing is not a perfect system; not every task can be crowd-sourced and quality control is usually an issue [12].

The reliability of the crowd-sourcing results are analyzed and discussed. The study contributes to the identification of challenges and research directions in crowd-sourcing for image retrieval tasks.

5.1 Crowd-Sourcing Platform

Several crowd-sourcing platforms appeared in the last years. *Amazon Mechanical Turk* is one of the most popular, although requester accounts are limited to users with a billing address located in United States. An alternative meta crowd-sourcing platform is *Crowdflower*. Jobs created in *Crowdflower* can be published in several crowd-sourcing channels, including *Amazon Mechanical Turk*. Channels usually vary in work force size and contributors' geographical location.

Although judgements can be ordered in different channels, not all functionalities are available via *Crowdflower*. For example, *Amazon Mechanical Turk* allows requesters selecting the workers who can access the job, manually rejecting low quality answers and republishing rejected assignments at no cost. Instead, *Crowdflower* uses an automatic quality control based on gold units. Gold units are unambiguous question for which an answer is provided by the requester. Contributors need to answer at least 4 gold questions with a minimum 70% accuracy to get their answers included in the results. In *Crowdflower* requesters can create jobs, which consist of a data file and units. Units contain the tasks to be performed and are instantiated using the data file. Before ordering, requesters can calibrate the number of judgements per unit, number of judgements per page, and worker pay per page.

5.2 Crowd-Sourcing Study Design

We designed two crowd studies: *Study 1* addresses the representativeness in images provided by the machine-analysis step, while *Study 2* addresses the diversification of the representative images extracted in *Study 1*. The studies were adapted to the monument retrieval scenario used for validating our approach.

Study 1 - representativeness. The representativeness task collected data on the variable locations (i.e., indoor, outdoor), relevance, and representativeness. In this study, relevant pictures contain, partially or entirely, the query monument. Representative images are prototypical outside views of a monument. Relevance is an objective concept indicating the presence/absence of the monument, or part of it, while representativeness is a more subjective concept that might depend on visual context or personal perception.

The task was divided in two parts: first, participants were familiarized with the task and provided with a contextualized visual example of the monument query (*Wikipedia* entry of the monument). Secondly, contributors were asked to answer questions on location, relevance, and representativeness for a given retrieval result.

Study 2 - diversification. The diversification task collected data on perceived diversity among the set of representative pictures of the same monument that were provided by Study 1. Participants assessed visual variation considering the use case scenario of constituting a monument photo album.

As for the previous task, first participants were given an introduction to the task accompanied with some visual examples including a link to the *Wikipedia* page of the monument. Afterwards, contributors were asked to answer whether they would include the provided pictures in the photo album.

6   Experimental Results and Discussion

To validate our approach we use a data set of more than 25,000 images depicting 94 Italian monument locations, from very famous ones (e.g., "Verona Arena") to lesser known to the grand public (e.g., "Basilica of San Zeno"[1]). Images were retrieved from *Picasa, Flickr* and *Panoramio* using both the name of the monument and GPS tags (with a certain radius). For each monument, when available, we retain the first 100 retrieved images per search engine, thus around 300 image in total per monument. To serve as ground truth for validation, each of the images was manually labeled (as being representative or not) by several experts with extensive knowledge of monument locations.

6.1   Validation of Machine Media Analysis

The proposed machine media analysis aims to refine the initial retrieval results by retaining for each query only a small set of around $P = 10$ images (see Step 6 in Section 4) that are both representative and diverse. To assess performance, we use the retrieval precision computed as $tp/(tp + fp)$, where tp is the number of true positives and fp are the false positives.

The first test consisted on determining the influence of the content descriptors on the precision of the results. We experimented with various state-of-the-art approaches:

• *MPEG-7 and related texture and color information*: we compute color autocorrelogram (autocorr.), Color Coherence Vector (CCV), color histogram (hist.), color layout (c.layout), color structure (c.struct), color moments (c.moment), edge histogram (edgehist.), Local Binary Patterns (LBP), Local Ternary Patterns (LTP) and color histogram in [16] (c.n.hist).

• *feature descriptors* that consist of Bag-of-Visual-Words representations of Histogram of oriented Gradients (HoG), Harris corner detector (Harris), STAR features, Maximally Stable Extremal Regions (MSER), Speeded Up Robust Feature (SURF) and Good Features to Track (GOOD). We use 4,000 words dictionaries, experimentally determined after testing different dictionary sizes with various descriptors.

Table 1 summarizes some of the results (we report the global average precision over all the results of the 94 monument queries; descriptors are fused using early fusion). Most interesting is that for this particular set-up very low complexity texture descriptors are able to provide results very close to the use of much more complex feature point

---

[1] Data set is available at http://www.cubrikproject.eu, FP7 CUbRIK project.

Table 1. Average precision for various descriptor combinations

autocorr.	CCV	hist.	c.layout	c.struct	**c.moment**	edgehist.	LBP	LTP
57.70%	58.80%	57.98%	57.70%	60.00%	**60.80%**	56.67%	56.58%	57.68%
Harris	STAR	MSER	SURF	GOOD	c.struct & GOOD	c.struct & HoG	c.moment &GOOD	all color desc.
57.84%	58.87%	59.17%	59.18%	60.39%	60.59%	59.02%	59.71%	58.43%
c.n.hist	HoG	all color desc. & GOOD						
58.33%	58.33%	58.73%						

Table 2. Performance comparison (average precision)

proposed	method in [17]	*Picasa*	*Flickr*	*Panoramio*
60.8%	46.8%	39.47%	53.51%	39.85%

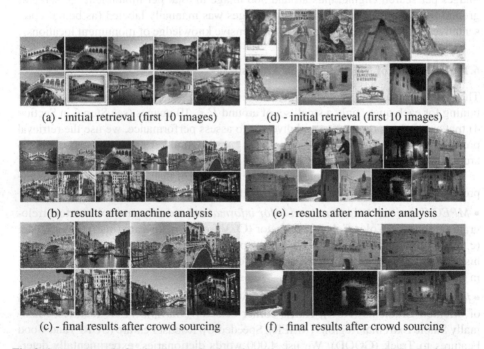

(a) - initial retrieval (first 10 images)   (d) - initial retrieval (first 10 images)

(b) - results after machine analysis   (e) - results after machine analysis

(c) - final results after crowd sourcing   (f) - final results after crowd sourcing

Fig. 2. Output example for the proposed approach: (a)(b)(c) results for query "*Rialto Bridge*" and (d)(e)(f) results for query "*Castle of Otranto*" (image sources: reference *Wikipedia* (image in red rectangle), others *Picasa*, *Flickr* and *Panoramio*).

representations. Therefore, for a higher computational speed, one may go along with a simpler approach without loosing performance. The highest precision is provided by color moments, 60.8%, which are further employed for the machine-crowd integration.

Another interesting result is the limited representative power provided by automatic content descriptors in this context. Regardless the feature used, the disparity of the performance is within a [56%;60%] precision interval. This basically shows the limitation of automated machine analysis and the need for addressing other information sources.

Nevertheless, results achieved by media analysis show significant improvement over the initial retrieval. Table 2 compares these results against the initial retrieval given by the three image search engines and an approach from the literature. The average improvement in precision over initial retrieval is more than 16%. In addition, we achieve an improvement of more then 23% over the approach in [17], that is very promising.

To have a subjective measure of performance, several example results are illustrated in Figure 2 (for visualization reasons, we also display a *Wikipedia* picture of the query - depicted with the red rectangle).

One may observe that compared to the initial images, the automatic machine analysis allows for significant improvement of representativeness and diversity. However, not all the results are perfect, the limited representative power of content descriptors may lead to misclassification, e.g., in Figure 2.(b) some of the initial duplicates are filtered - Figure 2.(a) images 4 and 8, but near-duplicates may appear in the final results, e.g., Figure 2.(b) images 2 and 5.

6.2 Validation of Crowd-Sourcing Analysis

This section reports the results of the representativeness and diversity studies on the images selected by the automated media analysis step. We discuss implications for media analysis and future work. The machine analysis step reduced the initial 25,000 images to only 915 for all the 94 monuments, data set that is further used in the crowd studies.

Crowd-sourcing results on representativeness. The representative study was conducted between 15th and 21st November 2012. Each unit contained one image, which was judged by at least three contributors. Contributors earned 0.07$ per unit. Contributors from 20 different countries located in diverse world regions were allowed to access the job. In total, 228 contributors judged 5,377 units. Due to platform's quality control method, 18% of the units were annotated more than three times. Contributors performed an average of 23.7 units (SD=25.5) with a minimum of 8 and maximum of 174 units. Most of the contributors were located in India (21%), Indonesia (18%), USA (15%), Germany (14%), Canada (9%), Italy (8%), and Morocco (6%).

Reliability analysis was calculated using Kappa statistics that measure the level of agreement among annotators discarding agreement given by chance (values from -1 to 1) [18]. As general guideline, Kappa values higher than 0.6 are considered adequate and above 0.8 are considered almost perfect [11]. There are several variations of Cohen's Kappa depending on categories and raters. For this study, fixed marginal multirater Kappa was used [18]. This variation of Cohen's Kappa is indicated when the number of categories is fixed and there are more than two raters. For analysis purposes, only images annotated exactly three times were considered. In total, 749 images were considered in the reliability analysis. Reliability among annotations achieve a Kappa value of 0.7 for location, 0.44 for relevance and 0.32 for representativeness.

In order to aggregate crowd-sourcing results, the average value among contributors' judgements for location, relevance and representativeness was calculated. First, categorical values (i.e., "Indoor", "Outdoor", "yes", "no") were coded into binary values. "Indoor" judgements for the variable location were coded with 1 and "Outdoor" judgements were coded with 0. For the variables relevance and representativeness, pictures annotated with "yes" were coded with 1 and those annotated with "no" took the value 0. Average values were calculated per image and variable, and mapped onto a binary distribution. For location, pictures with average value equal or above 0.5 were coded as indoor locations. For relevance and representativeness, pictures with average values equal or higher than 0.5 were considered both as related and representative.

Averaged results indicate that 81% of the images depict an outdoor location, 63% contain at least a part of the monument and 57% are representative. As the definition of representativeness used in this paper implies that a picture is representative if it depicts an outdoor scenario, distributions were also calculated considering just images annotated as outdoor. In this case, results show that the percentage of images containing the monument is up to 66%, the percentage of representative pictures is 60%.

Crowd-sourcing results on diversity. Images judged as representative were included in the diversity task, grouped by monument. The diversity study was conducted on November the 26th. In total, 499 images grouped in 82 units (i.e. monuments) were annotated. As the estimated time per unit was similar as in the Study 1, contributors also earned 0.07$ per unit. Units were judged by at least three contributors. In total, 62 contributors participated in 262 units. Number of performed units per contributor varied from 2 to 24 units. Most of the contributors were located in Indonesia (30%), Italy (26%), India (24%), Germany (5%), Brazil (4%), and United Kingdom (4%). Results were averaged and mapped onto a binary distribution using a similar procedure as in Study 1. Aggregated averaged results indicate that: 48% of the monument-grouped images contain all representative and diverse images; some 73% of the grouped images contain at least 75% representative and diverse images, 90% of the group images contain at least 50% representative and diverse images.

Crowd-sourcing is a promising approach for human validation. The results of this study support existing research which identify low costs and reduced annotation time as the advantages of using crowd-sourcing. However crowd-sourcing is not a perfect system, issues such as quality control and reliability of results need further investigation. Automatic quality control methods, as the gold units used in this study, can be a good option since data cleaning is a time-consuming and tedious task [14]. However, they do not always ensure high quality data; since answers can only be rejected at run-time, they may attract more spammers, malicious, and sloppy workers [12] [14].

Reliability of results may vary depending on the level of subjectiveness of the measured variable. For example, annotations for the variable location (i.e. indoor or outdoor) achieve an adequate level of reliability, while the reliability of the annotations for representativeness is quite low. These results suggest that representativeness is a concept subject of individual variations (e.g., visual perception, previous experiences, level of expertise), image features (e.g., perspective, color, and scene composition), or monuments' features (e.g., popularity, distinguishable features, and kind of monument). These issues should be further investigated and considered in the development of methods for aggregation of crowd-sourcing results, since current aggregation methods may underestimate the value of heterogeneous answers [19].

6.3 Validation of the Machine-Crowd

The final experiment consisted in analyzing the performance of the whole machine-crowd chain. The overall average precision after the crowd step is up to 78% which is an improvement over the machine analysis and initial retrieval results (see Table 2). Several examples are depicted in Figures 2.(c) and 2.(f). In general, if enough representative pictures are provided after the machine analysis, the crowd-sourcing step allows for increasing the diversity among them; while for the case when not enough representative pictures are available, the crowd-sourcing tends to increase the relevance (this is usually due to the fact that these pictures are already highly diverse, but not that relevant).

Overall, the use of machine and human validation allows for better performance than using solely the automated media analysis. The fact that crowd-sourcing acts like a human computational machine allows its integration in the processing chain. Depending on the complexity of the content descriptors (e.g., Bag-of-Visual-Words), crowd-sourcing may yield faster response than the machine analysis, but it is limited in the accuracy of the results for large data sets and therefore running it directly on the initial data is not efficient. Crowd-sourcing is not capable of providing perfect result, despite the use of human expertise. The main reason is that low monetary costs attracts people with limited experience to the task and results may be variable.

7 Conclusions

We addressed the problem of enforcing representativeness and diversity in noisy images sets retrieved from common social image search engines. To this end, we introduced a hybrid approach that combines a pre-filtering step carried out with an automated machine analysis with a crowd-sourcing study for final refinement. The motivation of using the media analysis is also to reduce the workload in the crowd-sourcing tasks for enabling better results. Experimental validation was conducted on more than 25,000 photos in the context of the retrieval of photos with monuments. Results show that automatic media analysis reached the point where further performance improvement requires the use of human intelligence, since regardless the image descriptors use, they are limited to reach only up to 60% precision. Instead, the further use of crowd-sourcing led to an improving of both representativeness and diversity in the final results. Thanks to the initial diversification of the results, after crowd-sourcing some 73% of the grouped images contained a promising number of at least 75% representative and diverse images. Future work will mainly consist on adapting the proposed approach to the large scale media analysis constraints.

Acknowledgment. This research is supported by the CUbRIK project, an IP funded within the FP7/2007-2013 under grant agreement n287704 and by the Romanian Sectoral Operational Programme POSDRU/89/1.5/S/62557.

References

1. Rae, A., Kelm, P.: Working Notes for the Placing Task at MediaEval. In: MediaEval 2012 Workshop, Pisa, Italy, October 4-5. CEUR-WS.org (2012) ISSN 1613-0073
2. Jordan, C., Watters, C.: Extending the Rocchio Relevance Feedback Algorithm to Provide Contextual Retrieval. In: Favela, J., Menasalvas, E., Chávez, E. (eds.) AWIC 2004. LNCS (LNAI), vol. 3034, pp. 135–144. Springer, Heidelberg (2004)
3. Elsas, J.L., Donmez, P., Callan, J., Carbonell, J.G.: Pairwise Document Classification for Relevance Feedback. In: TREC 2009 (2009)
4. Quinn, A.J., Bederson, B.B.: Human computation: a survey and taxonomy of a growing field. In: Proceedings of the SIGCHI Conference on Human Factors in Computing Systems (CHI 2011), pp. 1403–1412. ACM, New York (2011)
5. Soleymani, M., Larson, M.: Crowd-sourcing for affective annotation of video: Development of a viewer-reported boredom corpus. In: SIGIR Workshop on Crowd-sourcing for Search Evaluation (2010)
6. Taneva, B., Kacimi, M., Weikum, G.: Gathering and ranking photos of named entities with high precision, high recall, and diversity. In: ACM on Web Search and Data Mining, USA, pp. 431–440 (2010)
7. Zhu, X., Goldberg, A., Gael, J.V., Andrzejewski, D.: Improving Diversity in Ranking using Absorbing Random Walks, pp. 97–104 (2007)
8. Deselaers, T., Gass, T., Dreuw, P., Ney, H.: Jointly optimising relevance and diversity in image retrieval. In: ACM on Image and Video Retrieval, USA, pp. 39:1–39:8 (2009)
9. Huang, Z., Hu, B., Cheng, H., Shen, H., Liu, H., Zhou, X.: Mining near-duplicate graph for cluster-based reranking of web video search results. ACM Trans. Inf. Syst. 28, 22:1–22:27 (2010)
10. Radu, A.-L., Stöttinger, J., Ionescu, B., Menéndez, M., Giunchiglia, F.: Representativeness and Diversity in Photos via Crowd-Sourced Media Analysis. In: 10th International Workshop on Adaptive Multimedia Retrieval, AMR 2012, Copenhagen, Denmark, October 24-25 (2012)
11. Nowak, S., Rüger, S.: How reliable are annotations via crowd-sourcing? a study about inter-annotator agreement for multi-label image annotation. In: Int. Conf. on Multimedia Information Retrieval (2010)
12. Kittur, A., Chi, E.H., Suh, B.: Crowd-sourcing user studies with Mechanical Turk. In: SIGCHI Conf. on Human Factors in Computing Systems, Italy, pp. 453–456 (2008)
13. Crandall, D.J., Backstrom, L., Huttenlocher, D., Kleinberg, J.: Mapping the world's photos. In: Int. Conf. on World Wide Web, pp. 761–770 (2009)
14. Eickhoff, C., de Vries, A.P.: Increasing Cheat Robustness of Crowd-sourcing Tasks (2012)
15. Rudinac, S., Hanjalic, A., Larson, M.: Finding representative and diverse community contributed images to create visual summaries of geographic areas. In: ACM Int. Conf. on Multimedia, pp. 1109–1112 (2011)
16. Van De Weijer, J., Schmid, C.: Applying Color Names to Image Description. In: IEEE Int. Conf. on Image Processing, USA, p. 493 (2007)
17. Kennedy, L.S., Naaman, M.: Generating diverse and representative image search results for landmarks. In: Int. Conf. on World Wide Web, China, pp. 297–306 (2008)
18. Randolph, J.J., Bednarik, R., Myller, N.: Author Note: Free-Marginal Multirater Kappa (multirater kfree): An Alternative to Fleiss' Fixed - Marginal Multirater Kappa
19. Noble, J.A.: Minority voices of crowd-sourcing: why we should pay attention to every member of the crowd. In: ACM Conf. on Computer Supported Cooperative Work Companion, USA, pp. 179–182 (2012)
20. Li, J., Ma, Q., Asano, Y., Yoshikawa, M.: Re-ranking by multi-modal relevance feedback for content-based social image retrieval. In: Sheng, Q.Z., Wang, G., Jensen, C.S., Xu, G. (eds.) APWeb 2012. LNCS, vol. 7235, pp. 399–410. Springer, Heidelberg (2012)

Visual Saliency Weighting and Cross-Domain Manifold Ranking for Sketch-Based Image Retrieval

Takahiko Furuya and Ryutarou Ohbuchi

Graduate School of Medicine and Engineering, University of Yamanashi, Japan
{g13dm003,ohbuchi}@yamanashi.ac.jp

Abstract. A Sketch-Based Image Retrieval (SBIR) algorithm compares a line-drawing sketch with images. The comparison is made difficult by image background clutter. A query sketch includes an object of interest only, while database images would also contain background clutters. In addition, variability of hand-drawn sketches, due to "stroke noise" such as disconnected and/or wobbly lines, also makes the comparison difficult. Our proposed SBIR algorithm compares edges detected in an image with lines in a sketch. To emphasize presumed object of interest and disregard backgrounds, we employ Visual Saliency Weighting (VSW) of edges in the database image. To effectively compare the sketch containing stroke noise with database images, we employ Cross-Domain Manifold Ranking (CDMR), a manifold-based distance metric learning algorithm. Our experimental evaluation using two SBIR benchmarks showed that the combination of VSW and CDMR significantly improves retrieval accuracy.

Keywords: sketch-based image retrieval, visual saliency detection, cross-domain matching, manifold ranking.

1 Introduction

Querying modality is an issue central to image retrieval. Query-by-keywords may be the easiest among query modalities for users to perform, but its retrieval accuracy is not satisfactory. Images taken by phones and cameras generally do not have keyword tags, and if they do, specifying image content by keywords alone can be difficult. An alternative is content-based image retrieval, which uses an example image or a user-drawn sketch of some kind to query images. The first option, querying by image example, is not practical as a user often does not have an image appropriate for query. The second option, Sketch-Based Image Retrieval (SBIR), has become a popular modality of late. Earlier SBIR systems, mostly likely using a mouse as an input device, have drawn sketches by using simple geometrical primitives, such as straight lines, rectangles or circles for drawing, possibly combined with color specification. Recently, with prevalence of touch- and pen-based devices, line-drawing sketch has become the most popular querying modality for content-based image retrieval.

The issue in SBIR is an effective comparison between a line drawing sketch and a 2D image. The comparison should be *robust against background clutter* in the 2D

C. Gurrin et al. (Eds.): MMM 2014, Part I, LNCS 8325, pp. 37–49, 2014.

image. It should also be robust against "stroke noise" of the line drawing sketch, that is, wobbling and/or disconnected lines and difference in drawing styles.

A SBIR algorithm uses a common ground representation, either line/edge image [1-5] or gradient field image [6, 7], for sketch-to-image comparison. Then, one or more image features are extracted from the common ground representation for comparison. Saavedra et al. [3] uses a set of local histograms of Canny edge orientations as an image feature, while Eitz et al. [2] combines several orientation sensitive image features including a variant of HoG [14]. The BF-GFHoG by Hu et al. [6][7] employs gradient field image as the common ground and extracts HoG descriptors. Hundreds of HoG descriptors extracted from a gradient field image are integrated into a feature vector per image by using the Bag-of-Features (BF) approach. To our knowledge, BF-GFHoG is one of the best performing SBIR algorithms.

These algorithms perform well if database images do not have any background clutter and if all the strokes in sketch images are stable and well-connected. However, images contain background clutters, and sketch strokes are disconnected or wobbly. For example, images shown in Figure 1a contain leaves and clouds in their backgrounds, producing noisy Canny edge images in Figure 1c. A casual photograph taken by using phones would contain even more clutters. Also, hand-drawn sketches contain stroke noise. Comparison of a sketch having stroke-noise with an image having background-clutter would lead to low retrieval accuracy.

To suppress background clutter, visual saliency map is often employed. A visual saliency map, which approximates visual attention of human, is used for image segmentation [16], object detection [17], and other applications. Recently, Yang et al. [8] proposed a graph-based saliency detection method called Manifold Ranking-based Saliency Detection (MRSD). The MRSD yields better saliency maps than existing methods. Our proposed algorithm uses the MRSD algorithm to emphasize foreground object of an image for background-clutter-free comparison of a sketch with an image.

To perform robust sketch-to-image comparison under the presence of sketch stroke noise, distance metric learning has been employed [9][12][15]. For example, Weinberger et al. [15] compares handwritten digits in a subspace where distances among feature vectors are robust against stroke noise. Recently, we proposed Cross-Domain Manifold Ranking (CDMR) algorithm for the task of sketch-to-3D model "cross-domain" comparison [12]. The CDMR is based on Manifold Ranking proposed by Zhou et al. [9]. It learns distributions, or manifolds, of sketch features and 3D model features to improve distance (or similarity) computation among them. Our proposed algorithm adopts the CDMR algorithm for improved sketch-to-image comparison.

In this paper, we aim for a SBIR algorithm that is *robust against background clutters* in images and *robust against stroke noise* in sketches. Figure 2 illustrates the overall processing pipeline of the proposed algorithm.

To gain robustness against background clutters, the algorithm tries to emphasize a region presumed to be foreground, *i.e.*, the object sought for by the sketch query. The emphasis, called Visual Saliency Weighting, is done by multiplying, pixel-by-pixel, a Canny edge image with the visual saliency map computed by using the MRSD algorithm [8]. Figure 1d shows the examples of visual-saliency-weighted edge images. Background clutter, i.e., edges of leaves and clouds, are effectively suppressed.

To gain robustness against stroke noise in sketch, the algorithm employs data-adaptive distance metric learning. Relevance values from a sketch to images in the database are computed by using the CDMR algorithm [12]. The CDMR uses a Cross Domain Manifold (CDM) constructed by using similarity values due to multiple, heterogeneous features, each of which is may be optimized for a given comparison task. A manifold of database images is computed (without edge detection) by using the densely sampled SIFT [11] features. A manifold of sketches is computed by using the BF-fGALIF [12] feature. Then these two manifolds in two different domains are coupled into a CDM by using the BF-fGALIF feature. Once the CDM is constructed, relevance values from a sketch to images in a database are computed as diffusion distances over the CDM. The CDMR may be used in either semi-supervised, supervised or unsupervised mode. In this paper, we use the CDMR in unsupervised mode.

Note that the CDMR algorithm has a built-in ability to perform automatic query expansion if an unlabeled (or labeled) corpus of sketches is available. Relevance diffused from a sketch query turns sketches similar to it into secondary sources of relevance diffusion, creating an expanded set of queries. Such a corpus of sketches may be collected beforehand, or collected online over time as sketch queries are made.

We experimentally evaluated the proposed algorithm by using two sketch-based image retrieval benchmarks by Hu et al., that are, the Flickr160 [6] and the Flickr15k [7]. Small but consistent improvement in retrieval accuracy is observed for both benchmarks when the VSW is applied. The CDMR improved retrieval accuracy very significantly. For example, for the Flickr160 benchmark, the combination of the VSW and the CDMR produced MAP score of 72.3 %, which is about 18 % higher than 54.0 % of the BF-GFHoG reported in [6].

Contribution of this paper can be summarized as follows;

- Proposal of a novel sketch-based image retrieval algorithm. It employs Visual Saliency Weighting (VSW) to suppress background clutter in images. The features extracted from edge images processed by VSW are compared against the feature of a sketch query by using the Cross-Domain Manifold Ranking (CDMR), a distance metric learning algorithm adept at comparing heterogeneous feature domains.
- Experimental evaluation of the proposed algorithm using multiple benchmarks, which showed effectiveness of the proposed algorithm.

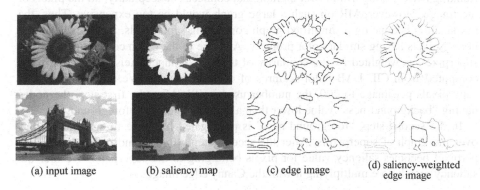

| (a) input image | (b) saliency map | (c) edge image | (d) saliency-weighted edge image |

Fig. 1. Examples of visual saliency weighting

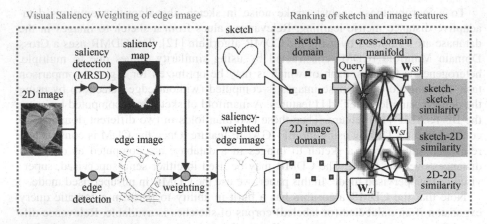

Fig. 2. Outline of the proposed method that employs Visual Saliency Weighting (VSW) of Canny edge images and Cross-Domain Manifold Ranking (CDMR) of image features

2   Proposed Method

2.1   Visual Saliency Weighting of Edge Image

Our algorithm converts (2D) images in a database into saliency-weighted edge images for comparison with sketches. The algorithm first computes Canny edge image from the database image. We used the parameters found in [2] for Canny edge detection; low threshold and high threshold are set to 0.05 and 0.2, respectively. Then, edges due to background clutters are suppressed by using *Visual Saliency Weighting* (*VSW*), which multiplies, pixel-by-pixel, a visual saliency map with the Canny edge image. Visual saliency map is computed by using the MRSD algorithm by Yang et al. [8]. The MRSD algorithm computes visual saliency in two steps, assuming periphery of an image as background.

In the first step, "background-ness" is propagated from image periphery at four sides of the image toward the center. The propagation is performed by using Manifold Ranking (MR) [9] algorithm over a graph that connects, conceptually, all the pixels of the image. However, MR on such a large graph would be too expensive. Thus, the propagation is done on a simplified graph connecting superpixels, a cluster of neighboring pixels having similar color property. An edge of the graph connecting a pair of superpixels is weighted by the similarity of the pair of superpixels. The similarity is computed from CIE LAB color features of each superpixel. We set the number of superpixels per image to 200, the number used in [8]. After the first step, the nodes having "background-ness" value lower than a threshold becomes foreground.

In the second step, "foreground-ness" is propagated from the foreground regions over the graph of superpixels. After the second step, "foreground-ness" of each superpixel becomes saliency value for pixels belonging to the superpixel. We blur the saliency map before multiplying it with the Canny edge image.

Figure 1d shows examples of saliency-weighted edge images. Edges on background (e.g., edges on leaves and edges on clouds) are effectively suppressed.

2.2   Cross-Domain Manifold Ranking for Sketch-Based Image Retrieval

Cross-Domain Manifold Ranking (CDMR)
The CDMR [12] consists of two stages; Cross-Domain Manifold (CDM) generation stage and relevance diffusion stage.

In the CDM generation stage, a CDM matrix $\mathbf{W}$ is generated. $\mathbf{W}$ is a graph whose vertices are the features from a sketch domain and an image domain. Given the number of sketches N_S and the number of images N_I, the matrix $\mathbf{W}$ has the size $(N_S + N_I) \times (N_S + N_I)$.

$$\mathbf{W} = \begin{pmatrix} \mathbf{W}_{SS} & \mathbf{W}_{SI} \\ \mathbf{W}_{IS} & \mathbf{W}_{II} \end{pmatrix} \tag{1}$$

The submatrix $\mathbf{W}_{SS}$ having size $N_S \times N_S$ is the manifold graph of sketch features generated by linking features of sketches produced by the BF-fGALIF [12] algorithm. We will describe the BF-fGALIF in the following section. An edge of the graph $\mathbf{W}_{SS}$ connecting vertices i and j is undirected and has a weight, which is a similarity $w(i, j)$ among the vertices. The similarity $w(i, j)$ is computed by using the equation (2) after normalizing the distance $d(i, j)$ of features i and j to range [0,1].

$$w(i, j) = \begin{cases} \exp(-d(i, j)/\sigma) & \text{if } i \neq j \\ 0 & \text{otherwize} \end{cases} \tag{2}$$

The submatrix $\mathbf{W}_{II}$ having size $N_I \times N_I$ is a manifold graph of image features. It is created in a similar manner as $\mathbf{W}_{SS}$. Features for image-to-image comparison are computed by using BF-DSIFT [13] algorithm.

The submatrix $\mathbf{W}_{SI}$ of size $N_S \times N_I$ couples two submanifolds $\mathbf{W}_{SS}$ and $\mathbf{W}_{II}$ that lie in different domains, that are, sketch feature domain and image feature domain. To compute a feature similarity $w(i, j)$ between a sketch i and an image j, BF-fGALIF features are extracted from a sketch image i and a saliency-weighted edge image generated from an image j. Finally, feature similarity $w(i, j)$ is computed by using equation (2).

Submatrix $\mathbf{W}_{IS}$ of size $N_I \times N_S$ is a zero matrix as we assume no diffusion of relevance occurs from image features to sketch features.

In the relevance diffusion stage, ranking of images to a query sketch is done by diffusing relevance value from the query to the images over the CDM by using MR [9]. We normalize $\mathbf{W}$ for $\mathbf{S}$ by using the following equation;

$$\mathbf{S} = \mathbf{D}^{-1/2} \mathbf{W} \mathbf{D}^{-1/2} \tag{3}$$

where $\mathbf{D}$ is a diagonal matrix whose diagonal element is $\mathbf{D}_{ii} = \sum_j \mathbf{W}_{ij}$. We use the following equation to find rank values in $\mathbf{F}$ given initial value, or "source" matrix $\mathbf{Y}$;

$$F = (I - \alpha S)^{-1} Y \qquad (4)$$

Y is a diagonal matrix of size $(N_S + N_I) \times (N_S + N_I)$ that defines source(s) of relevance value diffusion. If a vertex i is the source of diffusion $Y_{ii} = 1$, and, if not, $Y_{ii} = 0$. In our case, the vertex corresponding to the query sketch becomes the source of diffusion. F_{ij} is the relevance value of the image j given the sketch i. The higher the relevance value F_{ij}, the higher the rank of the image j in the retrieval result.

The ranking is very robust, as diffusion from the query sketch to the images occurs via multiple paths. For example, the relevance value may first diffuse quickly through sketches similar to the query before reaching to the images. In such a case, the CDMR embodies a form of query expansion.

The parameter σ in equation (2) controls diffusion of relevance value across the CDM. We use different values σ_{SS}, σ_{II}, and σ_{SI} for each of the submatrices W_{SS}, W_{II}, and W_{SI} as optimal value of σ depends on each submatrix. The parameter $\alpha = [0,1)$ in equation (4) controls regularization.

Computing Feature Similarities for Cross-Domain Manifold
Generation of the CDM W requires computation of similarities in the submatrices W_{SS}, W_{II}, and W_{SI}. In this section, we describe features used for computing each submatrix.

Computing Similarities for W_{SI}
Figure 3a shows a feature comparison pipeline for W_{SI}. Given an image in a database, it is resized so that the shorter edge of the image becomes 256 pixels. A saliency-weighted edge image is generated from the resized image by the VSW described in Section 2.1.

From each of the saliency-weighted edge images and the sketch images, we extract BF-fGALIF feature, which is shown to be among the best methods for sketch-based 3D shape retrieval [12]. GALIF features [10] are extracted densely at regular grid points on the image. GALIF feature captures orientation of lines and intensity gradient in the image by using Gabor filters. Parameters for GALIF we used in this paper are the same as those in [12]. 1,000 to 1,500 GALIF features are extracted per image.

The set of GALIF features extracted from an image is integrated into a feature vector per image by using a standard bag-of-features (BF) approach. This integration reduces cost of image-to-image matching significantly compared to directly comparing a set of features to another set of features. We used vocabulary size $k=3,500$ for the experiments. We used k-means clustering to learn the vocabulary, and used kd-tree to accelerate vector quantization of GALIF features into words of vocabulary.

A BF-fGALIF feature of a sketch image is compared against a BF-fGALIF feature of a saliency-weighted edge image by using Cosine distance.

Computing similarities for W_{SS}
Similarities for the submatrix W_{SS} is computed, again, by the BF-fGALIF algorithm (Figure 3b). For each sketch image, BF-fGALIF feature is extracted after the sketch

image is resized down to 256×256 pixels. A distance between BF-fGALIF features is computed using Cosine distance. All the parameters for the BF-fGALIF (i.e., parameters for Gabor filter, number of GALIF features per image, and vocabulary size) are the same as those used to compute $\mathbf{W}_{SI}$.

Computing Similarities for $\mathbf{W}_{II}$

Similarities for the submatrix $\mathbf{W}_{II}$ is computed by the BF-DSIFT algorithm [13] (Figure 3c). BF-DSIFT is extracted from grayscale images without edge detection. Given an image, it is resized so that the shorter edge of the image becomes 256 pixels as with computing $\mathbf{W}_{SI}$.

From each resized image, about 3,000 SIFT [11] features are extracted at densely and randomly placed feature points on the image. SIFT has invariance against scaling, rotation, illumination changes and minor changes in viewing direction. The set of about 3,000 SIFT features are integrated into a feature vector per image by using a BF approach. We used vocabulary size k=3,500 for the experiments. A distance between BF-DSIFT features is computed using symmetric version of Kullback-Leibler Divergence.

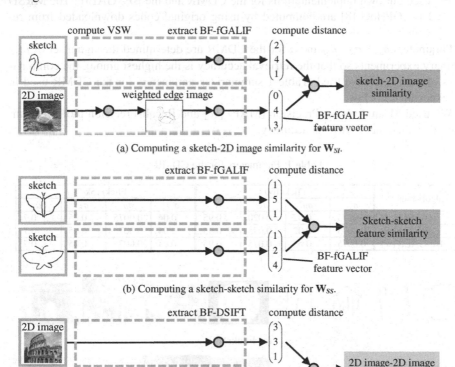

(a) Computing a sketch-2D image similarity for $\mathbf{W}_{SI}$.

(b) Computing a sketch-sketch similarity for $\mathbf{W}_{SS}$.

(c) Computing a 2D image-2D image similarity for $\mathbf{W}_{II}$.

Fig. 3. Feature comparison methods for generating CDM

3   Experiments and Results

We experimentally evaluated effectiveness of weighting by the VSW of Canny edge images and ranking by the CDMR for sketch-based image retrieval. We used two sketch-based image retrieval benchmark databases; the Flickr160 [6] and the Flickr15k [7], both by Hu et al. Figure 4 shows examples of sketch queries and retrieval target images for the two benchmarks.

The Flickr160 consists of a set of 25 sketch queries and a set of 160 retrieval target images. Each of the set of sketch queries and the set of target images is partitioned into 5 categories. The Flickr15k, which is a larger-scale version of the Flickr160, consists of a set of 330 sketch queries and a set of 14,660 retrieval target images. The set of sketch queries is partitioned into 33 shape categories (e.g., "round", "heart-shape", etc.). The set of retrieval target images is partitioned into 60 semantic categories (e.g., "pyramid", "bicycle", etc.). A query in the Flickr15k could belong to multiple semantic categories, for example, a round-shaped sketch query belongs to such semantic categories as "moon", "fire_balloon", and "london_eye".

We used our own implementations for the CDMR and the BF-fGALIF. The MRSD [7] and BF-GFHoG [8] are computed by using original codes downloaded from respective authors' websites.

Parameters σ_{SS}, σ_{II}, σ_{SI} and α for the CDMR are determined through a set of preliminary experiments so that the retrieval accuracy is the highest among the combinations of parameters we tried. Table 1 summarizes parameters for the CDMR used in the experiments below.

We used Mean Average Precision (MAP) [%] and Recall-Precision plot for quantitative evaluation of retrieval accuracy.

Table 1. Parameters for the CDMR

algorithms	Flickr160				Flickr15k			
	σ_{SS}	σ_{II}	σ_{SI}	α	σ_{SS}	σ_{II}	σ_{SI}	α
BF-GFHoG	0.0075	0.02	0.020	0.95	0.04	0.0075	0.05	0.7
BF-GALIF	0.0075	0.02	0,075	0.95	0.04	0.0075	0.05	0.7
BF-GALIF(w)	0.0075	0.02	0.075	0.95	0.04	0.0075	0.05	0.7

(a) Flickr160.

(b) Flickr15k.

Fig. 4. Examples of sketch queries and retrieval target images

3.1   Effectiveness of Edge Weighting by VSW

Figure 5 shows the relationship between vocabulary size k and retrieval accuracy for the Flickr160 and the Flickr15k benchmarks. In the figure, "BF-fGALIF" means BF-fGALIF is extracted directly from Canny edge images, while "BF-fGALIF(w)" with "(w)" means that the feature is computed from saliency-weighted edge images generated by the VSW. For both benchmarks, at almost all the vocabulary size, the BF-fGALIF(w) with saliency weighting produced MAP scores about 2 % higher than the BF-fGALIF without saliency weighting.

It can be concluded that edges due to background clutter are suppressed to certain degree by the VSW, resulting in small but consistent improvements in retrieval accuracy.

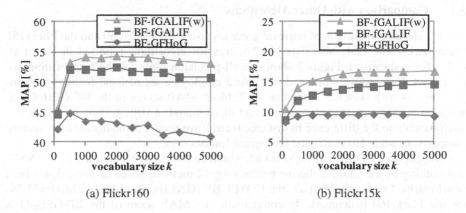

(a) Flickr160   (b) Flickr15k

Fig. 5. Vocabulary size and retrieval accuracy. (Please note the difference in MAP scales.).

3.2   Effectiveness of Ranking by CDMR

Table 2 shows ranking performance of the CDMR for the Flickr160 and the Flickr15k benchmarks. In this experiment, the sketch-to-image comparison algorithm for the submatrix $\mathbf{W}_{SI}$ of the CDMR is selected from the following three; the BF-GFHoG, the BF-fGALIF, and the BF-fGALIF(w). We fixed the sketch-to-sketch comparison algorithm for the submatrix $\mathbf{W}_{SS}$ to the BF-fGALIF. We also fixed the image-to-image comparison algorithm for the submatrix $\mathbf{W}_{II}$ to the BF-DSIFT.

For all the three sketch-to-image comparison algorithms we compared, that are, the BF-GFHoG, the BF-fGALIF, and the BF-fGALIF(w), the CDMR significantly improved their retrieval accuracy for the two benchmarks. In case of the Flickr160, the BF-fGALIF(w) feature using the CDMR produced the highest MAP score of 72.3 %. It is about 19 % better than the BF-fGALIF(w) feature without using the CDMR. In case of the Flickr15k, the BF-fGALIF(w) with the CDMR yielded MAP score of 22.4 %. This score is 6 % higher than the BF-fGALIF(w) without the CDMR.

Overall, gain in retrieval accuracy due to the CDMR is quite significant. Diffusion of relevance via multiple paths through the sketches and images of the CDM makes

Table 2. Feature selection and ranking accuracy by the CDMR

algorithms	Flickr160		Flickr15k	
	without CDMR	with CDMR	without CDMR	with CDMR
BF-GFHoG	44.8	64.1	9.5	15.4
BF-fGALIF	51.6	69.4	14.3	20.6
BF-fGALIF(w)	53.7	72.3	16.5	22.5

the ranking robust against stroke noise and other variations in sketches. Also, diffusion of relevance from the query sketch via sketches similar to the query toward target images pushes retrieval accuracy up, as it works like an automatic query expansion.

3.3 Comparison with Other Algorithms

Table 3 shows comparison of retrieval accuracy using the Flickr160 and the Flickr15k benchmarks. The table also lists MAP scores for algorithms reported in Hu et al. [6][7] for comparison. Figure 7 shows recall-precision plots for the 6 algorithms we compared. Every algorithm listed in Table 2 employs a set of local features to compare images with sketches. Note that, in Table 2, MAP scores of the BF-GFHoG are different among our own experiments and theirs found in [6][7]. This discrepancy is due probably to the difference in distance metric; we used Cosine distance for feature comparison, while Hu et al. used Histogram Intersection.

Our proposed CDMR-BF-fGALIF(w), which employs edge weighting by the VSW and ranking by the CDMR, did the best among 12 methods listed in Table 3, for both benchmarks. Our best performer, the CDMR-BF-fGALIF(w), produced MAP=72.3% for the Flickr160 benchmark. In comparison, the MAP score of the BF-GFHoG is 54% [6]. When compared using the Flickr15k benchmark, the CDMR-BF-fGALIF(w) produced MAP score of 22.5 %. This score is about 10 % higher than that of BF-GFHoG, 12.2%, reported in [7]. Most of the gain in retrieval accuracy comes from the CDMR ranking, but contribution due to visual saliency weighting is consistent.

Recall-precision curves of Figure 6 also shows the advantages in retrieval accuracy of the CDMR-BF-fGALIF(w) over the other algorithm.

Table 3. Comparison of MAP scores [%] among several algorithms

algorithms	Flickr160	Flickr15k
BF-GFHoG	44.8	9.5
BF-fGALIF	51.6	14.3
BF-fGALIF(w)	53.7	16.5
CDMR-BF-GFHoG	64.1	15.4
CDMR-BF-fGALIF	69.4	20.6
CDMR-BF-fGALIF(w)	72.3	22.5
BF-GFHoG [6][7]	54.0	12.2
BF-HoG [6][7]	42.0	10.9
BF-SIFT [6][7]	41.0	9.1
BF-SelfSimilarity [6][7]	42.0	9.6
BF-ShapeContext [7]		8.1
BF-StructureTensor [7]		8.0

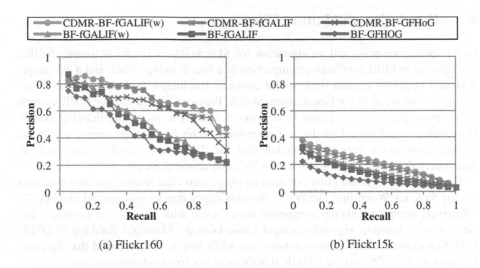

(a) Flickr160   (b) Flickr15k

Fig. 6. Recall-Precision plots

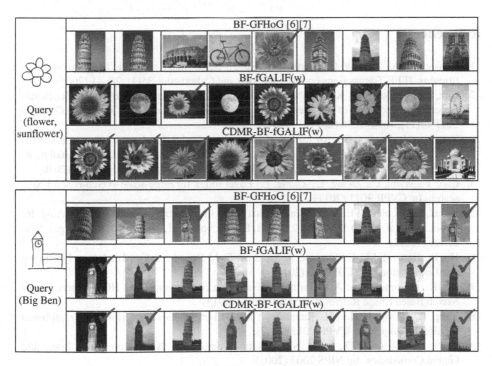

Fig. 7. Retrieval examples for the Flickr15k

4   Conclusion and Future Work

In this paper, we proposed an algorithm for Sketch-Based Image Retrieval (SBIR). A challenge in SBIR is effective comparison of a line drawing sketch and a 2D image. It should be robust against background clutter in the image. It should also be robust against stroke noise in the line drawing sketch. Previous algorithms for SBIR convert a 2D image into an edge image to compare it against a query line drawing sketch. However, unnecessary edges due to background clutter in an image interfere with the feature comparison between a sketch and the 2D image. Stroke noise such as disconnected or wobbly lines also makes the comparison difficult.

Our proposed algorithm first converts an image into edge image, and then performs Visual Saliency Weighting (VSW) to suppress edges due to background clutters. To effectively compare a sketch containing stroke noise with images, we employ a distance metric learning algorithm called Cross-Domain Manifold Ranking (CDMR) [12]. Our experimental evaluation using two SBIR benchmarks showed that the combination of the VSW and the CDMR significantly improves retrieval accuracy.

We are currently looking into the improvement of computational efficiency of the CDMR algorithm, as the CDMR is expensive to compute for a large database.

References

1. Chalechale, A., Naghdy, G., Mertins, A.: Sketch-based image matching using angular partitioning. IEEE Transactions on Systems, Man and Cybernetics 35(1), 28–41 (2005)
2. Eitz, M., Hildebrand, K., Boubekeur, T., Alexa, M.: Sketch-Based Image Retrieval: Benchmark and Bag-of-Features Descriptors. Visualization and Computer Graphics 2011 17(11), 1624–1636 (2011)
3. Saavedra, J., Bustos, B.: An improved histogram of edge local orientations for sketch-based image retrieval. In: Goesele, M., Roth, S., Kuijper, A., Schiele, B., Schindler, K. (eds.) Pattern Recognition. LNCS, vol. 6376, pp. 432–441. Springer, Heidelberg (2010)
4. Cao, Y., Wang, C., Zhang, L., Zhang, L.: Edgel index for large-scale sketch-based image search. In: CVPR 2011 (2011)
5. Bozas, K., Izquierdo, E.: Large scale sketch based image retrieval using patch hashing. In: Bebis, G., et al. (eds.) ISVC 2012, Part I. LNCS, vol. 7431, pp. 210–219. Springer, Heidelberg (2012)
6. Hu, R., Barnard, M., Collomosse, J.: Gradient Field Descriptor for Sketch based Retrieval and Localization. In: ICIP 2010 (2010)
7. Hu, R., Collomosse, J.: A Performance Evaluation of Gradient Field HOG Descriptor for Sketch Based Image Retrieval. In: CVIU 2013 (2013)
8. Yang, C., Zhang, L., Lu, H., Ruan, X., Yang, M.H.: Saliency Detection via Graph-based Manifold Ranking. In: CVPR 2013 (2013)
9. Zhou, D., Bousquet, O., Lal, T.N., Weston, J., Schölkopf, B.: Learning with Local and Global Consistency. In: NIPS 2003 (2003)
10. Eitz, M., Richter, R., Boubekeur, T., Hildebrand, K., Alexa, M.: Sketch-Based Shape Retrieval. ACM TOG 31(4), 1–10 (2012)
11. Lowe, D.G.: Distinctive Image Features from Scale-Invariant Keypoints. IJCV 60(2) (November 2004)

12. Furuya, T., Ohbuchi, R.: Ranking on cross-domain manifold for sketch-based 3D model retrieval. Accepted as Regular Paper, Cyberworlds 2013 (2013)
13. Furuya, T., Ohbuchi, R.: Dense sampling and fast encoding for 3D model retrieval using bag-of-visual features. ACM CIVR 2009, Article No. 26 (2009)
14. Dalal, N., Triggs, B.: Histograms of oriented gradients for human detection. In: CVPR 2005 (2005)
15. Weinberger, K.Q., Saul, L.K.: Distance Metric Learning for Large Margin Nearet Neighbor Classification. In: Journal of Machine Learning Research (JMLR) (2009)
16. Achanta, R., Estrada, F., Wils, P., Süsstrunk, S.: Salient region detection and segmentation. In: Gasteratos, A., Vincze, M., Tsotsos, J.K. (eds.) ICVS 2008. LNCS, vol. 5008, pp. 66–75. Springer, Heidelberg (2008)
17. Hou, X., Zhang, L.: Saliency detection: A spectral residual approach. In: CVPR 2007 (2007)

A Novel Approach for Semantics-Enabled Search of Multimedia Documents on the Web

Lydia Weiland and Ansgar Scherp

University of Mannheim, Germany
{lydia,ansgar}@informatik.uni-mannheim.de

Abstract. We present an analysis of a large corpus of multimedia documents obtained from the web. From this corpus of documents, we have extracted the media assets and the relation information between the assets. In order to conduct our analysis, the assets and relations are represented using a formal ontology. The ontology not only allows for representing the structure of multimedia documents but also to connect with arbitrary background knowledge on the web. The ontology as well as the analysis serve as basis for implementing a novel search engine for multimedia documents on the web.

1 Introduction

Multimedia search on the web is limited to keyword search today. Search engine giants like Google simply just index the textual information encoded in the multimedia documents and the incoming and outgoing hyperlinks. Thus, Google squeezes structured multimedia documents to fit into its Page Rank model for hypertext.[1] In contrast, search for structured multimedia documents such as Silverlight presentations, Flash documents, and Adobe's Edge documents in the new W3C format HTML 5 is still very limited. Structured multimedia documents are composed of media assets like images, videos, audio, and text [6,17]. The multimedia document obtains its structure by organizing the media assets coherently in time, space, and interaction [6,17]. However, this information is not used for indexing and retrieving the content today. In addition, the arrangements of media assets as well as the media assets themselves exhibit certain semantics that is typically not explicitly encoded in the multimedia documents. This makes it hard to search for and within structured multimedia documents, which is of high benefit for a variety of reasons. Multimedia documents that can be better searched by making its media assets accessible for retrieval are better visible in the web. In addition, it enables for a better reuse of media assets in order to save costs and time, e. g., in large enterprises that professionally produce multimedia documents for e-learning, advertisement, or for creating professional websites.

In order to improve search in structured multimedia documents, we conduct an analysis of a large multimedia corpus. For the purpose of analysis, we represent

---

[1] http://support.google.com/webmasters/bin/answer.py?hl=en&answer=72746, access: 3/10/2013

C. Gurrin et al. (Eds.): MMM 2014, Part I, LNCS 8325, pp. 50–61, 2014.

the documents using a generic multimedia document ontology (M2DO). Besides representing media assets and their temporal, spatial, and interaction relations, the M2DO allows for a seamless integration of semantic annotations in form of background knowledge provided from the Linked Open Data (LOD) cloud.[2] On the LOD cloud, data is interlinked and provides machine readable semantics. The M2DO has been designed in terms of a backwards analysis of the existing models [18].

The remainder of the paper is organized as follows: In the following section, we present an illustrative scenario motivating the need for a multimedia document search engine. The related work is discussed in Section 3. In Section 4, we describe the requirements to our ontology, which is presented in Section 5. The results of our analysis are shown and discussed in Section 6, before we conclude the paper in the last section.

2 Scenario

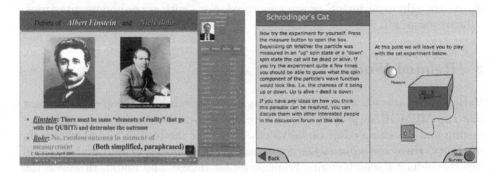

Fig. 1. Multimedia document with audio of a debate between Einstein and Bohr.[3]

Fig. 2. Animation of Schroedinger's Cat.[4]

The scenario illustrates features of a search engine for multimedia documents. We consider the physics teacher Mr. Particle who is preparing his next lecture. He wants to get his pupils interested in quantum mechanics. Thus, he uses as introductory story the famous dispute of Albert Einstein and Niels Bohr at the Solvay physics conference back in 1930, where Einstein doubted the completeness of Bohr's quantum mechanics model. To stress his doubt, Einstein invented a thought experiment called 'photon level'. Mr. Particle wants to visualize the story and topic with an interactive multimedia presentation. Thus, he

---

[2] http://www.w3.org/DesignIssues/LinkedData.html

[3] Taken from: http://nanohub.org/local/breeze/nt501/2005.04.01-Hess/viewer.swf, last accessed: April 2013

[4] Taken from: http://www.gilestv.com/tutorials/find1.swf, last accessed: April 2013

collects appropriate content for reuse in his presentation. He searches for images of the two physicists. Mr. Particle formulates a query to find all multimedia documents, where images appear simultaneously with the terms 'Einstein' and 'Bohr' (Fig. 1). Subsequently, he specializes his query to find only multimedia documents that contain animations, videos, or serious games dealing with 'photon level'. Mr. Particle wants to explain an additional thought experiment, called 'Schroediger's cat' later in the lecture. To find multimedia documents for this experiment, he searches for documents being annotated with the category 'thought experiment'. The resulting documents should either contain embedded videos or interactive elements like the example in Fig. 2.

Overall, we can observe that Mr. Particle conducts queries along the different dimensions of multimedia documents, i.e., time, space, and interaction. In order to find multimedia presentations about persons like Einstein and Bohr and events such as the Solvey conference in 1930, he makes use of higher-level semantics associated with the multimedia documents in form of background knowledge such as the annotations 'thought experiment'. Such a query on the background knowledge allows to find documents that are annotated with the category 'thought experiment' or are annotated, e. g., with synonymous categories, subcategories, etc.

3   Related Work

Various *semantic multimedia retrieval systems* have been developed in the past like the MEMORAe project [15], where ontological knowledge is used for indexing and searching educational videos. Breaking the barrier of a single media modality, there are approaches for semantic cross-media search and retrieval like the semantic search engine Squiggle [8] for images and audio. However, the images and audios do not originate from a multimedia document and thus no spatial, temporal, or interaction relations are considered in Squiggle. The FLash Access and Management Environment (FLAME) [23] is an approach where Flash files where converted to XML in order to extract and index media assets, events, and interaction features. Although the authors recognize the importance of temporal relations between media objects, they are not considered in their work. The video search engine Yovisto [22] and work by Diemert et al. [10] allow for a semantic enrichment and search of audio-visual media content. As such they do not propose a model for representing rich, structured multimedia documents. In the area of *multimedia document models*, we find HTML 5 and Flash and abstract document models such as MM4U [18], ZyX [4], and AHM [13]. These models are targeted towards the presentation of multimedia content to the users. They are not designed for serving as internal representation model for a multimedia retrieval engine. In addition, they do not allow for a seamless integration of LOD background knowledge such as today's very popular Linked Open Data. An extensive overview is beyond the scope of this work and has been conducted earlier (see, e. g., [19,5]). In the area of *multimedia query algebras*, we find the MP7QF query language for audio-visual media content encoded in the MPEG-7 format [11]. The Unified Multimedia Query Algebra (UMQA) aims at integrating different features for multimedia querying such as content-based features, spatio-temporal relations, and traditional metadata [7]. The existing algebras either focus on spatial or temporal relations. The semantics of multimedia documents is typically not considered. Only, the query algebra EMMA [24] allows to state queries against media assets stored in Enhanced Multimedia Meta Objects (EMMOs) and the typed edges, which allow for modeling background knowledge, between EMMOs. However, there is no support for temporal, spatial, and interaction relations between media assets.

Overall, the work on multimedia retrieval is still at its beginning. So far, there is no sophisticated solution available that allows for representing and querying structured multimedia content along time, space, and interaction relations as well as background knowledge provided by the LOD cloud.

4   Requirements to a Semantic Multimedia Search Engine

From this scenario and the related work, we extract the requirements for a multimedia retrieval engine. First, the multimedia documents and its characteristics need to be represented. Multimedia documents are composed of different types of media assets. We have to be able to identify each occurrence of a media asset and assign it to its document [6]. These requirements enable a user to formulate a query like "Show me all audios within a multimedia document containing the keyword 'Solvay' (cf. Scenario)". Second, a multimedia retrieval engine needs to represent the three central relations for time, space and interaction [6,19]. These relations enable queries containing keywords as mentioned in the scenario (cf. Scenario) like 'simultaneously, after, below, on click, etc.'. Third, support for semantic annotations is required to support use of background knowledge.

5   Representing Multimedia Content with the M2DO

As prerequisite for a multimedia search engine, we first need to investigate the nature of multimedia documents on the web. To this end, we consider multimedia documents as graphs and represent them using a multimedia representation ontology. Please note that the existing multimedia document models (see discussion in Section 3) are not sufficient to this purpose as they are aimed as exchange format for the purpose of presenting the content to the users. In addition, the existing models do not support representing background knowledge. Thus, we have developed the Multimedia Document Ontology (M2DO) for the specific purpose of representing the media assets and structure of multimedia documents and its association(s) with background knowledge. The M2DO is designed such that it can be easily used in a multimedia retrieval engine. It is defined in OWL[5] and

---

[5] http://www.w3.org/2001/sw/wiki/OWL, last accessed: July 2013

axiomatized in Description Logics (DL) [2]. This has various advantages: First, it allows for checking consistency of our model due to the DL axioms. Second, existing ontologies can be easier reused. For the development of the M2DO, we made use of a scenario-based methodology, similar to the NeOn methodology[6]. We use the foundational ontology DOLCE+DnS Ultralight (DUL) [14] as basis for M2DO. DUL has proven being well suited for designing ontologies in various domains [20]. DUL makes use of ontology design patterns, allowing for a modularization of the knowledge it formalizes [21]. The M2DO is depicted in Fig. 3 and consists of three patterns. Classes taken from DUL are shown in white. Newly defined classes in M2DO have pale blue background. While the Media Asset Pattern and the Media Occurrence Relation Pattern are newly defined in the M2DO, the Media Annotation Pattern is extended from the Annotation Pattern of the M3O [16].

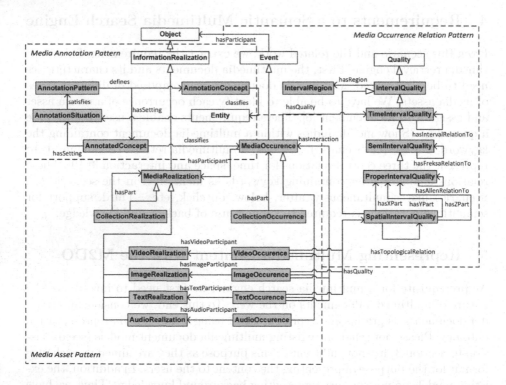

Fig. 3. M2DO Ontology

The *Media Asset Pattern* allows for representing the whole multimedia document and its assets. A Media Realization is an instance of a media type and represents a media asset. Media Occurrences are concrete occurrences of a specific

---

[6] http://www.neon-project.org, last accessed: July 2013

Media Realization. Thus, while one occurrence refers to exactly one realization, a realization can be associated to many occurrences.

The *Media Occurrence Relation Pattern* allows for modeling temporal, spatial, and interaction relations between media assets. The temporal relation is grouped into relations for semi-intervals and proper intervals by Freksa [12] and Allen [1], respectively. Within semi-intervals, either the starting point in time or the ending point in time is known. For proper intervals, both information is given. Spatial relations can be either topological [9], like disjoint, inside or covers, or rectangular [3]. The rectangular relations are based on the Allen calculus, where the 13 relations are pairwise connected to each other. The Cartesian product results in 169 possible combinations for spatial objects. The M2DO allows to represent interactions started by clicking on an image, video, or text occurrence. Interactions are defined using the property m2do:hasInteraction, where different types of interactions like a mouse click, hovering over an element, or hitting the enter key, are distinguished by specialization.

The *Media Annotation Pattern* allows for representing multimedia annotations such as the category "thought experiment" in the scenario. Annotations can be in principle any resources on the web, e. g., the Wikipedia article on thought experiments[7]. The Annotation Pattern of the M3O [16] serves as basis of this pattern. It has been connected to the M2DO via the concepts MediaRealization. With respect to our example in the scenario about thought experiments, the figure would have been connected to the AnnotatedConcept, because that is the asset which is annotated, and the AnnotationConcept is connected to the DBPedia-Link[8], because this is the annotation the asset gets.

The M2DO ontology has been created using the Protégé Ontology editor. We have checked its consistency with the FacT++ reasoner. The Ontology Pitfall Scanner (OOPS)[9] was used to detect errors in the ontology, which cannot be detected by a consistency check in order to remove missing inverse relationships or equivalent properties. The ontology together with its axiomatization in Description Logics is available as OWL file[10].

6   Analysis of a Large Corpus of Multimedia Documents

In 2012, we have crawled around 18.000 Flash files. From these, we have analyzed around 14.000 which we were able to extract with swfmill[11]. The Flash files versions range from 5 to 10 and sizes from 8 KB to 45 MB with an average of 484.75KB and a standard deviation of 1.7MB. For converting the binary flash files using swfmill an average duration of 4 seconds per file was needed.

---

[7] http://en.wikipedia.org/wiki/Thought_experiment, last accessed: April 2013
[8] http://de.dbpedia.org/page/Kategorie:Gedankenexperiment,
   last accessed: April 2013
[9] www.oeg-upm.net/oops, last accessed: July 2013
[10] http://dws.informatik.uni-mannheim.de/fileadmin/lehrstuehle/ki/
   research/M2DO
[11] http://swfmill.org/

Subsequently, we have analyzed the XML files to extract relevant information for a multimedia retrieval engine and represent the Flash file using our M2DO introduced above. The average duration for modeling a file with triples is 27 seconds, with a standard deviation of 483.

6.1   Basic Analysis of Flash File Characteristics

In order to provide insights about the nature of our data set, we have extracted the different media assets and their occurrences grouped by their media type. In Table 1, the averages and standard deviations for every media type is shown. The high standard deviation shows that there is not only a high difference in the sizes of the flash files, but also in the type of the content. Also the minimum and maximum amount of media occurrences (Table 2) per media asset grouped by type is shown. More specific characteristics of image assets within the flash files are shown in Tables 3. The assets are investigated for their widths and heights (in pixels) and their sizes (in KB).

Table 1. Mean, standard deviation, minimum, and maximum number of media assets per file

Amount of	μ	σ	min	max
Text Assets	10.78	30.87	0	276
Image Assets	6.57	15.69	0	173
Video Assets	0.08	0.27	0	1
Audio Assets	0.79	3.43	0	32

Table 2. Occurrences per media asset with mean, standard deviation, minimum, and maximum values

Amount of	μ	σ	max
Text Occurrences	0.78	1.00	5
Image Occurrences	1.25	1.61	8
Video Occurrences	0.12	0.44	3
Audio Occurrences	0.24	0.56	3

We have checked the consistency of 10% of the Flash files, which were randomly chosen from our dataset. Arguing with the assumption that 10% of 18.000 flash files covers a representative amount for all of the different types. There are around 2000 axioms per file with 1% logical axioms. The latter are those of equivalences or properties like transitivity. 30% of the axioms are inferred, e. g., the

Table 3. Characteristics of image asset with mean, standard deviation, minimum and maximum values

Characteristics of Image Occurrences	μ	σ	min	max
Width	297.5	238.73	9	1602
Height	203.1	149.07	4	1202
Size	225.97	343.61	4.59	2292.00

property `hasPart` which is defined in the ontology will automatically be added by its inverse, in this case the `isPart` relation. The consistency checks have been done as preparation to use the data in the multimedia retrieval engine. It allows to identify errors when transforming the XML structure of the flash content to an OWL graph based on the axiomatization of the ontology.

In Figure 4, the distribution of the media assets can be seen, which is given as a power-law distribution. This means that there are a lot of files with a few assets and only a few files with a huge amount of assets.

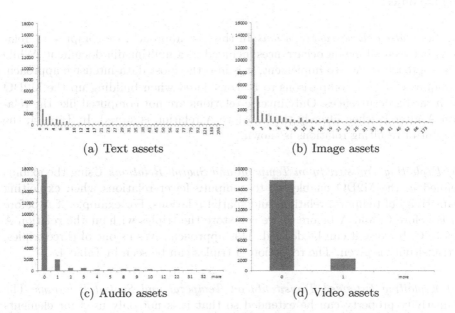

(a) Text assets   (b) Image assets

(c) Audio assets   (d) Video assets

Fig. 4. Distribution of assets of the four different media types, x-axis: Amount of asset, y-axis: Amount of files

6.2 Approaches for Representing Multimedia Documents

We investigate different strategies for representing multimedia documents using our M2DO ontology for the purpose of using this representation in a multimedia retrieval engine. The initial investigation focuses on the representation of time and space. The number of comparisons to be made between the media occurrences increases quadratic with the total number of assets in a document. If all relationships of every occurrence to all others is computed, a large data storage is required. For reasons of comparison calculating every relation is considered in variant (a). A lot of triples do not only lead to storage intensive use, but also to time-consuming queries. Therefore other approaches for representing multimedia documents without restricting or changing our ontology are investigated. We are exploring different strategies: We have decided to investigate a variant (b) that exploits the transitivity of properties such as `before` and `contains`. This idea is also considered in variant (c), where the transitivity of more complex relationships are exploited. Our last approach (d) is derived from the idea that the spatial relations in direction of the z-axis will not be needed. Following down the (negative) z-axis, one can only recognize if some object is behind some other. The user will not be able to distinguish if an object is covered by another objects that hold a before or a meets relation and that are smaller than the original object.

(a) Computing all possible relations In the first approach, we compute all relations between all media occurrences included in a multimedia document. While this approach is easy to implement, it is also the most data-intensive approach. It requires $\frac{n*(n-1)}{2}$ comparisons to be considered when building up the M2DO for n media occurrences. Only inverse relations are not computed like the relation A after B when already the B before A relation is stored. In Table 4, the amount of resulting relations is shown.

(b) Exploiting Transitivity on Temporal and Spatial Relations Using the axioms defined in the M2DO enables us to compute fewer relations when exploiting transitivity of temporal relations and spatial relations. For example, if A before B, B before C, and A before C, we can store the triples without the relation A before C, because it can be derived. This approach saves us one of three triples, if transitivity is given. The reduction of triples can be seen in Table 4.

(c) Exploiting Extended Transitivity on Temporal and Spatial Relations The transitivity property can be extended so that it is not only used for elements which are connected over the same relation, but for all of the 13 Allen relations which have logical similarities, in a sense that transitivity is given. For example, if A during B, B meets C, and A before C, we can again leave out the relation of A and C, because it can be derived by B meets C, which means that C is directly after B, and A is before C, as A is contained in B, which means that A is finished earlier than B. All possible combinations can be seen in [1].

(d) Reducing the Spatial Depth The x- and y-axis expand from 0,0 to the end of the screen. Thus, a user is able to see if two objects are next to each other or if the objects have some distance between them. That is why the spatial relations are important. In contrast, the z-axis has its extension in the direction to the user. This geometry leads to the fact that a user can only distinguish between three situations regarding the z-axis: A before B, A after B, and A equal B. Computing all spatial relations in direction of the z-axis would not give any benefit to the user, as he can only distinguish the three cases. Reducing the z-axis relation to the three cases reduces further the number of triples. In addition, before and after is a pair, which can be derived from each other. Thus, we only need to compute before and equal relations for the z-axis. Thus, only a few relations are needed to represent the whole z-axis relations.

In Table 4, the results for the different approaches described in the paragraphs *(a)* to *(d)* are shown, grouped by temporal and spatial relations. Comparing rows *(a)* and *(b)* shows a high improvement regarding storage capacity using the transitive property (mean of (a): 10,016.74 and of (b): 1,773.2). Comparing rows *(b)* to *(c)* shows a low improvement. As approach *(c)* leads to higher calculation costs while extracting and retrieving the data, the improvement of storing and the costs for conducting the calculations need to be compared. *(d)* has got benefits towards the other approaches such that the representation model can be computed efficiently (row Modeling Time in the table).

Table 4. Analysis results for different approaches (a)-(d)

Representation Approaches	μ	σ	min	max
(a) All relations				
Temporal Relations	10,016.74	95,454.66	0	1,209,364
Spatial Relations	645.33	7,238.10	0	126,318
Modeling Time	4.6	96.8	0	704.71
(b) Exploiting transitivity				
Temporal Relations	1,773.2	1,623.67	0	58,247
Spatial Relations	55.38	390.47	0	6,468
Modeling Time	7.71	558.3	0	2746.92
(c) Exploiting extended transitivity				
Temporal Relations	1,626.67	1,607.49	0	50,264
Spatial Relations	50.52	386.70	0	6,390
Modeling Time	12.5	648.62	0	3836.05
(b) Reducing Depth				
Temporal Relations	10,016.74	95,454.66	0	1,209,364
Spatial z-Relation	8.13	23.12	0	214
Modeling Time	3.24	41.79	0	296.44

7 Conclusion

We have shown with our M2DO a first step towards a whole multimedia retrieval engine. Implementing different approaches for representing the structural information of multimedia documents in an efficient way helps to improve the storage and query of documents in later steps. Taking various flash files into account, we can demonstrate that our approach is flexible w.r.t. the complexity of multimedia documents. In our future work, we will also consider interaction relations. In addition, the combination of reducing the z-axis information with the use of transitivity feature will be evaluated. We will develop a prototype using approaches (c) and (d) and conduct extensive evaluations.

Acknowledgments. This research is supported by the SeMuDocs project co-funded by the Ministry for Science, Research, and Arts, Baden-Wuerttemberg, Germany. We would like to thank D. Schmeiß, D. Mies, N. Kurz, and T. Thesing who implemented an early version of the multimedia extraction and representation engine.

References

1. Allen, J.F.: Maintaining knowledge about temporal intervals. Commun. ACM 26(11), 832–843 (1983)
2. Baader, F., Calvanese, D., McGuinness, D.L., Nardi, D., Patel-Schneider, P.F. (eds.): The Description Logic Handbook: Theory, Implementation, and Applications. Cambridge University Press (2003)
3. Balbiani, P., Condotta, J., del Cerro, L.F.: A model for reasoning about bidimensional temporal relations. In: Principles of Knowledge Representation and Reasoning, pp. 124–130. Morgan Kaufmann (1998)
4. Boll, S., Klas, W.: ZYX - A Multimedia Document Model for Reuse and Adaptation. IEEE Trans. on Knowledge and Data Engineering 13(3), 361–382 (2001)
5. Boll, S., Klas, W., Westermann, U.: Multimedia document models - sealed fate or setting out for new shores? In: Int. Conf. on Multimedia Computing and Systems, p. 9604. IEEE, Washington, DC (1999)
6. Candan, K.S., Sapino, M.L.: Data Management for Multimedia Retrieval. Cambridge University Press, New York (2010)
7. Cao, Z., Wu, Z., Wang, Y.: UMQL: A unified multimedia query language. In: Signal-Image Technologies and Internet-Based System, pp. 109–115. IEEE (2007)
8. Celino, I., Valle, E.D., Cerizza, D., Turati, A.: Squiggle: a semantic search engine for indexing and retrieval of multimedia content. In: Semantic Enhanced Multimedia Presentation Systems. CEUR-WS.org (2006)
9. Clementini, E., Sharma, J., Egenhofer, M.J.: Modelling topological spatial relations: Strategies for query processing. Computers & Graphics 18, 815–822 (1994)
10. Diemert, B., Abel, M.-H., Moulin, C.: Semantic audiovisual asset model. Multimedia Tools and Applications 63, 663–690 (2013)
11. Döller, M., Kosch, H., Wolf, I., Gruhne, M.: Towards an mpeg-7 query language. In: Damiani, E., Yetongnon, K., Chbeir, R., Dipanda, A. (eds.) SITIS 2006. LNCS, vol. 4879, pp. 10–21. Springer, Heidelberg (2009)

12. Freksa, C.: Temporal reasoning based on semi-intervals. Artif. Intell. 54(1-2), 199–227 (1992)
13. Hardman, L., Bulterman, D.C.A., van Rossum, G.: The Amsterdam hypermedia model: adding time and context to the Dexter model. Communications of the ACM 37(2) (February 1994)
14. Masolo, C., Borgo, S., Gangemi, A., Guarino, N., Oltramari, A., Schneider, L.: The wonderweb library of foundational ontologies and the dolce ontology. Technical report, ISTC-CNR (2007)
15. Merzougui, G., Djoudi, M., Behaz, A.: Conception and use of ontologies for indexing and searching by semantic contents of video courses. International Journal of Computer Science Issues 8(3) (2012)
16. Saathoff, C., Scherp, A.: M3o: The multimedia metadata ontology. In: Proceedings of the Workshop on Semantic Multimedia Database Technologies, 10th International Workshop of the Multimedia Metadata Community (SeMuDaTe 2009) (2009)
17. Scherp, A.: Authoring of Multimedia Content: A Survey of 20 Years of Research. In: Semantic Multimedia Analysis and Processing. CRC Press (2013)
18. Scherp, A., Boll, S.: MM4U - A framework for creating personalized multimedia content. In: Managing Multimedia Semantics. Idea Publishing (2005)
19. Scherp, A., Boll, S.: Paving the Last Mile for Multi-Channel Multimedia Presentation Generation. In: Chen, Y.-P.P. (ed.) Proc. of the 11th Int. Conf. on Multimedia Modeling, Melbourne, Australia, pp. 190–197. IEEE (January 2005)
20. Scherp, A., Saathoff, C., Franz, T., Staab, S.: Designing core ontologies. Applied Ontology 6, 177–221 (2011)
21. Suarez-Figueroa, M., Gomez-Perez, A., Motta, E., Gangemi, A.: Ontology Engineering in a Networked World. Springer (2012)
22. Waitelonis, J., Sack, H.: Towards exploratory video search using linked data. In: Multimedia Tools and Applications, pp. 1–28 (2011)
23. Yang, J., Li, Q., Wenyin, L., Zhuang, Y.: Searching for flash movies on the web: A content and context based framework. World Wide Web Journal (September 2005)
24. Zillner, S., Westermann, U., Winiwarter, W.: EMMA – A query algebra for enhanced multimedia meta objects. In: Meersman, R. (ed.) OTM 2004. LNCS, vol. 3291, pp. 1030–1049. Springer, Heidelberg (2004)

Video to Article Hyperlinking by Multiple Tag Property Exploration

Zhineng Chen[1], Bailan Feng[1], Hongtao Xie[2], Rong Zheng[1], and Bo Xu[1]

[1] Interactive Digital Media Technology Research Center, Institute of Automation,
Chinese Academy of Sciences, Beijing, China
[2] Institute of Information Engineering, Chinese Academy of Sciences, Beijing, China
{zhineng.chen,bailan.feng}@ia.ac.cn, xiehongtao@iie.ac.cn,
{rong.zheng,xubo}@ia.ac.cn

Abstract. Showing video and article on the same page, as done by official web agencies such as CNN.com and Yahoo!, provides a practical way for convenient information digestion. However, as the absence of article, this layout is infeasible for mainstream web video repositories like YouTube. This paper investigates the problem of hyperlinking web videos to relevant articles available on the Web. Given a video, the task is accomplished by firstly identifying its contextual tags (e.g., *who* are doing *what* at *where* and *when*) and then employing a search based association to relevant articles. Specifically, we propose a multiple tag property exploration (mTagPE) approach to identify contextual tags, where tag relevance, tag clarity and tag correlation are defined and measured by leveraging visual duplicate analyses, online knowledge bases and tag co-occurrence. Then, the identification task is formulated as a random walk along a tag relation graph that smoothly integrates the three properties. The random walk aims at picking up relevant, clear and correlated tags as a set of contextual tags, which is further treated as a query to issue commercial search engines to obtain relevant articles. We have conducted experiments on a large-scale web video dataset. Both objective performance evaluations and subjective user studies show the effectiveness of the proposed hyperlinking. It produces more accurate contextual tags and thus a larger number of relevant articles than other approaches.

Keywords: Web Video, Video to Article Hyperlinking, Random Walk, Contextual Tag.

1 Introduction

With the prosperity of social video websites such as YouTube and YouKu, massive amount of user and professional generated videos have been uploaded and accumulated online. As a consequence, the way that people acquire and digest information has been largely enriched. When people want to get news or knowledge about certain events or topics (e.g., making a hamburger). Instead of traditional search and reading relevant text articles, they may prefer to visit a social video website and watch a video about it, or do both of them, to obtain the desired information.

C. Gurrin et al. (Eds.): MMM 2014, Part I, LNCS 8325, pp. 62–73, 2014.
© Springer International Publishing Switzerland 2014

To meet the diverse taste and habits of different people, official web agencies such as CNN.com and Yahoo! tend to provide information sources of rich-media forms, i.e., besides showing text article, one to several images and/or videos are embedded in the same page as additional references. This layout indeed provides convenient information digestion. However, it is infeasible to do so for mainstream web video websites, where videos are uploaded by independent web users who are hesitant to provide sufficient text description [1]. The information in these websites thus has the peculiarity of rich in multimedia content but sparse in text description. Additionally, the text is often noisy and subjective [2, 3]. Consequently, in many cases the text is of little use in helping people understand the video content, especially for those who are interested in details. Nowadays web videos are one of the most important information sources in people's daily life. Thus, it becomes an urgent demand of hyperlinking web videos with rich and relevant text material for references.

Human's behavior is a useful guidance for algorithm design. When a person want to search relevant text articles given a video, generally he or she will select several keywords that accurately describe the context of video (e.g., *who what, where, when,* etc), and then turn to commercial search engines to seek the desired articles. Therefore, in order to enable hyperlinking from web video to articles, a key technique is by automatically mining the context of video, i.e., identifying keywords that well summarize the video. We define these keywords as contextual tags in this paper. Thus, how to extract contextual tags of a video, and then employ appropriate search based means to build the video to article hyperlinking becomes a timely issue.

In this paper, we investigate the problem of hyperlinking web videos to relevant articles available on the Web. Our study mainly focuses on: multiple tag property exploration (mTagPE) for contextual tags extraction, and search based association for video to article hyperlinking. In mTagPE, three tag properties, namely tag relevance, tag clarity and tag correlation are separately defined and measured by leveraging visual duplicate analyses, online knowledge bases and tag co-occurrence. Then, a tag relation graph is built to fuse the three properties, where tags are nodes whose intensity are represented by both tag relevance and tag clarity, and edges between the nodes are weighted by tag correlation. The extraction of contextual tags is then formulated as a random walk along the graph. Since properties describing different aspects of tags are appropriately fused, the mTagPE is able to pick up relevant, clear and correlated tags as contextual tags. To determine relevant articles, a search based video-article association is proposed, where the set of contextual tags are treated as a query submitted to Google Search and the relevance of returned documents is heuristically judged. We have conducted experiments on MCG-WEBV [23], a real-world large-scale web video dataset. Both objective performance evaluations and subjective user studies show the proposed hyperlinking is effective. It produces more accurate contextual tags and thus a larger number of relevant articles than other approaches.

2   Related Work

Despite relatively little studied, video to article hyperlinking is closely related to two topics widespread studied in multimedia domain, namely video annotation and cross-media analyses.

Video Annotation. We can broadly categorize existing video annotation methods into model-based, search-based and retagging-oriented approaches. Model-based approaches [4-8] adopt supervised or semi-supervised learners to learn a pre-defined, small set of concepts, and then apply the learned models to predict appearance of the concepts in videos. Nevertheless, the limited number of concepts and low prediction accuracy restrict the practical use of these approaches. In search-based approaches [9-10], content redundancy of video repositories has been utilized, where the annotation task is implemented by tag propagation along overlapping or (near) duplicate videos. For retagging-oriented approaches, motivated by the metadata (i.e., title and tag sequence) is often noisy and subjective in social multimedia websites [2, 3], a number of retagging approaches [11-16] that aims to improving the completeness, precision and ranking of metadata have been proposed. Although significant research efforts have been made, it is observed that most existing methods emphasize on extracting keywords to describe the content of videos, which may prominent in visual aspects but not discriminative from the perspective of video context. There is little work focus on extracting contextual tags of videos.

Cross-Media Analyses. There are several application-oriented cross-media analyses approaches similar to our work. In [17], the authors propose a method to link Wikipedia entries with broadcast videos explained the entries. While in [18], a method combining semantic inferencing and visual analysis for finding automatically photos and videos illustrating events has been proposed. In recent MediaEval competition, a Search and Hyperlinking task [19] is launched to first find a relevant video instance and then follow hyperlinks to other related video clips to further explore the video collection. Furthermore, Tan et al [20] investigate the task of synchronizing and leveraging different information sources including video and text for efficient video browsing. Compared with our work, [17-18] are text to video mapping and ours is video to text mapping. The Search and Hyperlinking task [19] is to seek for meaningful videos with respect to a text query, while the purpose of [20] is for efficient video browsing, both of which are also different with ours.

3   Video to Article Hyperlink

3.1   Overview

The proposed video to article hyperlinking consists of three components, i.e., visual duplicate analyses, mTagPE and search based association, as illustrated in Fig. 1. Given a web video, the first module is employed to find a set of duplicate videos as well as their associated metadata, which is further parsed as semantic entities (abbreviated as tags in this paper). In the mTagPE module, tag relevance, tag clarity and tag correlation, which describe tag properties from different but complementary angles, are separately defined and appropriately measured. To fuse the three properties, a tag relation graph is built and a random walk is applied to produce the contextual tags.

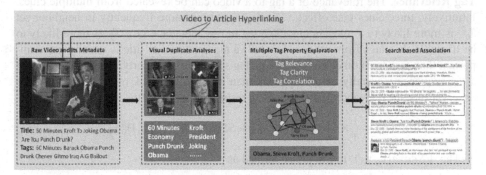

Fig. 1. Illustrative flowchart of the proposed video to article hyperlinking

In the third module, a search based video-article association is proposed to find relevant articles, where all contextual tags are treated as a query submitted to Google Search and the relevance of returned documents are appropriately judged.

3.2   Visual Duplicate Analyses

Given a web video v_0 and its metadata $\mathcal{T}_0$, the objective of this module is to find a set of videos from a web video corpus $\mathcal{D}$, which describe the same event or topic with v_0. As social video websites often have certain amount of content redundancy, especially for those popular ones [10, 23], we employ the method proposed in [9] to perform the visual duplicate analysis. Ideally, the process returns a set of videos duplicated, near duplicated or partially duplicated with v_0. These videos constitute a neighbor video set $\Theta_0^{\#} = \{v_1, v_2, ..., v_K\}$ for v_0, where $K = \left| \Theta_0^{\#} \right|$ is the cardinality of set $\Theta_0^{\#}$. Similarity, the collection of metadata associated with the neighbor videos are define as $\mathcal{T}_0^{\#}$, which is often a enriched but even noisy set with respect to $\mathcal{T}_0$.

It is observed that there are many named entities (e.g., person names, locations) in both $\mathcal{T}_0$ and $\mathcal{T}_0^{\#}$. However, as a named entity consists of one to several words and the metadata of web videos is lack of grammar structure, it becomes a nontrivial task to correctly recognize them. To address this problem, we implement a Wiki-based tag identification method to recognize named entities (tags). In the method, candidate tags are identified by stepwise testing whether a word or a succession of words in metadata could represent a wikipedia entry. If multiple entries found, we only keep the longest one (but not exceed 4 words). By this way, $\mathcal{T}_0$ and $\mathcal{T}_0^{\#}$ can be rewrite as two sets of tags, i.e., $\mathcal{T}_0 = \left\{ t_1, t_2, ..., t_{|\mathcal{T}_0|} \right\}$ and $\mathcal{T}_0^{\#} = \left\{ t_1^{\#}, t_2^{\#}, ..., t_{|\mathcal{T}_0^{\#}|}^{\#} \right\}$.

3.3   Multiple Tag Property Exploration (mTagPE)

In this module, we first give the measurement for tag relevance, tag clarity and tag correlation, and then introduce how the three properties are integrated into the built tag relation graph and produce contextual tags.

Tag Relevance. The relevance of a tag to a video can be revealed from multiple clues. Intuitively, three clues take effect: (1) a tag appearing more frequently in neighbor set $\mathcal{T}_0^{\#}$ is more likely to a relevant tag; (2) a tag in raw metadata, i.e., $\mathcal{T}_0$, is more likely to be trustable than tags from other videos; (3) a tag appeared in title is usually more important than tags from other metadata (e.g., tag sequence). Similarly, a tag in the front of a tag sequence is usually more important than those in the back ones. Based on these observations, the relevance of a tag t_i with respect to v_0 is defined as:

$$TRele(t_i) = \frac{1}{4} \times \left(VScore(t_i, v_0) + SScore\_Norm(t_i, \Theta_0^{\#}) \right) \tag{1}$$

The $VScore(t_i, v_0)$ in Eq. (1) is defined as

$$VScore(t_i, v_0) = tt_{i0} + 1 - \frac{r_i}{|v_0|} \tag{2}$$

where tt_{i0} is a binary value of 0 or 1, denoting the absence or presence of tag t_i in the title of v_0, r_i is the rank of tag t_i in the tag sequence of v_0, and $r_i = |v_0|$ if $t_i \notin \mathcal{T}_0$. $SScore\_Norm(t_i, \Theta_0^{\#})$ in Eq. (1) is the normalized $SScore(t_i, \Theta_0^{\#})$ given by

$$SScore(t_i, \Theta_0^{\#}) = \sum_{j=1}^{K} \frac{VScore(t_i, v_j)}{|v_j|} \times Sim(v_0, v_j) \tag{3}$$

where $Sim(v_0, v_j)$ is the visual duplicate score between v_0 and v_j obtained from the visual duplicate analysis step. The $SScore\_Norm(t_i, \Theta_0^{\#})$ is obtained by normalizing $SScore(t_i, \Theta_0^{\#})$ to interval $[0, 2]$, using the min-max normalization.

Tag Clarity. Tag clarity measures the extent of definitiveness of a tag. We have two observations on the tag clarity. (1) More words a tag has, clearer the tag is. For instance, "table cloth" is clearer than "cloth". (2) A clearer tag usually has more Wikipedia category labels in its Wikipedia entry than those less clear tags. This is reasonable as Wikipedia category labels are descriptive explanations for tags. More labels imply a tag is more clear and definite. Therefore, the clarity of tag t_i is defined as

$$TClar(t_i) = \frac{1}{2} \times \left(\frac{Len_i}{4} + WV_i \right) \tag{4}$$

where Len_i is the number of words tag t_i has, WV_i is a unit ramp function defined on the number of Wikipedia category tags (nWCT) of t_i. It equals to 1 when $nWCT \geq T$, and it is given by $\frac{|WCT_i|}{T}$ in cases that $nWCT < T$. The parameter T is empirically set as 20 in this paper. Note that unlike tag relevance, tag clarity is a video-independent property that needs to be computed only once.

Tag Correlation. Intuitively, the co-occurrence between tags somewhat reflects the correlation of tags as simultaneously being contextual tags. Therefore, we model the pair-wised tag correlation $TCor(t_i, t_j)$ between tag t_i and t_j as follows.

$$TCor(t_i, t_j) = CO_{\mathcal{T}_0}(t_i, t_j) + CO_{\mathcal{D}}(t_i, t_j) \tag{5}$$

where $CO_{T_0}(t_i, t_j)$ and $CO_D(t_i, t_j)$ are tag co-occurrence on $T_0 \cup T_0^\#$ and on the whole video corpus D, respectively. And the tag co-occurrence is defined as

$$CO_C(t_i, t_j) = \frac{|v_i \cap v_j|}{|v_i|} + \frac{|v_i \cap v_j|}{|v_j|} \tag{6}$$

where V_i is the set of videos with tag t_i found in metadata in corpus C, which is either $T_0 \cup T_0^\#$ or D. $TCor(t_i, t_j)$ balances tag co-occurrence in both the neighbor video set and the whole dataset. Two tags with larger $TCor(t_i, t_j)$ are more compatible and less likely to conflict with each other as contextual tags.

Random Walk Based Fusion. The three tag properties reflect tag characteristics from different angles. Therefore, we build a tag relation graph to integrate them together. In the graph, tags are nodes whose intensity, e.g., the intensity of tag t_i, is defined as

$$TI(t_i) = (1 - \beta) \cdot TRele(t_i) + \beta \cdot TClar(t_i), \ t_i \in T_0 \cup T_0^\# \tag{7}$$

where $\beta \in [0,1]$ is a parameter balancing tag relevance and tag clarity. Let $r_k(i)$ be the probability of tag t_i being contextual tags at iteration k and define $T = |T_0 \cup T_0^\#|$, the identification of contextual tags can be formulated as a random walk process over the built tag relation graph as follows,

$$\mathbf{r}_k = \alpha \mathbf{P} \cdot \mathbf{r}_{k-1} + (1 - \alpha) \cdot \mathbf{s} \tag{8}$$

where $\mathbf{r}_k \equiv [r_k(i)]_{T \times 1}$, $\mathbf{P} \equiv [p_{i,j}]_{T \times T}$, $\mathbf{s} \equiv [s_i]_{T \times 1}$. s_i and $p_{i,j}$ are the normalized tag intensity and the degree of tag t_j being identified as compatible, or saying correlated, given tag t_i, respectively. The two variables are defined as

$$s_i = \frac{TI(t_i)}{\sum_j TI(t_j)}, \ t_j \in T_0 \cup T_0^\# \quad \text{and} \quad p_{i,j} = \frac{TCor(t_i, t_j)}{\sum_k TCor(t_i, t_k)}, \ t_k \in T_0 \cup T_0^\# / t_i \tag{9}$$

Since with exactly the same formulation as in [21], Eq. (8) will converge to

$$\mathbf{r}_\pi = (1 - \alpha)(1 - \alpha \mathbf{P})^{-1} \cdot \mathbf{s} \tag{10}$$

where $\mathbf{r}_\pi$ is the converged vector whose element $r_\pi(i)$ shows the probability of tag t_i being a contextual tag. The random walk assigns large probabilities to tags with high tag intensity and co-occurred frequently [21, 22], both of which are actually attributed to tag relevance, tag clarity and tag correlation computed by Eq. (1), (4) and (5), respectively. Consequently, the information carried by the three tag properties are fused in $\mathbf{r}_\pi$. Finally, we pick up four tags with the largest $r_\pi(\cdot)$ as contextual tags.

3.4   Search Based Association

With the obtained contextual tags, which are deemed to accurately describe the video context, a search based video-article association is proposed to implement the hyperlinking. Specifically, a text query is formed by simply concatenating all contextual tags. Note that articles describing the same content as the video may available on the

Web, we submit the query to Google Search to find relevant articles. For each query, information of the top 32 returned documents, including title, Google provided description and URL, is crawled by using the Google API.

With these documents, we employ a two-step filtering strategy to automatically determine relevant articles. Firstly, the documents are verified from bottom to top, e.g., from the 32-th documents to the first one. The verification is stopped when it finds a document whose title and description contains all the query tags. The document and the ones ranked higher than it are temporarily marked as candidate relevant. Secondly, we parse website names based on URLs of the candidate relevant documents. A document is judged as a relevant article if it comes from an approved website. The determination of approved websites will be introduced in the experiments.

4   Experiments

4.1   Dataset and Ground-truth

The dataset employed in the experiments is a subset of MCG-WEBV [23], which has 19,934 YouTube videos with channel names "News&Politics". We select the subset as news is often reported by both videos and text articles. In the subset, 963 videos are "most viewed videos of the month" from Dec. 2008 to Nov. 2009, while the remaining 18,971 ones are their "related video" and "same author video", i.e., videos uploaded by the same author. By applying the Wiki-based tag identification, there are 4.64 and 10.66 tags in title and tag sequence per video on average, respectively.

We recruit two assessors to manually annotate contextual tags for the 963 videos. The assessors are asked to provide three or four contextual tags for each video independently. The two assessors then compare their annotations and observe that inconsistency between annotations is common, as a video could be described by different sets of contextual tags. We also note that both sets usually differ in only a few tags. Thus, to resolve the inconsistency, the union of the two annotation sets is adopted as the final annotation. The whole annotation takes each assessor nearly 40 hours. Finally, the annotation is formed by providing 3,483 contextual tags with 4,475 words for 873 videos, and leaving 90 videos unlabeled and removed for reasons of hard to understandable and difficult to verbalize. The contextual tags for the 873 videos are called HumanTag as they provide the performance that human can achieve.

To label the approved websites, contextual tags from different approaches (mentioned below) for the 873 videos are collected and submitted to Google Search, where URLs from candidate relevant articles are collected. Then, top websites are obtained by parsing the URLs to site names and ranking them in a descending order. The top 200 websites are manually annotated. Since the purpose of the paper is hyperlinking video to article, websites of video, image and music are removed while others are retained. The annotation is finished by providing 177 websites as approved websites.

4.2   Objective Performance Evaluations

Evaluation Metric. We use three metrics to evaluate the performance of the mTagPE approach in contextual tag identification as well as the performance of video to article hyperlinking. The first one is Tag Consistency (TC). It measures the degree of consistency between the extracted contextual tags and the HumanTag as follows.

$$TC(S^c, S^g) = |S^c \cap S^g| + \sum_{S_i^c \notin S^g} \max_{1 \leq j \leq |S^g|} EDist(S_i^c, S_j^g) \tag{11}$$

where S^c and S^g are the set of extracted contextual tags and HumanTag for the same video, S_i^c is the i-th tags in S^c, $EDist(S_i^c, S_j^g)$ is the edit distance between S_i^c and S_j^g.

The second and third metrics are the number of Relevant Articles (nRA) and rank of the first Relevant Article (rRA), which both evaluate the performance of video to article hyperlinking. Since the three metrics only shows the performance of a single video, we use mTC, mnRA and mrRA, simply the mean of TCs, nRAs and rRAs to evaluate average performance over a set of videos, respectively.

The Parameter Sensibility Analysis. The significance of different tag properties has been of great interest to researchers. In the tag relation graph, parameter β linearly weights tag relevance and tag clarity to produce node intensity. While parameter α balances the strength between node intensity and edge weights represented by tag correlation. We analyze the parameter influence on the 873 annotated videos.

We first analyze the sensibility of α. For simplicity, we set β to 0.5 and range α from 0 to 1 with an increasing step of 0.1. The upper part of Table 1 shows the results. The optimal performance is achieved by setting α around 0.2. This implies that both tag correlation and tag intensity are useful in identifying contextual tags. We also note that tag intensity plays a dominant role, and setting α to 0 can obtain much better results than setting α to 1, showing that tag intensity is more reliable than tag correlation in identifying contextual tags.

We then set α to 0.5 and range β from 0 to 1.0 with the same increasing step. The results are shown in the lower part of Table 1. It is seen that setting β in (0,1) is better than setting β to 0 or 1 individually, demonstrating the effectiveness of combining tag relevance and tag clarity. We also observed the optimal performance is achieved by setting β to 0.3, which shows tag relevance play a more important role than tag clarity. This is in accordance with what we anticipated as tag relevance is a video-dependent property while tag clarity is irrelevant with the given video.

Drawing from these analyses, we have the conclusion that all three properties are taken effects in producing contextual tags. The random walk has fused them smoothly. Hereafter in the experiments, α and β are set to 0.2 and 0.3, respectively.

Table 1. Performace of the mTagPE approach varing with α and β

β = 0.5	α	0	0.1	0.2	0.3	0.4	0.5	0.6	0.7	0.8	0.9	1.0
	mTC	3.063	3.168	**3.256**	3.12	3.076	3.006	2.903	2.731	2.5	2.28	1.936
	mnRA	5.546	**5.766**	5.751	5.355	5.043	4.889	4.749	4.509	4.406	3.855	3.757
	mrRA	3.231	3.101	2.888	**2.819**	2.955	2.93	3.086	3.005	3.131	3.192	3.26
α = 0.2	β	0	0.1	0.2	0.3	0.4	0.5	0.6	0.7	0.8	0.9	1.0
	mTC	3.195	3.258	3.345	**3.354**	3.327	3.256	3.103	2.788	2.528	2.323	2.109
	mnRA	6.399	6.659	**6.676**	6.66	6.173	5.751	4.973	4.171	3.452	3.077	2.792
	mrRA	3.092	2.942	2.98	2.619	**2.579**	2.888	3.114	3.29	3.323	3.309	3.424

Evaluations. To evaluate the performance of the mTagPE in producing contextual tags and the search based association, four baseline approaches are conducted for comparison, which are introduced as follows.

1) **Raw Tag (RTag)**: RTag is a baseline that simply selects the first four tags in metadata as contextual tags, where the metadata is parsed and ranked in a title first and tag sequence followed order.

2) **PoS Tagging Tag (PoSTTag)**: PoSTTag extracts contextual tags based on their part-of-speech labels. The first four tags in metadata labeled as nouns or cardinal number, provided by the Stanford's Part-of-Speech Tagger [26], are chosen with priority. A tag is labeled as noun if at least one of its words is labeled as noun. This idea has been used in [24].

3) **Wiki-Classified Tag (WCTag)**: WCTag first classifies tags to four categories including person names, locations, other proper nouns and unclassified tags, using the heuristic rule proposed in [25]. Then four tags belonging to person names, locations and other proper nouns are picked out in turn. Unclassified tags are selected only if all tags of the other three categories have already been chosen.

4) **Visual Duplicate Tag Voting (VDTagV)**: VDTagV first collects a set of duplicate videos as well as their tags, the tags are ranked according to their relevance computed in [9], which has similar equation with ours (i.e., Eq. (1)) but neglects the effect of tag position, i.e., whether it is a title tag and its rank in tag sequence. Finally, four tags with the largest relevance are picked out as contextual tags.

Table 2 lists the performance of the four baselines, HumanTag and our mTagPE on the 873 annotated videos. The main observations are:

- mTagPE performs better than the four baselines, and there is not much differences between the performance of mTagPE and HumanTag, demonstrating that mTagPE captures the video context more accurately compared to other syntax-based (i.e., PosTTag) and knowledge-based (i.e., WCTag) approaches.
- mrRAs of all six approaches are larger than 2.5, this is attribute to the top searched documents for a large portion of queries are come from YouTube, which has been ruled out from the list of approved websites. But even so, the first relevant article can usually be found from the top 3 returned documents.
- VDTagV is better than the other three baselines, indicating that the visual duplicate analysis is effective in discovering relevant tags. However, the mTagPE can further boost the performance by jointly considering other tag properties.

Table 2. Performance of different approaches on the 873 videos

	RTag	PoSTTag	WCTag	VDTagV	mTagPE	HumanTag
mTC	2.78	2.846	2.981	3.218	3.354	3.99
mnRA	5.609	5.514	5.709	6.251	6.66	7.109
mrRA	3.095	3.152	3.288	3.13	2.619	2.598

Noted that both the reported nRA and rRA may not be the exact performance our approach achieved. For instance, a document, despite containing all query words and coming from an approved website, may still an irrelevant article with respect to the video. However, we argue that such cases are statistically independent with the approaches **such** that the results reported by nRA and rRA are still meaningful.

We also evaluate the mTagPE approach on the remaining 18,971 videos. By filtering 197 videos with less than two contextual tags, 17,449 sets of different contextual tags (i.e., queries) are extracted for the 18,774 videos. We submit these queries to Google Search, 538,860 documents, where 112,144 ones are judged as relevant articles, are returned. The mTC, mnRA and mrRA for these videos are 3.241, 6.427 and 2.878, respectively. It is observed that the results are not much different with those on the small dataset, i.e., the 873 videos, showing that the parameters tuned on small dataset can be smoothly apply to large dataset with similar sources. It is also demonstrated that the random walk can stably capture the intrinsic complementary natures of multiple tag properties, generating discriminative contextual tags and thus high quality articles relevant to the video.

4.3   Subjective User Studies

We conduct user based studies to further compare the contextual tags generated by mTagPE, RTag, PoSTTag, WCTag, VDTagV and HumanTag. Specifically, ten web users are recruited in this study, including three females and seven males with age ranging from 22 to 34. They all have the experience of watching online videos. To avoid bias on this study, the mark of the queries, which indicates which approach the contextual tags is correspond to, is hided.

We randomly select 50 videos from the 873 annotated videos. The participants are required to watch the videos one-by-one, and to evaluate the degree of match between the video content and each of the six types of contextual tags. Specifically, they are asked to provide their judgments by choosing one from the four options: the query is "perfectly matched", "moderate matched", "little matched" or "unmatched" with the video content. As a result, there are 3,000 judgments for the participants in total.

The evaluation results are illustrated in Fig. 2. As can be seen, for contextual tags generated by mTagPE, the percentage of videos rated as "perfectly matched" or "

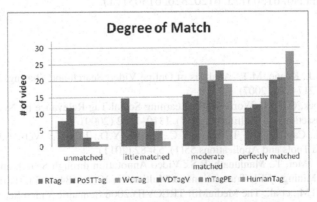

Fig. 2. Performance comparisons of different approaches from the 10 participants

mo-derate matched" is 87.6%, while the percentages for RTag, PoSTTag, WCTag and VDTagV are 54.8%, 57.6%, 78% and 79.8%, respectively, clearly showing the effectiveness of multiple tag property exploration. We also note that for HumanTag, 2.3 videos on average are judged as "little matched" or "unmatched". After viewing the samples and discussions with the participants, we find the inconsistency among humans, i.e., annotators and user study participants, is mainly because of unknown foreign languages and errors in the Wiki-based tag identification. It is also seen that the number of "unmatched" and "little matched" videos have risen to 6.2 on average for mTagPE. Besides the above two reasons, errors in visual duplicate analyses becomes another factor that results in increased number of mismatches.

In summary, although there are a few cases that the contextual tags are not well identified. Both objective performance evaluations and subjective user studies shows that, by exploring multiple tag properties and fusing them via a random walk process, we can extract contextual tags that basically summarize the context of video, and thus hyperlinking video to relevant articles that detailedly describe the video.

5   Conclusion

Observing that hyperlinking web videos to relevant articles provides convenient information digestion, in this paper we have presented the mTagPE approach to extract contextual tags of a video, and the search based video-article association to implement the hyperlinking. In mTagPE, multiple tag properties such as tag relevance, tag clarity and tag correlation are measured and fused by a random walk process. Both the objective and subjective experiments conducted on MCG-WEBV dataset basically validate our proposal. Performance improvements are also observed when compared to the RTag, PoSTTag, WCTag and VDTagV. In the future, we will test the validity of our approach on videos with other channel labels, e.g., Howto (a channel name in YouTube). We are also interested in extending the approach to video shot hyperlinking and exploring effective methods to evaluate the performance of hyperlinking.

Acknowledgments. This work is supported by National Nature Science Foundation of China (Grant No. 61303175, 61202326, 61303171).

References

1. Halvey, M.J., Keane, M.T.: Analysis of Online Video Search and Sharing. In: Proceeding of HT, pp. 217–226 (2007)
2. Li, X., Snoek, C.G.M., Worring, M.: Learning Social Tag Relevance by Neighbor Voting. IEEE Transactions on Multimedia 11(7), 1310–1322 (2009)
3. Chen, Z.N., Cao, J., Xia, T., Song, Y.C., Zhang, Y.D., Li, J.T.: Web Video Retagging. Multimedia Tools and Applications 55(1), 53–82 (2011)
4. Moxley, E., Mei, T., Manjunath, B.S.: Video Annotation through Search and Graph Reinforcement Mining. IEEE Transactions on Multimedia 12(3), 184–193 (2010)
5. Snoek, C.G.M., et al.: The MediaMill TRECVID 2008 Semantic Video Search Engine. In: Proceeding of TRECVID (2008)

6. Jiang, Y.G., Ngo, C.W., Yang, J.: Towards Optimal Bag-of-Features for Object Categorization and Semantic Video Retrieval. In: Proceeding of ACM CIVR, pp. 494–501 (2007)
7. Qi, G.J., Hua, X.S., Rui, Y.: Correlative Multilabel Video Annotation with Temporal Kernels. ACM Trans. Multimedia Comput. Commun. Appl. 5(1), 1–27 (2008)
8. Aradhye, H., Toderici, G., Yagnik, J.: Video2Text: Learning to Annotate Video Content. In: Proceeding of IEEE Data Mining Workshops, pp. 144–151 (2009)
9. Zhao, W.L., Wu, X., Ngo, C.W.: On the Annotation of Web Videos by Efficient Near-duplicate Search. IEEE Transactions on Multimedia 12(5), 448–461 (2010)
10. Siersdorfer, S., Pedro, J.S., Sanderson, M.: Content Redundancy in YouTube and its Application to Video Tagging. ACM Transactions on Information Systems, 301–331 (2011)
11. Liu, D., Hua, X.S., Yang, L., Wang, M., Zhang, H.J.: Tag Ranking. In: Proceeding of WWW, pp. 351–360 (2009)
12. Liu, D., Yan, S.C., Hua, X.S., Zhang, H.J.: Image Retagging using Collaborative Tag Propagation. IEEE Transactions on Multimedia 13(4), 702–712 (2011)
13. Liu, D., Hua, X.S., Zhang, H.J.: Content-based Tag Processing for Internet Social Images. Multimedia Tools and Applications 51(2), 723–738 (2011)
14. Sang, J., Xu, C.S., Liu, J.: User-aware image tag refinement via ternary semantic analysis. IEEE Transactions on Multimedia 14(3), 883–895 (2012)
15. Zhang, X.M., Huang, Z., Shen, H.T., Yang, Y., Li, Z.J.: Automatic tagging by exploring tag information capability and correlation. World Wide Web 15(3), 233–256 (2012)
16. Ballan, L., Bertini, M., Bimbo, A.D., Meoni, M., Serra, G.: Tag Suggestion and Localization in User-Generated Videos based on Social Knowledge. In: Proceeding of WSM, pp. 3–8 (2010)
17. Okuoka, T., Takahashi, T., Deguchi, D., Ide, I., Murase, H.: Labeling News Topic Threads with Wikipedia Entries. In: Proceeding of IEEE ISM, pp. 501–504 (2009)
18. Liu, X.L., Troncy, R., Huet, B.: Finding Media Illustrating Event. In: Proceeding of ICMR, pp. 1–8 (2011)
19. Eskevich, M., Jones, G.J.F., Aly, R.: Multimedia Information Seeking through Search and Hyperlinking. In: Proceeding of ACM ICMR, pp. 287–294 (2013)
20. Tan, S., Ngo, C.W., Tan, H.K., Pang, L.: Cross Media Hyperlinking for Search Topic Browsing. In: Proceeding of ACM Multimedia, pp. 1095–1098 (2011)
21. Hsu, W., Kennedy, L., Chang, S.F.: Video Search Reranking through Random Walk over document-level context graph. In: Proceeding of ACM Multimedia, pp. 971–980 (2007)
22. Yao, T., Ngo, C.W., Mei, T.: Circular Reranking for Visual Search. IEEE Transactions on Multimedia 12(5), 1644–1655 (2013)
23. Cao, J., Zhang, Y.D., Song, Y.C., Chen, Z.N., Zhang, X., Li, J.T.: MCG-WEBV: A Benchmark Dataset for Web Video Analysis. Technical Report, pp. 1-10 (2009)
24. Chen, Z.N., Cao, J., Song, Y.C., Guo, J.B., Zhang, Y.D., Li, J.T.: Context-Oriented Web Video Tag Recommendation. In: Proceeding of WWW, pp. 1079–1080 (2010)
25. Chen, Z.N., Cao, J., Song, Y.C., Zhang, Y.D., Li, J.T.: Web Video Categorization based on Wikipedia Categories and Content-Duplicate Open Resources. In: Proceeding of ACM Mutimedia, pp. 1107–1110 (2010)
26. Toutanova, K., Klein, D., Manning, C., Singer, Y.: Feature-Rich Part-of-Speech Tagging with a Cyclic Dependency Network. In: Proceeding of HLT-NAACL, pp. 252–259 (2003)

Rebuilding Visual Vocabulary
via Spatial-temporal Context Similarity
for Video Retrieval

Lei Wang[1], Eyad Elyan[1], and Dawei Song[2,3]

[1] School of Computing Science and Digital Media, Robert Gordon University,
Aberdeen, AB10 7GJ, United Kingdom
[2] Tianjin Key Laboratory of Cognitive Computing and Applications,
School of Computer Science and Technology, Tianjin University, Tianjin, China
[3] Department of Computing and Communications, The Open University,
Milton Keynes, United Kingdom

Abstract. The Bag-of-visual-Words (BovW) model is one of the most popular visual content representation methods for large-scale content-based video retrieval. The visual words are quantized according to a visual vocabulary, which is generated by a visual features clustering process (*e.g.* K-means, GMM, etc). In principle, two types of errors can occur in the quantization process. They are referred to as the *UnderQuantize* and *OverQuantize* problems. The former causes ambiguities and often leads to false visual content matches, while the latter generates synonyms and may lead to missing true matches. Unlike most state-of-the-art research that concentrated on enhancing the BovW model by disambiguating the visual words, in this paper, we aim to address the *OverQuantize* problem by incorporating the similarity of spatial-temporal contexts associated to pair-wise visual words. The visual words with similar context and appearance are assumed to be synonyms. These synonyms in the initial visual vocabulary are then merged to rebuild a more compact and descriptive vocabulary. Our approach was evaluated on the TRECVID2002 and CC_WEB_VIDEO datasets for two typical Query-By-Example (QBE) video retrieval applications. Experimental results demonstrated substantial improvements in retrieval performance over the initial visual vocabulary generated by the BovW model. We also show that our approach can be utilized in combination with the state-of-the-art disambiguation method to further improve the performance of the QBE video retrieval.

Keywords: Visual Vocabulary, Synonyms, Spatial-Temporal Context, Content based Video Retrieval, Bag-of-visual-Word.

1 Introduction

Large amount of videos are continuously produced, and there is an urgent need for more advanced video retrieval technology that helps users to access desired visual information more efficiently and effectively. The Bag-of-visual-Words model

C. Gurrin et al. (Eds.): MMM 2014, Part I, LNCS 8325, pp. 74–85, 2014.
© Springer International Publishing Switzerland 2014

(BovW) has become one of the most popular methods for video content representation. A bunch of "interesting" local regions are first extracted from the images/videos, and then each local region is described by a high dimensional vector. An unsupervised clustering method, e.g. K-Means, is then applied to the features, and each unseen or training feature is quantized to its nearest centroid. In this way, a centroid acts as a basic visual element, analogous to a word, which is often called a visual word. Each feature is called an instance of the corresponding visual word (in this paper, the term "an instance of a visual word" is interchangeably used with "a visual word", for convenience, except when they are explicitly distinguished). The images/videos in a data collection are then represented by bags of visual words, and two instances belonging to an identical visual word are considered as "matched" in the similarity measurement.

Despite the significant progress that has been made in term of retrieval performance, the BovW model has been considered not as effective as the textual words. The textual words always have relatively clear semantics, but the meaning of a quantized visual word is less stable, mainly due to two types of quantization errors that may hamper the retrieval performance.

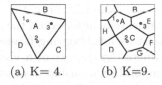

(a) K= 4.   (b) K=9.

Fig. 1. Two types of quantization error. K is the size of visual vocabulary.

The first type of error tends to occur when the quantization is too rough, and the features with different meaning may be quantized into the same visual word. The ambiguous visual words will result in false visual content matches. In this paper, we refer to this type of quantization error as *UnderQuantize*. As can be seen in Figure 1, the feature space is shown as a square and the feature points with same meaning are identified by the same color(e.g., features 1 and 2 in green, but feature 3 in purple). If the size of the vocabulary K equals 4, then all features 1, 2, and 3 are mis-quantized into the same visual word A in Figure 1(a).

A larger sized vocabulary may help to disambiguate the visual words, and then the granularity of individual visual words becomes finer. For example, in Figure 1(b), the size of vocabulary increases from 4 to 9, and the cells become smaller. This ensures distinction between features 1 and 3. However, increasing the vocabulary size could bring in the second type of quantization error, where features representing similar visual information will be quantized into different visual words. In this paper, we refer to this type of quantization error as *OverQuantize*, which tends to cause this "synonymy" problem, e.g. the features 1 and 2 are not matched in Figure 1(b). This *OverQuantize* error would cause a loss of the relevant visual information in the retrieval process, which will also hurt the performance of the BoVW model.

In previous research, intensive investigations were devoted to reducing the *UnderQuantize* error in order to achieve more accurate visual content match. Recently, large vocabularies [11] are often used to address the *UnderQuantize* problem. However, there tends to be a trade-off between the two types of errors, i.e., the smaller *UnderQuantize* error, the bigger *OverQuantize* error.

In this paper, we focus on addressing the *OverQuantize* problem. The idea is to integrate multiple synonymous visual words into an identical visual word, and then rebuild a more compact and descriptive visual vocabulary. For example, if the visual words A and C in Figure 1(b) are merged, then features 1 and 2 can be correctly matched. Because the appearance of features has been used by the initial visual vocabulary to classify the meaning of visual words, extra information should be used to detect the synonyms. We propose to leverage the context of visual words to address the problem of *OverQuantize*. Here, we define the context of an individual visual word as its co-occurring visual words.

It is assumed that the visual words with similar meaning tend to occur in the similar context. Furthermore, the video content is always a mixture of large amount of miscellaneous information. Wang et al. [14] proposed a method to quantitatively compute the correlation between the co-occurring instances of visual words for similarity measurement based on the spatial proximity and the temporally relative motion across neighbouring frames. It inspires the method we propose in this paper, where the context of a visual word is refined by its accumulated spatial and temporal relationship to other visual words. A context vector is then quantized for a visual word, w, and the context similarity between a pair of visual words is (inversely) measured by the distance between their context vectors. Nevertheless, the original feature information (i.e., appearance) of visual words should not be completely ignored in determining the similar visual words. It can be used for the appearance similarity verification, and for example, in Figure 1(b), A and G are far from each other in the feature space, and they may not be considered as synonyms, even though their contexts are similar.

The rest of the paper is organized as follows: Section 2 reviews the related work on visual vocabulary quantization and error reduction; Section 3 describes our vocabulary rebuilding method in detail; Section 4 presents the experiment settings and the results for an extensive evaluation on two typical QBE video retrieval tasks; Section 5 concludes the paper and points out future research directions.

2   Related Work

One of the popular disambiguation methods is to further verify the matched visual words with additional constraints. For example, Jegou et al. [4] proposed a visual words matching framework based on the so-called geometric weak consistency constraints (WGC), which utilized the additional information about the features, e.g. main direction and scales. This method and its variant Tight Geometric Constraint (TGC) achieved promising performance for applications such as near duplicate image/video search and video automatic annotation [17]. Other techniques such as Spatial Pyramid Matching and its variants [7] are also proposed to promote the accuracy of visual similarity measurement by incorporating the encoded spatial information.

The problem of losing relevant visual content attracted significant attention. Jiang et al. [6] have proposed to softly quantize a feature into multiple visual words to reduce the risk of missing relevant information. Another type of soft-matching equivalent schemes [2] were proposed to expand the query with the synonymous and related visual words to promote performance and indirectly tackled this problem.

Building contextual visual vocabulary has also attracted much attention. For example, the context aware visual word clustering method proposed in [16] and [13] directly utilized the co-occurrence information of visual features in the regularized clustering. However, this regularized and iterative clustering algorithm can not easily generalize to the unseen data. In this paper, we proposed to merge the synonyms in initial visual vocabulary, and the generalization becomes as straightforward as classical BovW framework.

Other than the spatially and temporally cross-relationship defined on the visual words level, pixels level spatial and temporal information is also proposed to be integrated into local feature [1] to represent the visual content in the video. In real world applications, the visual vocabulary could also be constructed with multiple features, either by concatenating multiple features into a long vector before the quantization [3] or aggregating the visual words after quantizing the different features separately [5]. However, in this paper, we only exploit vocabulary based on single feature (SIFT) to demonstrate our approach to rebuilding the vocabulary, and it can be easily applied to the visual vocabulary based on other single or combined features.

3   Visual Vocabulary Based on Spatial Temporal Context

The initial visual vocabulary is generated by the approximate K-Means clustering algorithm [11]. To focus our research on the $OverQuantize$ problem, the initial K of visual vocabulary could be as large as tens of thousands. Each video in the data collection D is represented as a sequence of key frames $v_d = \{f_d\}$. Each f_d is represented by a K dimensional vector: $f = \{w_i\}$ where each weight w_i is the Term Frequency (TF) of the ith visual words. The D is a video set $\{v\}$, and it is also a set of key-frames $\{f\}$. The data collection D can be generally represented by an $N \times K$ document-term matrix M, where N is the number of key-frames in the data-collection. Each row of M is a keyframe f.

In the query-by-example (QBE) video retrieval, a given query example, which is also a video, is represented as $v_q = \{f_l\}$. The key frame similarity $sim(f_d, f_q)$ can be measured by the Cosine function. In this paper, it is assumed that the video data are well segmented shots or the videos are short and consist of a relatively small number of shots. We then adopt the shot similarity measurement method proposed by Peng et al. [10], where the highest similarity among all possible pairs of key frames is used to measure the similarity between two video shots:

$$sim_{v_d,v_q} = \max_{f_d \in v_d, f_q \in v_q} sim(\boldsymbol{f_d}, \boldsymbol{f_q}) \tag{1}$$

3.1   The Spatial-temporal Context

As discussed in the first section, the context of a visual word in this paper is defined as the co-occurring relationship with other visual words. The co-occurrence context of a visual word is simply characterized by its co-occurring words in the same frame. For example, if an instance a of visual word w_i, co-occurs with b, an instance of visual word w_j, the context of w_i would consider its co-occurrence with w_j:

$$c_i^c(j) = \boldsymbol{f}(w_i) \times \boldsymbol{f}(w_j) = \sum_{a=1}^{\boldsymbol{f}(w_i)} \sum_{b=1}^{\boldsymbol{f}(w_j)} 1 \tag{2}$$

where each co-occurrence in the image is uniformly weighted. As a result, the $c_i^f(j)$ in Equation 2 represents the frame-level correlation, in which the spatial and temporal information is ignored.

Practically, an instances always co-occurs with a variety of visual information contained in the frame, and the other instances should not be equally related to its context. Similar to [14], the spatial context, which is defined based on the spatial correlation between two instances of visual word, can be approximately quantized by:

$$c_i^s(j) = \begin{cases} \sum_{a=1}^{\boldsymbol{f}(w_i)} \sum_{b=1}^{\boldsymbol{f}(w_j)} e^{-\kappa d_{a,b}} & \boldsymbol{f}(w_i) \& \boldsymbol{f}(w_j) \neq 0 \\ 0 & otherwise \end{cases} \tag{3}$$

where $d_{a,b}$ denotes the distance between the instance a of the ith visual word and the instance b of the jth visual word in the same frame. The parameter κ determines the degree of spatial context. Similarly, weighting the co-occurrence by quantized temporal motion coherence [14] between the instances extend the context model to temporal information:

$$c_i^t(j) = \begin{cases} \sum_{a=1}^{\boldsymbol{f}(w_i)} \sum_{b=1}^{\boldsymbol{f}(w_j)} e^{-\gamma \|\Delta \boldsymbol{m}_{a,b}\|} & \boldsymbol{f}(w_i) \& \boldsymbol{f}(w_j) \neq 0 \\ 0 & otherwise \end{cases} \tag{4}$$

where the parameter γ determines the degree of temporal context. Here, $\Delta \boldsymbol{m}_{a,b}$ denotes the relative motion between the two instances [14].

To model the context of visual words in the current data collection, the c for each frames should be then accumulated. The computation is formulated as follows:

$$C_i(j) = \sum_{f \in D} c_i(j) \qquad (5)$$

where the calculated values $C_i(j)$ for $0 \le j < K$ are combined into a K dimensional row vector C_i. Furthermore, the row vectors C_i for $0 \le i < K$ are computed, the correlation matrix C is then formulated to represent the contexts of all visual words. The co-occurrence, spatial and temporal correlation matrixes can be calculated by utilizing c^c, c^s and c^t in the Equation 5 respectively.

Furthermore, to make the modelled context comparable between visual words, the vector C_i is normalized as:

$$C_i(j) = \frac{C_i(j)}{\parallel C \parallel_1 * tf_D(w_j)} \qquad (6)$$

where $\parallel * \parallel_1$ denote the first order $Norm$ of matrix and $tf_D(w_j)$ denote the term frequency of w_j within the entire video collection.

3.2 Context Similarity and Rebuilding the Visual Vocabulary

As introduced in the previous section, the synonymous visual words will be detected by examining the visual words context similarity. The context vector of a visual word has been quantitatively computed and the context similarity between visual words can be measured by the distance between the vectors. The distance d^c between two co-occurrence context vectors is formulated as:

$$d_D^c(i,j) \approx \sum_{k=1}^{K} |C_i^c(k) - C_j^c(k)| \qquad (7)$$

where Manhattan Distance $|C_i^c - C_j^c|$ is used as an example, and other distance metrics can also be used for the context similarity measurement. Similarly, the distance d^s and d^t, which is based on the spatial correlation and temporal correlation respectively, can also be computed using Equation 7.

It is clear from Equation 7 that the larger the $d(i,j)$, the more similar are the two visual words in terms of the corresponding contexts. Thus, a set of pair-wise synonyms $W_s = \{(w_i, w_j)\}$ is selected. The selection procedure can be formulated as follows:

$$(w_i, w_j) \begin{cases} \in W_s & \text{if } d_D(i,j) < \epsilon \\ \notin W_s & otherwise \end{cases} \qquad (8)$$

where d_D can be replaced by either d_D^c, d_D^s or d_D^t, which would consider co-occurrence, and spatial or temporal context respectively. ϵ is an empirically selected threshold to determine the number of visual words that should be merged. In our experiment, the range of selected ϵ is from 0.1 to 0.5, and the results will be demonstrated in the next section.

To exclude the falsely selected synonyms, we utilize the appearance of the visual words to verify the similarity. The visual words that are close in the feature space represent the features of similar appearances. A set of nearest visual words $N_i = \{w_n\}$ of visual word w_i is detected using K-d tree algorithm. If the w_j does not belong to the set N_i, it can not be merged with w_i. This guarantees that the merged synonyms are not only similar with respect to the context, but their appearances are also similar. Based on this arrangement, the representation of the frames can be updated according to the rebuilt visual vocabulary as follows:

$$f'(w_i) = \begin{cases} f(w_i) + f(w_j) & \text{if } (w_i, w_j) \in W_s \And w_j \in N_i \\ f(w_i) & otherwise \end{cases} \tag{9}$$

After merging K_m pairs of visual words, the initial visual vocabulary is pruned as a new visual vocabulary, and the size of it becomes $K - K_m$. The similarity between the query and a video document is then measured based on the updated representation vectors as:

$$sim_{v_d, v_q} = \max_{f'_d \in v_d, f'_q \in v_q} sim(f'_d, f'_q) \tag{10}$$

4   Experiments

The main objective of our experiments in this paper is to evaluate if the rebuilt visual vocabulary can deal with the synonyms caused by the *OverQuantize* problem. The Hessian detector [9] and SIFT feature [8] are used.

We evaluated our approach by the performance of the Query-by-Example video retrieval, and here the video relevance was defined in terms of concept similarity and visual similarity. Accordingly, two retrieval tasks were used: (1) general topics QBE video retrieval task, for retrieving "conceptually" similar videos; (2)QBE near-duplicate video search task for searching visually similar videos. Common performance criteria utilized in IR research community were applied: Mean Average Precision (MAP) and Precision-Recall.

The classical BovW model is used as a baseline, which is denoted by **BovW**. The performances of three variations of visual vocabulary rebuilding approaches proposed in this paper are reported, namely the co-occurring, spatial, and temporal correlation based approaches (Equation 10), denoted by **c-corr** , **s-corr**, and **t-corr** respectively. Furthermore, we also compared the effectiveness of our approach with the state-of-the-art disambiguation methods, Tight Geometric Constraint (**TGC**) [17]. Rather than simple comparison, **TGC** is utilized in combination with the rebuilt visual vocabularies, and the effectiveness can also be referred to as the compatibility between our approaches and the disambiguate methods. The **TGC** method was implemented with a public toolkit SOTU [17].

4.1   Experimental Set-Up

Two commonly used video retrieval datasets were selected for the experiments:
    (1) TRECVID2002 [12] is selected to perform the general topic video retrieval. The videos in this data collection contain diversified video sources: old film, news, documentary and advertisement, and the topics cover various information needs.

The data collection consists of approximately 10K shots segmented from 133 videos based on the shot boundary ground truth provided by TRECVID2002. Totally, 71 quires video were performed by the experiment. The topics involve searching for specific person, object, action, scene or instances of a category of a person. On average, each topic is associated with 23 relevant shots.

(2) CC_Web_Video Near-duplicate video search is performed on data collection CC_Web_Video [15]. Most videos in this data-collection are 3-5 minutes long and not longer than 10 minutes. The videos are crawled from Youtube, Yahoo and Google Video. 4590 videos are randomly selected from the original data-collection to form the experimental dataset. In summary, 24 topics are utilized, with 69 queries selected from the ground-truth. On average,10 key frames were sampled from each query to measure the video similarity, and each topic is associated with 84.7 relevant videos.

4.2   Experiment 1: General Topics QBE Video Retrieval

The relevance in the general topics QBE video retrieval is defined at the concept level. The scale of initial vocabulary in this experiment was selected to be 5K, which is relatively small and suitable to the great visual differences between the relevances.

Improving Bag-of-Visual-Words Model. The overall performance in terms of precision-recall curve are demonstrated in the Figure 2(a). It can be seen that the rebuilt visual vocabulary generally outperformed the original visual vocabulary. It is shown that at the points where the recall ranges from 5% to 35%, the corresponding precision of **c-corr** is higher than the **BovW**, which means that more relevances are retrieved. Thus the **c-corr** appear to compensate the missing information by the classical Bag-of-visual-Word model in this experiment. Furthermore, while **c-corr**, **t-corr**, and **s-corr** performs closely to each other, which may be caused by two reason: i) the spatial and temporal concurrence is an extension of simple co-occurrence, and ii) the accumulation of context across the data collection blurred the distinction based on single frame level spatial and temporal information.

It is important to point out that the number of merged visual words of rebuild vocabularies varied across different ϵ values (Equation 8). In order to capture this variation, we compared the performances of visual vocabularies, which were rebuilt by merging different number of visual words. Figure 2(b) illustrates the performances regarding MAP criteria of a variety of rebuilt vocabularies by **c-corr**, **s-corr**, and **t-corr** approaches respectively. The MAP performances were arranged according to the number of merged visual words K_m.

It can be seen from Figure 2(b) that the performances of all three approaches generally improved along with the increasing of K_m. The best performance of three methods was achieved by merging 150-250 out of 5K visual words, and around 3%-5% visual words were *OverQuantize*-ed by original quantization process for the TRECVID2002 data-collection. However, it does not necessarily imply that merging more visual words would always improve performance.

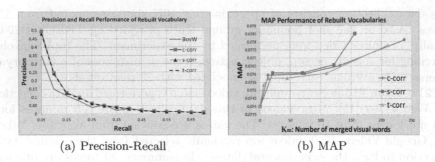

(a) Precision-Recall   (b) MAP

Fig. 2. Overall performance of rebuilt vocabulary for general topics QBE video retrieval

For example, if we increase the K_m up to 1538, the MAP of **t-corr** will drop to 0.06891 from the best performance 0.0781.

It is also clear from Figure 2(b) that the performance of **s-corr** is better than the **c-corr** and **t-corr**. This shows that the spatial context of visual words is more meaningful than simple co-occurring correlation. Furthermore, **s-corr** outperformed **t-corr** with respect to the MAP criteria. In this data-collection, the discovered temporal relationship across visual words between consecutive keyframes may not be meaningful enough.

The **s-corr** outperformed **BovW** on more topics (16 out of 22) and on average, **s-corr** significantly outperformed the initial vocabulary by 5.4% (P-value = 0.047). These results clearly demonstrate that the rebuilding process improved the initial vocabulary and partially solve the *OverQuantize* problem.

Compatibility with TGC Approach. Furthermore, to test whether the rebuilt vocabulary is compatible with other enhancement approach for the BovW framework, we incorporate it with the **TGC** approach as an example. **TGC** is one of the most common techniques used to disambiguate visual words with additional geometric information constraints. It has been proven to be effective to promote the accuracy of visual content similarity match.

As illustrated by Figure 3, **TGC** has obviously promoted the precision of the top ranked results of **BovW**. Afterward, we incorporate the **TGC** approach with the rebuilt visual vocabularies. The overall Precision-Recall curve is presented in Figure 3, and **c-corr**, **s-corr**, and **t-corr** outperformed the baseline **TGC** and **BovW**. The rebuilt vocabularies kept the high precision of the **TGC** for the top ranked results, and also improve the performance on the middle part of curve. It can be concluded that the rebuilt vocabularies based on context similarity could compensate the missing of relevances caused by *OverQuantize* and additional similarity constraints.

In summary, the evaluation on the QBE general topic video retrieval task shows that the rebuilt visual vocabulary based on spatial-temporal context effectively improve the performance of the classical BovW model, and it can be utilized in combination with the **TGC** approach to further improve the BovW model.

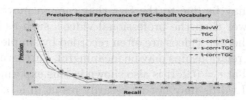

Fig. 3. Precision-Recall performance of TGC approaches based on the rebuilt vocabularies

4.3 Experiment 2: QBE Near-Duplicate Video Search

The near duplicate video search is another important application of QBE content based video retrieval. The search targets are always near identical with the queries. A 20K initial visual vocabulary is used to evaluate the performance of the rebuilding approach.

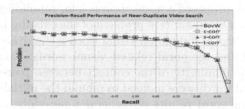

Fig. 4. Precision-Recall performance of based on rebuilt vocabulary based

The overall precision-recall curve is shown in Figure 4, and it demonstrates that **c-corr** outperformed **BovW**. The curve associated with **c-corr** is mostly located above the **BovW**. The curves associated with **c-corr**, **s-corr**, and **t-corr** are not very different with respect to the precision-recall criteria. However, all rebuilt vocabularies outperformed the initial visual vocabulary.

Table 1. MAP Comparison of rebuilt vocabularies

ϵ	c-corr	s-corr	t-corr
0		0.8144	
0.1	0.8243	0.8246	0.8346
0.2	**0.8358**	**0.8363**	**0.8364**
0.5	0.8181	0.8257	0.8360

The performances comparison regarding the MAP criteria is displayed in Table 1. The first column shows the different values of parameter ϵ used in Equation 8. It is demonstrated that **t-corr** slightly outperforms the other two rebuilt vocabularies. This may be due to the fact that camera movements in the web videos are often slower than the professional videos, such as movie and advertisements, because the web videos are often recorded by simple instruments. The discovered temporal relationships may be meaningful. Furthermore, the best performance is achieved at $\epsilon = 0.2$. It is also shown in this experimental results that merging more visual words may not necessarily result in a better performance. For example, vocabularies based on $\epsilon = 0.5$ performed worse than $\epsilon = 0.2$. Empirically, in this experiment, the number of merged redundant visual words should be around 40-70 (out of 20K visual words), which is less than the 150-200 merged visual words in the general topic video retrieval experiment. The main reason may be the higher requirement of precision by this application.

In summary, the evaluation on the QBE near-duplicate video search task shows that the rebuilt visual vocabulary based on spatial-temporal context can effectively improve the performance of initial visual vocabulary of the BovW.

5 Conclusions and Future Work

In this paper, we have proposed a novel approach for rebuilding the visual vocabularies to address the *OverQuantize* problem of the existing BovW models. The spatial-temporal context of individual visual word is discovered by weighting its co-occurring spatial and temporal related visual words. Furthermore, we proposed to measure the similarity of the context between pair-wise visual words, and the similar visual words were merged to rebuild the initial visual vocabulary.

A series of experimental results on general topic video retrieval and near-duplicate video search tasks showed that the rebuilt visual vocabulary generally promotes the performances of initial vocabulary. Experimental results also demonstrated that the rebuilt vocabulary is compatible with another state-of-the-art BovW enhancement approach TGC. The results clearly indicate that rebuilding technology based on the spatial-temporal context effectively reduces the redundant visual words in the initial visual vocabulary, and contribute to solving the *OverQuantize* problem.

A possible research direction is to investigate how to combine the spatial and temporal constraints to model the context of visual words.

Acknowledgement. This work is supported in part by the Chinese National Program on Key Basic Research Project (973 Program, grant No. 2013CB329304, 2014CB744604), the Natural Science Foundation of China (grant No. 61272265), and EU's FP7 QONTEXT project (grant No. 247590).

References

1. Cao, L., Tian, Y., Liu, Z., Yao, B., Zhang, Z., Huang, T.S.: Action detection using multiple spatial-temporal interest point features. In: ICME, pp. 340–345 (2010)
2. Chum, O., Mikulik, A., Perdoch, M., Matas, J.: Total recall ii: Query expansion revisited. In: 2011 IEEE Conference on Computer Vision and Pattern Recognition (CVPR), pp. 889–896 (June 2011)

3. Hsu, W.H., Chang, S.-F.: Visual cue cluster construction via information bottleneck principle and kernel density estimation. In: Leow, W.-K., Lew, M., Chua, T.-S., Ma, W.-Y., Chaisorn, L., Bakker, E.M. (eds.) CIVR 2005. LNCS, vol. 3568, pp. 82–91. Springer, Heidelberg (2005)
4. Jégou, H., Douze, M., Schmid, C.: Improving bag-of-features for large scale image search. Int. J. Comput. Vision 87, 316–336 (2010)
5. Jegou, H., Perronnin, F., Douze, M., Sánchez, J., Perez, P., Schmid, C.: Aggregating local image descriptors into compact codes. IEEE Trans. Pattern Anal. Mach. Intell. 34(9), 1704–1716 (2012)
6. Jiang, Y.-G., Ngo, C.-W.: Visual word proximity and linguistics for semantic video indexing and near-duplicate retrieval. Comput. Vis. Image Underst. 113, 405–414 (2009)
7. Lazebnik, S., Schmid, C., Ponce, J.: Beyond bags of features: Spatial pyramid matching for recognizing natural scene categories. In: Proceedings of the 2006 IEEE Computer Society Conference on Computer Vision and Pattern Recognition (CVPR 2006), vol. 2, pp. 2169–2178. IEEE Computer Society, Washington, DC (2006)
8. Lowe, D.G.: Distinctive image features from scale-invariant keypoints. Int. J. Comput. Vision 60, 91–110 (2004)
9. Mikolajczyk, K., Tuytelaars, T., Schmid, C., Zisserman, A., Matas, J., Schaffalitzky, F., Kadir, T., Gool, L.V.: A comparison of affine region detectors. Int. J. Comput. Vision 65(1-2), 43–72 (2005)
10. Peng, Y., Ngo, C.-W.: EMD-based video clip retrieval by many-to-many matching. In: Leow, W.-K., Lew, M., Chua, T.-S., Ma, W.-Y., Chaisorn, L., Bakker, E.M. (eds.) CIVR 2005. LNCS, vol. 3568, pp. 71–81. Springer, Heidelberg (2005)
11. Philbin, J., Chum, O., Isard, M., Sivic, J., Zisserman, A.: Object retrieval with large vocabularies and fast spatial matching. In: IEEE Conference on Computer Vision and Pattern Recognition (CVPR 2007) (June 2007)
12. Smeaton, A.F., Over, P., Kraaij, W.: Evaluation campaigns and trecvid. In: Proceedings of the 8th ACM International Workshop on Multimedia Information Retrieval (MIR 2006), pp. 321–330. ACM Press, New York (2006)
13. Wang, H., Yuan, J., Tan, Y.-P.: Combining feature context and spatial context for image pattern discovery. In: Proceedings of the 2011 IEEE 11th International Conference on Data Mining (ICDM 2011), pp. 764–773. IEEE Computer Society, Washington, DC (2011)
14. Wang, L., Song, D., Elyan, E.: Improving bag-of-visual-words model with spatial-temporal correlation for video retrieval. In: Proceedings of the 21st ACM International Conference on Information and Knowledge Management (CIKM 2012), pp. 1303–1312. ACM, New York (2012)
15. Wu, X., Ngo, C.-W., Hauptmann, A.G., Tan, H.-K.: Real-time near-duplicate elimination for web video search with content and context. Trans. Multi. 11, 196–207 (2009)
16. Yuan, J., Wu, Y.: Context-aware clustering. In: IEEE Conference on Computer Vision and Pattern Recognition (CVPR 2008), pp. 1–8 (2008)
17. Zhao, W.-L., Wu, X., Ngo, C.-W.: On the Annotation of Web Videos by Efficient Near-Duplicate Search. IEEE Transactions on Multimedia 12(5), 448–461 (2010)

Approximating the Signature Quadratic Form Distance Using Scalable Feature Signatures*

Jakub Lokoč

SIRET research group, Department of Software Engineering,
Faculty of Mathematics and Physics, Charles University in Prague
lokoc@ksi.mff.cuni.cz

Abstract. The feature signatures in connection with the signature quadratic form distance have become a respected similarity model for effective multimedia retrieval. However, the efficiency of the model is still a challenging task because the signature quadratic form distance has quadratic time complexity according to the number of tuples in feature signatures. In order to reduce the number of tuples in feature signatures, we introduce the scalable feature signatures, a new formal framework based on hierarchical clustering enabling definition of various feature signature reduction techniques. As an example, we use the framework to define a new feature signature reduction technique based on joining of the tuples. We experimentally demonstrate our new feature signature reduction technique can be used to implement more efficient yet effective filter distances approximating the original signature quadratic form distance. We also show the filter distances using our new feature signature reduction technique significantly outperform the filter distances based on the related maximal component feature signatures.

Keywords: Similarity Search, Approximate Search, Content-based Retrieval, Signature Quadratic Form Distance, Scalable Descriptor.

1 Introduction and Related Work

The content-based multimedia retrieval [6] has become an integral part of various information systems managing multimedia data (e.g., e-shops, image banks, industry and medical systems), providing users an alternative to the keyword-based retrieval approaches. In order to search the multimedia data in the content-based way, the systems often employ a similarity model enabling ranking of the database objects according to a query object, where the similarity model comprises multimedia data descriptors and a suitable similarity measure defined for the utilized descriptors. The selection of a proper similarity model then belongs among key tasks when designing an effective and efficient content-based multimedia retrieval system. During the last decades, many types of similarity models

* This research has been supported by Czech Science Foundation project GAČR P202/12/P297.

C. Gurrin et al. (Eds.): MMM 2014, Part I, LNCS 8325, pp. 86–97, 2014.

have been designed and even standardized for a particular multimedia retrieval tasks (e.g., the MPEG-7 standard [15]). One of the most popular similarity models investigated during the last decade is the bag of visual words (BoVW) model [18], utilizing a statically-created vocabulary of codewords, so called codebook. In the BoVW model, each object is represented as a frequency histogram of codewords present in the object, where all the objects in a database share one codebook. Such representation enables efficient retrieval using inverted files, a well established technique for the text-based retrieval area. Whereas the efficiency of the BoVW model is sufficient for large scale multimedia retrieval, the practical effectiveness of the model is still an open problem. While recent works have tried to improve the effectiveness of the BoVW model using semantic preserving models [19], Hamming embedding [10], compressed Fisher vectors [16] or vectors of locally aggregated features [11], several new approaches have relaxed from a common static vocabulary and investigated more general similarity models based on the feature signatures [17] and the adaptive distance measures (e.g., Signature Quadratic Form Distance [3] or Signature Matching Distance [1]). The signature-based models utilize an object specific vocabulary and thus can flexibly represent the contents of an object. Hence, the feature signatures can capture more disparities in the data, which can be beneficial in dynamic databases rapidly changing in content (e.g., multimedia streams). As recently shown, several signature-based models can outperform the BoVW approaches in the terms of effectiveness [1], however, the efficiency of the signature-based models is still a challenging task, especially for feature signatures comprising a high number of tuples. In [9], the authors employ metric/ptolemaic indexing to improve the efficiency of the retrieval, however, the approach is restricted only to distances satisfying metric/ptolemaic postulates. In [13], the authors show signature-based models can be utilized for effective re-ranking when obtaining a candidate result set using an efficient model based on a subset of the MPEG-7 descriptors. In this paper, we focus on new feature signature reduction techniques enabling more efficient yet still effective retrieval. Furthermore, we consider also scalability of the reduction techniques enabling adjusting the size of the feature signatures according to the actual system load. Let us now recall several basic concepts and definitions referred in this paper.

1.1   Feature Signatures and Signature Quadratic Form Distance

Feature signatures [17] have been introduced to flexibly aggregate and represent the contents of a multimedia object mapped into a feature space $\mathbb{F}$. Whether the requested feature space $\mathbb{F}$ comprises color, position, texture information, SIFT gradient vectors or other complex features [7,14], the feature signatures are often obtained by an adaptive variant of the k-means clustering selecting the most significant centroids. In Figure 1, we depict an example of image feature signatures according to a CPT feature space[1]. The feature signatures were extracted using the GPU extractor [12] employing an adaptive k-means clustering

---

[1] Color $\langle L, a, b \rangle$, position $\langle x, y \rangle$ and texture information $\langle contrast, entropy \rangle$, $\mathbb{F} \subseteq \mathbb{R}^7$.

Fig. 1. Example of feature signatures

algorithm, where the extraction of the first image in Figure 1 has put stress on the color, while the extraction of the second image in the figure has put stress on the position. The representatives $r_i \in \mathbb{F}$ corresponding to the selected centroids are depicted by circles in the corresponding position and color, while the weights $w_i \in \mathbb{R}^+$ corresponding to the size of the cluster determine the diameter of the circles (texture information is not depicted). Formally, the feature signatures are defined as:

Definition 1 (Feature Signature). *Given a feature space $\mathbb{F}$, the feature signature S^o of a multimedia object o is defined as a set of tuples $\{\langle r_i^o, w_i^o \rangle\}_{i=1}^n$ from $\mathbb{F} \times \mathbb{R}^+$, consisting of representatives $r_i^o \in \mathbb{F}$ and weights $w_i^o \in \mathbb{R}^+$*

The number of tuples in a feature signature can vary depending on a complexity of a corresponding multimedia object and the parameters used for the extraction. As a consequence, a feature signature can comprise tens or hundreds of tuples, which significantly affects the time for similarity computations. In [2], the authors propose a simple feature signature reduction technique based on maximal components of a feature signature O, where the maximal component feature signature O_{MC} with c components is defined as: $O_{MC} \subseteq O, |O_{MC}| = c$, such that $\forall \langle r_i^o, w_i^o \rangle \in O_{MC}, \forall \langle r_j^o, w_j^o \rangle \in O - O_{MC} : w_i^o \geq w_j^o$. In other words, the maximal component feature signature contains c tuples with the highest weights. The authors also define a signature quadratic form filter distance, applicable for approximate filter and refine retrieval, where the filter distance just evaluates the signature quadratic form distance using maximal component feature signatures. In Figure 2, we depict an example of maximal component feature signatures with 10 and 20 components. We may observe the maximal component feature signatures can omit representative tuples from the original feature signatures (the rightmost signatures in Figure 2) when few maximal components are utilized.

Let us now shortly recall the Signature Quadratic Form Distance [3], an effective adaptive distance measure generalizing the quadratic form distance.

Definition 2 (SQFD). *Given two feature signatures $S^o = \{\langle r_i^o, w_i^o \rangle\}_{i=1}^n$ and $S^p = \{\langle r_i^p, w_i^p \rangle\}_{i=1}^m$ and a similarity function $f_s : \mathbb{F} \times \mathbb{F} \to \mathbb{R}$ over a feature space $\mathbb{F}$, the signature quadratic form distance SQFD_{f_s} between S^o and S^p is defined as:*

$$\mathrm{SQFD}_{f_s}(S^o, S^p) = \sqrt{(w_o \mid -w_p) \cdot A_{f_s} \cdot (w_o \mid -w_p)^T},$$

where $A_{f_s} \in \mathbb{R}^{(n+m) \times (n+m)}$ is the similarity matrix arising from applying the similarity function f_s to the corresponding feature representatives, i.e., $a_{ij} = f_s(r_i, r_j)$. Furthermore, $w_o = (w_1^o, \ldots, w_n^o)$ and $w_p = (w_1^p, \ldots, w_m^p)$ form weight vectors, and $(w_o \mid -w_p) = (w_1^o, \ldots, w_n^o, -w_1^p, \ldots, -w_m^p)$ denotes the concatenation of weight vectors w_o and $-w_p$.

To determine similarity values between all pairs of representatives from the feature signatures, the Gaussian similarity function $f_{gauss}(r_i, r_j) = e^{-\alpha L_2^2(r_i, r_j)}$ or the Heuristic similarity function $f_{heuristic}(r_i, r_j) = 1/(\alpha + L_2(r_i, r_j))$ can be utilized, where α is a parameter for controlling the precision, and L_2 denotes the Euclidean distance. If we utilize similarity function $f_{L_2}(r_i, r_j) = -L_2^2(r_i, r_j)/2$, we obtain the L_2-Signature quadratic form distance [4] suffering from worse effectiveness but computable in linear time.

The rest of the paper is structured as follows: we present the scalable feature signatures and our new reduction technique in the following section, then we experimentally demonstrate in section 3 our new reduction technique can be employed for effective approximate search with the signature quadratic form distance, and finally we conclude the paper and point on the future work in section 4.

2   Scalable Feature Signatures

In this section, we introduce the *scalable feature signatures* – a formal framework based on hierarchical clustering enabling definition of sophisticated reduction strategies for feature signatures. The framework extends and generalizes the maximal component feature signatures [2] primarily designed for approximate filter and refine retrieval. As we experimentally demonstrate, the filter signature quadratic form distance employing the maximal component feature signatures does not approximate the original distance (or its lower bound) well, and thus we focus on new filter distances using new feature signature reduction techniques providing better approximations of the original feature signatures. Unlike the maximal components approach that just removes tuples with small

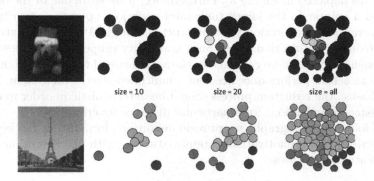

Fig. 2. Maximal component feature signatures

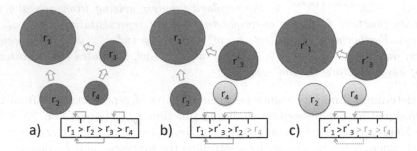

Fig. 3. Scaling feature signature

weights, our new approach aggregates the tuples during the reduction of the feature signatures. Our new approach is motivated by the feature signature extraction process [12], where the adaptive k-means clustering removes centroids with small weights, while the points distributed within the removed centroids are assigned to the remaining centroids. However, after the extraction process is finished and new feature signatures are stored, the points are no longer available and thus only information in the stored tuples can be used for the reduction of feature signatures.

Before we proceed to formal definitions, let us describe a motivation example depicted in Figure 3 where the feature signature FS in Figure 3a is consecutively reduced to the half of the original size in Figure 3c. If the maximal components approach was used, the reduced feature signature would contain only two blue tuples, which would not correspond to the original image. Therefore, instead of removing tuples, we can join them using an aggregation function τ to keep the original information at least in the aggregated form. To determine which tuples are joined in each step, we expect a total ordering $>$ defined over all the tuples in $\mathbb{F} \times \mathbb{R}^+$ and a mapping function ϕ^{FS} defined for all tuples in FS depicted as gray arrows in Figure 3. Using a suitable $>$, ϕ^{FS} and τ, we may observe the reduced feature signature can be a good approximation of the original feature signature (as depicted in Figure 3). Furthermore, if we store one of the original tuples and a pointer to the joined tuple after each join operation, we can later utilize a reverse split operation to reconstruct the original feature signature (or just less reduced feature signature). Such scalability property of the descriptor can be beneficial because we can balance the actual size of the feature signatures according to actual performance needs of a multimedia retrieval system. Let us also emphasize, the reduction process should be deterministic in order to enable preprocessing optimizations for a particular distance functions.

In the following paragraphs we provide definitions formalizing the key concepts described in the motivation example, starting with the definition of the scalable feature signatures.

Definition 3 (Scalable Feature Signature). *Given a feature signature FS over a feature space $\mathbb{F}$, a total ordering $>$ defined over all tuples in $\mathbb{F} \times \mathbb{R}^+$, a total* *mapping function* $\phi^{FS} : FS \to FS$ *and an aggregation function* $\tau : (\mathbb{F} \times \mathbb{R}^+)^2 \to \mathbb{F} \times \mathbb{R}^+$, *then the tuple* $(FS, >, \phi^{FS}, \tau)$ *is called scalable feature signature.*

In the following paragraphs, we show an example of the total ordering and several examples of mapping and aggregation functions. Let $\mathbb{F}$ be the euclidean space over field $\mathbb{R}^n$ and FS be a feature signature over that feature space. We can define a total ordering $>$ using weights and the lexicographic ordering $>_{lex}$ over vectors in $\mathbb{R}^n$ as: $\forall \langle r_i, w_i \rangle, \langle r_j, w_j \rangle \in \mathbb{F} \times \mathbb{R}^+ : \langle r_i, w_i \rangle >_{wl} \langle r_j, w_j \rangle$ if and only if $w_i > w_j \vee (w_i = w_j \wedge r_i >_{lex} r_j)$.

The mapping function ϕ^{FS} can utilize the total ordering $>_{wl}$ and can be defined for each tuple $\langle r_i, w_i \rangle \in FS$ as:

$$\phi^{FS}_{min}(\langle r_i, w_i \rangle) = \langle r_i, w_i \rangle \text{ for } \langle r_i, w_i \rangle = \max\nolimits_{>_{wl}} FS, \text{ and otherwise as:}$$

$$\phi^{FS}_{min}(\langle r_i, w_i \rangle) = \min\nolimits_{>_{wl}} \{\langle r_j, w_j \rangle : \langle r_j, w_j \rangle \in FS \wedge \langle r_j, w_j \rangle >_{wl} \langle r_i, w_i \rangle\}.$$

The mapping function ϕ^{FS}_{min} just maps each tuple from FS to the first greater tuple in FS, except for the maximal tuple that is mapped to itself. The mapping function can consider also a Minkowski distance L_p between the representatives in FS as follows:

$$\phi^{FS}_{L_p}(\langle r_i, w_i \rangle) = \langle r_i, w_i \rangle \text{ for } \langle r_i, w_i \rangle = \max\nolimits_{>_{wl}} FS, \text{ and otherwise as:}$$

$$\phi^{FS}_{L_p}(\langle r_i, w_i \rangle) = \langle r_j, w_j \rangle \text{ such that } \langle r_j, w_j \rangle \in FS \wedge \langle r_j, w_j \rangle >_{wl} \langle r_i, w_i \rangle \wedge$$
$$(\forall \langle r_k, w_k \rangle \in FS, k \neq i \neq j : \langle r_k, w_k \rangle >_{wl} \langle r_i, w_i \rangle \implies (L_p(r_j, r_i) < L_p(r_k, r_i) \vee$$
$$(L_p(r_j, r_i) = L_p(r_k, r_i) \wedge \langle r_k, w_k \rangle >_{wl} \langle r_j, w_j \rangle)))).$$

The aggregation operation τ can be defined trivially as a projection:

$$\tau_{first}(\langle r_i, w_i \rangle, \langle r_j, w_j \rangle) = \langle r_i, w_i \rangle,$$

or as a more complex aggregation:

$$\tau_{avg}(\langle r_i, w_i \rangle, \langle r_j, w_j \rangle) = \langle r_i \cdot w_i / (w_i + w_j) + r_j \cdot w_j / (w_i + w_j), w_i + w_j \rangle.$$

Having defined scalable feature signature $(FS, >, \phi^{FS}, \tau)$, we can now define an unary reduction operation that replaces the minimum $\langle r, w \rangle$ in FS and the corresponding tuple $\phi^{FS}(\langle r, w \rangle)$ by the join tuple $\tau(\langle r, w \rangle, \phi^{FS}(\langle r, w \rangle))$.

Definition 4 (Scalable Feature Signature Reduction). *Given a scalable feature signature* $SFS = (FS, >, \phi^{FS}, \tau)$, *let* $\langle r, w \rangle = \min_> FS$, *let* $FS' = (FS - \{\phi^{FS}(\langle r, w \rangle), \langle r, w \rangle\})$ *and let* $\langle r_k, w_k \rangle = \tau(\phi^{FS}(\langle r, w \rangle), \langle r, w \rangle)$, *then the reduction of scalable feature signature* SFS *denoted as* $\otimes SFS$ *is defined as* $\otimes SFS = (FS_r, >, \phi^{FS_r}, \tau)$, *where* $FS_r = FS' \cup \{\langle r_k, w_k \rangle\}$ *for* $\langle r_k, w_k \rangle \notin FS'$, *and* $FS_r = FS' - \{\langle r_k, w_k \rangle\} \cup \{\langle r_k, 2 \cdot w_k \rangle\}$ *otherwise.*

Let as denote ϕ^{FS_r} is defined in the same way as ϕ^{FS}, but in the context of new feature signature FS_r. We provide also a simple lemma to emphasize the unary reduction operation creates another scalable feature signature. The lemma is without proof because it is a direct consequence of the previous definitions.

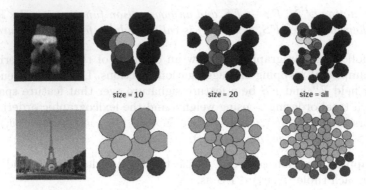

size = 10   size = 20   size = all

Fig. 4. Scalable feature signatures using $\phi_{L_2}^{FS}$ and τ_{avg}

Lemma 1. *Let $(FS, >, \phi^{FS}, \tau)$ be a scalable feature signature over a feature space $\mathbb{F}$, then $\otimes(FS, >, \phi^{FS}, \tau)$ is also a scalable feature signature over the feature space $\mathbb{F}$.*

So far we have provided a formal framework enabling definition of various feature signature reduction techniques. Using the framework, we can simply define our new reduction technique based on joining of the tuples as a quintuplet $(FS, >_{wl}, \phi_{L_2}^{FS}, \tau_{avg}, \otimes)$, consisting of the scalable feature signature $(FS, >_{wl}, \phi_{L_2}^{FS}, \tau_{avg})$ and the reduction operation $\otimes$. In Figure 4, we may observe our new reduction technique can approximate the distribution of the tuples in the original feature signature well even for smaller number of tuples.

Let us now provide several notes, for the lack of the space without proofs. First, the $\otimes(FS, >, \phi^{FS}, \tau)$ can create new scalable feature signature with $|FS|$, $|FS| - 1$ or $|FS| - 2$ tuples depending on the result of ϕ^{FS} and τ. Second, in case $|FS| = 1$, the reduction operation does not have to be identity, for example, if the aggregation function τ penalizes one of the arguments. Third, $(FS, >_{wl}, \phi_{min}^{FS}, \tau_{first})$ with the reduction operation $\otimes$ corresponds, except for minor differences[2], to the maximal component feature signatures. However, in the text we will strictly use the label *maximal component feature signatures* in order to distinguish the related work from the scalable feature signatures based on joining of the tuples.

Having defined an operation reducing the size of a scalable feature signature, we can now proceed to the definition of a new signature quadratic form filter distance, generalizing the filter signature quadratic form distance $SQFD_{filter}$ defined in [2], where the filter distance is utilized for the approximate search in a filter and refine architecture. In order to express multiple superpositions of the unary operation $\otimes$, we use in the following definition $\otimes^n(FS, >, \phi^{FS}, \tau)$ notation as a shortcut for $\underbrace{\otimes \cdots \otimes}_{n-times}(FS, >, \phi^{FS}, \tau)$.

---

[2] The maximal component feature signatures do not assume a total ordering of the tuples.

Definition 5 (Signature Quadratic Form Filter Distance). *Given two reduced scalable feature signatures* $(FS_{r_1}, >, \phi^{FS_{r_1}}, \tau) = \otimes^{|FS_1|-n}(FS_1, >, \phi^{FS_1}, \tau)$ *and* $(FS_{r_2}, >, \phi^{FS_{r_2}}, \tau) = \otimes^{|FS_2|-n}(FS_2, >, \phi^{FS_2}, \tau)$ *over a feature space* $\mathbb{F}$, *and let* $SQFD$ *be the signature quadratic form distance, then the distance* $SQFD_f^n = SQFD(FS_{r_1}, FS_{r_2})$ *is called the signature quadratic form filter distance according to* $SQFD(FS_1, FS_2)$.

The filter distance just simply reduces the original scalable feature signatures to a requested size and evaluates the original distance measure for the two reduced feature signatures. If we define the scalable feature signatures using $>_{wl}$, ϕ^{FS}_{min} and τ_{first} (which corresponds to the maximal component feature signatures), then the signature quadratic form filter distance $SQFD_f^n$ corresponds to the filter distance $SQFD_{filter}$ presented in [2]. The new signature quadratic form filter distance can be also utilized for the approximate search in a filter and refine schemes, where the reduced scalable feature signatures can be either cached or evaluated every time the filter distance is requested. Furthermore, such retrieval system can decide to temporarily use a reduced version of the scalable feature signatures also for the refinement step. In order to prevent from storing multiple versions of the scalable feature signatures, we can implement the reduction operation as a reversible update of the original scalable feature signatures enabling to keep just one actual version of the scalable feature signatures.

For example, the reduction operation in Figure 3ab replaces $\phi^{FS}_{L_2}(\langle r_4, w_4 \rangle) = \langle r_3, w_3 \rangle$ by $\langle r_{3'}, w_{3'} \rangle = \tau_{avg}(\langle r_3, w_3 \rangle, \langle r_4, w_4 \rangle)$, removes $\langle r_4, w_4 \rangle$, inserts pair $(\langle r_4, w_4 \rangle$, pointer to $\langle r_{3'}, w_{3'} \rangle)$ into a stack and sorts the tuples. The corresponding reverse operation removes pair $(\langle r_4, w_4 \rangle$, pointer to $\langle r_{3'}, w_{3'} \rangle)$ from the stack, inserts $\langle r_4, w_4 \rangle$ into the feature signature, replaces $\langle r_{3'}, w_{3'} \rangle$ by $\tau^{rev}_{avg}(\langle r_{3'}, w_{3'} \rangle, \langle r_4, w_4 \rangle) - \langle r_3, w_3 \rangle$ and sorts the tuples, where τ^{rev}_{avg} is derived from τ_{avg} as:

$$\tau^{rev}_{avg}(\langle r_i, w_i \rangle, \langle r_j, w_j \rangle) = \langle (r_i - r_j \cdot w_j/w_i) \cdot w_i/(w_i - w_j), w_i - w_j \rangle.$$

3 Experimental Evaluation

For the experiments, we make use of the three different datasets, each with different source of ground truth. Specifically, we use a subset of the ALOI dataset [8] comprising 12,000 images divided into 1,000 classes, each class contain 12 images of a 3D object rotated by 30 degrees; a subset of the Profimedia dataset [5] comprising 21,993 images divided into 100 classes, where the ground truth was collected semi-automatically and verified by users; the TWIC dataset [13] comprising 11,555 images forming 197 classes, where each class represents images obtained by a keyword query to the google images search engine. Each TWIC class was further manually filtered by users. The feature signatures were extracted using a GPU extractor tool [12]. For all three datasets we have used the same extractor parameters except the multiplicative vector that was adjusted to each dataset separately. The average number of tuples in feature signatures was 33 for ALOI dataset and 66 for TWIC and Profimedia datasets. As the query

Table 1. The time (in milliseconds) needed to evaluate the filter distances using various number of tuples (1, 2, 4, ..., 64) and the signature quadratic form distance (all)

	1	2	4	8	16	32	64	all
Gaussian	0.0006	0.0007	0.0015	0.0039	0.013	0.050	0.193	0.205
Heuristic	0.0004	0.0006	0.0013	0.0031	0.010	0.038	0.149	0.159

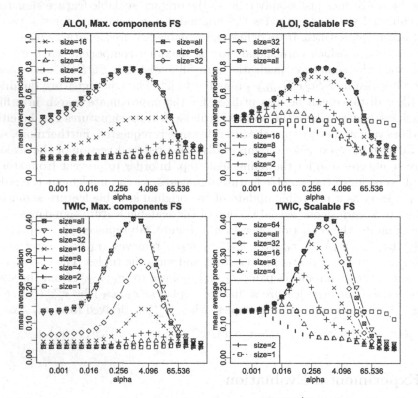

Fig. 5. Mean average precision of the filter distance $SQFD_f^{size}$ utilizing Gaussian similarity function

objects, one representative from each class was selected for all three datasets[3], resulting in 1000 query objects for ALOI, 100 query objects for Profimedia and 197 query objects for TWIC. The experiments have run on 64-bit Windows Server 2008 R2 Standard with Intel Xeon CPU X5660, 2.8 GHz.

In the experiments, we use $(FS, >_{wl}, \phi_{min}^{FS}, \tau_{first})$ as an implementation of the maximal component feature signatures and compare them to the scalable feature signatures using joining of the tuples $(FS, >_{wl}, \phi_{L_2}^{FS}, \tau_{avg})$. For each reduction technique, we utilize six variants of the filter distance $SQFD_f^{size}$ using $size \in \{1, 2, 4, 8, 16, 32, 64\}$ and compare them to the original distance denoted

---

[3] Profimedia dataset is already provided with a set of query objects.

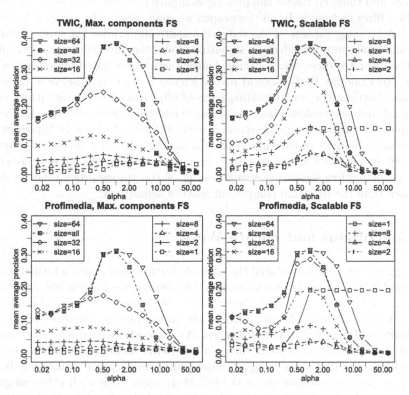

Fig. 6. Mean average precision of the filter distance $SQFD_f^{size}$ utilizing Heuristic similarity function

as *size = all*. Before we proceed to the experiments comparing the two reduction techniques, we present a table of average times (in milliseconds) needed to evaluate the utilized filter distances and the original distance, measured for the TWIC dataset for both Gaussian and Heuristic similarity functions. In Table 1, we may observe the Heuristic similarity function is slightly faster then the Gaussian similarity function. We may also observe the expected quadratic time dependency of the signature quadratic form filter distance on the number of tuples in the reduced feature signatures.

Let us now proceed to the following two figures, where the filter distances utilizing the Gaussian similarity function are depicted in Figure 5 and the filter distances utilizing the heuristic similarity function are depicted in Figure 6. In both figures, we have focused on the mean average precision (y-axis) measured for varying parameter α (x-axis). The figures are organized into two columns, where the first column contains results for the maximal component feature signatures (denoted as Max. components FS), while the second column contains the results for the scalable feature signatures using joining of the tuples (denoted simply as Scalable FS). Let us also denote, we have unified the y-axis scaling for each row and thus the reader can directly compare the effectiveness of two corresponding filter distances. In all the graphs we may observe the similar behavior – when decreasing the size of the reduced feature signatures, the filter distances using maximal component feature signatures loose the effectiveness more rapidly than the filter distances using scalable feature signatures based on joining of the tuples. For example, in the second row of Figure 5, we may observe a markable difference between the corresponding pairs of filter distances for signatures comprising 32 and less tuples, where for scalable feature signatures using joining of the tuples the mean average precision is over 30% even for just 16 tuples, while for the same number of tuples and the maximal component feature signatures the mean average precision is just 15%. From the experiments, we may conclude the scalable feature signatures using joining of the tuples provide better filter distances than the maximal component feature signatures.

4   Conclusions and Future Work

In this paper, we have introduced the scalable feature signatures, a formal framework enabling definition of various reduction techniques for feature signatures. As an example, we have defined a new feature signature reduction technique employing joining of the tuples and utilized the technique for definition of effective signature quadratic form filter distances. We have also experimentally demonstrated the filter distances using our new reduction technique significantly outperform the filter distances using maximal component feature signatures. In the future, we plan to examine the scalable feature signatures with other adaptive distance measures and measure the effectiveness of the corresponding similarity models. We would also like to design more complex mapping and joining functions in order to provide more options for the reduction of the scalable feature signatures. We also plan to investigate the performance of the scalable feature signatures on various different features extracted from the images (e.g., SIFT or color SIFT descriptors). We would also like to utilize the scalable feature signatures for more efficient retrieval using new filter and refine schemes or metric/ptolemaic access methods.

References

1. Beecks, C., Kirchhoff, S., Seidl, T.: Signature matching distance for content-based image retrieval. In: Proc. ACM International Conference on Multimedia Retrieval (ICMR 2013), Dallas, Texas, USA, pp. 41–48. ACM, New York (2013)
2. Beecks, C., Uysal, M.S., Seidl, T.: Efficient k-nearest neighbor queries with the signature quadratic form distance. In: Proc. 4th International Workshop on Ranking in Databases (DBRank 2010) in Conjunction with IEEE 26th International Conference on Data Engineering (ICDE 2010), Long Beach, California, USA, pp. 10–15. IEEE, Washington (2010)
3. Beecks, C., Uysal, M.S., Seidl, T.: Signature quadratic form distance. In: Proc. ACM CIVR, pp. 438–445 (2010)

4. Beecks, C., Uysal, M.S., Seidl, T.: L2-signature quadratic form distance for efficient query processing in very large multimedia databases. In: Lee, K.-T., Tsai, W.-H., Liao, H.-Y.M., Chen, T., Hsieh, J.-W., Tseng, C.-C. (eds.) MMM 2011 Part I. LNCS, vol. 6523, pp. 381–391. Springer, Heidelberg (2011)
5. Budikova, P., Batko, M., Zezula, P.: Evaluation platform for content-based image retrieval systems. In: Gradmann, S., Borri, F., Meghini, C., Schuldt, H. (eds.) TPDL 2011. LNCS, vol. 6966, pp. 130–142. Springer, Heidelberg (2011)
6. Datta, R., Joshi, D., Li, J., Wang, J.Z.: Image retrieval: Ideas, influences, and trends of the new age. ACM Comput. Surv. 40(2), 5:1–5:60 (2008)
7. Deselaers, T., Keysers, D., Ney, H.: Features for image retrieval: an experimental comparison. Information Retrieval 11(2), 77–107 (2008)
8. Geusebroek, J.-M., Burghouts, G.J., Smeulders, A.W.M.: The Amsterdam Library of Object Images. IJCV 61(1), 103 112 (2005)
9. Hetland, M.L., Skopal, T., Lokoč, J., Beecks, C.: Ptolemaic access methods: Challenging the reign of the metric space model. Inf. Syst. 38(7), 989–1006 (2013)
10. Jegou, H., Douze, M., Schmid, C.: Hamming embedding and weak geometric consistency for large scale image search. In: Forsyth, D., Torr, P., Zisserman, A. (eds.) ECCV 2008, Part I. LNCS, vol. 5302, pp. 304–317. Springer, Heidelberg (2008)
11. Jegou, H., Perronnin, F., Douze, M., Sánchez, J., Perez, P., Schmid, C.: Aggregating local image descriptors into compact codes. IEEE Trans. Pattern Anal. Mach. Intell. 34(9), 1704–1716 (2012)
12. Kruliš, M., Lokoč, J., Skopal, T.: Efficient extraction of feature signatures using multi-GPU architecture. In: Li, S., El Saddik, A., Wang, M., Mei, T., Sebe, N., Yan, S., Hong, R., Gurrin, C. (eds.) MMM 2013, Part II. LNCS, vol. 7733, pp. 446–456. Springer, Heidelberg (2013)
13. Lokoč, J., Novák, D., Batko, M., Skopal, T.: Visual image search: Feature signatures or/and global descriptors. In: Navarro, G., Pestov, V. (eds.) SISAP 2012. LNCS, vol. 7404, pp. 177 191. Springer, Heidelberg (2012)
14. Mikolajczyk, K., Schmid, C.: A performance evaluation of local descriptors. IEEE Transactions on Pattern Analysis and Machine Intelligence 27(10), 1615–1630 (2005)
15. MPEG-7. Multimedia content description interfaces. Part 3: Visual. ISO/IEC 15938-3:2002 (2002)
16. Perronnin, F., Liu, Y., Sánchez, J., Poirier, H.: Large-scale image retrieval with compressed fisher vectors. In: 2010 IEEE Conference on Computer Vision and Pattern Recognition (CVPR), pp. 3384–3391 (2010)
17. Rubner, Y., Tomasi, C.: Perceptual Metrics for Image Database Navigation. Kluwer Academic Publishers, Norwell (2001)
18. Sivic, J., Zisserman, A.: Video google: A text retrieval approach to object matching in videos. In: Proceedings of the Ninth IEEE International Conference on Computer Vision (ICCV 2003), vol. 2, p. 1470. IEEE Computer Society, Washington, DC (2003)
19. Wu, L., Hoi, S.C.H., Yu, N.: Semantics-preserving bag-of-words models and applications. Trans. Img. Proc. 19(7), 1908–1920 (2010)

A Novel Human Action Representation via Convolution of Shape-Motion Histograms

Teck Wee Chua and Karianto Leman

Institute for Infocomm Research,
A*STAR (Agency for Science, Technology and Research), Singapore
{tewchua,karianto}@i2r.a-star.edu.sg

Abstract. Robust solutions to vision-based human action recognition require effective representations of body shapes and their dynamics. Combining multiple cues in the input space can improve the recognition task. Although conventional method such as concatenation of feature vectors is straightforward, it may not sufficiently encapsulate the characteristics of an action. Inspired by the success of convolution-based reverb application in digital signal processing, we propose a novel method to synergistically combine shape and motion histograms via convolution operation. The objective is to synthesize the output (action representation) which carries the characteristics of both source inputs (shape and motion). Analysis and experimental results on the Weizmann and KTH datasets show that the resultant feature is more efficient than other hybrid features. Compared to other recent works, the feature that we used has much lower dimension. In addition, our method avoids the need for determining weights manually during feature concatenation.

1 Introduction

There has been a surge, in recent years, towards the study of human action recognition because it is fundamental to many computer vision applications such as video surveillance, human-computer interface, and content-based video retrieval. The search results retrieved from the search engine upon the keyword 'human action recognition' is astonishing. For instant, Google search engine returned 27,800,000 results as on 9 August, 2013. While human can recognize an action in a seemingly effortless fashion, the solutions using computer have, in many cases, proved to be immensely difficult. One open problem is the choice of optimal representations for human actions. Ideally, the representation should be robust against inter/intra variations, noises, temporal variations, and sufficiently rich to differentiate huge number of possible actions. Practically, such representation does not exist.

Recent approaches can be categorized into local and global representations. Local representation encodes the image or video frames as a collection of local patches. Common local representation includes spatio-temporal interest points such as 3D Harris, cuboid, Hessian, Dense etc. Usually the interest points are extracted at different spatial and temporal scales. Laptev and Lindeberg [1] proposed to extend Harris corner detector to the third dimension. Dollár *et al.* [2]

C. Gurrin et al. (Eds.): MMM 2014, Part I, LNCS 8325, pp. 98–108, 2014.

cuboid detector is based on temporal Gabor filter. Willems *et al.* [3] measured the saliency with the determinant of the 3D Hessian matrix. For global representation, the region-of-interest (obtained by tracking or background subtraction) is encoded as a whole. In other words, the entire human figure is considered. Silhouette, optical flow, edge and space-time volume fall into this category. Efros *et al.* [4] used blurred optical flows to recognize the actions of small human figures. Blank *et al.* [5] stacked silhouettes over a sequence to form space-time volumes. Poisson equation was used to compute local space-time saliency and orientation features. Ikizler *et al.* [6] extended the motion descriptor of Efros by using spatial and directional binning and then combined it with line shape descriptor. Following that, they proposed to use histogram of oriented rectangles as the shape descriptor [7]. Likewise, Lin *et al.* [8] used silhouettes as the shape descriptor by counting the number of foreground pixels and motion-compensated optical flow as motion descriptor. Ikizler *et al.* [7] pointed out that human actions can be encoded as spatial information of body poses and dynamic information of body motions. As a matter of fact, some actions cannot be distinguished using shape or motion feature alone. For example, as shown in Fig. 1 a *skip* action may look very similar to a *run* action if only the pose of the body is observed. The classification task would be easier if the motion flow of the body is considered simultaneously. One would expect that *skip* action generates more vertical flows (upward/downward). Besides, actions such as jogging, walking and running can be easily confused if only pose information is used due to similar postures in the action sequences. Likewise, there are some actions which cannot be fully described by motion feature alone. Combining both cues potentially provides complementary information about an action. Conventionally, motion and shape feature vectors are concatenated to form a super vector [8,9]. However, the super vector may not explicitly convey the underlying action. Moreover, the super vector is unnecessarily long and require feature dimension reduction techniques. In this regard, an efficient representation of action is highly desirable. Motivated by the idea of convolution-based reverb in digital signal processing (DSP), we propose to encode the human action by convolving shape and motion histograms. This novel representation extracts rich information from actions.

Fig. 1. Similar poses observed in *skip* (left) and *run* (right) action sequences

The paper is organized as follows. In Section 2, we describe the details of shape and motion feature extraction. Next, we define the atomic action and explain how it can be represented as the convolved shape-motion histogram. Section 5 sets the backdrop for the experimental evaluation on Weizmann and KTH datasets while Section 6 shows the results. Finally, Section 7 gives the conclusion remarks of the paper.

2   Motion and Shape Histogram Binning

Observing shape and motion is a very natural way to recognize an action. Jhuang *et al.* [10] pointed out that the visual cortex in the brain has two pathways to process shape and motion information. Motivated by the robustness of the histogram of feature, we use histogram-of-oriented gradient (HOOG) and histogram-of-oriented optical flow (HOOF) as the shape and motion descriptors respectively. We adopt the histogram formation method which was originally introduced by Chaudhry *et al.* [11]. The method is more robust against scale variation and the change of motion direction. The method is illustrated in Fig. 4(a) with example of creating a 4-bin histogram. The main idea is to bin the vectors according to their primary angles from the horizontal axis. Therefore, the vectors are symmetry about the vertical axis. As a result, the histogram of a person moving from left to right will be same as the one with a person moving in the opposite direction. The contribution of each vector is proportional to its magnitude. The histogram is normalized to sum up to unity to make it scale-invariant. Therefore, we do not normalize the size of the bounding box. We further enhance Chaudhry *et al.* algorithm by including spatial information. This is done by dividing the bounding box of the subject into 4×4 regions as shown in Fig. 4(c).

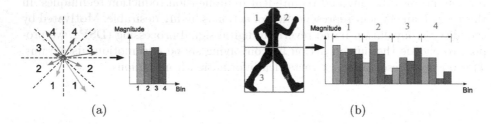

(a)   (b)

Fig. 2. (a) Histogram binning, (b) the bounding box is divided into 4×4 grid and the resultant histograms from each region are concatenated

3   Atomic Action Representation – Convolution of Shape-Motion Histrograms

3.1   Defining 'Atomic Action'

Formally, a complex action can be decomposed to into a sequence of elementary building blocks, known as 'atomic actions'. For example, Fig. 3 shows a walking action can be decomposed into several atomic actions – *right-leg stepping, two-leg crossing* and *left-leg crossing*. In this study, an atomic action is defined as the action performed at video frame t. It is represented by a shape histogram extracted at frame t and optical flow histogram computed at frames $(t-1)$ and t. Therefore, a T-frame action video has $(T-1)$ number of atomic actions.

Fig. 3. A walking sequence can be decomposed into sequence of atomic actions: *right-leg stepping* (left most), *two-leg crossing* (middle two), and *left-leg crossing* (right most)

3.2 Convolving Shape-Motion Histograms

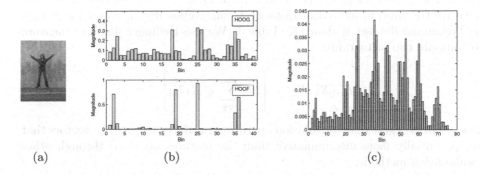

(a)   (b)   (c)

Fig. 4. (a) A jumping-jack action, (b) the corresponding shape histogram (upper) and motion histogram (lower), (c) the convolved histogram

Inspired by the idea of convolution-based reverb in digital signal processing (DSP), we propose to encode atomic actions by convolving shape and motion histograms. In DSP, convolution is a mathematical way of combining two source signals to form an output signal. The output signal bears the characteristics of both sources. One important application of convolution is convolution-based reverb, a process for digitally simulating the reverberation of a virtual or physical space. Given the impulse response of a space which can be obtained by recording a short burst of a broad-band signal, we can convolve any "dry" signal (little room or space influence) with that impulse response. The results is that the sound appears to have been recorded in that space. Analogously, knowing that an action is characterized by both shape and motion information, we can obtain an atomic action histogram $A[i]$ by convolving the corresponding shape histogram $X_s[i]$ and motion histogram $X_m[i]$:

$$A[i] = X_s[i] * X_m[i] = \sum_{k=-\infty}^{k=+\infty} X_s[k].X_m[i-k] \tag{1}$$

where the asterisk '*' denotes the convolution operator and square bracket [ ] indicates the signal is discrete. Since histograms are discrete in nature, for notation consistency the square bracket is omitted in the latter parts of this paper (i.e., x_i is equivalent to $x[i]$). Convolution operation is commutative meaning that it does not mathematically matter the order of the inputs. Thus, convolving X_s with X_m opposed X_m with X_s does not affect the result of the output. The length of output is given by the $\|X_s\| + \|X_m\| - 1$. Fig. 4 shows the convolution process of a jumping-jack action.

The proposed representation has two major advantages: First, the action histogram is more robust against noises. This is because each bin in the action histogram is influenced by bins in the shape histogram weighted by the motion histogram or vice versa (commutative property of convolution), therefore the effect of abrupt changes in the histogram magnitude can be minimized; second, the action histogram produced using convolution is more discriminative. We measure the ratio of inter-class distance to intra-class distance and the results on Weizmann dataset is shown in Table 1. We use Hellinger distance measure to compare two histograms:

$$D_h(X_1, X_2) = \left[1 - \sum_{\forall x} \sqrt{X_1 X_2}\right]^{\frac{1}{2}} \tag{2}$$

The results suggest that convolution operation produces the feature vectors that are potentially more discriminative than the features obtained through other combination methods.

4   Compact Video Representation: Distance Weighted Bag-of-Atomic-Actions

Over the past decade, a large body of work on human action recognition using local representation has been focusing on the bag-of-visual-words model [2, 3, 12–14]. In local representation, a huge collection of independent patches (e.g.: spatial-temporal features) is extracted from the training data. A codebook is then created from these local patches using some clustering algorithms such as K-means. Following that, an image or short sequence of images can be represented as a histogram which corresponds to the frequency of the visual-words. Apparently, this technique may not be directly applicable to global representation. Therefore, we propose an extension scheme to encode a video that is represented by the global features. In the proposed scheme, an action video is represented as a collection of repetitive atomic actions. Recall that an atomic actions is fully characterized by the convolved shape-motion histogram as described in the previous section. A visual codebook can be created by performing K-means clustering on all atomic actions from the training data. The cluster centroids, which are essentially some normalized histograms, serve as the visual codewords. Next, each atomic action in the video is compared against those codewords and the distances are recorded accordingly. The distance between the atomic action and its nearest codeword is used to weight the histogram bin. The histogram for all relevant codewords in a video is computed by aggregating their respective distances. This final representation allows any lengthy video to be 'compressed' into a compact histogram. The histogram is normalized such that sum of the bins is unity. The normalization ensures that the histogram distribution is invariant to the video length. For instant, given a particular action class, we expect to see the codewords (i.e., key atomic actions) frequencies for a variable length video remains relatively stable.

Table 1. Comparison of normalized inter-intra class distance ratio on Weizmann dataset for different types of feature combination methods. Higher value indicates that the feature is potentially more discriminative.

Combination Strategies	Ratio
Convolution (*Conv*)	1.0000
Summation (*Sum*)	0.8535
Product (*Prod*)	0.8489
Concatenation (*Concat*)	0.8743

5   Experiments

We performed various experiments to evaluate the proposed action recognition framework on two publicly available datasets (see Fig. 5):

- **Weizmann.** The dataset was originally introduced in [5]. The dataset contains 90 low-resolution (180×144 pixels) video sequences with 9 subjects performing 10 actions: bend (*bend*), jumping-jack (*jack*), jump-forward (*jump*), jump-in-place (*pjump*), run (*run*), gallop-sideways (*side*), jump-forward-one-leg (*skip*), walk (*walk*), wave-one-hand (*wave1*), and wave-two-hands (*wave2*)[1]. We used the silhouettes provided to compute the bounding boxes for the subjects. HOOG and HOOF features are extracted from the silhouettes.

- **KTH.** The dataset was introduced in [15]. There are 25 subjects performing 6 actions: boxing, handclapping, handwaving, jogging, running, and walking. The low resolution (160×120) videos were recorded under four scenarios (s1-outdoors, s2- outdoors with scale variation, s3- outdoor with different clothes,

---

[1] Note that there are two versions of Weizmann dataset, the original one has 9 actions while the augmented version has 10 which includes *skip* action.

s4- indoors with lighting variation) and each video was split into 4 sub-clips. Originally, the dataset has (4 settings) × (25 subjects) × (6 actions) × (4 sub-clips) = 2400 clips. However, only 2391 clips are available because 8 clips were missing. We used the bounding box provided by Lin *et al.* [8] to locate the subject. Nevertheless, we did not compute the silhouette because object segmentation is not our focus in this work. Therefore, HOOG and HOOF features are extracted directly from the raw grayscale video frames.

In the literature, KTH dataset has been regarded either as one large set with strong intra-subject variations (*all-in-one*) or as four independent scenarios. In the latter case, each scenario is trained and tested separately. In this work, we only focus on the *all-in-one* case.

Leave-one-out cross validation (LOOCV) protocol is used in all evaluations. We use multiclass Support Vector Machine (SVM) as the classifier.

Fig. 5. Examples of different actions from databases Weizmann (left) and KTH (right)

6   Results

Table 2 shows the LOOCV recognition rate for Weizmann dataset. With only using 5 clusters (codewords), the convolved feature yielded a much higher accuracy (96.67%) compared to other features. When the number of clusters is increased further, the convolved feature consistently gives perfect classification accuracy (100%). It is worth noting that using only shape feature (HOOG) or motion feature (HOOF) resulting poorer results. On average, by using the proposed method we gain about 11.29% overall improvement (4% for *sum*, 5.33% for *prod*, 4.44% for *concat*, 23.56% for *HOOG* and 19.11% for *HOOF*). Since KTH is a more challenging dataset with strong intra-class variations, it would be interesting to find out if the proposed method can still perform well as for the Weizmann dataset. The results for KTH dataset is tabulated in Table 3. Obviously, similar to Weizmann dataset, for all number of clusters, higher accuracies are attained from the convolved feature. Although classification task on KTH dataset is more difficult, the advantage of using the convolved feature is more prominent on this dataset. The average improvement over all other five features is 19.56% (10.61% for *sum*, 10.58% for *prod*, 6.49% for *concat*, 42.72% for *HOOG* and 27.41% for *HOOF*). On this dataset, both HOOG and HOOF features fail to provide discriminative information which contribute to the poor classification results. One important observation from the results is that our method consistently requires a much smaller number of clusters (codewords) to give higher accuracy.

Table 2. Weizmann dataset: LOOCV classification accuracy using different number of clusters

No. of Clusters \Features	Convol	Sum	Prod	Concat	HOOG	HOOF
5	96.67	87.78	87.78	86.67	66.67	73.33
10	**100**	94.44	94.44	93.33	74.44	78.89
15	**100**	97.78	94.44	98.89	77.78	81.11
20	**100**	97.78	96.67	97.78	78.89	84.44
25	**100**	98.89	96.67	97.78	81.11	83.33

Table 3. KTH dataset: LOOCV classification accuracy using different number of clusters

No. of Clusters \ Features	Convol	Sum	Prod	Concat	HOOG	HOOF
10	83.94	70.25	72.22	75.58	45.88	57.73
25	91.63	79.94	79.92	83.30	51.90	63.88
40	**92.46**	82.44	81.43	87.64	45.24	64.88
55	91.46	84.43	83.62	86.97	45.58	63.37

For example, with only 10 clusters our method achieves comparable accuracy with the product feature which uses 40 clusters. This confirms the finding that the convolved feature is significantly more discriminative. Moreover, we notice that the product-based combination method is slightly inferior to the sum-based approach, probably due to the information lost when multiplying two features with elements equal to zero. The bottom part of Fig. 4(b) shows an optical flow histogram with many bins equal to zero. When multiplying this histogram with the gradient histogram, the output will bias to the motion input while some useful information from the shape input is discarded.

The confusion matrix for Weizmann and KTH datasets are given in Fig. 6(a) and 6(b) respectively. For KTH dataset, errors mainly occur when classifying *boxing*, *handwaving* and *handclapping*. The misclassification of boxing action could be due to the erroneous bounding box extracted which is off-centered from the body axis when the person punches to the side. Nevertheless, our method is able to discriminate *jogging-running-walking* classes very well. This is remarkable given that these three actions share substantial similarity in motion and shape cues.

We have compared the results with state-of-the-art action recognition approaches. Table 4 shows that our method achieved the same perfect accuracy as [9], [16] for Weizmann dataset. As for KTH dataset, it may be argued that the results are not directly comparable as different authors employed different evaluation protocol (splits vs. LOOCV). While not definitive, Table 5 still provides some indicative comparison. The result shows that our method achieves one of the highest accuracies (only slightly lower than [8]) as far as the LOOCV protocol is concerned. One possible reason is that Lin *et al.* used silhouette for feature

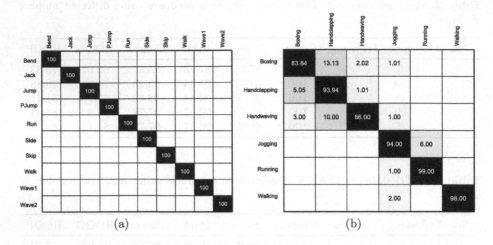

Fig. 6. Confusion matrix for (a) Weizmann (accuracy = 100%), and (b) KTH dataset (accuracy = 92.46%)

Table 4. Comparison of Recognition Rates for Weizmann Dataset

Method	Accuracy (%)
Our method	**100.00**
Fathi [16]	100.00
Schindler [9]	100.00
Blank [5]	99.64
Jhuang [10]	98.80
Wang [17]	97.78
Chaudhry [11]	94.44
Niebles [14]	90.00

extraction while in our KTH experiment we only used the original grayscale image containing inside the bounding box. It is well known that silhouette-based approach is more robust but it requires good background modelling which is more restrictive than the bounding box-based approach. From the result, it can be deduced that our method can perform very well even without using silhouette. One biggest advantage of our approach is the simplicity in implementation as opposed to their complicated prototype trees generation procedure. Moreover, the feature used in our method has much lower dimension (length = 79) than those used in [8](length = 512) and [9](length = 1000). Most importantly, the tedious task of determining the optimal weight to control the relative importance of shape and motion cues during concatenation, is no longer required.

Table 5. Comparison of Recognition Rates for KTH Dataset

Method	Protocol	Accuracy (%)
Our method	LOOCV	**92.46**
Lin [8]	LOOCV	93.43
Schindler [9]	Splits	92.70
Fathi [16]	Splits	90.50
Ahmad [18]	Splits	88.83
Willems [3]	Splits	84.26
Niebles [14]	LOOCV	83.33
Dollár [2]	LOOCV	81.17
Ke [19]	LOOCV	80.90
Schüldt [15]	Splits	71.72

7   Conclusion

This paper presents a novel method to encode human actions by convolving shape-motion histograms. The inspiration comes from the success of convolution reverb application in digital signal processing. The main idea is to produce an output signal (i.e., action histogram) from the source signals (i.e., shape and motion histograms) so that the output shares the characteristics of both sources. The experimental results demonstrate that the proposed method is very efficient compared to other combination strategies such as sum, product and concatenation. Moreover, our results are compared to the state-of-the-art results.

References

1. Laptev, I., Lindeberg, T.: Space-time interest points. In: ICCV, Nice, France, pp. 432–439 (2003)
2. Dollár, P., Rabaud, V., Cottrell, G., Belongie, S.: Behavior recognition via sparse spatio-temporal features. In: VS-PETS, Beijing, China, pp. 65–72 (2005)
3. Willems, G., Tuytelaars, T., Van Gool, L.: An efficient dense and scale-invariant spatio-temporal interest point detector. In: Forsyth, D., Torr, P., Zisserman, A. (eds.) ECCV 2008, Part II. LNCS, vol. 5303, pp. 650–663. Springer, Heidelberg (2008)
4. Efros, A.A., Berg, A.C., Mori, G., Malik, J.: Recognizing action at a distance. In: ICCV, Nice, France, pp. 726–733 (2003)
5. Blank, M., Gorelick, L., Shechtman, E., Irani, M., Basri, R.: Actions as space-time shapes. In: ICCV, Beijing, China, pp. 1395–1402 (2005)
6. Ikizler, N., Cinbis, R.G., Duygulu, P.: Human action recognition with line and flow histograms. In: ICPR, Tampa, FL, pp. 1–4 (2008)
7. Ikizler, N., Duygulu, P.: Histogram of oriented rectangles: A new pose descriptor for human action recognition. Image and Vision Computing 27, 1515–1526 (2009)
8. Lin, Z., Jiang, Z., Davis, L.S.: Recognizing actions by shape-motion prototype trees. In: ICCV, Kyoto, Japan (2009)
9. Schindler, K., Van Gool, L.: Action snippets: How many frames does human action recognition require? In: CVPR, Anchorage, Alaska, pp. 1–8 (2008)

10. Jhuang, H., Serre, T., Wolf, L., Poggio, T.: A biologically inspired system for action recognition. In: ICCV, Rio de Janeiro, Brazil, pp. 1–8 (2007)
11. Chaudhry, R., Ravichandran, A., Hager, G., Vidal, R.: Histograms of oriented optical flow and binet-cauchy kernels on nonlinear dynamical systems for the recognition of human actions. In: CVPR, Miami, FL, USA, pp. 1932–1939 (2009)
12. Poppe, R.: A survey on vision-based human action recognition. Image and Vision Computing 28, 976–990 (2010)
13. Wang, H., Ullah, M.M., Kläser, A., Laptev, I., Schmid, C.: Evaluation of local spatio-temporal features for action recognition. British Machine Vision Conference (BMVC), 127 (2009)
14. Niebles, J.C., Wang, H., Fei-fei, L.: Unsupervised learning of human action categories using spatial-temporal words. Int'l J. Computer Vision 79, 299–318 (2008)
15. Schüldt, C., Laptev, I., Caputo, B.: Recognizing human actions: A local SVM approach. In: ICPR, Cambridge, United Kingdom, pp. 32–36 (2004)
16. Fathi, A., Mori, G.: Action recognition by learning mid-level motion features. In: CVPR, Anchorage, Alaska, pp. 1–8 (2008)
17. Wang, L., Suter, D.: Recognizing human activities from silhouettes: Motion subspace and factorial discriminative graphical model. In: CVPR, Minnesota, USA, pp. 1–8 (2007)
18. Ahmad, M., Lee, S.: Human action recognition using shape and CLG-motion flow from multi-view image sequences. Pattern Recognition 41, 2237–2252 (2008)
19. Ke, Y., Sukthankar, R., Hebert, M.: Spatio-temporal shape and flow correlation for action recognition. In: 7th Int. Workshop on Visual Surveillance (2007)

How Do Users Search with Basic HTML5 Video Players?

Claudiu Cobârzan and Klaus Schoeffmann

Alpen-Adria-Universität Klagenfurt
9020 Klagenfurt, Austria
{claudiu,ks}@itec.aau.at

Abstract. When searching within a video for a specific scene most non-expert users employ a basic video player. The main advantage of such a player over more advanced retrieval tools lies in its ease of use and familiar controls and mode of operation. This means that the available navigation controls (play, fast forward, fast reverse, seeker-bar) will be used for interactive search and browsing. We compare the search behavior by type of interaction and speed of interactive search of two groups of users, each numbering 17 participants. Both groups performed the same tasks using an HTML5 video player but in different setups: the first group performed Known Item Search tasks, while the second performed Description Based Search tasks. The goal of this study is twofold. One: better understand the way users search with a basic video player, so that useful insights can be taken into consideration when designing professional video browsing and search tools. Two: evaluate the impact of the different setups (Known Item Search vs. Description Based Search tasks)

Keywords: video search, video browsing, user behavior, HTML5 video player.

1 Introduction

The amount of video data made available on the Internet is continuously increasing thanks in part to social media and sharing platforms, as well as to the wide availability and popularity of video recording devices in consumer electronics. A large portion of the videos are recorded by non-professionals that are driven by the most diverse of motives [5]. Those non-professionals will mostly employ simple video players, not only for viewing the content, but also for searching for specific sequences. This is because the available navigation features like play, pause, fast-forward, fast-reverse as well as random access using a seeker-bar are familiar and easy to use. Those simple features are in fact preferred by non-expert users as reported in [10]. However, they limit the user experience and interaction metaphors especially in mobile setups. Recent research has begun to investigate more appropriate controls for such situations [3], [4].

Popular web based sharing platforms like YouTube often provide only the most basic video player functionality (play/pause buttons and a seeker-bar) and

C. Gurrin et al. (Eds.): MMM 2014, Part I, LNCS 8325, pp. 109–120, 2014.

dump some of the typical VCR controls like the fast-forward and the fast reverse buttons. Current implementations of standard HTML5 video players also tend to favor those basic controls. For example, the Safari implementation offers the fast-forward and fast-reverse functionalities only when the player runs in full-screen mode.

This paper investigates the way those basic controls are used for video browsing (a combination of video playback and video search) within long videos when searching for a specific sequence. It continues the work in [8], which presents evaluation results from a user study performed with 17 participants that had to solve "Known Item Search" (KIS) tasks used for the Video Browser Showdown [7]. We add the result from a second user study, also with 17 participants, which had to solve "Description Based Search" (DBS) tasks with the same long videos as in the Video Browser Showdown [7], but with the target videos replaced by their textual descriptions. We compare and discuss the results from the two studies.

We are aware of only another study that evaluates search strategies, but concentrates on VCR-like (play, pause, stop, fast-forward, fast-reverse) controls [1] and not typical HTML5 navigation features.

2   Related Work

User interaction with VCR-like controls has been studied in [1]. The authors performed a user study in which the tasks were of type "Known Item Fact Retrieval" or "I'll know it when I see it" [2]. They consisted of finding video segments within a small archive based on a semantic question. The available navigation controls were *play, stop, pause, fast rewind, fast forward, step reverse* and *step forward*. Four search strategies were identified within the study: (1) *incremental linear search* (55%), (2) *decremental linear search* (10%), (3) *educated guess* (29%), and (4) *random selection* (6%). Within one file, the browsing behavior consisted of: (1) *straight viewing* (21%), (2) *speed switching*: linear viewing with switching back and forth between playback and fast-forward (46%), (3) *inaccurate shuttle determination*: fast-forward too far, fast-reverse, then play – or if too far back, fast-forward again (13%), (4) *accurate shuttle determination*: similar to (3) but with step-forward and step-backward (7%), (5) *halt and refine*: step-forward and play but pause sometimes to reflect on where they are (13%). The fastest approach proved to be *speed switching* and *halt and refine, accurate shuttle determination, straight viewing*, and *inaccurate shuttle determination*.

In contrast, the study in [8] on user interaction by the means of a basic HTML5 video player, focused on the use of the default controls in non-full-screen mode: *play* and *pause* buttons and the *seeker-bar*. The users had to perform "Known Item Search" tasks: a short video sequence of 20 seconds was initially presented and then it had to be located within an hour long video. As far as we are aware of, this is the only recent study on interactive search and behavior while using modern players employing mainly *play, pause* and the *seeker-bar* as main navigation aid. This is somewhat surprising since in recent years numerous video browsing tools have been proposed (a detailed review can be found in [9]).

Five navigation methods while interacting with the HTML5 video player were identified: (1) *Playback* (36%), (2) *Forward@Playback* (12%), (3) *Forward* (36%), (4) *Reverse@Playback* (7%) and (5) *Reverse* (10%). The best results in terms of completion time were obtained by users applying *linear forward search* with *seeker-bar dragging* in *non-playback state*.

3   Navigation Patterns: "Description Based Search" (DBS) vs. "Known Item Search" (KIS) Tasks

In [8], a preliminary report on the behavior of 17 users performing "Known Item Search" tasks within hour long videos, can be found. The tool used during the tests was a simple HTML5 video player (see Figure 1d). Following, we compare those findings with those of a newly performed user test also with 17 participants. The same HTML5 video player was used, but in a different setup: the users had to identify exactly the same sequences as in the KIS tasks but were presented only with the textual description of the video sequences that had to be located and not with the actual footage.

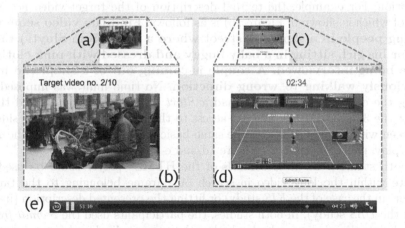

Fig. 1. The interface used in the *KIS* study. (a) and (b) the HTML5 video player during the first stage of a trial with the automatic playback of the target scene. (c) and (d) the HTML5 video player during the second stage of a trial: the effective search. (e) close up of the provided interaction possibilities of the video player.

3.1   User Studies

The data used in both studies is identical and consists of one-hour Dutch news videos. It comes from the public dataset of the Video Browser Showdown [7] in 2013 and it is currently available on its website. The same hardware was used in both studies: an 17-inch MacBook Pro laptop with the resolution set at 1920×1200 pixels to which a wired optical mouse was connected. The applications' interfaces were presented in a Safari web-browser window in full-screen mode.

The first study in [8] (which we will call the *KIS* study throughout the rest of the paper) had 17 participants (2 female, 15 males) with ages between 23 and 52 years, all of which were daily computer users. Each had to complete 10 search tasks consisting of finding a 20 seconds video sequence within an hour long news video. For each task, the short sequence was played back once within an automatic playback player on the left side of a full screen window as shown in Figure 1a.

All interaction elements were removed during the playback (see close-up in Figure 1b). Once the playback ended, the users had up to 3 minutes to find the presented scene within the corresponding long video which was presented on the left side of the screen within a player with basic controls (play, pause, seeker-bar - Figure 1d and Figure 1e for a close-up).

The second study (which we will call the *DBS* study throughout the rest of the paper) also had 17 participants (11 female, 6 males) with ages between 20 and 36 years and all daily computer users. They had to find the exact video sequences as the users in the *KIS* study but instead of the playback of the target scene, they were presented also on the left side of the screen with its textual description. For example, the textual description of the target video no. 2 (a frame of which is shown in Figure 1b) is as follows: **Find the video sequence showing people in a shopping street where a man (Tom Sluyts) takes care for his girl, sitting in a baby buggy and dressed with pink clothes, before being interviewed. A couple is coming out of a shop; the man is obviously walking to wrong direction**. No time limit was imposed on reading the description. The users had a *Start test* button that allowed them to start the 3 minutes timer and get access to the long video on the left side of the screen within the player with the same basic controls as the one in the *KIS* study (play, pause, seeker-bar - Figure 1d and e for a close-up).

In both studies, the basic controls of the HTML5 player could be used to navigate within the long video in search of a frame belonging to the target segment (in the case of the *KIS* study) or fitting the presented description (in the case of the *DBS* study). In both studies, the participants used the *Submit frame* button below the player to check whether the current displayed frame actually belonged to the target video or it fitted the description. False submissions were signaled by setting the background red for 4 seconds, while correct ones were signaled by a green background and the achieved score being displayed for 10 seconds. The next trial was available after a successful submission or after the allowed 3 minutes search time period expired.

3.2   Discussion

In the following we will discuss and compare the interaction types as well as the interaction speed we have observed during the two discussed tests. Also some interesting particular cases that came up during the two tests will be presented.

Search Start Interaction. To our surprise, the users in both the *KIS* and *DBS* studies approached the tasks almost in the same manner, as can be seen in Figure 2. The users in the *DBS* study (see Figure 2a) started in 71% of all 170 tasks with playback from the beginning of the video performing linear search forward by forward navigation. The users in the *KIS* study (see Figure 2b) employed the same approach in 64% of their 170 tasks. In 29% of the tasks in the *DBS* study respectively 33% of the tasks in the *KIS* study, the users preferred to start the search with a jump within the video. For *DBS*, a 30 second jump was recorded for 19% of the tasks, while for the *KIS* study, the same jump of 30 seconds appeared for 20% of the tasks. A 60 seconds jump was recorded for 10% of *DBS* tasks and 13% of *KIS* tasks. Random positioning, which usually marks some kind of educated or instinctive guess, was seldom employed. The users trusted their luck only for 1% of the *DBS* tasks and 3% of the *KIS* tasks.

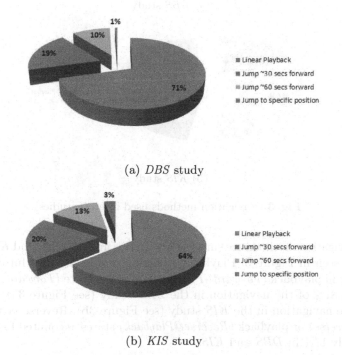

(a) *DBS* study

(b) *KIS* study

Fig. 2. Interaction methods used to start the search

Interaction Types. The application logs revealed for both our studies two classes of user behavior:

– *Click & Play*: the play button of the HTML5 video player is pressed and then multiple clicks are performed on the seeker-bar towards the end (*forward*) or towards the beginning (*reverse*) of the video. The pause button is sometimes used (usually when a certain frame is examined or when it is submitted for evaluation)
– *Dragging*: the player's seeker-bar is used to browse the content in both directions. The play and pause buttons are rarely used.

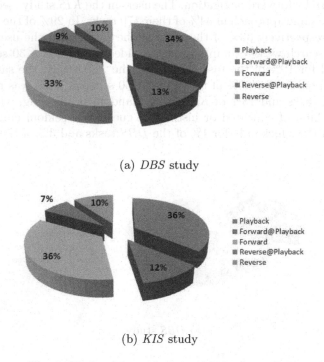

(a) *DBS* study

(b) *KIS* study

Fig. 3. Navigation methods used in both studies

The navigation methods are surprisingly similar in both *DBS* and *KIS* studies, as can be seen in Figure 3. Playback, dragging or clicking to a future point in time while in playback (*Forward@Playback*) or pause state (*Forward*) accounted for aprox. 81% of the navigation in the *DBS* study (see Figure 3a) and aprox. 83% of the navigation in the *KIS* study (see Figure 3b). Reverse positioning in pause (*Reverse*) or playback (*Reverse@Playback*) states accounted for only 19%, respectively 17% in *DBS* and *KIS*.

Navigation Strategies. Individual users showed in both studies varied strategies (see Figure 4). Overall, the users in the *DBS* study were significant slower than the ones in the *KIS* study. The approaches attempted in the *DBS* study appear to be more consistent (Figure 4a), while the ones in the *KIS* study appear a little bit more diverse (Figure 4b). Participant 17 in the *DBS* study and participant 1 in the *KIS* study did not use playback at all, while participants 14 in the *DBS* and participants 10 and 12 in *KIS*, used playback for most of the search time. Reverse positioning was used by some of the participants of the *KIS* study almost equally long as forward positioning (participants 2 and 16). In contrast, all the participants in the *DBS* study showed a strong preference for forward navigation. All the participants in the two studies, preferred positioning in paused mode over positioning in playback mode. If we compare Figure 4 with Figure 5, it becomes apparent that users that employed a lot of *Dragging* (e.g. participants 1, 4 and 9 in the *KIS* study) had significant less frames than those preferring *Click&Play* (e.g. participants 7, 10 and 12 also from the *KIS* study) and they were also a lot faster - in fact, they had the fastest overall submission in the *KIS* study. The most effective approach proved to be interactive search using the seeker-bar without playback.

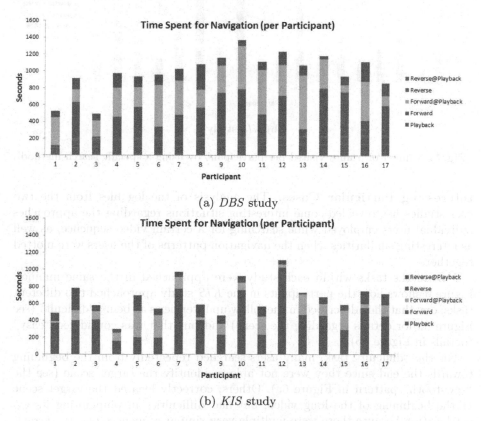

(a) *DBS* study

(b) *KIS* study

Fig. 4. Time spent for a specific search method (per user)

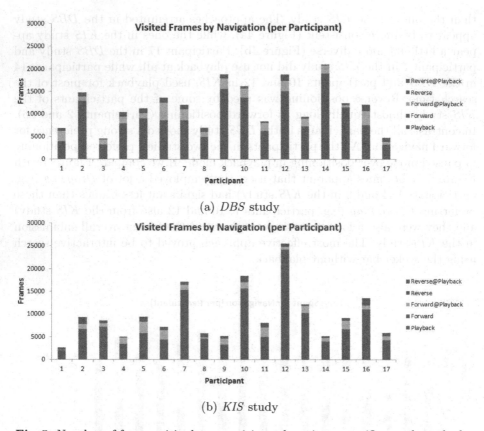

(a) *DBS* study

(b) *KIS* study

Fig. 5. Number of frames visited per participant by using a specific search method

Interesting Particular Cases. The analysis of the log files from the two user studies has revealed some interesting situations regarding the approaches individual users employed while searching for a certain video sequence, as well as interesting similarities when the navigation patterns of the users were plotted together.

Not all the tasks within each study were approached in the same manner. Figure 6 shows how the participants in the *KIS* study approached two different tasks: one that they described in the follow up interviews as being "difficult" (see Figure 6a for details regarding the scene) and one that was considered "easy" (details in Figure 6b).

For the "difficult" task, many users searched repeatedly from the beginning towards the end since they were not able to identify the target scene (see the "saw-tooth" pattern in Figure 6a). Others, correctly located the target scene at the beginning of the long video, but had difficulties in pinpointing its exact location because there were multiple very similar sequences (see the erratic movement between 0 and 10000 on the $0y$ axis also in Figure 6a).

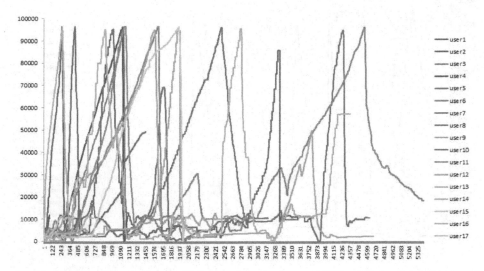

(a) **Navigation diagram of all participants for a "difficult" task in the *KIS* study**: in split screen a reporter talks to an anchor person in a studio. Almost identical scenes with the two people appear multiple times within the news program in and outside the target area. This made it hard to locate the exact scene.

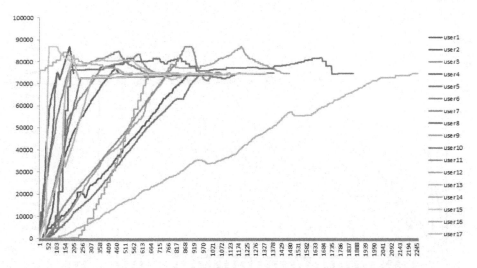

(b) **Navigation diagram diagram of all participants for an "easy" task in the *KIS* study**: a goal scene from a soccer match is shown. The player scoring the goal wears a white t-shirt and blue shorts. The coach (Ariel Jacobs) from Anderlecht gives an interview in front of an UEFA Europa League sponsor wall.

Fig. 6. *KIS* study: "Difficult" vs. "Easy" tasks

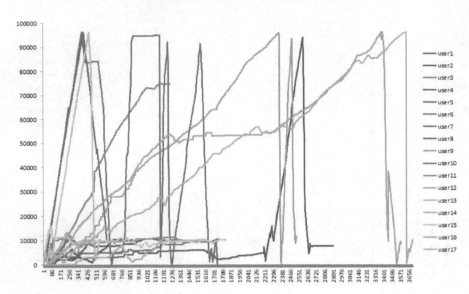

(a) **Navigation diagram of all participants for a "difficult" task in the _DBS_ study**: in split screen a reporter talks to an anchor person in a studio. Almost identical scenes with the two people appear multiple times within the news program in and outside the target area. This made it hard to locate the exact scene.

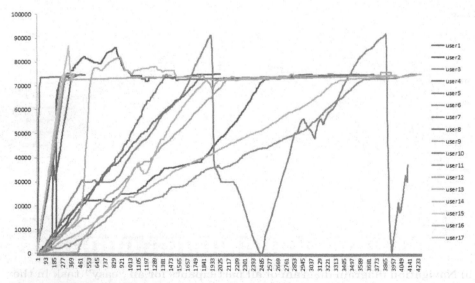

(b) **Navigation diagram diagram of all participants for an "easy" task in the _DBS_ study**: a goal scene from a soccer match is shown. The player scoring the goal wears a white t-shirt and blue shorts. The coach (Ariel Jacobs) from Anderlecht gives an interview in front of an UEFA Europa League sponsor wall.

Fig. 7. _DBS_ study: "Difficult" vs. "Easy" tasks

For the "easy" task, most of the users made an "educated guess" that the target scene has to be located towards the end of the video and acted accordingly. The majority moved very quickly to the end and then concentrated on finding the exact scene (the lines which are the closest to the Oy axis in Figure 6b). The others also navigated towards the end, but at a slower pase (the next group of lines more towards the middle in Figure 6b). A single user (**user 13**) chose to ignore the "obvious" hint provided by the target scene preview and applied linear search for the entire duration of the test. This is a user who changed his strategy in mid-test session and switched from an awkward approach in which he searched from the beginning to the end of the video and then reverted the direction from the end towards the beginning. After recognizing it is a failing strategy, he switched for the rest of the session to linear search, which he stubbornly applied even when there were indications that other approaches might be more appropriate.

In Figure 7 we present the approaches made by the participants in the *DBS* study for solving the exact same pair of "difficult" and "easy" tasks. The similarities between the approaches taken by the participants can easily be seen. We have basically the same behavior for both of the tasks. For the "difficult" one, some of the users struggle to locate the scene, hence the repeated dragging towards the end of the video and back to the beginning (the same "saw-tooth" patterns in Figure7a similar to the ones in Figure 6a. Others recognized the scene in the beginning, hence the movement between 0 and 20000 on the Oy axis also in Figure 7a).

For the "easy" task most of the users in the *DBS* study also made the "guess" that the target scene lies somewhere towards the end of the long video. Some moved quickly (the lines which are most close to the Oy axis in Figure 7b), while others applied linear search while dragging and finally concentrated on the last part of the video. The study also had a participant who had a hard time in locating the target scene in this test (the single "saw-tooth" line in Figure 7b corresponding to **user 6**). He had to start two times from the beginning after unsuccessful browsing two times through the video.

4   Conclusion

We have presented and compared the results from two user studies which focused on interactive search using basic HTML5 video players with limited navigation features. The adopted strategies vary quite significantly, especially in the case of the *KIS* study, but they also share some common characteristics. Most users (and especially the ones participating in the *DBS* study) favored linear forward search with seeker-bars positioning, as it helped alleviate the fact that the target scenes were introduced only by their textual description. Reverse search was seldom used and most of the times did not lead to success. The alternative to reverse search that almost all the participants adopted was to start fresh from the beginning. Linear search with and without playback was preferred since it helps to remember visited segments. This is especially important when the user does not clearly recollect the content in the target scene.

Our two studies also show that linear forward search with seeker-bar dragging in non-playback state is the most efficient in terms of search time, since the users can concentrate on solving the actual task. This means that video search tools do not necessarily have to provide playback feature and instead can confidently employ static images for interactive search.

Acknowledgments. This work was funded by the Federal Ministry for Transport, Innovation and Technology (bmvit) and the Austrian Science Fund (FWF): TRP 273-N15 and by Lakeside Labs GmbH, Klagenfurt, Austria, and funding from the European Regional Development Fund (ERDF) and the Carinthian Economic Promotion Fund (KWF).

References

1. Crockford, C., Agius, H.: An empirical investigation into user navigation of digital video using the vcr-like control set. International Journal of Human-Computer Studies 64(4), 340–355 (2006)
2. Huang, A.-H.: Effects of multimedia on document browsing and navigation: an exploratory empirical investigation. Information & Management 41(2), 189–198 (2003)
3. Hudelist, M., Schoeffmann, K., Böszörményi, L.: Mobile Video Browsing with a 3D Filmstrip. In: Proceedings of the 3rd ACM International Conference on Multimedia Retrieval, pp. 299–300. ACM, New York (2013)
4. Hudelist, M., Schoeffmann, K., Böszörményi, L.: Mobile Video Browsing with the ThumbBrowser. In: Proceedings of the 21st ACM Conference on Multimedia. ACM, Barcelona (accepted for publication 2013)
5. Lux, M., Huber, J.: Why did you record this video? An exploratory study on user intentions for video production. In: 13th International Workshop on Image Analysis for Multimedia Interactive Services (WIAMIS), pp. 1–4 (2012)
6. Over, P., Awad, G., Michel, M., Fiscus, J., Sanders, G., Shaw, B., Kraaij, W., Smeaton, A.-F., Quénot, G.: Trecvid 2012 – an overview of the goals, tasks, data, evaluation mechanisms and metrics. In: Proceedings of TRECVID 2012, NIST, USA (2012)
7. Schoeffmann, K., Bailer, W.: Video browser showdown. SIGMultimedia Rec. 4(2), 1–2 (2012)
8. Schoeffmann, K., Cobârzan, C.: An Evaluation of Interactive Search with Modern Video Players. In: IEEE International Conference on Multimedia and Expo (ICME), pp. 1–4. IEEE, San-Jose (2013)
9. Schoeffmann, K., Hopfgartner, F., Marques, O., Böszörményi, L., Jose, J.-M.: Video browsing interfaces and applications: a review. SPIE Reviews 1(1), 018004 (2010)
10. Scott, D., Hopfgartner, F., Guo, J., Gurrin, C.: Evaluating novice and expert users on handheld video retrieval systems. In: Li, S., El Saddik, A., Wang, M., Mei, T., Sebe, N., Yan, S., Hong, R., Gurrin, C. (eds.) MMM 2013, Part II. LNCS, vol. 7733, pp. 69–78. Springer, Heidelberg (2013)

Visual Recognition by Exploiting Latent Social Links in Image Collections

Li-Jia Li[1,*], Xiangnan Kong[2,*], and Philip S. Yu[2]

[1] Yahoo! Research, USA
[2] Department of Computer Science, University of Illinois at Chicago
lijiali@yahoo-inc.com, {xkong4,psyu}@cs.uic.edu

Abstract. Social network study has become an important topic in many research fields. Early works on social network analysis focus on real world social interactions in either human society or animal world. With the explosion of Internet data, social network researchers start to pay more attention to the tremendous amount of online social network data. There are ample space for exploring social network research on large-scale online visual content. In this paper, we focus on studying multi-label collective classification problem and develop a model that can harness the mutually beneficial information among the visual appearance, related semantic content and the social network structure simultaneously. Our algorithm is then tested on CelebrityNet, a social network constructed by inferring implicit relationship of people based on online multimedia content. We apply our model to a few important multimedia applications such as image annotation and community classification. We demonstrate that our algorithm significantly outperforms traditional methods on community classification and image annotation.

1 Introduction

Visual recognition research by using image content has achieved promising progress in recent years. There emerges robust object detectors [1,2], large scale image classification methods [3,4], efficient image retrieval algorithms [5,6] and advanced image annotation algorithms [7,8], most of which are developed by using image content alone. At the same time, social network analysis has been playing an important role in many fields such as recommendation [9] and search [10]. In recent years, it has attracted significant amount of interests in the multimedia community [11,12,13]. Not much work has been done on exploiting the rich structural information in social networks, which cannot be captured by the traditional visual models, for the multimedia research. Such mutually beneficial information is very common in the large online multimedia data. For example, in Fig. 1, we can observe the consistency between the photos (visual content) and the tags (semantic information). In addition, people who are socially connected share similar photos and tags. Furthermore, photos and tags belonging to the same social group (indicated by the circle) are very characteristic. An interesting question is what is the benefit of introducing social network to high level visual recognition tasks. In this paper, we advocate that by modeling the mutual benefit of the visual content,

---

* Indicates equal contributions.

C. Gurrin et al. (Eds.): MMM 2014, Part I, LNCS 8325, pp. 121–132, 2014.

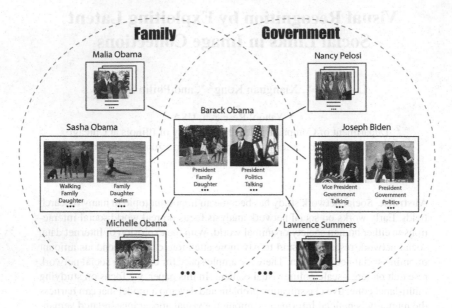

Fig. 1. Example to show the mutually beneficial information among photos, related tags and social structure. Each person has a set of photos and their related photo tags. Circles correspond to communities. Links between people indicate that they are connected.

semantic content and relationship structure in a social network, better visual recognition algorithm can be developed. Inspired by the ICA algorithm [14], we develop a collective classification algorithm to learn the mutually beneficial information modeled in this joint model. Fundamentally different than pioneering social network algorithms [11,12,13], our approach serves as generic annotation or classification algorithms which are not limited to face recognition or friendship prediction. We apply our algorithm to CelebrityNet, an implicit social network, as the source of visual, semantic and social structural information. It is constructed from the co-occurrence statistics of celebrities who appear in millions of professionally produced news images. It encodes visual, semantic and social structural information, making it a valuable resource for developing structural learning algorithms which jointly models such information. Experimental results demonstrate the effectiveness of our model built upon social network structure encoding multi-modal information resources for applications such as classification and image annotation by providing informative tags for unseen people or images. Specifically, we make the following contributions in this paper:

- Develop a principle model for modeling the mutually beneficial information among visual, semantic and implicit social structure from large scale online image dataset.
- Derive a collective classification algorithm for learning this model.
- Demonstrate significant result improvement by incorporating social structure in visual recognition.

It is worth noting that our model is a generic one. Although we demonstrate its effectiveness by using CelebrityNet in this paper, our algorithm is not limited to the celebrity social network. It can be directly applied to other information sources, such as the users of a general social network or the objects of an online photo sharing website.

2 Related Works

Social network analysis has achieved substantial progress recently with the emergence of large scale online structural data [9,10]. While much of the research has been done on textual documents or hyper links, social network analysis on visual content has also made promising progress these years [15,11,12,13]. With the emergence of photo sharing websites such as Flickr! and Facebook, pioneer research [11,12] has been conducted to tackle visual recognition problems by incorporating social network structure. For example, [12] models the types of relationships based on face features such as face size ratio, age difference and gender distribution. [11] leverages the social network structure to improve face recognition in a collection of face photos by using a MRF model. Both models focus on face recognition and analysis, which are not directly applicable to generic recognition tasks. Little has been done to construct an implicit social network from visual data, uncover the structure embedded and applying it for generic multimedia tasks e.g. image annotation and classification. In this paper, we propose a model to harness the mutual information embedded in the social network encoding implicit relationship inferred from co-occurrence of people in photos.

At the mean time, visual recognition algorithms [1,2,3,4,7,8] have shown effectiveness in recognizing objects and classifying images. Most of them are using only the visual content of images. Interesting research such as the multi-label classification approaches [16,17] further explore the correspondence among the tags related to the images. Sophisticated models are developed to model the relationship between the visual content and the semantic meaning of the images. Our algorithm, on the other hand, takes the visual content, related semantic information and social network structure into account. We aim to emphasize the impact of social network structure in visual tasks as a new knowledge resource for multimedia tasks.

3 Implicit Social Network

Before we describe our model, we first introduce the social network constructed from an online image dataset and explain the motivation of developing our model. Intuitively, photos encode implicit relationship: People who are related to each other usually appear together and are photographed in some occasions; The more photos they appear in together, the stronger the relationship might be. Inspired by this observation, we construct a social network called CelebrityNet from large scale online celebrity image dataset collected from a professional photograph website [1] named Getty [18]. In this social network, a network link is placed between two persons if they appear in images together. The strength of the link is simply the frequency of their occurrence in the dataset.

---

[1] Person names are manually labeled in this image dataset. Specifically, we use 2 million images to construct our implicit social network.

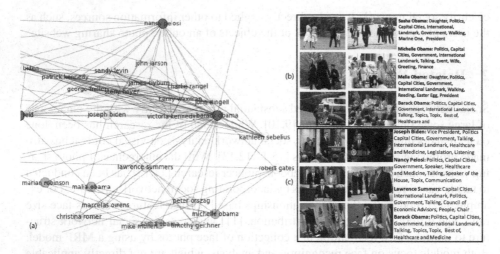

Fig. 2. Sample overlapped communities, related photos and popular tags. (a) Overlapped communities of 'Barack Obama'. Nodes represent people in the social network. Lines denote connection between people. Nodes highlighted in the same color refers to people belonging to the same community. Persons assigned to multiple communities are highlighted in red. (b) Example images and frequent tags of each person in the 'Obama Family' community. (c) Example images and popular tags of each person in a 'Obama Government' community.

Human beings, as social creatures, naturally form communities due to similar profession, location, and hobbies etc., which can be reflected by photos they take together. Let's take a deep dive in the constructed social network and uncover such phenomena. Our assumption is that the group of people belonging to a community often appears together much more often than people who are not part of the group. At the same time, one person could belong to multiple communities. For example, a person with computer vision research as profession could have the hobby of cooking and he/she then simultaneously belongs to both the computer vision researcher community and the cooking community. Therefore, we adopt the Clique Percolation Method (CPM) proposed by Palla et al. [19] to discover the overlapping communities.

In Fig. 2, we show example of overlapped communities related to 'Barack Obama' with the images and the most frequent tags for the persons in the community. From Fig. 2, our first observation is the consistency between visual appearance and the tags. Traditional image annotation algorithms [1,2,3] are developed based on this observation. By modeling the correspondence between the visual and semantic content, these algorithms are capable of automatically predicting related tags for unknown test images. In addition, people who are linked to each other in the social network usually share visual and semantic content. Furthermore, we observe that images and tags are very characteristic within each community. This applies not only to isolated communities but also overlapped communities. Social network encodes informative structure for learning the visual and textual data. Inspired by this observation, we propose a model jointly models the visual, textual and social information in Section 4.

4 Models

In this paper, we formulate the image annotation and community classification tasks as multi-label classification problems. For each unknown test image, our algorithm needs to provide a list of tags related to it. Similarly, each person in the social network will be assigned to one or multiple communities. We first introduce some notations that will be used throughout. Suppose we have a multi-label dataset $\mathcal{D}\,(\mathcal{X},\mathcal{Y}) = \{(\boldsymbol{x}_i, \boldsymbol{Y}_i)\}_{i=1}^n$ and a network $G(\mathcal{V}, \mathcal{E})$ among the samples of the dataset. Here $\mathcal{X} = \{\boldsymbol{x}_i\}_{i=1}^n$, and $\boldsymbol{x}_i \in \mathbb{R}^d$ denotes the feature vector of sample $\boldsymbol{x}_i$ in the d-dimensional input space. $\mathcal{Y} = \{\boldsymbol{Y}_i\}_{i=1}^n$, where $\boldsymbol{Y}_i = \left(Y_i^1, \cdots, Y_i^q\right)^\top \in \{0,1\}^q$ denotes the multiple labels assigned to sample $\boldsymbol{x}_i$. Let $\mathcal{C} = \{\ell_1, \cdots, \ell_q\}$ denote the set of q possible label concepts. In the network G, $\mathcal{V} = \{v_1, \cdots, v_n\}$ is a set of nodes, which corresponds to the samples in $\mathcal{D}$. $\mathcal{E}$ is the set of links/edges in $\mathcal{V} \times \mathcal{V}$. Assume that we have a training set $\mathcal{X}_\mathcal{L} \subset \mathcal{X}$ where the values $\mathcal{Y}_\mathcal{L}$ are known. Here $\mathcal{L}$ denotes the index set for training data, i.e., $\mathcal{Y}_\mathcal{L} = \{\boldsymbol{Y}_i = \boldsymbol{y}_i | \boldsymbol{x}_i \in \mathcal{X}_\mathcal{L}\}$. $\boldsymbol{y}_i = \left(y_i^1, \cdots, y_i^q\right)^\top \in \{0,1\}^q$ is a binary vector representing the observed label set assigned to sample $\boldsymbol{x}_i$. $y_i^k = 1$ if the k-th label is in $\boldsymbol{x}_i$'s label set.

Multi-label collective classification corresponds to the task of predicting the values of all $\boldsymbol{Y}_i \in \mathcal{Y}_\mathcal{U}$ for the testing set collectively ($\mathcal{X}_\mathcal{U} = \mathcal{X} - \mathcal{X}_\mathcal{L}$), where the inference problem is to estimate $\Pr(\mathcal{Y}_\mathcal{U} | \mathcal{X}, \mathcal{Y}_\mathcal{L})$. Conventional supervised classification approaches usually has i.i.d. assumptions, i.e., the inference for each sample is independent from other samples, i.e., $\Pr(\mathcal{Y}_\mathcal{U} | \mathcal{X}, \mathcal{Y}_\mathcal{L}) \propto \prod_{i \in \mathcal{U}} \Pr(\boldsymbol{Y}_i | \boldsymbol{x}_i)$. Moreover, in multi-label classification, the simplest solution (i.e., one-vs-all) assumes that the inference of each label is also independent from other labels for an sample, i.e., $\Pr(\boldsymbol{Y}_i | \boldsymbol{x}_i) = \prod_{k=1}^q \Pr(Y_i^k | \boldsymbol{x}_i)$. However, in many real-world classification tasks, there are complex dependencies not only among different samples but also among different labels.

In order to solve the multi-label collective classification problem more effectively, we explicitly consider three types of relationships. We adopt the multi-kernel learning framework (MKL) [20] and build one kernel on each type of relationship. SVMs have been widely used for classification problems in recent years. Different than traditional SVM, MKL incorporates multiple kernels and can learn a convex combination of these kernels (i.e., the kernel weights) simultaneously $\mathbf{K} = \sum_i \beta_i \mathbf{K}_i$. Specially, we build three different kernels that can capture three different types of relationship in the data.

4.1 Content Relationship

The first type of relationships we consider is about the visual content features of the samples. Conventional image annotation approaches focus on using the image content features to build inference models. In order to capture the content/visual information of different samples, we build the *content kernel* based upon the input visual feature vector of different samples. $K_{content}(i, j) = \phi(\boldsymbol{x}_i, \boldsymbol{x}_j)$. Here, any conventional kernel function can be used for $\phi(\cdot, \cdot)$. Intuitively, the content kernel denotes the relationship that if two images share similar visual features, they are more likely to have similar labels.

Input:
  $\mathcal{G}$: a network, $\mathcal{X}$: attribute vectors for all instances.
  $\mathcal{Y}_{\mathcal{L}}$: label sets for the training instances, A: a base learner for multi-kernel learning model, T_{max}: maximum # of iteration (default=10)
Training:
  - Learn the MKL model f:
    1. Construct q extended training sets $\forall 1 \leq k \leq q, \mathcal{D}_k = \left\{ (\mathbf{x}_i^k, y_i^k) \right\}$ by converting each instance $\boldsymbol{x}_i$ to $\mathbf{x}_i^k$ as follows:
      $$\boldsymbol{x}_i^k = (\boldsymbol{x}_i, \text{LabelSetFeature}(\ell_k, \boldsymbol{Y}_i), \text{NetworkFeature}(i, \mathcal{Y}_{\mathcal{L}}))$$
    2. Computer the corresponding kernels for each label: $\Phi, \Phi_{Labelset}$, and $\Phi_{network}$
    3. Calculate kernel weights and train MKL models on each label. Let $f_k = A(\mathcal{D}_k)$ be the MKL model trained on $\mathcal{D}_k$.
Bootstrap:
  - Estimate the label sets, for $i \in \mathcal{U}$: produce an estimated values $\hat{\boldsymbol{Y}}_i$ for $\boldsymbol{Y}_i$ as follows: $\hat{\boldsymbol{Y}}_i = f((\boldsymbol{x}_i, 0))$ using attributes only.
Iterative Inference:
  - Repeat until convergence or #iteration> T_{max}
    1. Construct the extended testing instance by converting each instance $\boldsymbol{x}_i$ to $\boldsymbol{x}_i^k$'s $(i \in \mathcal{U})$ as follows:
      $$\boldsymbol{x}_i^k = \left(\boldsymbol{x}_i, \text{LabelsetFeature}(\ell_k, \hat{\boldsymbol{Y}}_i), \text{NetworkFeature}(i, \mathcal{Y}_{\mathcal{L}} \cup \{\hat{\boldsymbol{Y}}_i | i \in \mathcal{U}\}) \right)$$
    2. Update the estimated value $\hat{\boldsymbol{Y}}_i$ for $\boldsymbol{Y}_i$ on each testing instance $(i \in \mathcal{U})$ as follows: $\forall 1 \leq k \leq q, \hat{Y}_i^k = f_k(\boldsymbol{x}_i^k)$.
Output:
  $\hat{\mathcal{Y}}_{\mathcal{U}} = \left(\hat{\boldsymbol{Y}}_1, \cdots, \hat{\boldsymbol{Y}}_{n_u} \right)$: the label sets of testing instances $(i \in \mathcal{U})$.

Fig. 3. The MKML algorithm

4.2 Label Set Relationship

The second type of relationships we consider is the label correlations within the label set of each sample. Different labels are inter-dependent in multi-label classification, thus should be predicted collectively. For example, in image annotation tasks, an image is more likely to have the tag 'sports' if we know the image has already been assigned with the tag 'NBA' or 'basketball'. The image is less likely to be annotated as 'sports', if we already know the image contains the label 'academy awards'.

Conventional multi-label classification approaches focus on exploiting such label correlations to improve the classification performances, which model $\Pr(Y_i^k | \boldsymbol{x}_i, \boldsymbol{Y}_i^{\{-k\}})$. $\boldsymbol{Y}_i^{\{-k\}}$ represents the vector of all the variables in the set $\{Y_i^p : p \neq k\}$. Hence, we have $\Pr(\boldsymbol{Y}_i | \boldsymbol{x}_i) = \prod_{k=1}^q \Pr(Y_i^k | \boldsymbol{x}_i, \boldsymbol{Y}_i^{\{-k\}})$. Based upon the above observation, we build the *label set kernel* encoding the correlations among different labels. $K_{labelset}(Y_i^k, Y_j^k) = \phi\left(\boldsymbol{Y}_i^{\{-k\}}, \boldsymbol{Y}_j^{\{-k\}} \right)$. Intuitively, the label set kernel denotes the relationship that if two images share similar label sets, they are more likely to have similar values in any label variable.

4.3 Network Relationship

The third type of relationships we consider is the correlations among label sets of the related samples that are inter-connected in the network. The label sets of related samples are usually inter-dependent in a network. For example, in our CelebrityNet network, the probability of an image having the label 'politics' should be higher if we already know the image contains the same people appearing in some other images with a label set of {'government', 'politics'}.

Conventional collective classification approaches focus on exploiting this type of dependencies to improve the classification performances, which models $\Pr(Y_i^k | \boldsymbol{x}_i, \boldsymbol{Y}_{j \in \mathcal{N}(i)})$. Here $\boldsymbol{Y}_{j \in \mathcal{N}(i)}$ denotes the set containing all vectors $\boldsymbol{Y}_j$ $(\forall j \in \mathcal{N}(i))$, and $\mathcal{N}(i)$ denotes the index set of related samples to the i-th sample, *i.e.*, the samples directly linked to the i-th sample. Hence, we will have $\Pr(Y_{\mathcal{U}}^k | X) = \prod_{i \in \mathcal{U}} \Pr(Y_i^k | \boldsymbol{x}_i, Y_{j \in \mathcal{N}(i)})$. Based upon the above observation, we build the *network kernel* encoding the correlations among related samples that are connected in the network. $K_{network}(Y_i^k, Y_j^k) = \phi\left(Y_{l \in \mathcal{N}(i)}, Y_{l \in \mathcal{N}(j)}\right)$. Intuitively, the network kernel denotes the relationship that if the neighbors of the two images share similar label sets, these two images are more likely to have similar label sets.

The general idea is as follows: We build one kernel on each type of the relations mentioned above, and then use MKL method to learn the weights of the multiple kernels (i.e., the importance of different kernels). We model the joint probability based upon the Markov property: if sample $\boldsymbol{x}_i$ and $\boldsymbol{x}_j$ are not directly connected in network G, the label set Y_i is conditional independent from Y_j given the label sets of all $\boldsymbol{x}_i$'s neighbors. The local conditional probability on label k can be modeled by a MKL learner with aforementioned kernels. The computation of these kernels depends on the predicted Y_j ($j \in \mathcal{N}(i)$) and the predicted $Y_i^{\{-k\}}$. Then, the joint probability can be approximated based on these local conditional probabilities by treating different labels as independent and the samples as *i.i.d.*. To simply demonstrate the effectiveness of our approach, we use linear kernels for all relations here.

Motivated by the ICA framework [14,21], we proposed the following inference procedure of our MKML method as shown in Figure 3. (1) At the beginning of the inference, the label sets of all the unlabeled samples are unknown. The *bootstrap* step is used to assign an initial label set for each sample using the content feature of each sample. In our current implementation, we simply initialize the label set features and the network features for unlabeled samples with all zero vectors. Other strategies can also be used for bootstrapping, e.g, training SVM (single kernel) on training data using content feature only, and then we use these models to assign the initial label sets of unlabeled samples. (2) In the *iterative inference* step, we iteratively update the label set features/kernels and network features/kernels based upon the predictions of MKL models and update the prediction of MKL models using the newly updated kernels. The iterative process stops when the predictions of all MKL models are stabilized or a maximum number of iteration has been reached.

5 Experiment

5.1 Compared Methods

In this subsection, we compared a set of methods exploring different information resources:

- BSVM (binary SVM): This baseline method uses binary decomposition to train one classifier on each label separately, which is similar to [22]. BSVM assumes all the labels and all instances are independent. It is based on visual content alone.
- MKL (Multi-kernel learning): We directly apply multiple kernel learning algorithm on the joint information of the visual, semantic and social network without iterative inference steps.

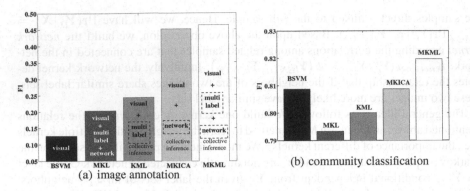

(a) image annotation   (b) community classification

Fig. 4. Overall performances of the compared methods

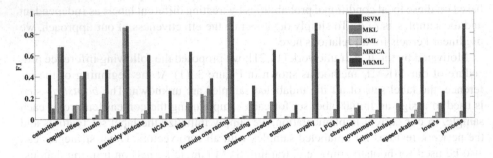

Fig. 5. F1 scores on example labels in image annotation task

•KML (visual kernel + multi-label kernel): This baseline method trains one multi-kernel learner on each label, using two different kernels visual feature kernel and multi-label kernel. KML not only models the correspondence between the visual content and the tags, but also models the correlation among the tags.

• MKICA (visual kernel + network kernel): In this baseline method, the multi-label dataset is first divided into multiple single-label datasets by one-vs-all binary decomposition. For each binary classification task, we use a multi-kernel version of ICA [14], as the base classification method. MKICA combines the social structure with visual modeling of the tags. However, it ignores the relationship among the tags.

•MKML (Multi-kernel Multi-label Collective classification): Our proposed method for multi-label collective classification based upon multi-kernel learning, which jointly models the visual, semantic and social network information.

For a fair comparison, we use LibLinear [23] as the base classifier for BSVM and LibLinear MKL as the base learner for all the remaining methods. The maximum number of iterations in the methods KML, MKICA, and MKML are all set as 10 based on observation from the validation experiment.

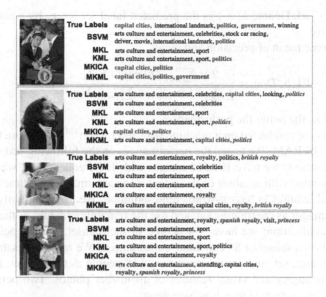

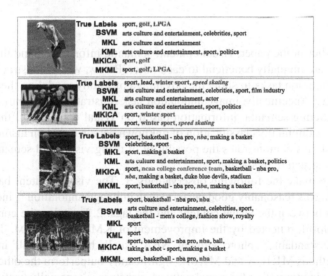

Fig. 6. Annotation Examples of different algorithms. BSVM, MKL, KML, MKICA, MKML represent binary SVM, traditional multi-kernel learning method, method built upon visual kernel + network kernel, Multi-kernel Multi-label Collective classification method respectively. Tags exclusively recognized by our methods are highlighted in color.

5.2 Evaluation Metrics

We use some evaluation criteria in [24,25,26] to verify the image annotation performance. Suppose a multi-label dataset $\mathcal{D}_U$ contains n instances $(\boldsymbol{x}_i, \boldsymbol{Y}_i)$, where $\boldsymbol{Y}_i \in$

$\{0, 1\}^q$ $(i = 1, \cdots, n)$. Denote $h(\boldsymbol{x}_i)$ as the predicted label set for $\boldsymbol{x}_i$ by a multi-label classifier h, we have

- *F1*: is the harmonic mean of precision and recall.

$$\text{F1}(h, \mathcal{D}_U) = \frac{2 \times \sum_{i=1}^{n} \|h(\boldsymbol{x}_i) \cap \boldsymbol{Y}_i\|_1}{\sum_{i=1}^{n} \|h(\boldsymbol{x}_i)\|_1 + \sum_{i=1}^{n} \|\boldsymbol{Y}_i\|_1}$$

The larger the value, the better the performance.

All experiments are conducted on a machine with Intel Xeon™ Quad-Core CPUs of 2.26 GHz and 24 GB RAM. We tested the performances on the following tasks:
1) image annotation task: we have 102,565 images with 159 frequent tags, where each image can be annotated with a subset of these tags. On each image we extracted 5000 dimensional visual features in bag-of-words representation. We then randomly sample two thirds of the images as the training set, and use the remaining images as the test set.
2) community classification: we have 554 people in the dataset, where each person can be classified into a subset of 80 candidate communities. We randomly sample 436 people into the training set, and use the remaining 118 people as the test set. For each person, we use the aggregated visual features of all his/her photos. Two persons are linked together if they appeared in at least one photo.

5.3   Results

As mentioned earlier in the paper, visual content, semantic information and the social network structure are mutually beneficial to each other. Below, we demonstrate results of two visual recognition tasks to show the advantage of jointly modeling these three information sources. Specifically, image annotation task illustrates the potential of our approach for predicting semantic information based on visual content and the social network structure. Community classification of unknown person based on his/her set of photos and related tags demonstrates the possibility of using visual and semantic tags for social network structure prediction.

In Fig. 4(a), we make the following observations: 1. The visual content based approach BSVM achieves reasonably good performance in image annotation [2], indicating strong correlation between the visual content and the tags. 2. Learning the correlation among tags is helpful, reflected by the improvement of KML over BSVM. This improvement is understandable: a photo with tag 'NBA' usually has 'basketball' in the tag list as well. 3. Methods MKICA and MKML significantly outperform the other methods indicating that incorporating the social network structure is especially useful. From the analysis of social network in Section 3, we learn that images and tags belonging to the same person and same community are very characteristic. Therefore, modeling the social network structure naturally improve the tag prediction performance of unknown images. 4. The significant improvement of MKML over the traditional MKL shows the power of iterative prediction and error correction in our proposed method. 5. Finally, jointly modeling the visual, semantic and social structure (MKML) provides additional improvement over combining visual and semantic information (MKL), demonstrating the effectiveness of social network structure.

---

[2] Random approach achieves only 0.03 by using the F1 measure.

In Fig. 4(b), we show the community classification results of different algorithms. In this experiment, tag correlation (KML) is not as useful as it is in the image annotation task. This is interpretable: as long as we know the person is related to the tag 'NBA', we can already do a good community classification without knowing other tags. On the other hand, if we know whom the unknown person is connected to, it is fairly easy to predict his/her community. This leads to the good performance of social network based algorithms.

To provide more details of the annotation result, we show the F1 scores of example labels in Fig. 5. While we observe similar pattern as in Fig. 4(a) with clear advantage of the social network based algorithms over the other methods, the social network based algorithms usually perform much better on specific labels with social meaning such as 'kentucky wildcats' and 'royalty'. Such social meaning can not be inferred from the visual content. The observation aligns well with our motivation of incorporating social network structure as a source of complimentary information for high level visual recognition tasks.

Finally, we show example results of image annotation in Fig. 6. Visual only method provides conservative prediction of common tags correlated to the visual content. Incorporating social network upon the visual and semantic modeling enables the algorithms to be more accurate in image annotation. MKML further enrich the tag annotation accurately by exploring the tag correlation upon jointly modeling the three sources of information. For example, in the 4th picture, the tags 'princess' and 'Spanish royalty' can only be inferenced correctly by combining information from Social netowrks and correlations with other tags (such as 'royalty').

6 Conclusion

In this paper, we propose a model to jointly model the visual content, semantic information and relationship structure for a few multimedia tasks. Our algorithm has been tested on a social network constructed from large scale images. We demonstrate significant improvement on community classification and image annotation of celebrity images over related algorithms. Our algorithm is a generic algorithm for modeling mutual information of content and relational data. In the future, we would like to explore its potential on generic online user photos such as those available on Flickr! for automatically prediction of missing tags. Another interesting direction is to develop advanced algorithms upon the proposed one for high level visual recognition tasks such as large scale object detection by combining detailed visual content information and objects' relational structure.

References

1. Dalal, N., Triggs, B.: Histograms of oriented gradients for human detection. In: CVPR, p. 886 (2005)
2. Felzenszwalb, P., Girshick, R., McAllester, D., Ramanan, D.: Object Detection with Discriminatively Trained Part Based Models. JAIR 29 (2007)
3. Lin, Y., Lv, F., Zhu, S., Yang, M., Cour, T., Yu, K., Cao, L., Huang, T.: Large-scale image classification: fast feature extraction and svm training. In: CVPR, pp. 1689–1696 (2011)

4. Deng, J., Berg, A.C., Li, K., Fei-Fei, L.: What does classifying more than 10,000 image categories tell us? In: Daniilidis, K., Maragos, P., Paragios, N. (eds.) ECCV 2010, Part V. LNCS, vol. 6315, pp. 71–84. Springer, Heidelberg (2010)
5. Wu, Z., Ke, Q., Isard, M., Sun, J.: Bundling features for large scale partial-duplicate web image search. In: CVPR, pp. 25–32 (2009)
6. Jégou, H., Douze, M., Schmid, C.: Improving bag-of-features for large scale image search. IJCV 87, 316–336 (2010)
7. Cao, L., Yu, J., Luo, J., Huang, T.: Enhancing semantic and geographic annotation of web images via logistic canonical correlation regression. In: ACM MM (2009)
8. Weston, J., Bengio, S., Usunier, N.: Large scale image annotation: Learning to rank with joint word-image embeddings. Machine learning 81, 21–35 (2010)
9. Konstas, I., Stathopoulos, V., Jose, J.: On social networks and collaborative recommendation. In: SIGIR, pp. 195–202. ACM (2009)
10. Brin, S., Page, L.: The anatomy of a large-scale hypertextual web search engine. Computer Networks and ISDN Systems 30, 107–117 (1998)
11. Stone, Z., Zickler, T., Darrell, T.: Toward large-scale face recognition using social network context. Proceedings of the IEEE 98, 1408–1415 (2010)
12. Wang, G., Gallagher, A., Luo, J., Forsyth, D.: Seeing people in social context: Recognizing people and social relationships. In: Daniilidis, K., Maragos, P., Paragios, N. (eds.) ECCV 2010, Part V. LNCS, vol. 6315, pp. 169–182. Springer, Heidelberg (2010)
13. Zhuang, J., Mei, T., Hoi, S., Hua, X., Li, S.: Modeling social strength in social media community via kernel-based learning. In: ACM MM (2011)
14. Lu, Q., Getoor, L.: Link-based classification. In: ICML (2003)
15. Ding, L., Yilmaz, A.: Learning relations among movie characters: A social network perspective. In: Daniilidis, K., Maragos, P., Paragios, N. (eds.) ECCV 2010, Part IV. LNCS, vol. 6314, pp. 410–423. Springer, Heidelberg (2010)
16. Read, J., Pfahringer, B., Holmes, G., Frank, E.: Classifier chains for multi-label classification. In: Buntine, W., Grobelnik, M., Mladenić, D., Shawe-Taylor, J. (eds.) ECML PKDD 2009, Part II. LNCS, vol. 5782, pp. 254–269. Springer, Heidelberg (2009)
17. Liu, X., Shi, Z., Li, Z., Wang, X., Shi, Z.: Sorted label classifier chains for learning images with multi-label. In: ACM MM (2010)
18. http://www.gettyimages.com/
19. Palla, G., Derényi, I., Farkas, I., Vicsek, T.: Uncovering the overlapping community structure of complex networks in nature and society. Nature 435, 814–818 (2005)
20. Vishwanathan, S.V.N., Sun, Z., Theera-Ampornpunt, N.: Multiple kernel learning and the smo algorithm. In: NIPS (2010)
21. McDowell, L.K., Gupta, K.M., Aha, D.W.: Cautious inference in collective classification. In: AAAI, Vancouver, Canada, pp. 596–601 (2007)
22. Boutell, M.R., Luo, J., Shen, X., Brown, C.M.: Learning multi-label scene classification. Pattern Recognition 37, 1757–1771 (2004)
23. Fan, R.E., Chang, K.W., Hsieh, C.J., Wang, X.R., Lin, C.J.: LIBLINEAR: a library for large linear classification (2008)
24. Ghamrawi, N., McCallum, A.: Collective multi-label classification. In: CIKM, Bremen, Germany, pp. 195–200 (2005)
25. Kang, F., Jin, R., Sukthankar, R.: Correlated label propagation with application to multi-label learning. In: CVPR, New York, NY, pp. 1719–1726 (2006)
26. Liu, Y., Jin, R., Yang, L.: Semi-supervised multi-label learning by constrained non-negative matrix factorization. In: AAAI, Boston, MA, pp. 421–426 (2006)

Collections for Automatic Image Annotation and Photo Tag Recommendation*

Philip J. McParlane, Yashar Moshfeghi, and Joemon M. Jose

University of Glasgow,
School of Computing, Glasgow, G12 8QQ, UK
p.mcparlane.1@research.gla.ac.uk,
{Yashar.Moshfeghi,Joemon.Jose}@glasgow.ac.uk

Abstract. This paper highlights a number of problems which exist in the evaluation of existing image annotation and tag recommendation methods. Crucially, the collections used by these state-of-the-art methods contain a number of biases which may be *exploited* or *detrimental* to their evaluation, resulting in misleading results. In total we highlight *seven* issues for *three* popular annotation evaluation collections, i.e. Corel5k, ESP Game and IAPR, as well as *three* issues with collections used in *two* state-of-the-art photo tag recommendation methods. The result of this paper is two freely available Flickr image collections designed for the fair evaluation of image annotation and tag recommendation methods called Flickr-AIA and Flickr-PTR respectively. We show through experimentation and demonstration that these collection are ultimately fairer benchmarks than existing collections.

Keywords: evaluation, collection, annotation, tag recommendation.

1 Introduction

Given the increase in popularity of photo sharing websites, there has been a recent research focus on the indexing and retrieval of such content. A recent study, showed that 65% of images uploaded to popular image sharing website, Flickr[1], contain less than four tags [18], this in turn makes retrieval difficult. Therefore, one of the major challenges in the field involves predicting the objects and concepts present within an image in order to allow for such retrieval. Last decade, a number of research works focussed on the automatic image annotation (AIA) of images and the semi-automatic process of photo tag recommendation (PTR) in order to extract meaning from an image.

Despite the amount of work taken out in these fields, a comparison of approaches is difficult due to the lack of a unified evaluation framework and collection. A review of the 20 most popular automatic image annotation papers[2]

---

* This research was supported by the the the European Community's FP7 Programme under grant agreements nr 288024 (LiMoSINe).

[1] http://www.flickr.com/
[2] Selected by searching http://citeseerx.ist.psu.edu/ for "automatic image annotation". Order by descending citation count (Dec'12).

C. Gurrin et al. (Eds.): MMM 2014, Part I, LNCS 8325, pp. 133–145, 2014.
© Springer International Publishing Switzerland 2014

showed that at least 15 different collections were tested upon[3]. These collections vary in characteristics and hence introduce biases of their own into the evaluation, highlighting the need for a single test collection which is *representative* of images uploaded to image sharing websites. Additionally, the most prominent works in photo tag recommendation all use their own collections [18,11,5].

Aside from the large number of collections used to benchmark annotation models, we have identified *seven* flaws which may result in misleading performance measures and therefore the incomparability of state-of-the-art (SOTA) models. The problems are as follows: (i) *class ambiguity*, in the form of synonyms e.g. testing for ocean vs sea (ii) *testing on unnormalised collections*, where SOTA models are able to boost annotation performance by promoting popular tags (iii) *low image quality* (iv) *lack of image meta-data* (v) *lack of image diversity* (vi) *using location tags as ground truth* (vii) *copyright restrictions*.

For photo tag recommendation, we have identified the *three* problems with the collections used in [18] and [5]: (i) *using crowdsourced ground truths*; only the photographer of an image understands the true content and context of an image (ii) *synonyms in the ground truth*; models which promote synonyms in their suggestions are promoted over those models which suggest diverse recommendations, (iii) *lack of distribution;* currently tag recommendation works test on their own private collection.

There are two major contributions in this paper: firstly, the identification and elaboration of the problems in current evaluation test sets, and secondly, the introduction of two new freely available image evaluation collections, namely Flickr-AIA and Flickr-PTR[4], which aim to overcome these discussed issues for the fair evaluation of image annotation and tag recommendation models. The rest of this paper is as follows. Section 2 collates the related work in AIA, PTR and its evaluation. In Section 3 we detail further the problems associated in automatic image annotation evaluation before introducing Flickr-AIA. In Section 4 we detail further the problems associated in photo tag recommendation evaluation before introducing Flickr-PTR. Finally, we conclude in Section 5.

2   Background Work

Automatic Image Annotation: The area of AIA has been a well researched area in the last decade [3,10,12]. Firstly, Duygulu *et al.* [3] used a machine translation approach, between image contents and annotations, which was tested on the Corel5k image collection. Joen *et al.* [10] adopted the Cross-Media Relevance Models (CMRM) to predict the probability of generating a word given blobs in an image in the training set. More recently, Makadia *et al.* [12] showed that five existing models could be outperformed by adopting a K-nearest neighbour approach (KNN) trained on colour and texture image features. Despite the

---

[3] The collections were: Corel5k, Corel30k, ESP Game, IAPR, Google Images, LabelMe, Washington Collection, Caltech, TrecVid 2007, Pascal 2007, MiAlbum & 4 other small collections.

[4] Both collections are available for download at http://dcs.gla.ac.uk/~philip/

progress made in the field, all of these models are evaluated on *small, unrealistic* and *unnormalised* image collections [2,20,8,1]. In this work we introduce the Flickr-AIA collection which aims to overcome these, and other, issues.

AIA Evaluation: A number of issues associated with the evaluation of AIA models have been identified in a number of previous works: Westerveld *et al.* [20] highlighted a number of problems with the Corel collection, such as the fact that images are grouped into *coherent themes*, resulting in misleadingly high performance measures. Athanasakos *et al.* [1] compared two existing models showing that the high performance reported was more to do with the evaluation scheme and test set instead of the approach itself. Müller *et al.* [16] highlighted issues with using the Corel image collection, in that many models test on a different subset of this collection resulting in different performance measures. In this paper we discuss new biases, resolving them in new evaluation collections.

Recently, AIA models have been tested on much larger image collections. Deng *et al.* [2] introduced the ImageNet collection, consisting of 3.2M images (which is constantly being extended), structured into synonym sets of the lexical databse, WordNet [15]. Huiskes *et al.* [13] introduced two Flickr collections of 25K [8] and 1M images [13]. However, these collections are not setup with defined train/test subsets for annotation or tag recommendation evaluation. Further, these collections fail to address a number of the issues presented in this paper such as tag ambiguity and normalisation. Despite the increase in the availability of computation power, in the forms of clusters and multi-core machines, the computationally intensive task of image annotation on this volume of images is out of the reach of many, and therefore a more manageable collection is desirable for most. Additionally, the large size of these collections increases the amount of *noise* and *synonyms* present, ultimately increasing the potential bias in evaluation, as well as the difficulty in its *distribution*. In the Flickr-AIA set, we reduce a large collection of 2M images, to a much smaller collection of high quality, well tagged images, free of *synonyms*. Thus, maintaining the diversity of these large, online collections, whilst allowing for their easy distribution.

Photo Tag Recommendation: Tag recommendation systems have been proposed in literature, which recommend tags based on those tags already present within an image. Sigurbjornsson *et al.* proposed a tag recommendation strategy to support users annotating photos on Flickr [18]. The relationships between tags were exploited to suggest highly co-occurring tags. Garg *et al.* offered personalised tag recommendations [5] in their approach which looked to combine suggestions made from personalised and global tag co-occurrence matrices. In this work we identify flaws in the evaluation procedure of these works, leading to the introduction of Flickr-PTR, a freely available image collection designed for the fair evaluation of tag recommendation models.

3   Automatic Image Annotation Evaluation

The purpose of an image annotation evaluation collection is to benchmark a given annotation method, for a number of image classes or scenes, based purely

Table 1. Comparison of the Collections (i) Ambiguity: % of tags where there exist at least one synonym (ii) Size: average dimension in pixels (iii) Time/Loc: whether time taken and location details are included (iv) I/T: average # images per tag

Collection	Images	Tags	Ambiguity	Time/Loc	Free	Size	Train	Test	I/T
Corel	5k	374	9.6%	✗	✗	160px	4.5k	0.5k	88
ESP	22k	269	9.7%	✗	✓	156px	20k	2k	377
IAPR	20k	291	12.7%	✓	✓	417px	18k	2k	386
Flickr-AIA	312k	420	0%	✓	✓	719px	292k	20k	2,304

on its visual discriminatory power. Therefore, these classes should be distinct (and not ambiguous) and easily identifiable by a human being based purely on their appearance. The images in this collection should reflect real, user images and should cover a diverse range of images for each class; alternatively, the images should be taken in different locations, by different users, in a number of different lighting conditions, on a range of devices. By doing so, annotation models would be benchmarked for as close to a real world scenario as possible. In the following sections, we first introduce three popular annotation collections and the problems they pose for fair evaluation. We introduce an experimental setup which upholds our hypotheses before detailing our new collection which aims to tackle the issues presented.

Existing Collections: We consider the following collections: *Corel* [3], *ESP Game* [19] and *IAPR* [6]. These collections are selected as they have been used to benchmark many AIA models of recent years [12]. We use the same *methods, training* and *test* subsets as used in [12]. These collections, along with the collection introduced in this paper (Flickr-AIA), summarized in Table 1.

Although this list does not cover all evaluation collections, they are amongst some of the most popular collections [12,10,1]. One popular collection which has been omitted and is related to this work is the MIR-Flickr 25k [8] and 1M [13] collections. We have not considered these collections as they are not setup with annotation evaluation in mind; they contain user tags rather than high level, visual, classes. However, these collections have been used in the ImageCLEF 2009 annotation task, where the referred 25k collection was annotated using a crowdsourced experiment. Despite this, the collection was only made available for the participants in this task and is no longer publicly available. Therefore, researchers are unable to compare new annotation approaches on this testbed. Additionally, a collection of 25k images, is too small by modern standards. In this work we introduce a larger collection for AIA evalation which is freely available.

3.1   Annotation Model

To demonstrate the issues with the given collections, we conduct a number of experiments using the annotation model described in [12]. The method models the problem of image annotation as that of image retrieval using a KNN

($K = 10$, as used in [12]) approach. Seven features are extracted from images, three colour histograms in three channels (RGB, HSV and LAB), two texture descriptors (HAAR and Gabor filters) and two quantized versions of the texture features. Each feature vectors is normalised, with visual similarity between images computed using the average of the seven distances (for each feature pair).

Each distance is scaled by its *maximum distance*, for the given feature, within the training set. The L_1 distance is used for all features, apart from the LAB descriptor, where the K-L divergence measure is used. N tags ($N = 5$, as used in [12]) are transferred from the nearest neighbour (ordered by frequency in the training set). If the number of tags in the nearest neighbour is $< N$, tags are transferred from the surrounding neighbourhood. The top tags, ranked by the *product* of tag occurrence in the neighbourhood and co-occurrence with the nearest neighbour, are selected. This model is used to highlight problems with testing on unnormalised collections. Firstly we introduce the problems with existing collections in the following section.

3.2   Problems

(1) Tag Ambiguity: One of the major problems with these collections concerns the classes they use as ground truth. All three collections contain synonyms (e.g. america/usa) or visually identical classes (e.g. sea/ocean). For the purposes of generic image annotation, a model should not have to differentiate between synonyms, as often (from analysing the visual contents), this is impossible e.g. consider, as a human, differentiating between an image of the sea or the ocean. To illustrate this problem, we use WordNet [15] to classify keyword pairs as synonyms i.e. those keywords which contain a common synonym set. After a list of potential synonyms is generated, pairs which are seen to be incorrect by an assessor (e.g. ball/globe) are removed.

Using this approach we identify 36, 26 and 37 *ambiguous tags* (i.e those tags which have at least one synonym) for the Corel, ESP and IAPR collections, respectively. Figure 1 highlights the percentage of *ambiguous tags* present in each collection. Around *one* in *ten* tags in each collection is deemed ambiguous. This equates to 15% of all photo annotations in the IAPR collection meaning a model may under perform by up to 15%, as for each *ambiguous* annotation in the ground truth, the model may predict the synonym. Therefore evaluating on these collections may result in misleading performance measures. For example, if an image's ground truth is [home, sea] and it is annotated with the tags [house, ocean] it will achieve precision and recall scores of 0. This is clearly a bias experimental framework as luck plays a *major* role in the scoring of evaluation measures. Table 2 summarises the most occurring synonyms pairs.

(2) Unnormalised Collections: One of the main issues with the evaluation of existing annotation models lies in the unbalanced nature of collections. By nature, the classes used in image collections follow a long tail distribution i.e. there exist a few *popular* tags and many *unpopular* tags. For the evaluation of annotation models, this leads to a bias experimental setup for two reasons: (i) *Selection Bias*: Popular tags exist in more training and test images. Therefore, annotation

Table 2. Top synonyms for each collections

Collection	Top Synonym Pairs *(Instances)*
Corel	field/lawn, field/plain, polar/arctic, ice/frost, ocean/sea
ESP	circle/ring, home/house, rock/stone, baby/child, child/kid
IAPR	woman/adult, building/skyscraper, rock/stone, bush/shrub

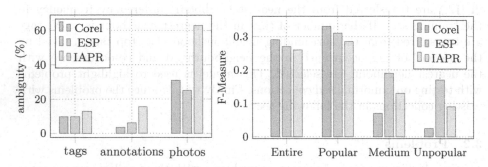

Fig. 1. *(Left)* Ambiguous tags: those tags which have at least one synonym. Ambiguous annotations: those assigned to images which have at least one synonym. Ambiguous photos: photos containing at least one ambiguous tag. *(Right)* Normalised annotation for each collection.

models are more likely to test their annotation model on these keywords, purely because a popular tag is more likely to exist in a random test image than an unpopular tag. (ii) *Prediction Bias*: Due to the wealth of training data available for popular keywords, annotations models are more likely to annotate images with these tags, as they are more likely to be correct. The unbalanced nature of collections therefore allows for potential *"cheating"* where models promote popular tags over less popular tags. To fairly measure a model's annotation accuracy based *purely on visual content*, models should not be able to exploit attributes of collections, such as tag popularity.

To demonstrate the hypothesis that popular keywords can be exploited to increase annotation accuracy, we split each collection into three vocabulary subsets representing the *popular*, *medium frequency* and *unpopular* tag sets. We denote the full vocabulary as *entire*. We select each subset so that each contains *approximately* the same number of keywords (i.e. one third), from the overall vocabulary. Using the annotation model described in Section 3.1, we annotate the images in each collection three times, annotating only with tags in each tag subset. Precision and recall measures are then computed against the tags in the ground truth, which *exist* in the given subset.

Figure 1 shows the results of this experiment. We observe that *popular* keywords are easier to annotate than *less popular* tags. Additionally, when we annotate the images purely with popular tags, we achieve higher results than the collection as a whole. Therefore, models may exploit this collection characteristic by promoting popular tags, leading to higher than expected measures for precision and recall. This annotation trend is observed across all collections.

It may be argued that by normalising, we are creating an *unrealistic* test set. However, if AIA models are benchmarked *purely* on visual features, we are measuring a model's *true* discriminative visual annotation power, without the bias of promoting popular tags. In our test collection, we propose two ground truths, an unnormalised (real life) and normalised version. We hypothesise that by improving annotation accuracy on the normalised ground truth, we will improve a model's visual discriminatory power, thus increasing accuracy on a real life collection. We encourage researchers to report evaluation metrics on both ground truths to ensure a model is not exploiting the long tail distribution and is annotating well on visual content.

(3) Quality of Images: The small size and poor quality of images in many collections often make it difficult to extract semantics from the visual contents of images, due to the lack of resolution and visual artefacts present. Despite this, the images contained in modern evaluation collections are often very small (see Table 1). The quality and size of images used in evaluation collections must increase to reflect those images taken on modern smart-phones and digital cameras.

(4) Lack of Meta-data: AIA is being more recently viewed from an information retrieval perspective, rather than that of content analysis, where time and location [14] are being exploited in the image annotation process. Despite this, all the collections used fail to include time, location and user meta-data. Therefore to allow deeper contextual analysis of images in the annotation process, every detail of an image's meta-data should be made available.

(5) Lack of Diversity: Images in the described collections are often taken by the same user, in the same place, of the same scene/object, using the same camera [20]. This leads to natural clustering in image collections, making annotation easier due to high inter-cluster visual similarity. This also causes problems such as duplicate images in the test and train set, making annotation easier.

(6) Identifying Location: As highlighted by Huiskes *et al.*, identifying a location from an image is often impossible [8]. Despite this, two of the three image collections contain ground truth classes which are locations (e.g. scotland).

(7) Copyright: The most popular baseline collection, Corel, is not freely available and is bound by copyright. To allow for the easy comparison of annotation models, a collection should be at least *free* and *distributable*.

3.3   Flickr-AIA

In this following section we detail the process used to build the Flickr-AIA collection, which aims to resolve these problems. In total, we present two test collection ground truths for 20k images, one with a normalised ground truth (i.e. where the image classes contain roughly the same number of test images), and one without (i.e. a real life scenario). We refer to how we address each problem by referencing the problem number in parenthesis e.g. (1).

Building the collection: The dataset is collected by querying Flickr for 2k popular nouns extracted from WordNet [15] (categorised as `animal, artifact, body, food, plant, substance`). The top 2k images, which contain the creative commons license, (7) location, user and time meta-data (4) and at least one tag, for each search are then considered for use in our collection. Using this approach, we collect images covering a wide range of topics (5). We download the "largest" available size version (not the original) for each image (3), ensuring high resolution and small file size.

Initially we collect 2M images before a number of pre-processing stages are taken out to resolve the discussed issues. As ground truth we use the tags assigned by the Flickr users; this has a number of advantages and disadvantages. By using user annotations, we are able to collect a *large* number of images, in comparison to the manually collated ground truths used in the Corel and IAPR collections. However, user tagging is often *noisy*, where tags do not refer to the visual contents of an image. In order to remove these tags deemed irrelevant for image annotation we use the following approach:

Removing Noise: Firstly, we manually removed (using three assessors) those tags which fell into the following categories: camera meta data (e.g. `d60`), Flickr awards (e.g. `excellentphotograph`) and Flickr groups (e.g. `5photosaday`), from the top 1,000 most frequently occurring tags. After removal of these redundant keywords we consider only the top 500 tags, ranked by descending number of users, for use in the collection. This removes tags which are used by only a few users (i.e. noise) and keeps popular classes which are more likely to be well known objects/concepts (i.e. potential image classes). We use WordNet to classify the remaining tags. Only *nouns* which are *not* categorised as the `noun.time` or `noun.location` sub-categories are used in the collection (6). By selecting nouns, we consider only visual objects, ignoring concepts difficult to identify e.g. verbs such as `talk`. Time and location tags are omitted as they are also difficult or impossible to annotate based purely on visual content [13] e.g. `Romania` or `2010`.

Promoting Diversity: As identified by [20], previous collections, such as Corel, often cluster images into coherent themes, where image similarity is high. This makes it easier for AIA models as, for every test image, there are likely to be many images in the training set which are almost visually identical. We therefore limit the number of images taken by a user to 20 to promote visual diversity (5).

Removing Synonyms: We remove synonyms in the remaining tag set using the same method as described in Section 3.2, by grouping tags which co-exist in a common WordNet synonym set. 49 synonym pairs are identified and merged (1). The details of the final collection are shown in Table 1.

Test Sets: From this collection, we remove 20k random images for testing purposes, leaving the rest for training. We offer two ground truths to test against for these images (i) *full ground truth* i.e. image contain all the classes remaining after preprocessing (ii) *normalised ground-truth* i.e. only those middle frequency classes are selected (2). Specifically, we select only those tags which occur in the middle third of tags ordered by frequency i.e. tags #140 to #280. By offering this normalised ground truth, we are able to test annotation models based purely on their visual discriminative power, removing the bias from offering popular tags.

4 Photo Tag Recommendation Evaluation

In photo tag recommendation, the typical evaluation approach is to take a small number of tags from an image and attempt to predict the other tags. As predictions are made based on textual features, the range of ground-truth classes can take a larger number of classes than those used in AIA. Differing to that of AIA evaluation, ground truth tags can also refer to both an image's visual content (e.g. an object within the scene) or its context (e.g. its location). In the following sections, we first highlight problems with test collections used by two existing tag recommendation methods. Finally we detail our new collection, Flickr-PTR, which is built for the purposes of tag recommendation evaluation in mind.

Table 3. Comparison of the Collections (i) I/T = average # images per tag (ii) T/I = averages # tags per image

Collection	# Training	# Test	Tags	Freely Available	Ground Truth
Sigurbjornsson	52M	331	3.7M	✗	Crowdsourced
Garg	50M	9k	-	✗	User Tags
Flickr-PTR	2M	1k	1M	✓	Clustered User Tags

4.1 Existing Collections

In this work we consider the evaluation collections for tag recommendation used by Sigurbjornsson et al. [18] and Garg et al. [5]. Unfortunately these collections are not freely available making any analysis or comparison with our collection difficult; however, we detail what is described in the respective papers, along with details of our new collection, Flickr-PTR, in Table 3. Firstly, we identify a number of problems with these collections:

4.2 Problems

(1) Crowdsourced ground-truths: The test collection used in [18] compares predictions against a crowdsourced ground truth for 331 images. We agree with [5], that the ground truth of an image can only be identified by the user whom the photograph is taken by. For example, consider an image taken by a father at their son's soccer game: only the father will know the location, team name etc. Therefore, an approach which tags images using a crowdsourced experiment will result in substandard annotations. Garg et al. follow this notion by adopting user tags as image ground truth, however, we identify an issue with this approach which may give mis-leading results, as described in the following subsection.

(2) Synonyms: One of the issues with using user tags is that, by nature, users tend to tag images with multiple synonyms in order make their image searchable for the various versions of the same entity. For example, instead of tagging an image solely `newyork`, many images also include a number of synonym tags e.g. `ny`, `nyc` and `newyorkcity`. In our collection, 52%, 43% and 35% of images tagged with `newyork` are also tagged with `nyc`, `ny` and `newyorkcity`, respectively.

This poses evaluation problems where models which simply promote synonyms achieve higher precision/recall scores than a model which promotes tag *novelty* and *diversity* in their rankings. For example, a recommendation model which suggests [`nyc, newyorkcity, ny`] may achieve a higher recommendation accuracy than a model which suggests [`taxi, street, centralpark`], due to the number of synonyms in the image ground-truth. However, we consider the recommendations made by the later more useful to the user as they offer novel tags instead of synonyms of already defined concepts. In this paper, we address this problem by clustering the tags in user images into related aspects, allowing for intent-aware metrics to be computed (e.g. αnDCG) instead of the traditional precision/recall metrics which ignore diversity.

(3) Free Distribution: One of the largest problems with these collections is that they are not available for distribution, making comparison with new recommendation method difficult. In our work, we download a manageable number of Flickr images which use the creative commons license, allowing for distribution.

4.3 Flickr-PTR

In this following section we detail the process used to build the Flickr-PTR collection, which aims to resolve these problems. As before, we refer to how we address each problem by referencing the problem number in parenthesis e.g. (1). The *training* collection contains the same 2M creative commons (3) images as in Flickr-AIA, before any preprocessing is taken out, using user tags as ground truth (1). The role of a training set in tag recommendation differs from image annotation, in that images can be categorised with a wide range of tags, whereas images in an AIA training set are categorised for a small number of visual classes. Therefore, for Flickr-PTR, we chose *not* to remove the noisy tags from the collection. Therefore, PTR models can be evaluated for a real-life scenario. Our main contribution, however, lies in our test collection, where tags are clustered into coherent aspects. In order to overcome the discussed problems with synonyms, we cluster tags which describe the same aspect of 1000 random images using a crowdsourced experiment. By doing so, we are able to build a test collection where the ground truth describes *aspects* for each image (2), rather than tags, as required for diversification evaluation.

Crowd-sourcing (i.e. outsourcing a task to a network of online workers) experiments have grown in popularity in recent years [7,4] and have been adopted to carry out tasks which are often difficult for computers but easy for humans e.g. image classification [17,2]. Recently, Nowak *et al.* showed that by using a majority voting scheme for an image annotation task, the quality of Turker judgements were in-line with those made by experts [17]. The ImageNet collection was also built using a crowd-sourced experiment where internet images were mapped to WordNet nodes [2]. In our work, we instead use the crowd to cluster related tags which are already assigned to images.

Task Description: We conduct this experiment on the popular Amazon Mechanical Turk[5] platform. On this platform, human intelligence tasks (HITs) are taken out by workers called 'Turkers'. In our experiment, only those Turkers with the *Master Qualification*[6] are able to accept our HIT. On acceptance of our HIT, users are presented with the following task description:

- **What is required of you:** You will be presented with an image with the tags describing its contents. You must group the *synonyms* or the tags which refer to the *same aspect* of the image.
- **Details:** You will be presented with 20 images. You may skip up to 3 images. You have a maximum of 45 minutes to finish the experiment. To group tags, simply click and drag them into the displayed boxes, then click submit. All of the tags must belong to one group, and every group must contain at least one tag. This experiment is supported for Firefox and Chrome (Res 1024+).
- **Finally:** You must judge at least 17 images and be a *fluent English* speaker. You can do the experiment multiple times, although you must make sure to login if you are coming back.

The interface allows for users to easily click and drag each tag into a number of clusters. The user is able to define one of more clusters (up to the number of tags) for each image. A *video* tutorial and two *example images* accompany the task description, allowing the worker to fully understand what is expected of them before accepting the HIT. Turkers are paid if they agreed to and carried out the conditions of the experiment. On acceptance of these terms, the worker is presented with a registration questionnaire asking for the following details: TurkID, age, sex, occupation, education level and proficiency in English.

Ensuring Quality: One of the major problems with crowdsourcing, however, is that workers often spam or try to complete tasks with as little effort as possible in order to maximize their profits [7,4]. This can lead to poor quality submissions. Many existing works have resolved this problem by introducing a number of *'honeypots'* [7,4] i.e. tasks where the correct 'answers' are already known. In our experiments we therefore introduce a number of honeypot images, which aimed to identify spamming users. Specifically, for every 20 images, we present the user with *three* images where the tags had been pre-grouped by an expert. Care is taken in creating these clusters to ensure that there is no ambiguity in the groupings. Creating these honeypots allows us to indicate the users whom completed the HIT without *reasonable effort.* Any user which grouped the tags of these *honeypot* images differently than the expert is blocked and their work is discarded. Further, the work of any Turker whom describes their English level

---

[5] https://www.mturk.com/
[6] 'Workers who have demonstrated excellence in a type of HIT, for instance categorization, are awarded the Master Qualification'.

as less than fluent is also removed. Finally, each image has its tags clustered by three *different* workers, allowing clusters to be computed using an aggregation scheme (as described in the following Section), thus minimizing spam.

Cluster Aggregation: As three different Turkers cluster the tags of each image, the votes from each are aggregated using a majority voting scheme, as adopted by [17,9]. Two tags are grouped if they are grouped in the *same cluster* by the *majority* of the three users i.e. two or more. The clusters are iteratively built, where clusters are merged if they contain a *common co-occurring* tag.

Workers: In total 197 different Turkers accepted the hit, with 20 users failing to pass the honeypot test. Therefore, work is accepted from 177 Turkers. Each HIT (20 images) is completed in 23 minutes and 40 seconds, on average. Turkers are paid between $1 and $3 for their work, which equates to $5.98/hour on average. From the entry questionnaire, around 70% of users say English is their *first language* and around 30% describe their English proficiency as *fluent*. Further, 49% of Turkers are female and 51% male, with an average age of 34. Finally, 80% of workers describe their education level as 'college' or higher.

Summary: After aggregation, each image in our test collection contains around 9.87 clusters (with each containing around 2.18 tags), on average. Considering that images in our test set are annotated with 21.5 tags on average, this indicates that more than half of the tags in our test collection are deemed *redundant* (if we assume that each tag in a cluster describes a single aspect of an image).

5   Conclusion and Future Work

This paper highlighted a number of problems which exist in using *three* popular image annotation and *two* popular photo tag recommendation evaluation collections. Most importantly, synonyms exist in annotation ground truths for all collections, which may result in misleading performance measures. Aside from this, we highlight six additional problems with annotation collections and two additional problems for tag recommendation collections. As a result, we introduce two new collections, namely Flickr-AIA and Flickr-PTR, which aim to overcome these issues and are created with fair evaluation in mind. For each collection, we also include extensive meta data relating on an image's photographer, location and time taken. Future work aims to include state-of-the-art image features and to increase the size of each collection.

References

1. Athanasakos, K., Stathopoulos, V., Jose, J.M.: A Framework for Evaluating Automatic Image Annotation Algorithms. In: Gurrin, C., He, Y., Kazai, G., Kruschwitz, U., Little, S., Roelleke, T., Rüger, S., van Rijsbergen, K. (eds.) ECIR 2010. LNCS, vol. 5993, pp. 217–228. Springer, Heidelberg (2010)
2. Deng, J., Dong, W., Socher, R., Li, L.-J., Li, K., Fei-Fei, L.: Imagenet: A Large-Scale Hierarchical Image Database. In: IEEE CVPR 2009 (2009)

3. Duygulu, P., Barnard, K., de Freitas, J.F.G., Forsyth, D.: Object Recognition as Machine Translation: Learning a Lexicon for a Fixed Image Vocabulary. In: Heyden, A., Sparr, G., Nielsen, M., Johansen, P. (eds.) ECCV 2002, Part IV. LNCS, vol. 2353, pp. 97–112. Springer, Heidelberg (2002)
4. Eickhoff, C., Vries, A.P.: Increasing Cheat Robustness of Crowdsourcing Tasks. Inf. Retr. 2013 (2013)
5. Garg, N., Weber, I.: Personalized, Tnteractive Tag Recommendation for Flickr. In: ACM RecSys 2008 (2008)
6. Grubinger, M., Clough, P., Mller, H., Deselaers, T.: The IAPR TC-12 Benchmark - A New Evaluation Resource for Visual Iinformation Systems (2006)
7. Hirth, M., Hoßfeld, T., Tran-Gia, P.: Cheat-detection Mechanisms for Crowdsourcing. Technical Report, University of Würzburg, 8 (2010)
8. Huiskes, M.J., Lew, M.S.: The MIR Flickr Retrieval Evaluation. In: MIR 2008 (2008)
9. Vuurens, A.P.D.V.J., Eickhoff, C.: How much Spam can you take? An Analysis of Crowdsourcing Results to Increase Accuracy. In: ACM SIGIR 2011 (2011)
10. Jeon, J., Lavrenko, V., Manmatha, R.: Automatic Image Annotation and Retrieval using Cross-media Relevance Models. In: ACM SIGIR 2003 (2003)
11. Liu, D., Hua, X.-S., Yang, L., Wang, M., Zhang, H.-J.: Tag ranking. In: WWW 2009 (2009)
12. Makadia, A., Pavlovic, V., Kumar, S.: Baselines for image annotation. In: IJCV 2010 (2010)
13. Mark, B.T., Huiskes, J., Lew, M.S.: New trends and Ideas in Visual Concept Detection: The MIR Flickr Retrieval Evaluation Initiative. In: MIR 2010 (2010)
14. McParlane, P.J., Moshfeghi, Y., Jose, J.M.: On Contextual Photo Tag Recommendation. In: SIGIR 2013 (2013)
15. Miller, G.A.: Wordnet: A lexical database for english. Communications of the ACM (1995)
16. Müller, H., Marchand-Maillet, S., Pun, T.: The Truth about Corel - Evaluation in Image Retrieval. In: Lew, M., Sebe, N., Eakins, J.P. (eds.) CIVR 2002. LNCS, vol. 2383, pp. 38–49. Springer, Heidelberg (2002)
17. Nowak, S., Rüger, S.: How reliable are annotations via crowdsourcing: a study about inter-annotator agreement for multi-label image annotation. In: MIR 2010 (2010)
18. Sigurbjörnsson, B., van Zwol, R.: Flickr Tag Recommendation based on Collective Knowledge. In: WWW 2008 (2008)
19. von Ahn, L., Dabbish, L.: Labeling images with a computer game. In: CHI 2004 (2004)
20. Westerveld, T., de Vries, A.P.: Experimental Evaluation of a generative probabilistic image retrieval model on 'easy' data. In: ACM SIGIR 2003 (2003)

Graph-Based Multimodal Clustering for Social Event Detection in Large Collections of Images

Georgios Petkos, Symeon Papadopoulos, Emmanouil Schinas,
and Yiannis Kompatsiaris

Information Technologies Institute, Centre for Research and Technology Hellas

Abstract. A common approach to the problem of SED in collections of multimedia relies on the use of clustering methods. Due to the heterogeneity of features associated with multimedia items in such collections, such a clustering task is very challenging and special multimodal clustering approaches need to be deployed. In this paper, we present a *scalable graph-based multimodal clustering* approach for SED in large collections of multimedia. The proposed approach utilizes example relevant clusterings to learn a model of the "same event" relationship between two items in the multimodal domain and subsequently to organize the items in a graph. Two variants of the approach are presented: the first based on a batch and the second on an incremental community detection algorithm. Experimental results indicate that both variants provide excellent clustering performance.

Keywords: Social media, Social event detection, Multimodal clustering.

1 Introduction

The wide availability of low cost media capturing devices along with the advent and massive adoption of online social publishing platforms have significantly transformed the behaviour of casual online users, turning them to a large extent into media content producers. Considering the huge growth that social media have had in the recent years, it is clear that there is a growing amount of diverse content that covers a huge range of real-world activities and that can be used to "sense the world". For instance, web content obtained from social media has been used in applications such as detecting breaking news [24], landmarks [12] or more recently social events [20,19]. Nevertheless, such content is often noisy, presents large heterogeneity and is often not well-structured and therefore presents many challenges to analysts.

In this paper we discuss the problem of discovering social events in collections of web multimedia. By social events, we mean events that are attended by people and are represented by multimedia content shared online. Instances of such events could include concerts, sports events, public celebrations or even protests. Formally, given a photo collection denoted by $\mathbb{P} \triangleq \{p\}$, where p stands for a photo and its metadata (owner, time, tags, location if available, etc.), we consider methods that produce a set of photo clusters $\mathbb{C} \triangleq \{c\}$, each cluster c comprising only photos $\mathbb{P}_c$ associated with a single event. Various approaches

C. Gurrin et al. (Eds.): MMM 2014, Part I, LNCS 8325, pp. 146–158, 2014.

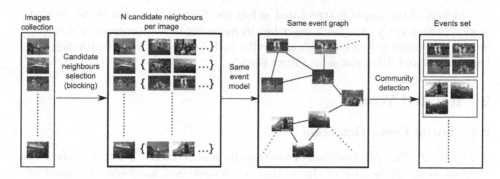

Fig. 1. Schematic representation of the proposed approach

have been proposed for tackling the problem of Social Event Detection (SED) in collections of multimedia. Some of them rely on auxiliary information obtained from online sources and directories, such as LastFM, EventFul and DBPedia [4,13]. Such approaches retrieve information about actual events or venues and attempt to match this information with items in the collection. A second class of methods do not utilize external information sources; instead, they attempt to cluster the items of the collection, so that the resulting clusters represent events [2,22,26,23]. This is clearly a more general approach than the first one, as it is does not rely on potentially unavailable external information.

In this paper we present a novel scalable approach for *multimodal* clustering that we apply to the problem of SED in collections of multimedia. The proposed approach utilizes a model that has been trained to predict whether a pair of items (images) belong to the same cluster (event), using as input the set of per modality distances between the pair of items (images). In the rest of the paper, we will refer to this as a "same event" (SE) model. We compute the predictions of this SE model for pairs of items in our collection and we organize the items in a graph. This graph has a node for each image of the collection and an edge between two nodes indicates that the prediction of the SE model for the corresponding pair of images was positive. Additionally, similarly to [26], we adopt an appropriate strategy to avoid having to predict the SE relationship between each possible pair of images, thereby making the approach applicable to large datasets. This involves computing the output of the SE model for each item only against its nearest neighbours according to each modality. Finally, we apply a community detection algorithm on the resulting graph. The proposed approach comes in two variants; in the first, batch community detection is applied on the graph, whereas in the second an incremental algorithm is applied. Figure 1 presents an overview of the proposed approach. Interestingly, the proposed approach is applicable to a variety of other multimodal clustering algorithms, as long as an example clustering that can be used for training the SE model is available. To the best of our knowledge, the presented *incremental* approach is one of the first to tackle the problem of incremental multimodal clustering.

The rest of this paper is structured as follows. Section 2 presents some related work. Subsequently, Section 3 describes in more detail the proposed approach. Section 4 presents some empirical results and finally Section 5 provides the conclusions and discusses some future work.

2   Related Work

2.1   Social Event Detection

The task of SED in collections of multimedia has attracted a lot of interest in the last years. Indicative of this is that a relevant task has been organized as part of the MediaEval Benchmark in 2011 [20] and 2012 [19], the data of which is used in the current work for evaluation. Most approaches commonly apply a sequence of clustering or filtering operations in order to obtain the required set of events, e.g. [21].

Of particular interest for this work are approaches that treat the problem as a clustering task. Since items in this problem are typically multimodal, this is a *multimodal clustering* task. Multimodal clustering is a challenging task, for which various approaches have been proposed (see Section 2.2). The essence of the problem is that appropriate similarity measures are required that take into account heterogeneous modalities. Such similarity measures in which the contribution of each individual modality is appropriately weighted may be determined either by prior knowledge or a search process. A solution to this problem that was utilized by previous approaches to event detection [2,26,23] uses an auxiliary example clustering, in which clusters are known to represent events, as a supervisory signal. Such a known clustering (collection of events) has been used in two different ways, in conjunction with different clustering procedures.

In the first, which will be refered to as the **item-cluster** approach, a part of the data is used to obtain an SE model that predicts whether an item belongs to some event, using as input the set of distances between the item and a prototype (aggregate) representation of the event. The auxiliary data can be used to obtain the training data for such a model by computing aggregate representations for part of the items that belong to the collection of events and then sampling from the rest of the items. This is the approach followed by [2] and [26]. Subsequently, this model can be used for clustering the incoming items in an incremental fashion, by computing the prediction of this SE model for the sets of similarities between the new item and the prototype representations of the already identified clusters. The set of predictions is either used to assign the item to the best matching cluster/event or to generate a new event. The prototype representation of each event is the average of the items assigned to it.

The second way that such an example clustering can be used will be refered to as the **item-item** approach. It involves learning a similar SE model that predicts whether two items, instead of an item and a cluster prototype, belong to the same cluster, again taking as input the corresponding set of similarities [23,2]. An item-item approach avoids the need to maintain a prototype representation, but requires a different clustering procedure. In the case of [2], the predictions of the model between an item and all other items that were assigned to some cluster are used to obtain an estimate of cluster membership by simple averaging. In [23], the SE model is applied on all pairs of items in the collection. For each item, a vector is maintained that contains the SE relationship of that item to all other items in the collection. Finally, it is assumed that items that belong to the same cluster will have similar sets of SE relationships to the set of items in the collection and therefore a final assignment to events is obtained by clustering the corresponding vectors. However, this approach is clearly not scalable as it has quadratic complexity to the collection size; it requires the evaluation of the SE model for all pairs of images. Moreover, it requires maintaining vectors of which the size increases with the dataset size. In this work, we propose a scalable item-item approach that deals with the problem of quadratic complexity by utilizing a candidate neighbour selection step and with the problem of large size neighbourhood vectors by utilizing a graph to store the SE relationships.

2.2 Multimodal Clustering

The task of clustering items which are expressed through heterogeneous modalities has commonly been treated using fusion approaches, of which two common classes exist: early and late fusion [29]. Early fusion approaches combine the features/modalities in some specific representation before the main analysis process - in this case clustering - is applied. Late fusion methods apply the main processing function to each modality separately and instead combine the results obtained for each modality. A recent spectral clustering approach that combines modalities at an intermediate level is presented in [5]. An important set of methods includes probabilistic approaches that utilize graphical models representations [11,3]. Finally, it is worth noting the connection between multimodal clustering and ensemble clustering [7], in which the goal is to combine a set of clusterings (produced e.g. from a set of different algorithms) in some optimal manner. A multimodal clustering problem can be cast into an ensemble clustering problem by performing a set of clusterings - e.g. one per modality - and then combining the individual clusterings. This is also one of the proposed approaches for event detection in [2].

2.3 Community Detection

Community detection is used to cluster the SE graph that is constructed as part of the proposed approach. There is a large body of work in this area, and a few comprehensive surveys on the topic [6,17]. The bulk of existing work examine the problem in the context of static graphs (i.e. batch mode). Recently, the problem of dynamic community detection (i.e. online mode) is increasingly gaining importance. The use of community detection for event detection is not new. For instance, [22] applies community detection on image similarity graphs for the extraction of landmarks and events in large image collections.

3   Approach Description

The proposed approach, has already been outlined in the introduction and in Figure 1. It consists of three components: The first retrieves for each image in the collection a set of images that are candidates for belonging to the same event. Subsequently, an SE model is employed to predict which of these candidate SE relationships are likely to hold. This leads to the generation of a graph that represents the SE relationships between the images of the collection. At the final step, a community detection algorithm is applied on this graph to extract clusters of images that constitute the detected events. We examine two variants, using either a batch or an incremental community detection algorithm. In the following, we provide more details on each of these components.

Candidate Selection. An important component of the proposed approach relies on the capability for fast retrieval of same-event candidates per incoming item for each of the different modalities, so that the SE model can be applied only on the candidate items. This is a technique, which in the context of related work, was first utilized by [26], but was used before in the field of record linkage and is commonly refered to as *blocking* [25]. Blocking greatly reduces the number of SE predictions required and can result in scalable implementations. To this end, all items in the collection are indexed in appropriate structures. For instance, textual metadata (title, tags) are inserted in a full-text index (e.g. Lucene) for rapidly retrieving documents with high textual similarity to a query document. Other metadata such as time and location are indexed by use of B-Trees, whereas visual similarity can be indexed by a variety of content-based retrieval approaches (in our implementation, we opted for the use of Product Quantization in tandem with Asymmetric Distance Computation [9]).

Multimodal "Same-Event" Model. The item-cluster comparison has the disadvantage that it relies on the prototype representation of the cluster being an accurate representation of the underlying event. Thus, if there are incorrectly assigned items, this may significantly affect the accuracy of the prototype representation. Due to the potential problem with averaging incorrectly assigned items into a prototype cluster representation, we adopt an item-item approach. Formally, given two images, p^i and p^j, which are expressed through a set of k features $p^i_1...p^i_k$ and $p^j_1...p^j_k$ respectively, we compute the vector that contains the per-modality distances $d(p^i, p^j) = [d(p^i_1, p^j_1), d(p^i_2, p^j_2)...d(p^i_k, p^j_k)]$. The SE model is then a function of $d(p^i, p^j)$:

$$SEM(p^i, p^j) = f(d(p^i, p^j))$$

and predicts if the images p^i and p^j belong to the same event, i.e. it is a classifier and for the following we assume that a predicted value of $+1$ denotes that the two images belong to the same event, whereas a value of -1 denotes the opposite. Having a separate example clustering, in which each cluster represents an event, it is straightforward to obtain training data for the SE model. In our scenario, where we use images from the social media, Flickr in particular, we use the following set of features and similarity measures:

1. *Uploader identity.* The identity of the Flickr user who uploaded the picture. We utilize a binary similarity measure where the value 0 is used for pairs of pictures uploaded by different users whereas the value 1 is used for pairs of pictures uploaded by the same user.
2. *Actual image content.* From each image we extract one global descriptor, GIST [16] and a set of local descriptors, SURF [1], which we aggregate using the VLAD scheme [10]. The distance between a pair of GIST descriptors or SURF-VLAD descriptors is computed using Euclidean distance.
3. *Textual information.* Images uploaded by users are typically accompanied by a title, a description and a set of tags. We utilize both the Term Frequency - Inverse Document Frequency and BM25 [14] weighting schemes, resulting in six textual-based distance elements per pair of items.
4. *Time of media creation.* Instead of using the difference of the time of creation between a pair of items, we use three binary distances. The first takes the value 0 if the difference in time of creation is larger than 6 hours and the value 1 if it is smaller. The other two are similar but set the threshold to 12 and 24 hours respectively. This is an approach that empirically led to better results when training the classifier compared to using the absolute time distance (various time scales were tried).
5. *Location.* Not all items in the collection come with location data. For pairs of items that do come with location information though, we compute two distance measures. The first is the geodesic distance in kilometers and the second is a boolean indicator which is 1 if the geodesic distance is smaller than 1 kilometer and 0 if it is larger.

In order to handle the case of missing location information we train two models. The first takes as input all the aforementioned distances and is used for pairs of items that both come with location information. The second takes as input the same set of distances except the location-based and is used when at least one of the two items for which SE model is evaluated does not have location information. In other scenarios, different types of data may be available or used, however, the presented approach is applicable without significant changes.

Graph-Based Clustering. The final part of the algorithm clusters the items in the collection. Our algorithms organize the items in a graph, in which the existence of an edge indicates that the SE model has predicted that the corresponding items belong to the same event. More formally, given a set of images to be clustered P, we generate a graph $G = (P, E)$, whose set of vertices is P, i.e. the set of photos to be clustered. The set of edges E, contains the pairs of images (p^i, p^j) for which $SEM(p^i, p^j) = +1$ and either p^i is a candidate neighbour of p^j or p^j is a candidate neighbour of p^i.

A community detection algorithm is then applied on the graph. The proposed approach comes in two variants, which utilize either a batch or an incremental community detection algorithm and are described below.
Batch community detection. In the first variant of our approach, a batch community detection algorithm is applied. The selected algorithm is the Structural Clustering Algorithm for Networks (SCAN) [30]. This choice was motivated by

some desirable properties of the algorithm: (a) computational efficiency, (b) possibility to leave spuriously connected nodes out of the clustering, (c) it can identify not only communities, but also hubs and outliers. This last property, is particularly interesting for the following reason. The predictions of the SE model will inevitably be imperfect. If they were perfect, then it would suffice to find the connected components of the graph. Therefore, in the case of noisy edges, one has to take into account the case of nodes that are sparsely connected to some cluster(s) due to incorrect predictions of the SE model. Such nodes are likely to be classified as outliers and can therefore be assigned to a separate cluster/event rather than to their erroneously adjacent clusters. Outliers on the other hand are assigned to the adjacent community to which it has the largest number of connections. SCAN is controlled by two parameters, μ and ϵ, which determine the minimum number of nodes and the minimum "tightness" in a community.

Incremental community detection. In the second variant of our approach, an incremental algorithm is applied. There are few incremental community detection algorithms in the literature. We opted for Quick Community Adaptation (QCA) [15], due to its efficiency and simplicity of implementation. QCA is an expansion of a previous physics-inspired, non-incremental approach [31]. It maintains the detected community structures up-to-date by appropriate processing operations in the event of four different graph changes: a) new node creation, b) new edge addition, c) node removal and d) edge removal. In short, in all operations, forces that attempt to pull a node inside adjacent communities are computed and each node is pulled to the community that applies the strongest attracting force. In all operations, the forces are appropriately computed for all affected nodes. For more details please see [15]. In our scenario we do not consider node or edge removals: neither is it likely that an item is removed from the collection, nor would an SE relationship between a pair of items be evaluated a second time. Finally, it should be noted that QCA is parameter free.

4 Experimental Evaluation

The proposed graph-based multimodal clustering method was tested on data from the 2012 MediaEval SED task [19]. This consists of three challenges that call for the detection of social events of specific types (there are three target types of events: soccer events, technical events and events related to the Indignados movement) that took place in specific geographic locations in a collection of approximately 167,000 images collected from Flickr. Out of those, 7,779 images did indeed belong to one of the 149 target events of the ground truth (79 soccer, 18 technical and 52 Indignados events). For examples of images and events of these classes, please see [18].

The SE models are trained using a Support Vector Machine classifier (SVM), notably the Weka [8] implementation. Various other classifiers were tested but the SVM classifier produced the best results. Its average classification accuracy on a sequence of sets on test data was 98.58%. A close second in accuracy among the tested algorithms was a decision tree, which resulted in an average accuracy score of around 96.62%.

Table 1. NMI for the graph-based batch and incremental methods, as well as an item-cluster based method (only event images are used in these runs)

	Batch	Incremental	Item-cluster [2]
Avg.	0,924	0,934	0,898
Std.	0,019	0,021	0,027

Table 2. NMI using both event and non-event images. The set of images is randomly selected from the complete 2012 challenge dataset.

Labelling acc.	# images	Batch	Incremental	Item-cluster [2]
0,95	15352	0,4824	0,5164	0,3954
0,90	22876	0,3121	0,3683	0,2899

As mentioned, to avoid evaluating the SE relationship between all possible pairs of items, a candidate selection mechanism is utilized. For each item, we retrieve the items with which it has the largest similarity with respect to the textual features (50 items chosen), time (150 items chosen), location (50 items chosen, if location is available), GIST (50 items chosen) and VLAD/SURF (50 items chosen). Items are indexed according to the different modalities in appropriate structures, so that the nearest neighbours for each modality can be obtained very fast. For the batch procedure, all items are indexed before the main processing is carried out, whereas for the incremental procedure items are indexed as they arrive. The following index implementations were used: Lucene for the textual features, a MySQL database for time and location features and the approximate nearest neighbor search method of [9] for the visual features.

As mentioned above, the SCAN algorithm used by the batch method has two parameters that may affect the quality of the final result. For these experiments, the ϵ parameter was empirically set to 0.7 and the μ parameter was set to 3. This experimentation was carried out on a separate random sample of events.

In a first set of experiments, the images that belonged to a random subset of 30 events (out of the 149 events in the collection) were used to train the SE model, whereas the images in the remaining events were used for testing the proposed methods. We generated 10 random tasks in this manner. The utilized measure of clustering quality is Normalized Mutual information (NMI). The two variants of the proposed algorithm are tested against the item-cluster approach of [2], where a threshold of 0.5 was used on the output of the SE model. That is, an incoming item was assigned to the best matching cluster if the probability output by the model was above 0.5, otherwise it was used to generate a new event. Table 1 presents the obtained results. The results demonstrate that both variants of the graph-based approach achieve very high clustering accuracy with an NMI clearly above 0.9. An important thing to notice is that the performance of the incremental algorithm is by no means inferior to the performance of the batch algorithm. Contrary, on average, the performance of the incremental algorithm is slightly higher than that of the batch algorithm, even though the difference is not statistically significant. The performance of both graph-based approaches is

Table 3. NMI of proposed methods using limited sets of features

	Batch	Incremental
Visual	0,8020 ± 0,0193	0,8179 ± 0,0151
Textual	0,7925 ± 0,0255	0,7792 ± 0,0310
Visual + time	0,9244 ± 0,0195	0,9360 ± 0,0183
Textual + time	0,9016 ± 0,0173	0,9049 ± 0,0209

higher than the performace of the tested item-cluster approach and the difference is statistically significant at a 0.95 confidence level.

In the previous experiments, we clustered sets of images that do belong to the target events. In the following, we examine the scenario that non-event pictures are also included in the collection to be clustered. More particularly, we use the data from the 2011 challenge for training. The test set is obtained from the complete collection of 167,000 images of the 2012 challenge as follows. Each image is randomly labeled as event or non-event with the probability of being labeled correctly being p and use only the images that have been labeled as events. This experiment will provide further comparison of the methods and will also test the robustness of the method in the presence of many spurious images. We test two different values for p: 0,95 and 0,9 which almost double and triple respectively the number of images to be clustered (originally 7,779 images belong to some event). It should be noted also, that in some recent work [27] we attempt to classify images as either representing a social event or not, we obtain accuracy values similar to these values (in particular 0,8962). The results can be seen in Table 2. The NMI has dropped significantly, however, we still observe that the item-item approaches are superior to the item-centroid approaches and that the parameter free incremental method is superior to the batch method. Finally, it should be noted that although these experiments are closer to the initial MediaEval SED challenge scenario, the results cannot be compared to results reported for the challenge. In order to do this, one would have to also perform further processing, e.g. to filter events or images as representing an event of a particular type. For an application of this method to the challenge together with these steps please see [28].

We also investigate the importance of the set of features used for computing the predictions of the SE model. We conducted the same run of experiments on limited sets of features. The results can be seen in Table 3. These surprisingly good results, especially for the case that only visual or textual features are used, are due to the fact that blocking is still applied. In the same experiments executed without blocking, the average NMI obtained using only visual features was 0,030 and 0,7148 for the textual features. From Table 3 it can also be seen that time is a crucial feature for the performance of the SE model.

Moreover, we examine whether the type of events used for training the SE model is important for detecting events of a different type. In the previous run of experiments, the 30 randomly sampled events were randomly chosen from the three categories (soccer, technical and Indignados). We now use all available

Table 4. NMI achieved by training and testing on different types on events

Batch			
	Soccer	Technical	Indignados
Soccer	-	0,8658	0,8494
Technical	0,7967	-	0,8977
Indignados	0,9645	0,8456	-
Incremental			
	Soccer	Technical	Indignados
Soccer	-	0,8892	0,8667
Technical	0,7661	-	0,7735
Indignados	0,9845	0,8482	-

events from one type of events to learn the SE model, which we then use to cluster the items that belong to one of the other two types of events. The results can be seen in Table 4. Clearly, the type of event used for training does have an effect on the quality of the produced clustering for different types of events. Most notably, using technical events for training the SE model and then using it for producing clusters that represent other types of events produces lower quality results using either variants of the graph-based method. However, it is noteworthy that in some cases training with a completely different type of events can lead to very high performance, e.g. in the case that Indignados events are used for training and soccer events are detected with an NMI of over 0.96 for both variants of the approach.

5   Conclusions and Future Work

This paper proposed two variants of a novel multimodal clustering approach and presented an application on the problem of SED in collections of multimedia. The proposed method utilizes the so-called "same event" (SE) model, which predicts whether a pair of items belong to the same cluster or not. The SE model is used to organize the collection in a graph, on which a community detection algorithm is applied. The two flavors of our method utilize either a batch or an incremental community detection algorithm. Empirical results indicate that the proposed algorithms achieve high quality clusterings. Interestingly, the performance of the incremental algorithm is not inferior to that of the batch algorithm.

Compared to the related approaches in [2] and [26], our approach computes the predictions of the SE model on pairs of images rather than on pairs of an image and an event. In these approaches the event representation is the result of an averaging of the features of the items that have been assigned to the event. Nevertheless, in the case of incorrectly assigned items, it is possible that the representation of the event may be significantly erroneous in some features. This may subsequently result in more items being erroneously assigned to the event, further affecting the representation of the event and leading to a progressive deterioration of the quality of the clusters. On the other hand, an item-item approach does not suffer from this issue and it is likely that the empirical advantage of the examined approaches over the item-cluster approach is due to this reason. Also, compared to the aforementioned approaches, the proposed approach does not require learning an additional model for determining whether a new event needs to be added (as in [26]) or setting a relevant threshold for the best matching event (as in [2]). Instead, the incremental approach can automatically determine whether a new item should join an existing cluster or not.

Compared to the approach in [23], where a SE model was also used between pairs of items, the graph-based methods are much more scalable and require far less resources. For instance, the approach in [23] requires that all pairwise same event relationships are maintained. Thus, all N^2 same event relationships need to be computed and the results need to be stored. Moreover, the sets of SE relationship of nodes are compared using Euclidean distance, which may make items with very few irrelevant neighbours appearing close to each other. Contrary, the proposed approaches do not require to either compute or store all the SE relationships.

In the future we intend to test the framework with other community detection algorithms as well as to apply the framework to other multimodal clustering tasks. Moreover, we plan to extend our recent work on distinguishing between event and non-event images. Regarding evaluation, we plan to explore the effect that the number of candidate neighbours has on the results.

Acknowledgments. This work is supported by the SocialSensor FP7 project, partially funded by the EC under contract number 287975.

References

1. Bay, H., Ess, A., Tuytelaars, T., Van Gool, L.: Speeded-Up Robust Features (SURF). Comp. Vis. Image Underst. 110(3), 346–359 (2008)
2. Becker, H., Naaman, M., Gravano, L.: Learning similarity metrics for event identification in social media. In: Proceedings of the Third ACM International Conference on Web Search and Data Mining, WSDM 2010, pp. 291–300. ACM, New York (2010)
3. Bekkerman, R., Jeon, J.: Multi-modal clustering for multimedia collections. In: CVPR (2007)
4. Brenner, M., Izquierdo, E.: Mediaeval benchmark: Social Event Detection in collaborative photo collections. In: MediaEval. CEUR Workshop Proceedings (2011)
5. Cai, X., Nie, F., Huang, H., Kamangar, F.: Heterogeneous image feature integration via multi-modal spectral clustering. In: 2011 IEEE Conference on Computer Vision and Pattern Recognition, CVPR, pp. 1977–1984 (June 2011)
6. Fortunato, S.: Community detection in graphs. Physics Reports 486(3-5), 75–174 (2010)
7. Goder, A., Filkov, V.: Consensus clustering algorithms: Comparison and refinement. In: Ian Munro, J. (ed.) Proceedings of the Workshop on Algorithm Engineering and Experiments, ALENEX 2008, San Francisco, California, USA, pp. 109–117. SIAM (January 19, 2008)
8. Hall, M., Frank, E., Holmes, G., Pfahringer, B., Reutemann, P., Witten, I.H.: The weka data mining software: an update. SIGKDD Explorations Newsletter 11(1), 10–18 (2009)

9. Jegou, H., Douze, M., Schmid, C.: Product quantization for nearest neighbor search. IEEE Transactions on Pattern Analysis and Machine Intelligence 33(1), 117–128 (2011)
10. Jégou, H., Douze, M., Schmid, C., Pérez, P.: Aggregating local descriptors into a compact image representation. In: 23rd IEEE Conference on Computer Vision & Pattern Recognition, CVPR 2010, pp. 3304–3311. IEEE Computer Society, San Francisco (2010)
11. Khalidov, V., Forbes, F., Horaud, R.P.: Conjugate mixture models for clustering multimodal data. Neural Computation 23(2), 517–557 (2011)
12. Li, Y., Crandall, D.J., Huttenlocher, D.P.: Landmark classification in large-scale image collections. In: IEEE 12th International Conference on Computer Vision, ICCV 2009, Kyoto, Japan, September 27 - October 4, pp. 1957–1964. IEEE (2009)
13. Liu, X., Troncy, R., Huet, B.: Using social media to identify events. In: ACM Multimedia 3rd Workshop on Social Media, WSM 2011, Scottsdale, Arizona, USA, November 18-December 1, p. 11 (2011)
14. Manning, C.D., Raghavan, P., Schütze, H.: Introduction to Information Retrieval. Cambridge University Press, New York (2008)
15. Nguyen, N.P., Dinh, T.N., Xuan, Y., Thai, M.T.: Adaptive algorithms for detecting community structure in dynamic social networks. In: 30th IEEE International Conference on Computer Communications, Joint Conference of the IEEE Computer and Communications Societies, INFOCOM 2011, Shanghai, China, April 10-15, pp. 2282–2290. IEEE (2011)
16. Oliva, A., Torralba, A.: Modeling the shape of the scene: A holistic representation of the spatial envelope. Int. J. Comput. Vision 42(3), 145–175 (2001)
17. Papadopoulos, S., Kompatsiaris, Y., Vakali, A., Spyridonos, P.: Community detection in social media. Data Mining and Knowledge Discovery 24(3), 515–554 (2012)
18. Papadopoulos, S., Schinas, E., Mezaris, V., Troncy, R., Kompatsiaris, I.: The 2012 Social Event Detection dataset. In: 4th ACM Multimedia Systems, Dataset Session, MMSys 2013, Oslo, Norway, February 27-March 1 (2013)
19. Papadopoulos, S., Schinas, E., Mezaris, V., Troncy, R., Kompatsiaris, Y.: Social Event Detection at MediaEval 2012: Challenges, Dataset and Evaluation. In: MediaEval 2012 Workshop, Pisa, Italy, October 4-5 (2012)
20. Papadopoulos, S., Troncy, R., Mezaris, V., Huet, B., Kompatsiaris, I.: Social Event Detection at Mediaeval 2011: Challenges, dataset and evaluation. In: MediaEval. CEUR Workshop Proceedings (2011)
21. Papadopoulos, S., Zigkolis, C., Kompatsiaris, Y., Vakali, A.: CERTH@Mediaeval 2011 social event detection task. In: MediaEval. CEUR Workshop Proceedings (2011)
22. Papadopoulos, S., Zigkolis, C., Kompatsiaris, Y., Vakali, A.: Cluster-based landmark and event detection for tagged photo collections. IEEE Multimedia 18(1), 52–63 (2011)
23. Petkos, G., Papadopoulos, S., Kompatsiaris, Y.: Social event detection using multimodal clustering and integrating supervisory signals. In: Proceedings of the 2nd ACM International Conference on Multimedia Retrieval, ICMR 2012, pp. 23:1–23:8. ACM, New York (2012)
24. Phuvipadawat, S., Murata, T.: Breaking news detection and tracking in twitter. In: IEEE/WIC/ACM International Conference on Web Intelligence and Intelligent Agent Technology, vol. 3, pp. 120–123 (2010)
25. Rendle, S., Schmidt-Thieme, L.: Scaling record linkage to non-uniform distributed class sizes. In: Washio, T., Suzuki, E., Ting, K.M., Inokuchi, A. (eds.) PAKDD 2008. LNCS (LNAI), vol. 5012, pp. 308–319. Springer, Heidelberg (2008)

26. Reuter, T., Cimiano, P.: Event-based classification of social media streams. In: Proceedings of the 2nd ACM International Conference on Multimedia Retrieval, ICMR 2012, pp. 22:1–22:8. ACM, New York (2012)
27. Schinas, E., Mantziou, E., Papadopoulos, S., Petkos, G., Kompatsiaris, Y.: CERTH @ Mediaeval 2013 Social Event Detection Task. In: MediaEval. CEUR Workshop Proceedings (2013)
28. Schinas, E., Petkos, G., Papadopoulos, S., Kompatsiaris, Y.: CERTH @ Mediaeval 2012 Social Event Detection Task. In: MediaEval. CEUR Workshop Proceedings, vol. 927 (2012)
29. Snoek, C.G.M., Worring, M., Smeulders, A.W.M.: Early versus late fusion in semantic video analysis. In: Proceedings of the 13th Annual ACM International Conference on Multimedia, MULTIMEDIA 2005, pp. 399–402. ACM, New York (2005)
30. Xu, X., Yuruk, N., Feng, Z., Schweiger, T.A.J.: Scan: a structural clustering algorithm for networks. In: Proceedings of the 13th ACM SIGKDD, KDD 2007, pp. 824–833. ACM, NY (2007)
31. Ye, Z., Hu, S., Yu, J.: Adaptive clustering algorithm for community detection in complex networks. Physical Review E 78(4) (2008)

Tag Relatedness Using Laplacian Score Feature Selection and Adapted Jensen-Shannon Divergence

Hatem Mousselly-Sergieh[1,2], Mario Döller[3], Elöd Egyed-Zsigmond[2],
Gabriele Gianini[4], Harald Kosch[1], and Jean-Marie Pinon[2]

[1] Universität Passau, Innstr. 43, 94032 Passau, Germany
[2] Université de Lyon, 20 Av. Albert Einstein, 69621 Villeurbanne, France
[3] FH Kufstein, Andreas Hoferstr. 7, 6330 Kufstein, Austria
[4] Università degli Studi di Milano, via Bramante 65, 26013 Crema, Italy
firstname.lastname@uni-passau.de, firstname.lastname@insa-lyon.fr,
mario.doeller@fh-kufstein.ac.at, gabriele.gianini@unimi.it

Abstract. Folksonomies - networks of users, resources, and tags allow users to easily retrieve, organize and browse web contents. However, their advantages are still limited according to the noisiness of user provided tags. To overcome this problem, we propose an approach for identifying related tags in folksonomies. The approach uses tag co-occurrence statistics and Laplacian score feature selection to create probability distribution for each tag. Consequently, related tags are determined according to the distance between their distributions. In this regards, we propose a distance metric based on Jensen-Shannon Divergence. The new metric named AJSD deals with the noise in the measurements due to statistical fluctuations in tag co-occurrences. We experimentally evaluated our approach using WordNet and compared it to a common tag relatedness approach based on the cosine similarity. The results show the effectiveness of our approach and its advantage over the adversary method.

Keywords: Folksonomies, Tag Relatedness, Laplacian Score, JSD.

1 Introduction

In the current internet era, collaborative tagging systems become ubiquitous tools which allow users to add contents to the web, annotate them using keywords called tags, and share them with each other. This results in a complex network of users, resources and tags which is commonly referred to as a folksonomy. According to the degree of user collaboration, folksonomies are classified in two main categories: broad and narrow [1]. In broad folksonomies, e.g., del.icio.us[1], multiple users tag the same resources with variety of terms. In narrow folksonomies, the tagging activity is mainly performed by the content creators. Image folksonomies like Flickr[2] belong to this category.

---

[1] www.delicious.com
[2] www.flickr.com

C. Gurrin et al. (Eds.): MMM 2014, Part I, LNCS 8325, pp. 159–171, 2014.
© Springer International Publishing Switzerland 2014

Tags simplify resource retrieval and browsing. Additionally, tagging allow users to annotate the same resources with several terms, thus, they can organize their resources in multiple categories. However, the unsupervised way of tag creation makes them suffer from noise, such as redundancy (different tags vs. same meaning) and ambiguity (same tag vs. different meanings) [2]. To overcome these problems, researches worked on techniques for identifying related tags in folksonomies (e.g. [3–5]). The proposed solutions help to identify redundant tags and to resolve tag ambiguity by providing the needed context through groups of related tags. Generally, the main research directions of most contributions are on investigating and proposing efficient clustering algorithms to determine similar tags. However, little research has focused on the (dis)similarity measure which is used to create the tag dissimilarity matrix (the main input for clustering algorithms). Most approaches follow a simple procedure for creating the tag dissimilarity matrix based on the cosine similarity of tag co-occurrence vectors. Although the cosine method seems to be efficient, we believe that more sophisticated measures can help to boost the performance of the tag clustering algorithms.

In this paper we propose a tag relatedness measure which deal with tags as probability distributions. Initially, a probability distribution is generated for each tag in the folksonomy. This is done based on the co-occurrence of each tag with a subset of tags called the feature set. To determine the feature set, we present a solution based on the idea of Laplacian score for feature selection [6]. Next, related tags are identified by calculating the distance between the corresponding probability distributions. For this purpose, we propose a new distance measure based on the well-known Jensen-Shannon Divergence (JSD). The new measure, called Adapted Jensen-Shannon Divergence (AJSD), can efficiently deal with fluctuations in the samples from which the probability distributions are created. We experimentally evaluated the proposed approach and compared it to a common method for tag relatedness based on the cosine similarity. The results are promising and show the advantage of our approach.

Section 2 surveys related work. In section 3, a definition for folksonomies is provided. In section 4 the proposed approach is presented. Section 5 shows experimental results. Section 6 provides a conclusion and discusses future work.

2   Related Work

Tag relatedness is essential component for applications that depend on mining knowledge from collective user annotations. Conventionally, a tag relatedness measure is used to create the tag dissimilarity matrix, which is used in a next step as input for a clustering algorithm to identify related tag groups.

The work in [3] proposes a tag relatedness measure which is based on tag co-occurrence counts. In that approach, the co-occurrence of each tag pair is computed and a cut-off threshold is used to decide whether two tags are related. The cut-off threshold is determined using the first and the second derivatives of the tag co-occurrence curve. Finally, tag clusters are built by using the computed tag similarity matrix as input to a spectral bisection clustering algorithm.

Gemmell et al. [4, 7] propose an agglomerative approach for tag clustering. For that purpose, they presented a tag relatedness measure based on the idea of term frequency-inverse document frequency (TF.IDF). Correspondingly, resources are considered as documents while the tags are considered as terms. Consequently, each tag is represented as a vector of tag-frequency-inversed resource frequency. Finally, the similarity between two tags is determined by the cosine similarity between the tag vectors. For their tag clustering approach, the authors of [8] propose a tag relatedness measure based on tag co-occurrence counts. First, the tags are organized in a co-occurrence matrix with the columns and the rows corresponding to the tags. The entries of the matrix represent the number of times two tags were used together to annotate the same resource. Next, each tag is represented by a co-occurrence vector and the similarity between two tags is calculated by applying the cosine measure on the corresponding vectors. Simpson et al. [9] propose a tag relatedness approach which uses Jaccard measure to normalize tag co-occurrences. After that, the tags are organized in a co-occurrence graph, which is then fed to an iterative divisive clustering algorithm to identify clusters of related tags. The tag relatedness measure presented in [5] is based on the notion of $(\mu, \epsilon) - cores$. Thereby, tags are organized in a graph with the edges weighted according to the structural similarity between the nodes. That means, tags that have a large number of common neighbors are considered related.

The presented works exploit tag co-occurrence counts to derive their tag relatedness measures. Additionally, either a simple threshold for tag co-occurrences [3, 9] or the cosine measure are used to identify similar tags [4, 7, 8]. The work in this paper, aims at addressing two important aspects which are less investigated in literature on tag relatedness. First, tags are dealt with as probability distributions and a new distance measure is proposed based on the well-known Jensen-Shannon Divergence. Second, to best of our knowledge, this work is the first to deal with the problem of feature selection for building tag co-occurrence vectors. In this regard, we propose a solution based on the method of Laplacian score for feature selection and demonstrate its advantage for tag relatedness measures.

3   Folksonomies and Tag Relatedness

A folksonomy F can be defined as a tuple $F = \{T, U, R, A\}$ [10]. T is the set of tags that are contributed by a set of users U to annotate a set of resources R. Two tags $t_1, t_2 \in T$ occur together if they are used by one or more user to describe a resource $r \in R$. This is captured by the assignment relation, $A \in U \times T \times R$. According to this definition a tag can be described in three different kinds of vector space representations: $\mathbb{R}^T$, $\mathbb{R}^U$ and $\mathbb{R}^R$ with respect to each of the three dimensions of the folksonomy T, U and R [11]. In $\mathbb{R}^T$ representation, which is called *tag-context*, a tag is represented as a vector, $v(t) \in \mathbb{R}^T$. The entries of the vector correspond to the co-occurrences of t with the other tags $t' \in T$. In the *user-context* representation each entry of $v(t) \in \mathbb{R}^U$ corresponds to a specific user and represents the number of times in which t was used by that user to annotate some resource. Finally, in the *resource-context* representation an entry of $v(t) \in \mathbb{R}^R$, corresponding to a specific resource, represents the number of times t was used to annotate that resource.

Approaches for tag relatedness use one (or more) of the presented vector space representations to identify related tags. This is done by generating the corresponding vectors and calculating the cosine similarity between them. The importance of each of the mentioned vector space representations differs according to the nature of the folksonomy (narrow vs. broad) and the goal of the tag relatedness task (e.g. retrieval, recommendation, etc.). In narrow folksonomies, e.g. image folksonomies, where a limited user collaboration is observed, tag relatedness approaches are mainly based on the tag-context representation.

4 Our Approach

We propose a tag relatedness approach based on the tag-context representation. The approach consists of two-steps. 1) For each tag $t \in T$ a probability distribution is built based on the co-occurrence of t with a set of tags $T_f \subseteq T$. We call T_f the feature set. 2) For two tags $t_1, t_2 \in T$ their relatedness is determined according to the distance between their corresponding probability distributions. To compute the feature set T_f, we propose a feature selection approach for tag relatedness based on the Lapalcian score (LS) method [6]. To calculate the distance between two tag probability distributions, we apply the well-known Jensen-Shannon Divergence (JSD) [12] and propose an extension thereof called AJSD. The main characteristic of AJSD is its ability to deal with statistical fluctuations in the generated probability distributions.

4.1 Tag Probability Distribution

In folksonomy F, an empirical probability distribution for a tag $t \in T$ can be created by quantifying the co-occurrences with each of the tags of the feature set $f \in T_f$ by counting the number of times $\#(t, f)$ in which t was used together with f to annotate the same resources. We can use this set of counts to create a histogram in the variable f. Then, by normalizing this histogram with the total number of co-occurrences of t with the elements of the set T_f, we obtain the empirical co-occurrence probability distribution $P_t(f)$ for the tag t with the elements $f \in T_f$: $P_t(f) = \frac{\#(t,f)}{\sum_{f \in T_f} \#(t,f)}$.

In this equation $P_t(f)$ represents the value of the distribution at the histogram channel which corresponds to the feature f. The empirical probability distribution of the tag t over the complete set of features T_f is denoted as $P_t(T_f)$.

4.2 Feature Selection for Tag Similarity

Identifying similar tags in a folksonomy is an all pairs similarity search problem (APSS). Given the set of $|T|$ tags and considering that each tag is represented by a d dimensional vector, the naive approach will compute the similarity between all tag pairs in $O(|T|^2.|d|)$ time. In the case of tag-context approach where $d = |T|$ the algorithm will have a complexity of $O(|T|^3)$.

For large folksonomies, performing such computations is impractical. However, the computational cost can be reduced if the tags are represented in reduced vector space, i.e, $\mathbb{R}^{T_f}$ where $T_f \subset T$ and $|T_f| \ll |T|$. A key requirement of the set T_f is that it should have no (or minimal) impact on the quality of the tag relatedness measure. This gives rise to the problem of feature selection for tag relatedness.

Laplacian Score Feature Selection. A simple approach to build the feature set T_f, is to select a subset of the most occurring (frequent) tags in the folksonomy (e.g. [11, 13]). This technique seems to be effective; however, most frequent tags can have uniform co-occurrence patterns with all other tags in the folksonomy. In that case, all tags would be considered related to each other since they will have very similar patterns of distribution over the set of most frequent tags. Therefore, a more sophisticated approach for identifying T_f is required. A possible solution for this problem is provided by the Laplacian score (LS) feature selection method [6]. LS is an unsupervised process for identifying good features for clustering problems. Therefore, it is also suitable for tag relatedness approaches, since the focus is also on finding clusters, i.e., groups of related tags.

The basic idea of LS is to evaluate the features according to their locality preserving power. To achieve that, the data points are organized in a weighted indirect graph in which the nodes correspond to the data points. An edge is drawn between two nodes if they are mutually "close" to each other. Furthermore, the edges are weighted according to the similarity between the connected data points. Now, the importance of a feature can be determined according to which extent it respects the graph structure. Specifically, a feature is considered "good" if and only if for every two data points, which are close based on this feature, there is an edge between these points. This can be formulated as a minimization problem with the following objective function:

$$L_f = \frac{\sum_{ij}(f_i - f_j)^2 S_{ij}}{Var(f)} \tag{1}$$

f_i and f_j correspond to the values of a feature f at the data points i and j respectively, while S_{ij} is the corresponding similarity. $Var(f)$ is the variance of the of feature f. The minimization of the objective function (equation 1) implies preferring features of larger variances. This conforms to the intuition that features with higher variance are expected to have more expressive power.

The feature selection algorithm and estimation for the solution of the objective function are summarized in the following steps (a mathematical justification can be found in [6]):

1) For the set of n data points a nearest neighbor graph is generated. In that graph, an edge between two data points x_i and x_j is drawn if the points are close to each other. That is, x_i belongs to the set of k nearest neighbors of x_j and vice versa.

2) The edges between close nodes are weighted according to a similarity functions. A widely used function is the Gaussian similarity $S_{ij} = e^{-\frac{\|x_i - x_j\|^2}{t}}$, where t is free parameter that can be determined experimentally. Pairwise similarities are then combined in a similarity matrix S.

3) For a feature f that is defined as a vector over the data points let:

$$\bar{f} = f - \frac{f^T D \mathbb{1}}{\mathbb{1}^T D \mathbb{1}} \mathbb{1} \tag{2}$$

$\mathbb{1} = [1...1]^T$ is the identity matrix. $D = diag(S\mathbb{1})$ is a diagonal matrix, the entries of which d_{ii} correspond to the sum of the entries of the column i in S.

4) Let $L = D - S$ be the Laplacian matrix of the similarity graph [14]. The Laplacian score of the feature f is then calculated as:

$$L_f = \frac{\bar{f}^T L \bar{f}}{\bar{f}^T D \bar{f}} \tag{3}$$

Accordingly, the final feature set will contain features with the top scores.

In our case, the data points as well as the features correspond to the tags of the folksonomy. In other words, we consider each element $t \in T$ as a data point, i.e., a multi-dimensional vector $v(t)$. The components of the vector correspond to the complete set of tags and the values correspond to the co-occurrence counts. On the other hand, the features (corresponding also to the tags) are represented as vectors over the data points.

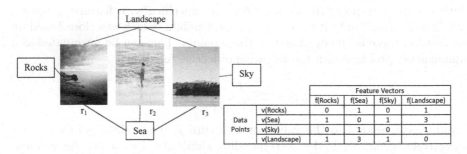

		Feature Vectors			
		f(Rocks)	f(Sea)	f(Sky)	f(Landscape)
Data Points	v(Rocks)	0	1	0	1
	v(Sea)	1	0	1	3
	v(Sky)	0	1	0	1
	v(Landscape)	1	3	1	0

Fig. 1. Simple folksonomy with the corresponding data points and feature vectors

For better understanding consider the simplified folksonomy (user links omitted) shown in Fig. 1. In that example, three resources (images) $R = \{r_1, r_2, r_3\}$ are annotated with tags form the set $T = \{Rocks, Sea, Sky, Landscape\}$. The table on the right shows the co-occurrence counts of the tags. The data points correspond to the row of the table while each column of the same table corresponds to a single feature vector. In the next step, the generated data points and features vectors can be processed according to the LS method to identify the final feature set.

4.3   Distance of Tag Probability Distributions

To determine if two tags are related, the distance between their corresponding empirical co-occurrence probability distributions must be calculated. In the literature, Jensen-Shannon Divergence (JSD) [12] is a widely used measure which has shown to outperform other measures [15]. It is based on Kullback-Leibler Divergence (D_{KL}); however, it is symmetric and has always a finite value. In our previous work, we proposed an extension of JSD called AJSD [13] which deals with the statistical fluctuations due to the finiteness of the sample.

Before discussing the new metric, first, we explain how to calculate JSD between two tag probability distributions. Given two tags $t_1, t_2 \in T$ and the corresponding empirical co-occurrence probability distributions $P(T_f)$ and $Q(T_f)$ respectively, over the feature set $T_f = \{f_1, ..., f_m\}$ (To avoid mathematical cluttering, we will omit the feature set from the notation). The values of P and Q at a specific feature $f_i \in T_f$ are given by $P(f_i)$ and $Q(f_i)$, respectively. Now, the JSD between P and Q is given by:

$$
\begin{aligned}
D_{JSD}(P||Q) &= \frac{1}{2}D_{KL}(P||M) + \frac{1}{2}D_{KL}(Q||M) \\
&= \frac{1}{2}\sum_{f\in T_f}\left(P(f)\log\frac{2P(f)}{P(f)+Q(f)} + Q(f)\log\frac{2Q(f)}{P(f)+Q(f)}\right)
\end{aligned} \quad (4)
$$

Adapted Jensen-Shannon Divergence (AJSD). If, as in our case, the probabilities P and Q are not available, rather we have an estimate of them through a finite sample represented in the form of a histogram for P and a histogram for Q, then the divergence computed on the histograms is a random variable. This variable, under appropriate assumptions, can be used to compute an estimate of the divergence between P and Q using error propagation under a Maximum Likelihood (ML) approach, as illustrated hereafter.

For P and Q consider that the channels at a point (feature) f of the corresponding histograms are characterized by the number of counts k_f and h_f respectively. We define the following measured frequencies $x_f \equiv k_f/n$ and $y_f \equiv h_f/m$, where $n = \sum_f k_f$ and $m = \sum_f h_f$ are the sum of counts for the first and second histogram, respectively. When the number of co-occurrences is high enough (large n and m), the quantities x_f and y_f can be considered to have normal distributions around the true probabilities $P(f)$ and $Q(f)$ respectively. As a consequence the *measured* JSD, denoted as d, can be considered as a stochastic variable defined as a function of the two normal variables x_f and y_f. By substituting x_f and y_f in equations 4 we get:

$$
d = \frac{1}{2}\sum_f\left(x_f\log\frac{2x_f}{x_f+y_f} + y_f\log\frac{2y_f}{x_f+y_f}\right) \quad (5)
$$

The value of this expression is not in general the maximum likelihood estimate of JSD. That is, due to the unequal variances of the terms in the sum. In order to find the maximum likelihood estimate $\hat{d}$ of the divergence we need to proceed through error propagation as in the following steps.

a) Thanks to the normality condition stated above, the ML estimate of $P(f)$ correspond to $x_f = k_f/n$ with the variance given in first approximation by $\sigma^2_{P(f)} = k_f/n^2$. Similarly, the ML estimate of $Q(f)$ is $y_f = h_f/m$ with the variance given by $\sigma^2_{Q(f)} = h_f/m^2$

b) Consider the individual addendum term in the sum expression of equation 5:

$$z_f \equiv x_f \log \frac{2x_f}{x_f + y_f} + y_f \log \frac{2y_f}{x_f + y_f} \tag{6}$$

If the two variables x_f and y_f are independent, the variance propagation at the first order is given by:

$$\sigma^2(z_f) \simeq \left(\frac{\partial z_f}{\partial x_f}\right)^2 \sigma^2(x_f) + \left(\frac{\partial z_f}{\partial y_f}\right)^2 \sigma^2(y_f) \tag{7}$$

$$\simeq \log^2 \frac{2x_f}{x_f + y_f} \sigma^2(x_f) + \log^2 \frac{2y_f}{x_f + y_f} \sigma^2(y_f) \tag{8}$$

$\sigma^2(z_f)$ can be easily calculated by substituting the quantities of step (a) in equation 8.

c) Define the (statistical) precision w_f (to be used later as a weight) as: $w_f \sim \frac{1}{\sigma^2(z_f)}$. Then, the maximum likelihood estimate of the quantity d of equation 5 is given by the following weighted sum:

$$\hat{d} = \frac{\sum_f w_f z_f}{\sum_f w_f}; \text{with } \sigma^2(\hat{d}) = \frac{1}{\sum_f w_f} \tag{9}$$

We use $\hat{d}$ as adapted Jensen-Shannon Divergence (AJSD). Note that AJSD, due to the statistical fluctuations in the samples, gives, in general, values greater than zero even when two samples are taken from the same distribution, i.e. even when the true divergence is zero. However, by weighting the terms according to their (statistical) precision AJSD provides a ranking for the terms that are correlated with the true ranking in a stronger way than JSD.

5   Experimental Results

Dataset. To evaluate the performance of the proposed tag relatedness approach we performed several experiments on a folksonomy extracted from Flickr. The folksonomy corresponds to images taken in the area of London[3]. To avoid bulk tagging we restricted the dataset to one image per user. The final dataset contains around 54,000 images with 4,776 unique tags occurring more than 10 times and a total of 544,000 tag assignments.

---

[3] Dataset and code: https://sites.google.com/site/hmsinfo2013/home/software

Qualitative Insight. For each of the 4,776 unique tags in the dataset, we identified its most related tags. Table 1 shows sample tags (first column) with the corresponding related tags ordered according to their degree of relatedness from left to right. The related tags are obtained by the cosine (COS), JSD and AJSD measures, respectively, and by using the top 2000 Laplacian features. First, one can notice the overlap among the groups of related tags corresponding to the same initial tag. That is, because the tag relatedness measures use the same context, namely the tag-context. Second, we have recognized that, in general, the groups of related tags which are identified by AJSD have a higher cardinality than their counterparts which are identified using JSD and the cosine approaches (e.g. Car, Garden in Table 1). That is, because AJSD generates non-zero similarity even two tags have different sample distributions (section 4.3).

Table 1. Sample tags with the corresponding most related tags

Initial Tag	Method	Related Tags
Airport	COS	Heathrow, KLM, duty, check, airports, runway
	JSD	Heathrow, runway, African, international, ramp
	AJSD	Heathrow, ramp, departures, president, restaurants
Car	COS	automobile, Citroen, driving, rolls, pit, wreck
	JSD	cars, classic, motor, Sunday, Ford, Mini, BMW, driving
	AJSD	cars, classic, Sunday, Ford, Mini, BMW, driving, Caterham, pit
Garden	COS	Covent, jardin, ING
	JSD	flower, gardens, rose, Covent, jardin
	AJSD	flower, gardens, Covent, jardin, pots, Nicholson, rocks
Thames	COS	path, Kingston, river, mud, embankment, Sunbury, shore
	JSD	river, path, Kingston, riverside, Greenwich, ship, embankment
	AJSD	river, water, riverside, path, Kingston, Greenwich, embankment
Music	COS	musician, bands, records, fighting, acoustic
	JSD	concert, rock, stage, festival, pop, jazz, song, records
	AJSD	concert, rock, festival, stage, pop, jazz, Simon, song
Olympics	COS	triathlon, men's
	JSD	Olympic, men's, arena, venue, women's, athlete
	AJSD	Olympic, men's, center, athlete, women's, venue, game, triathlon

To investigate the effect of feature selection, we applied the Laplacian score method on the dataset to identify the most important tags. To generate the tag graph we set the number of nearest neighbors to 10 and used the Gaussian similarity function with $t = 1$.

Fig. 2 shows a plot of the top tags according to LS against the number of occurrences of the tag (frequency). Additionally, the plot illustrates the most frequent tags in the folksonomy (italic). According to LS, the importance of a tag is determined according to its graph-preserving power and not according to its frequency. For example a tag like *potter* which is much less frequent than the tag *england* has a higher Laplacian score, thus, considered as more important. This can be explained since the folksonomy contains images taken in London, thus, it is very likely that most images will be tagged with the word *england* disregarding their contents. Correspondingly, *england* should have a kind of uniform

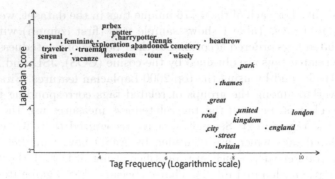

Fig. 2. Tags importance (Laplacian Score) vs. tag frequency

co-occurrence with all other tags in the folksonomy. Therefore, it is less discriminative (has a low LS) than a more specific tag like *potter* which expected to have non-uniform tag co-occurrence distribution.

Evaluation Using WordNet. To provide a quantitative evaluation, we performed additional experiments using WordNet[4]. WordNet has been used by several works as a tool for semantically grounding tag relatedness measures [11, 16, 17]. The goal is to assess how a given tag relatedness measure approximates a reference measure. For our study, we used the Jinag & Conrath (JCN) distance measure as a reference since it showed a high correlation with human judgment [18]. Initially, a gold standard dataset was created by extracting most similar tag pairs from our dataset according to WordNet and by applying JCN measure. After that, the relatedness between the tag pairs of the gold standard is calculated according to our tag relatedness approach as well as the cosine method. To evaluate the effectiveness of LS feature selection, we performed several experiments using different thresholds on the number of top LS features. Furthermore, we compared the performance of LS to frequency based features selection (FRQ).

The performance of the tag relatedness measures is determined according to the average JCN distance over the collection of most related tag pairs as identified by each of the investigated methods. Fig. 3 shows the average JCN distance for the most similar tag pairs (y-axis). The x-axis corresponds to the number of the features. The compared methods include the three measures JSD, AJSD and Cosine (COS) combined with the two features selections approaches, namely the Laplacian score (LS) and the frequency based approach (FRQ). The number of tag pairs which have correspondences in WordNet varies according to the applied similarity method. The average number of recognized WordNet pairs is 975 per method with a standard deviation of 81,6. The standard error in estimating the average JCN distance depends also on the similarity method. However, we observed close values in the range [0.15,0.19].

---

[4] http://wordnet.princeton.edu/

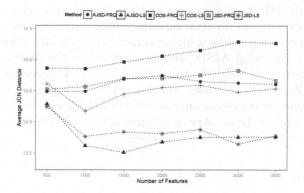

Fig. 3. Average JCN

LS leads to lower average JCN distance than FRQ for all similarity measures and disregarding the number of features (Fig. 3). Moreover, LS enables reducing the dimension of co-occurrence vector/probability distribution while preserving the quality of the identified similar tag pairs. For instance, a minimum JCN distance can be achieved when the top 1,500 Laplacian features (around 31% of total unique tags) are used to perform the calculation. Finally, regarding the distance measures, AJSD produces shorter JCN distances than JSD which in turn performs better than the cosine measure (Fig. 3).

Since the distributional properties of the investigated measures can be different, we followed the evaluation method described in [16]. In this approach, the performance of two tag relatedness measure can be compared according to the order of the ranks of the set of most similar tag pairs generated by each of them. This is done by calculating the correlation between the rankings of each tag relatedness approach and the corresponding rankings using WordNet. A suitable measure is provided by *Kendall* τ rank correlation coefficient.

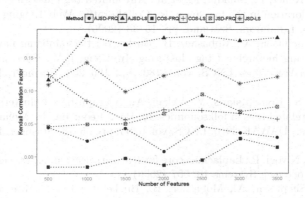

Fig. 4. Kendall Correlation

The performance of the tag relatedness measures based on Kendall correlation is in correspondence with our observations when JCN is used for the evaluation. AJSD combined with LS provides a higher correlation with WordNet than JSD and COS (Fig. 4). By Using AJSD, we can even reduce the dimension of the probability distribution to 80% (the top 1,000 LS tags) while getting the best correlation with WordNet. Moreover, the frequency features selection have a much negative impact on the cosine approach. COS-FRQ is negatively correlated with WordNet as long as the number of features is below 3,000. In contrast, LS leads to a positive correlation factor in all cases.

6   Conclusion

In this paper, a tag relatedness approach based on the Laplacian score feature selection (LS) and an adaptation of Jensen-Shannon Divergence (JSD) was presented. LS allows reducing the dimension of tag co-occurrence vectors without affecting the quality of the applied distance measure. The adapted JSD measure (AJSD) discovers tag pairs of smaller WordNet (JCN) distances and of higher correlation with WordNet than the original JSD measure. Furthermore, both AJSD and JSD performs better than cosine measure. In future work, we will work on improving the performance of our approach by determining the best parameter values for LS. Also, we aim at evaluating the performance of our approach by integrating it into a tag recommendation system.

References

1. Vander Wal, T.: Explaining and showing broad and narrow folksonomies (June 2005), www.vanderwal.net/random/entrysel.php?blog=1635 (accessed July 30, 2013)
2. Bischoff, K., Firan, C.S., Nejdl, W., Paiu, R.: Can all tags be used for search? In: Proceedings of the 17th ACM Conference on Information and Knowledge Management, CIKM 2008, pp. 193–202. ACM, New York (2008)
3. Begelman, G., Keller, P., Smadja, F., et al.: Automated tag clustering: Improving search and exploration in the tag space. In: Collaborative Web Tagging Workshop at WWW 2006, Edinburgh, Scotland, pp. 15–33 (2006)
4. Gemmell, J., Shepitsen, A., Mobasher, B., Burke, R.: Personalizing navigation in folksonomies using hierarchical tag clustering. In: Song, I.-Y., Eder, J., Nguyen, T.M. (eds.) DaWaK 2008. LNCS, vol. 5182, pp. 196–205. Springer, Heidelberg (2008)
5. Papadopoulos, S., Kompatsiaris, Y., Vakali, A.: A graph-based clustering scheme for identifying related tags in folksonomies. In: Bach Pedersen, T., Mohania, M.K., Tjoa, A.M. (eds.) DAWAK 2010. LNCS, vol. 6263, pp. 65–76. Springer, Heidelberg (2010)
6. He, X., Cai, D., Niyogi, P.: Laplacian score for feature selection. Advances in Neural Information Processing Systems 18, 507 (2006)
7. Gemmell, J., Shepitsen, A., Mobasher, B., Burke, R.: Personalization in folksonomies based on tag clustering. Intelligent Techniques for Web Personalization & Recommender Systems 12 (2008)

8. Specia, L., Motta, E.: Integrating folksonomies with the semantic web. In: Franconi, E., Kifer, M., May, W. (eds.) ESWC 2007. LNCS, vol. 4519, pp. 624–639. Springer, Heidelberg (2007)
9. Simpson, E.: Clustering Tags in Enterprise and Web Folksonomies. HP Labs Techincal Reports (2008)
10. Hotho, A., Jäschke, R., Schmitz, C., Stumme, G.: Information retrieval in folksonomies: Search and ranking. In: Sure, Y., Domingue, J. (eds.) ESWC 2006. LNCS, vol. 4011, pp. 411–426. Springer, Heidelberg (2006)
11. Cattuto, C., Benz, D., Hotho, A., Stumme, G.: Semantic grounding of tag relatedness in social bookmarking systems. In: Sheth, A.P., Staab, S., Dean, M., Paolucci, M., Maynard, D., Finin, T., Thirunarayan, K. (eds.) ISWC 2008. LNCS, vol. 5318, pp. 615–631. Springer, Heidelberg (2008)
12. Manning, C., Schütze, H.: Foundations of statistical natural language processing. MIT press (1999)
13. Mousselly-Sergieh, H., Egyed-Zsigmond, E., Gianini, G., Döller, M., Kosch, H., Pinon, J.M.: Tag Similarity in Folksonomies. In: INFORSID 2013 (May 2013)
14. Chung, F.R.: Spectral Graph Teory, vol. 92. Amer Mathematical Society (1997)
15. Ljubešić, N., Boras, D., Bakarić, N., Njavro, J.: Comparing measures of semantic similarity. In: 30th International Conference on Information Technology Interfaces, Cavtat (2008)
16. Markines, B., Cattuto, C., Menczer, F., Benz, D., Hotho, A., Stumme, G.: Evaluating similarity measures for emergent semantics of social tagging. In: Proceedings of the 18th International Conference on World Wide Web, pp. 641–650. ACM (2009)
17. Srinivas, G., Tandon, N., Varma, V.: A weighted tag similarity measure based on a collaborative weight model. In: Proceedings of the 2nd International Workshop on Search and Mining User-Generated Contents, pp. 79–86. ACM (2010)
18. Jiang, J.J., Conrath, D.W.: Semantic similarity based on corpus statistics and lexical taxonomy. arXiv preprint cmp-lg/9709008 (1997)

User Intentions in Digital Photo Production: A Test Data Set

Mathias Lux, Desara Xhura, and Alexander Kopper

Klagenfurt University, Klagenfurt Austria
{mlux,dxhura,akopper}@itec.uni-klu.ac.at

Abstract. Taking a photo with a digital camera or camera phone is a process triggered by a certain motivation. People want for instance to document the progress of a task, others want to preserve a moment of joy. In this contribution we present an openly available dataset with 1,309 photos along with annotations specifying the intentions of the photographers. This data set is the result of a large survey on Flickr and shall provide a common basis for joint research on user intentions in photo production. The survey data was validated using Amazon Mechanical Turk. Besides discussing the process of creating the data set we also present information of the structure and give statistics on the data set.

Keywords: User Intentions, Digital Photos.

1 Introduction

Researchers in multimedia information systems, visual information retrieval and information retrieval in general have lately put more and more emphasis on research regarding users' context. A common definition of context is: *Context is any information that can be used to characterize the situation of an entity. An entity is a person, place, or object that is considered relevant to the interaction between a user and an application, including the user and applications themselves.* [1]. A user's intention – defined as *a thing intended; an aim or plan*[1] – therefore is part of a user's context.

In multimedia information systems user intentions are manifold. In search scenarios users might want to find multimedia data to gain knowledge, or to entertain themselves. In publishing scenarios users might intend to communicate ideas or share feelings with others. To learn about intentions, the users have to answer *why* they want to search, share, or store a video or image. Figure 1 shows two images. Image (A) has been taken to preserve a bad feeling. The photographer noted: "because i [sic] was feeling sad at that time and everything seems as sharp and hard to me as the endins [sic] of this plant.". Image (B) on the other hand was taken for a functional reason. The photographer claimed: "I'm origami folder, and I took this photo to archive my work and share it with other origami folders."

---

[1] Oxford Dictionaries, http://oxforddictionaries.com/

C. Gurrin et al. (Eds.): MMM 2014, Part I, LNCS 8325, pp. 172–182, 2014.

(A)   (B)

Fig. 1. Two sample photos from our test data set taken with different intentions

In this sense we have created a data set, where 1,309 images, shared on the internet, have been annotated by their owners to indicate why these images have been taken. The images were randomly selected from the Flickr web site and their publishers have been contacted to take part in a survey. An additional, crowd-sourced verification step was done with the help of Amazon Mechanical Turk. The data set is publicly available for scientific use under Creative Commons Attribution License[2]. Note at this point that this is not a test data set in the common sense of multimedia retrieval. There are neither queries and topics given for the data set, nor can it considered being a ground truth. Its value is (i) its nature of being a first data source for research on user intentions in multimedia, and (ii) that the data set provides a basis common to different research groups due to its open nature. Its nature is comparable to the infamous *AOL search log data*, where also no topics were given, but the data set was appreciated (in terms of availability of data not in terms of releasing it without asking the users) by the research community. However, in our case we asked the photographers for permission to release the data.

This paper describes the data set starting with a short overview on related work and research on user intentions. Then the acquisition process is outlined and basic statistics and information on the data set are given. We conclude the paper with a discussion on the impact of the data set and give an outlook on future work.

2 Related Work

Data sets for multimedia retrieval and computer vision have quite a long history, as it is commonly agreed that building on each others research results can only work if methods and data are made available. A discussion on the Corel data set, which was employed often in sub sets, and its implications are presented

---

[2] Note that the URL is not given to the double blind review process.

in [2]. Today, a well-known and well-received data set is for instance the MIR-Flickr [3] data set. Since 2010 it provides 1,000,000 images from Flickr along with metadata including tags, title, license and EXIF. Other examples – just to pick a few out of many – are the Caltech-256 Object Category Data set [4], which consists of more than 30,000 images in 256 categories, and the PASCAL data set, which was developed for the PASCAL Visual Object Classes (VOC) Challenge [5].

The problem of capturing the intention of multimedia information system users is diverse, so different approaches have been tried. A preliminary survey on the creation of videos has been presented in [6]. Similarly [7], [8], [9] investigate the intentions people have for capturing photos with phone cameras and [10] investigates intentions for capturing photos independently from the camera used. Intentions for watching online videos have been investigated in [11]. As a part of a survey on user (sub-)groups in multimedia information systems, the goal-directedness of users is investigated in [12]. User intentions for searching images are discussed in [13], where also a taxonomy of user intentions for image retrieval is presented. A taxonomy on intention classes for online video search is discussed in [14]. An application of the research on user intentions for image search is discussed in [15], where the result view of Flickr is adapted to the automatically detected search intention class.

3   Methodology and Acquisition

To collect the data set, we developed an online survey tool, which is able to (i) download recently uploaded images along with the associated metadata from Flickr, (ii) create a questionnaire for every image and (iii) invite the photographers to the survey to fill in their individual questionnaire.

In the survey, the photographers were asked

1. if their image and associated metadata could be used for non-commercial, scientific research,
2. to give additional tags describing the image content, and
3. to provide information about their intentions for taking the image.

To capture the photographers' intentions we asked the photographers to write a free text description on their motivation for taking the image and to rate their intention on some predefined intention classes.

To determine the predefined intention classes we analyzed the results presented in [7], [8], [9] and [10]. What all these studies have in common is that they distinguish between images captured for emotional or functional reasons, between images captured for personal use, to be shared between a known group of people and those to be made public, as well as between images produced for archiving memories or to be kept for a short time. This led to a list of five main intentions,

- preserve an emotion
- support a task
- recall a situation
- share with family and friends.
- publish online

We used the class *preserve an emotion* as a mean to evaluate if the given ratings were not randomly assigned. This was done by decomposing this class in two contradictory subclasses *preserve good feeling* and *preserve bad feeling*. The resulting six predefined intentions were presented to the photographers with the following textual descriptions:

- I took the photo to capture the moment or recall a specific situation later on.
- I took the photo to preserve a good feeling (luck, joy, happiness etc.).
- I took the photo to publish it online.
- I took the photo to show it to my family and friends.
- I took the photo to support a task of mine (archive or document work or task, communicate work progress etc.).
- I took the photo to capture a bad feeling (sadness, anger, depression etc.).

The photographers were asked to rate these intentions based on a five point Likert scale including {*strongly disagree, disagree, neutral, agree, strongly agree*}

The survey was active from June to September 2011. During this time we downloaded 17,119 images and sent 13,583 invitations to photographers. Out of those, 1,309 were completed, which results in a return rate of 9,6%. The invitations were delivered using the public Flickr API and posted as comments to the images. For this task Flickr user accounts were created. Flickr's anti spam restrictions deleted these accounts three times.

To increase the quality of the collected data and to gather more information about the images, we validated the results using the Amazon Mechanical Turk[3] (AMT) marketplace, where people, called AMT workers or *turkers*, fill out surveys or solve small tasks, called human intelligence tasks or *HITs*, for a small amount of money. For each image, a HIT was created, which was then presented to 5 different turkers. In the HIT *turkers* were shown the image. In a first step they had to remove tags that have no relation to the image content, then add additional tags that describe the picture and then rate the degree of manipulation of the image (natural to artifical image). After that, the free text description of phototgrapher's intention was displayed. The turkers had to rate the readability and if an intention can be inferred from it. Also, turkers had to consider the free text description and think about why the photographer captured the photo. Then they had to rate the same six predefined intention classes like the photographers did. The turkers did not see the intention ratings by the photographers.

---

[3] http://www.mturk.com

The creation of the HIT was an iterative process including several pretests. A first offline pretest employed a convenience sample of 5 people for the comprehensibility of the HITs' questions. Further pretests were undertaken on AMT with actual turkers.

The evaluation of the data set was conducted in February 2012 and lasted one month. 6,545 HITs (five for each photo) were successfully completed by 177 turkers. By manual quality control (i) 321 HITs were rejected and republished for other workers to complete and (ii) eight Turkers were blocked due to their bad working performance. The completion of a HIT was rewarded with 0.05$. In total, we spent about 360$ for the validation including the 10% fee issued by AMT. Expenses for additional pretests are not included in this calculation.

To maintain quality in the results of turkers, (i) a large portion of the HITs were reviewed manually, (ii) each HIT was completed by five different turkers and (iii) turkers could only work on HITs if their approval rate of turkers was 95% and above and they had at least 100 HITs approved in their history as AMT workers.

4   Data Set

The resulting data set consists of 1,309 samples. Each sample contains information about the image collected from three main sources: (i) taken from Flickr's API including EXIF metadata, (ii) added from the photographer in the course of the survey and (iii) added by the turkers in the HITs. An example for the

Fig. 2. Sample image from the data set. The photographer described the intention for taking the photo as "a reminder of the beautiful Island were [sic] my father came from".

Table 1. Example of a data item from the data set giving the rating on the image shown in Figure 2. A value of -2 corresponds to *strongly disagree* on the Likert scale, while a value of 2 denotes *strongly agree*.

	Photogr.	Turkers
Recall situation:	2	2, 0, 1, 0, 2
Preserve good feeling:	2	-2, 1, 0, 0, 1
Publish online:	2	0, 0, 0, 1, 2
Show to family & friends:	2	1, 2, 1, 1, 0
Support task of mine:	0	-2, 1, 1, 1,-2
Preserve bad feeling:	-2	0,-2, 0,-2,-2
Degree of manipulation:	-	-1,-2,-1, 0,-2
Readability:	-	-2, 2, 0, 0, 2
Infer intention:	-	0, 2, 1, 0, 0

ratings of an instance from the data set is given in Table 1. The ratings were given to the image shown in Figure 2.

Using the IP addresses of the survey participants and turkers logged by our web server, we were able to assign locations to survey participants and turkers and therefore, to get a rough idea about the originating country. The survey participants – the actual photographers – are spread over 95 different countries. Around 38% of the participants were from English speaking countries like USA, UK and Australia.

In contrast to the widespread distribution of photographers from all over the world, the majority of turkers – the people doing the validation on AMT – were from India and only a small percentage from other countries. Figure 3 gives an overview on the absolute number of participants from the six top countries (on top) and an overview on the turkers' locations (bottom). A trend of an increase of Indian turkers on AMT was already noticed by Ross et al. [16] in 2009. They observed that the share of Indian workers went from 5% in November 2008 to 36% in November 2009, so the distribution of turkers in our survey is not too unusual.

A first and pressing question was to what degree the employed intention classes were redundant. Therefore we investigated if the 6 classes were correlated in a pair wise manner. Table 2 shows the correlation matrix. Most interesting correlations are to be found between the intentions *show to family and friends, recall situation* and *preserve good feeling*, and that the highest correlation is between *preserve good feeling* and *recall situation* with a value of 0.45. The rest of the correlations coefficients are too small to talk of a reasonable correlation. However, the actual values indicate that with the given data sets the 6 classes of intentions are not pair wise redundant and therefore, cannot be removed.

4.1   Inter Rater Agreements

With the validation on AMT by 5 turkers for each instance the question whether the turkers agree is obvious. For quantizing inter-rater agreement we chose

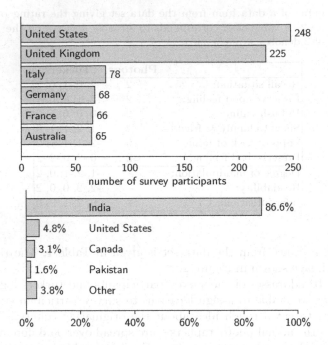

Fig. 3. Locations of the survey participants (photographers, top graph) and the turkers (workers on AMT in the validation step, bottom graph)

Table 2. Intentions Correlation Matrix

Attr.	recall	good	pub.	show	task	bad
recall	1	**0.45**	0,01	**0.29**	-0.08	-0.05
good		1	0.04	**0.29**	-0.05	-0.01
pub.			1	0.19	0.21	0.01
show				1	-0.08	-0.08
task					1	0.14
bad						1

Krippendorff's Alpha α, specifically the R implementation provided by the *irr* package.

Krippendorff's Alpha α is a reliability coefficient that measures the agreement between raters by taking into consideration the agreement for randomly assigned ratings. The following formula describes how the α coefficient is calculated:

$$\alpha = 1 - \frac{Observed\ disagreement}{Expected\ disagreement\ for\ random\ assignements} \quad (1)$$

Values very close to 0 indicate that the inter-rater agreement reliability is low as their ratings are very similar to random assignments and values close to 1 indicate a high agreement between raters. Negative values might occur as well, this often indicates that raters are systematically disagreeing with each other. Detailed description of how to calculate the *observed disagreement* and the *random disagreement* can be found in [17].

We used Krippendorff's Alpha α coefficient to investigate in how far the raters agreed on the rating for the six intention classes and compared that to the inter-rater agreement on (i) readability, (ii) possibility to infer intention classes based on the text description, and (iii) the degree of manipulation of the image. Figure 4 shows the histogram of Krippendorff's Alpha α for all 1,309 instances. In the left graph, showing a histogram of α for the six intention classes, α is generally lower with a large part of the values in $[0.0, 0.4]$. The agreement on the other classes is better as there are more values in $[0.4, 0.8]$. Also the plot in Figure 5 shows that the inter rater agreement follows a different distribution for the intention classes and the other ones.

Table 3. Descriptive statistics for α indicating the inter rater agreement for the six intention classes compared to the three other classes

	Intentions	Other
mean	0.1467	0.2321
variance	0.0316	0.0693
minimum	-0.2361	-0.2291
maximum	0.7096	0.9437

Table 3 shows statistics on α. It is easy to see that the turkers' agreement is worse for the intention classes. Of course these ratings can be considered highly subjective as the turkers had to rate the intention of the photographer, which is in many cases only partially expressed in the free text given by the photographers. Still, the inter rater agreement on the other three classes is somewhat discouraging. All in all the inter rater agreement of the turkers is low. Therefore, we assume to take a closer look at data cleaning methods before employing the data set to sort out instances that do not have the necessary information quality.

Table 4 shows statistics about the agreement of the five pretesters in the offline pretest. Compared to the Turkers' results, the participants of the pretest show a higher agreement. This intuitively sounds right as the people participating in the pretest were more likely to do a good job as they were watched and moderated while they were doing the tasks. Therefore, these values give us an

Table 4. Descriptive statistics for α from the pretest

	Intentions	Other
mean	0.5707	0.5104
variance	0.0157	0.1337
minimum	0.4330	-0.0499
maximum	0.7710	0.8571

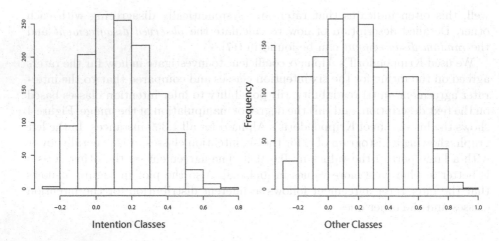

Fig. 4. Histograms of α indicating the inter rater agreement for the six intention classes compared to the three other classes

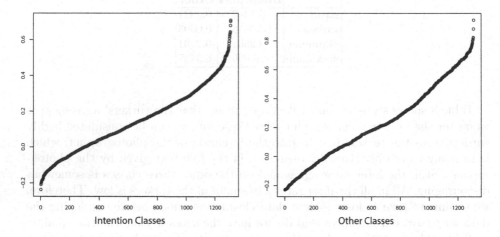

Fig. 5. Plot of the ranked values for α with the rank in the x-axis and α on y indicating the inter rater agreement for the six intention classes compared to the three other classes

idea an agreement on what issues might be possible regardless of the subjectivity in the nature of the task.

We hypothesize three main possible causes for the low inter rater agreement of the turkers.

1. **Subjective interpretation.** The turkers can have a different opinion on the the ratings.
2. **Carelessness.** Cases have been reported in literature that some turkers take short cuts and insert random values. This generates noise and lowers agreement scores.

3. **Judging on an unknown intention.** As it is already a hard task for the photographer to give an abstraction of the cause for taking the photo, a person who does not even know the photographers and their context has a hard time to judge upon intentions just knowing the photo and a few lines of text.

5   Conclusions

In this paper we have presented a data set of 1,309 photos. These photos were collected from Flickr and the photographers participated in a survey that tried to find out why the images have been taken. The data set is to this date – to the best of our knowledge – the only openly available data set[4] dealing with user intentions in multimedia. We consider this as one of the first steps towards joint research in user intentions in multimedia information systems on a common basis. While providing anecdotal evidence on actual intentions for taking photos, also text mining and pattern analysis on the data might lead to insights on why people actually take photos and put them online. Ultimately this understanding will help in providing better tools and algorithms for multimedia search, retrieval, distribution, storage and communication.

While the data set is a great tool to leverage understanding of user intentions in creating digital photos, there are several shortcomings. First of all the data set only includes photos that have already been shared and are available to the public. Hence, the data set is biased towards a sharing intention. Also the actual intention is hard to find, even for the original photographer or uploader of the image. This additional step of abstraction is something users do not appreciate. In face to face interviews we often heard the answer "I don't know". We assume that those, that were not willing to formulate their explicit intention for taking the photo either aborted the study or did not even start it. Still, there are multiple answers, that do not define the intention, but explain the content of the image. Furthermore the data set is rather noisy. Instances with rich information are mixed with instances that are most likely fakes or random answers.

In the near future we want to investigate the data set in full detail. First steps towards using the data set to infer photographers' intentions have shown promising results. Also manual selection of a sub set with richly annotated instances is a next, crucial step.

References

1. Abowd, G.D., Dey, A.K.: Towards a better understanding of context and context-awareness. In: Gellersen, H.-W. (ed.) HUC 1999. LNCS, vol. 1707, pp. 304–307. Springer, Heidelberg (1999)
2. Müller, H., Marchand-Maillet, S., Pun, T.: The truth about corel - evaluation in image retrieval. In: Lew, M., Sebe, N., Eakins, J.P. (eds.) CIVR 2002. LNCS, vol. 2383, pp. 38–49. Springer, Heidelberg (2002)

---

[4] Note that the URL is not given due to the double blind review process.

3. Huiskes, M.J., Lew, M.S.: The mir flickr retrieval evaluation. In: Proceedings of the 2008 ACM International Conference on Multimedia Information Retrieval (MIR 2008). ACM, New York (2008)
4. Griffin, G., Holub, A., Perona, P.: Caltech-256 object category dataset. Technical report, California Institute of Technology (2007)
5. Everingham, M., Gool, L., Williams, C., Winn, J., Zisserman, A.: The pascal visual object classes (voc) challenge. International Journal of Computer Vision 88, 303–338 (2010)
6. Lux, M., Huber, J.: Why did you record this video? an exploratory study on user intentions for video production. In: 2012 13th International Workshop on Image Analysis for Multimedia Interactive Services (WIAMIS), pp. 1–4 (May 2012)
7. Kindberg, T., Spasojevic, M., Fleck, R., Sellen, A.: The ubiquitous camera: An in-depth study of camera phone use. IEEE Pervasive Computing 4(2), 42–50 (2005)
8. Mäkelä, A., Giller, V., Tscheligi, M., Sefelin, R.: Joking, storytelling, artsharing, expressing affection: a field trial of how children and their social network communicate with digital images in leisure time. In: Proceedings of the SIGCHI Conference on Human Factors in Computing Systems, pp. 548–555. ACM (2000)
9. Van House, N., Davis, M., Ames, M., Finn, M., Viswanathan, V.: The uses of personal networked digital imaging: an empirical study of cameraphone photos and sharing. In: CHI 2005 Extended Abstracts on Human Factors in Computing Systems, pp. 1853–1856. ACM (2005)
10. Lux, M., Kogler, M., del Fabro, M.: Why did you take this photo: a study on user intentions in digital photo productions. In: Proceedings of the 2010 ACM Workshop on Social, Adaptive and Personalized Multimedia Interaction and Access (SAPMIA 2010), pp. 41–44. ACM, New York (2010)
11. Lagger, C., Lux, M., Marques, O.: Which video do you want to watch now? development of a prototypical intention-based interface for video retrieval. In: Workshop on Multimedia on the Web, pp. 45–48 (2011)
12. Kemman, M., Kleppe, M., Beunders, H.: Who are the users of a video search system? classifying a heterogeneous group with a profile matrix. In: 2012 13th International Workshop on Image Analysis for Multimedia Interactive Services (WIAMIS), pp. 1–4 (May 2012)
13. Lux, M., Kofler, C., Marques, O.: A classification scheme for user intentions in image search. In: CHI 2010 Extended Abstracts on Human Factors in Computing Systems (CHI EA 2010), pp. 3913–3918. ACM, New York (2010)
14. Hanjalic, A., Kofler, C., Larson, M.: Intent and its discontents: The user at the wheel of the online video search engine. In: Proceedings of the ACM 21 International Conference on Multimedia 2012, Nara, JP (November 2012)
15. Kofler, C., Lux, M.: Dynamic presentation adaptation based on user intent classification. In: Proceedings of the 17th ACM International Conference on Multimedia, MM 2009, pp. 1117–1118. ACM, New York (2009)
16. Ross, J., Irani, L., Silberman, M.S., Zaldivar, A., Tomlinson, B.: Who are the crowdworkers?: shifting demographics in mechanical turk. In: Proceedings of the 28th of the International Conference Extended Abstracts on Human Factors in Computing Systems, CHI EA 2010, pp. 2863–2872. ACM, New York (2010)
17. Krippendorff, K.: Computing krippendorff's alpha reliability. Departmental Papers (ASC) 43 (2007)

Personal Media Reunion: Re-collecting Media Content Scattered over Smart Devices and Social Networks

Mohamad Rabbath and Susanne Boll

rabbath@offis.de, Susanne.Boll@informatik.uni-oldenburg.de

Abstract. With the rapidly growing use of smart phones it becomes easier for people to document social events. Parts of the multimedia content are shared on social networks like Facebook, while others stay private on the users' phones. Friends also add content of the same event to Facebook. The media content of social events is becoming scattered over different accounts and becomes increasingly difficult for the user to re-collect, especially with the absence of the EXIF-header in Facebook. In this paper, we introduce an approach that supports the users in automatically detecting distributed event content after selecting very few photos and videos on the phone. We use the metadata of the content on the phone to increase the initial seeds, and then we exploit a novel face recognition approach using the social context, and a probabilistic fusion model to detect the distributed media content between several social contacts on Facebook to the respective events. Our event content detection significantly outperforms other approaches that give poor result in metadata-unfriendly environment such as Facebook.

Keywords: social networks, Facebook, events, multimedia retrieval, smart phones.

1 Introduction

The last few years witnessed an exponentially growing number of videos and photos taken with mobile phone cameras, due to the increase of camera resolution and the availability of mobile devices almost in every life activity. At the same time, the online-shared percentage of these multimedia contents is growing. The users especially share media contents of the social events like birthday parties, family events or holiday trips when they take videos or photos each with his mobile phone or camera and upload them directly or later to their social network accounts. For example [1] shows that videos and photos sharing apps are in the top-ten list of the most used applications. Facebook constitutes one of the largest sites where these contents are shared, with more than 20 million uploaded videos and 3 billion new shared photos a month [2].

The platforms and systems today, hence, allow to easily share content from personal devices with other users over the Web and Social Media. However, with the increasing sharing activities by instagram for example, there is no easy way to connect private videos and photos from the user's mobile phone, with photos added by friends or the users themselves in social network, e.g., in Facebook.

Till now, most of the work related to organizing and clustering media contents is concerned with a single source like smart personal cameras [7] or dedicated photo sharing sites like Flickr. However, there is no easy way to connect private videos and photos

C. Gurrin et al. (Eds.): MMM 2014, Part I, LNCS 8325, pp. 183–194, 2014.

from the user's mobile phone, with photos shared with and by friends and distributed between different social contacts in social network, e.g., in Facebook.

In our work, we aim to connect the distributed media contents in the social contacts of the user and those remaine private in the mobile device. We propose an approach in which the user can easily select as few as one photo or video and obtain the related content distributed both in her smart phone and between her social contacts in Facebook. We treat videos as photos by extracting representative scenes from them (Section 3.1). The automatic linking between photos in Facebook and mobile phones is very challenging and a face recognition approach is essential in this phase (Section 3.2). We exploit the tags in Facebook but also we use unsupervised social context novel approach to annotate the user's non-tagged photos on the phone and in Facebook. In the next step and based on the recognized people we use a mixture of features including content-based, social-based and if existed the metadata-based features to connect the photos of the same event in the mobile and in Facebook (Section 3.3). The results and the evaluation of our approach is presented in Section 4.

2   Related Work

Although we did not explicitly find related work that directly deals with the problem of connecting the media content distributed between smart phones and social networks such as Facebook, we divide the related work into face recognition as an important part of our approach and multimedia event clustering.

Face recognition is an active research area. [15] developed a scalable approach of using the visual features of the face as bag-of-words to retrieve and rank the faces later when receiving a query. The closest to our face recognition approach is [17] where the visual features of the face and the upper body capturing the clothes are used, however the social context of the user was not taken into consideration. The social context of the user may highly affect the prediction, because the other people appearing in a photo may increase or decrease the certainty of a recognized person. [9] introduced a robot system which is connected online to Facebook, and uses the images of Facebook to recognize the faces of the people dealing with the robot in an interactive way. The work of [14] introduced a supervised approach to use the social context where the person's name, age, and social relation features with the other people are known. However in real life applications all this information is very hard to collect. The closest work that explicitly uses the social context for face recognition is [12], where a propagation graph-based approach using several features including the metadata is introduced. In our approach of exploiting the social context, we consider that both the explicit relation between the users and metadata are completely unknown. We employ the appearance of the users together in different times and the tags in Facebook to annotate the faces in the mobile phone, and the non-tagged photos in the Facebook account of the user, to later use these annotations in detecting the media contents of the same event. Unlike other previous approaches, our recognition approach does not need a lot of data, and starts with small number of tagged photos in Facebook.

A lot of previous work deals with the event clustering of media contents. In the work of [10] features including metadata, textual and visual features are used in a fusion model to cluster the photos of events in Flickr. [13,5] introduced ontology-base approaches to link the events to the media contents. The work of [11] dealt with connecting the photos of the same event in Facebook where the geo-temporal metadata does not exist and the Facebook features were evaluated to link the distributed photos of an event. However it dealt only with photos with manually added Facebook tags, while the majority of photos are still not tagged.

3 Approach

To achieve our goal of connecting the distributed shared content the user in our approach selects as few as one photo or video or very few, and gets the related photos in the phone and those distributed in Facebook between the social contacts. The process consists of several steps: (1) efficiently extracting the representative video scenes from mobile videos (2) face recognition using the Facebook tags to annotate faces in mobile video scenes and photos as well as the non-tagged photos in Facebook by exploiting the social context(3) Detecting the related media content both on the mobile phones and Facebook. We call the detected related photos (and video frames) on the phone extended seeds, as they are easier to detect because the temporal metadata is usually available for the mobile phone photos. We then extend those seeds by the near-duplicate photos in Facebook using only very few high entropy SIFT features. At the last stage we use a probabilistic approach to detect related photos on Facebook by exploiting the recognized faces and other visual and social features in Facebook. In the detection process the recognized faces play a major role, because the features of the areas describing the clothes and the people around the person replace the metadata which is not available in Facebook. Figure 1 summarizes the steps of our approach.

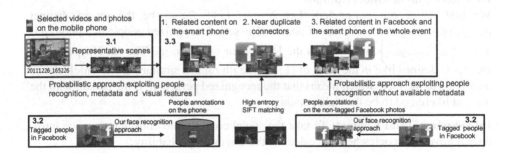

Fig. 1. The summary of the steps of our approach

3.1 Representative Scene Extraction from Personal Videos

In the first step, the user selects one or few videos and photos from on the phone. The representative scenes of the selected videos are automatically detected. Each distinct scene is a set of frame $\{f_r, ..., f_k\}$ and one representative frame is selected for each scene. The approach is light and suitable to be performed on the smart phone. For this task we adopt a simple approach that was originally used by [16]. However, instead of using pixel difference and color histograms we use an approach that is more invariant and robust to luminance and lighting conditions. The boundaries between the frames are detected by calculating the distance between the CEDD (colour edge directivity descriptor) [3] of each consecutive frames. A new scene begins if $\|f_r, f_{r+1}\| = \frac{d_{f_r}^T d_{f_{r+1}}}{d_{f_r}^T d_{f_r}^T + d_{f_{r+1}}^T d_{f_{r+1}} - d_{f_r}^T d_{f_{r+1}}} > T_f$ where d_{f_r} is the CEDD descriptor vector of frame f_r and $d_{f_r}^T$ is the transpose vector of d_{f_r}. T_f is a threshold that we empirically choose as 0.2. Then, from each scene the frame with the maximum details is chosen as a representative image. We select the frame with the maximum edge-map as described by [16].

The result of this step is a set of images representing each video. If the user selects a video on the phone the representative images of that video are used as seeds to find the related photos to the same event.

3.2 Face Recognition to Detect People Related to the Event

As our goal is to find as much related photos from both the mobile phone content and from social media content, we assume that an important factor is the people that participated in the same event. In this subsection, we introduce a novel approach that exploits the *social context* to increase the accuracy of the face recognition. Knowing the people who appear with a person of undecided annotation may increase the probability of a specific annotation over another. The types of relationships between the people are unlike the work of [14] unknown. We use the manually added tags of the user's social zone in Facebook, to annotate the non-tagged photos both on the smart phone and in the Facebook account. If the face belongs to an unknown person (with no previously connecting manually added tag) a unique id is assigned to this person.

Face Recognition Model Notation
We first define our parameters. Let $X = \{x_1, ..., x_N\}$ be the set of distinct people appearing in a collection of photos. The set of recognized people in a photo p is $Y = \left\{ y_1^{(i)}, ..., y_n^{(i)} \right\}$, where y_k^i is the Hit-person i of face y_k. The ith Hit-person of face y_k is defined like in the work of [17] as the ith ranked similar person which can be annotated to the face, which means that the recognized person y_k^0 is the person with the highest likelihood to be the right person.

Introducing Social Context: To take the social context into consideration, we introduce a novel approach where we calculate the Bayesian probability:

$$p\left(\Omega_{x_i} | \Delta F, \Lambda_{x_{r0}}, ..., \Lambda_{x_{rn}}\right) = \frac{p\left(\Delta F | \Omega_{x_i}\right) * \prod_{x_{rk}=x_{r0}}^{x_{rn}} p\left(\Lambda_{x_{rk}} | \Omega_{x_i}\right) * p\left(\Omega_{x_i}\right)}{p\left(\Delta F, \Lambda_{x_{r0}}, ..., \Lambda_{x_{rn}}\right)}$$

$$(1)$$

$$p\left(\Delta F, \Lambda_{x_{r0}}, ..., \Lambda_{x_{rn}}\right)$$
$$= p\left(\Delta F | \Omega_{x_i}\right) * \left(\Omega_{x_i}\right) * \prod_{x_{rk}=x_{r0}}^{x_{rn}} p\left(\Lambda_{x_{rk}} | \Omega_{x_i}\right) + p\left(\Delta F | \overline{\Omega_{x_i}}\right) * \left(\overline{\Omega_{x_i}}\right) * \prod_{x_{rk}=x_{r0}}^{x_{rn}} p\left(\Lambda_{x_{rk}} | \overline{\Omega_{x_i}}\right)$$

$$(2)$$

Where $\Delta F = \{f_1, ..., f_n\}$ is the set of the difference facial and body features introduced in the work of [4], and $p\left(\Delta F | \Omega_{x_i}\right) = \prod p\left(f_j | \Omega_{x_i}\right)$. Ω_{x_i} denotes that the current person belong to the same inter-variation class of person x_i, $\overline{\Omega_{x_i}}$ is the complement inter variation indicating another class of person. $\Lambda_{x_{rk}}$ denotes the co-existence of person x_{rk} in the same photo, where the relationship between person x_{rk} and x_i is unknown. Unlike [4] we use the eigenface value for the face features due to the low dimensionality resulting after applying the PCA in comparison to the block wavelet features. For the body features, we use the color correlogram of 256 bins because the color of the clothes is most effective factor [4,11]. Some of the features $f_j \in \Delta F$ related to the upper body might be missing or filtered due to the overlapping tags; and in this case we replace $p(f_j|\Omega_{x_i})$ by $p(\Omega_{x_i})$.

Initializing Our Parameters: We first initialize each $p(\Lambda_{x_{rk}}|\Omega_{x_i})$. The ratio of the together appearance of two people can be estimated ideally using the manual tags in Facebook, where the smart phone application client communicates with the server using the access token generated by the Facebook app.

$$p(\Lambda_{x_{rk}}|\Omega_{x_i}) = \frac{\#photos(x_{rk}, x_i)}{\#photos(x_i)} \tag{3}$$

where $\#photos(x)$ is the number of photos where x is tagged , and $\#photos(x_{rk}, x_i)$ is the number of photos where x_{rk}, x_i are tagged together. There are still people who are not tagged or rarely manually tagged, and to handle this we first perform face recognition using the the features in the set ΔF , where the social context is first ignored, and then $p(\Lambda_{x_{rk}}|\Omega_{x_i})$ is initialized as following

$$p(\Lambda_{x_{rk}}|\Omega_{x_i}) = \frac{\sum_{ph} p(\Omega_{x_i}|\Delta F_{x_i}(ph)) * p(\Omega_{x_{rk}}|\Delta F_{x_{rk}}(ph))}{\sum_{ph} p(\Omega_{x_i}|\Delta F_{ph_{x_i}})} \tag{4}$$

where $\Delta F_{x_i}(ph) = Argmax_{\Delta F_{ph_k}} p(\Omega_{x_i}|\Delta F_{ph_k})$, meaning the body and face features of the most relevant face to person x_i in the photo ph, if any face is within the 5-Hit of the person x_i (ΔF_{ph_k} denotes the features of face k in the photo ph). If the first Hit of the person did not exceed a certain probability threshold (empirically 50%) a new tag with a unique id is assigned to this people as a first Hit and the first Hit person is shifted to be the second Hit. In this initialization phase, we preserve all T-Hit people who appear in each photo and we will use them in our algorithm as will be described. The set of the T-Hit people in a photo p is $Y_p = \{y_{r0}^{0..T}, ..., y_{rn}^{0..T}\}$

Face Recognition Algorithm

1. Updating the predicted people: Using our current parameters in the Bayesian probabilities 1,2 we update each y_i in the set Y_p of present Hits in the photo p as follows
$$y_i^0 = Argmax_{y_i} p\left(\Omega_{y_i}|\Delta F, \Lambda_{y_{r0}^{0..T}}, ..., \Lambda_{y_{rn}^{0..T}}\right)$$
where $\Lambda_{y_{rk}}^{0..T}$ denotes the presence of person y_{rk} with Hit $h \in [0..T]$. This means all the Hits $0..T$ of the annotated people $y_{rk}^{0..T}$ who appear with the current inter-variation class of person y_i are passed to the probability, and the person with the maximum probability is chosen as the updated y_i^0. We choose the $T = 3$ and therefore if we for example have three people appearing with the current y_i person, the

maximum between $3^3 = 27$ probabilities is chosen. In the first update we use the initialization values explained before. For each y_i not only the 0- Hit is updated but the first T-Hits to use them in updating the other face annotations iteratively.

2. Updating the probabilities: With the new updated set of recognized people $Y = \bigcup_i Y^i$, we update the probabilities $p\left(\Omega_{x_i} | \Delta F\right)$, $p\left(\Lambda_{x_{rk}} | \Omega_{x_i}\right)$ in 4.

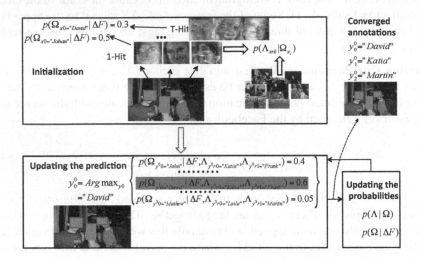

Fig. 2. The steps of our face recognition approach

Both steps are repeated till convergence, as shown in Figure 2 where our approach is applied on an example of photo containing three faces. Both steps can be looked at as EM steps of the Expectation-Maximization algorithm, however the pre-processing in the initialization step, makes our approach semi-supervised.

Evaluating the Face Recognition: To evaluate our face recognition approach we use the users data in Facebook accessed by one of our applications [Smilebooks]. The total amount of photos we have access to is 1 million photos from around 1500 Facebook users and their friends, 40k of these photos are manually tagged with at least one person. We used the 40k manually tagged photos for evaluation. We separate the data collection of the tagged photos according to each user and his friendship circle to sub collections, such that if two people have no social connectivity (they are not friends of each other) they belong to two different collections. In each of the sub collections we apply 10 fold cross validation where each time 1/10 of the collection is used as testing data and the rest is used as training data. We calculated the average accuracy of our face recognition approach exploiting our social context approach. We compare the accuracy of our approach to the baseline (the work of [17]) where it is the state-of-the-art when only face and upper body features are used. The other approaches where metadata is known is not applicable in our case. The accuracy is compared with 1-Hit where only one person is predicted. The performance of the baseline is between 0.59 and 0.63. Our approach starts to improve when more people are on the photos. On average our approach achieves 0.64 slightly better than the baseline when the photo contain two

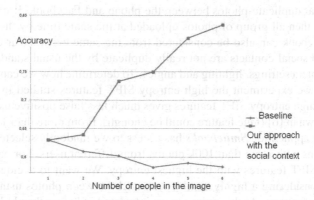

Fig. 3. A comparison between our face recognition approach adding non-supervised social context, and the baseline which uses both the face and the upper body features only

people, while the accuracy reaches 0.83 on average when the photo contains 5 people, a 40% increase of the accuracy as Figure 3 shows.

3.3   Detecting the Media Content Related to the Same Event

Detecting related photos distributed among friends in a social network such as Facebook after selecting as few as one photo or video on the smart phone is our main target. This is a very challenging task especially that the metadata is stripped out in Facebook. Our detection approach makes it easy to the user to browse the content of social events: that which remains private on the smart phone, that which is shared by her on Facebook and that shared by friends on their social account, by using only her smart phone. Our previously described approach in people recognition plays a very important role in the detection process because the manually tagged faces in Facebook are till now below 10%. The process of finding out the related content is divided into three steps.

1. Related Media Contents on the Mobile: First, we obtain the related contents on the phone from the representative video scenes and the selected initial photos. Unlike Facebook photos, the phone contents usually contain useful temporal information. We apply the mean-shift time clustering on the contents made within 1 week of the selected elements, due to the remark of [8] where the vast majority of the media elements of a specific event are uploaded within 5 days of that event. We measure the distance between the time stamps T_{ph_i}, T_{ph_j} of two photos ph_i, ph_j as:

$$\Delta f_t = exp\left(\frac{T_{ph_i} - T_{ph_j}}{K}\right) \tag{5}$$

where we choose K=100. We increase the selected seeds with all the photos and videos that belong to the same cluster. We use the newly detected content of the social event on the phone as a extended seeds to further detect the distributed content in Facebook.´

2. Obtaining High Confidence Near-Duplicate Photos in Facebook: Before applying our probabilistic approach that exploits features from the recognized people, we connect the near duplicate-photos between the phone and Facebook. If one duplicate-photo is found then all group of photos uploaded at the same time by the same social contact on Facebook can also be considered from the same event. Near-duplicate between different social contacts are not really duplicate by the usual standards because of different camera settings, lighting and angles. To determine how we can obtain those near-duplicate we experiment the high entropy SIFT features studied in the work of [6]. Matching high entropy SIFT features gives much less false positive using very few features as shown in [6] (One feature could be enough). From more than 1 Million user photos that our application **Smilebooks** has access to we randomly selected 10k different event albums with more than 100k photos from different users. For each photo we only store 20 SIFT features with the highest entropy. We want to decide the distance threshold of considering a highly confidence match between photos using only those highest entropy features. At this point we are interested only in a threshold that makes very low false positive, because increasing the recall comes in the next step. In our experimental dataset we consider two photos to be true positive if they are in the same album. We denote $DMatch_{ij}$ as the distance between two matched features in two photos. For matching we use the well known FLANN matcher. The matching process is very fast due to matching only 20 Features. We define the function

$$Match\,(p1, p2, T) = \sum_{ij}^{N} \frac{DMatch_{ij}}{N} : DMatch_{ij} < T \qquad (6)$$

which represents the average matches distance $DMatch_{ij}$ for matches with a distance smaller than a threshold T. If no match ij was found with a $DMatch_{ij} < T$ then the function $Match$ is not defined and the two photos $p1, p2$ cannot be matched under T.

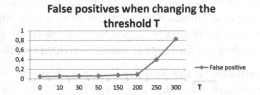

Fig. 4. The false positive when changine the threshold in the function $match$

Choosing a threshold T lower than 150 retrieves less than 6% false positive, while the number increases dramatically when T goes above 200 as shown in Figure 4. We select the match $p2$ in Facebook of each of the content $p1$ resulted from the previous step where $Match\,(p1, p2, 150)$ is defined. This way of using few high entropy SIFT features performas significantly better than matching all SIFT features or using the average distance instead of the $Match$ function.

3. Related Media Contents in Facebook: We employ each of the obtained contents in the previous two steps to get the related Facebook contents. From the previously automatically annotated faces we extract the same features introduced in [11] for the manually tagged photos on Facebook. We denote $\Delta f_v, \Delta f_{tag}, \Delta f_{fr}$ as the vectors of visual, tag-based and friendship-based features respectively. Extracting the tag-based and friendship-based features is now possible only after our face recognition approach that exploits the social-context. As a quick reminder summary of these features: tag-based features is the difference in color, texture and edges histograms of the areas around the face representing the clothes of the same person, because people wear the same clothes during the same event. Friendship-based features between two photos i, j measure the the shared people factor and penalizes the too often together-appearing people in different events. For simplicity we assign $f_{ij} = < \Delta f_v, \Delta f_{tag}, \Delta f_{fr} >$ as the joint vector of the features between two photos i, j. The probability of a photo to be from the same event of a selected representative scene or photo can be written as

$$p\left(event|f_{ij}\right) = \frac{cert_{ij}*p(f_{ij}|event)*p(event)}{p(f_{ij}|event)*p(event)+p\left(f_{ij}|\overline{event}\right)*p\left(\overline{event}\right)} \quad (7)$$

Where $event$ denotes that both photos belong to the same event, and $\overline{event}$ is its complement meaning that the two photos do not belong to the same event. $cert$ is a factor representing the average certainty of the recognized people in the photo $cert_{ij} = \sum_{k=1}^{N} \frac{p_{y_k^0}}{N}$ where N is the number of recognized faces in both photos and $p_{y_k^0}$ is the probability of the 1-Hit recognized person annotating the k face resulted from our face recognition approach. The posterior probability is

$$p\left(f_{ij}|event\right) = \prod_{l=1}^{N} p\left(\Delta f_l|event\right) \quad (8)$$

We use the posterior probabilities in the lookup table introduced by [11], where the features were evaluated in a supervised way, to come up with the optimal initial values. We consider the two photos to be from the same event if the calculated probability in 7 is higher than 60%.

4 Results

To evaluate our approach we perform a lab study to explore the performance of our system in term of the precision and recall for the users who are interested in creating photo products directly using their smart phone.

4.1 Study

To evaluate our system, we invited 15 users who have Facebook accounts and previously used our Facebook application. 9 of the users are males . The youngest participant is 12 years old while the oldest is 72 where the average age is 36. The users are students or from various occupations such as teacher, secretary, electrical engineer, etc. None of them are a photo expert, but all of them are interested in creating photo products such as photo books. Table 1 shows the type of 15 social events each belonging to users U1 to U15, the number of photos and videos remained private on the user smart phone, the number of photos and videos distributed between the user friends in Facebook, and the number of friends who shared the media contents in Facebook.

Table 1. The description of the multimedia contents of our 15 participants

Event		Photos on the smart phone	Videos on the smart phone	Photos distributed in Facebook	Videos distributed in Facebook	Number of people
U1	Vacation	53	5	15	2	2
U2	Vacation	43	1	12	0	4
U3	Vacation	56	4	43	1	3
U4	Vacation	25	2	6	1	3
U5	Vacation	121	2	41	0	3
U6	Vacation	61	4	3	1	2
U7	Vacation	45	4	6	0	4
U8	Wedding	26	1	6	0	3
U9	Festival	35	3	28	0	5
U10	Festival	46	0	56	0	4
U11	Birthday	50	1	16	0	3
U12	Christmas	26	1	6	0	3
U13	Party	34	1	11	0	3
U14	Party	6	0	107	1	9
U15	Party	16	0	82	0	11

4.2 Evaluating the Retrieval of Related Photos

We calculate the recall and the precision of the retrieved distributed multimedia contents of each event. In our evaluation the users of each data collection in Table 1 may select one video or very few photos (maximally three) on the phone. The users use the interface of the application to mark the photos that are indeed in the same event among those retrieved from Facebook and the phone. To measure the recall, the user may then manually select the missing photos from Facebook though the application.

We compare the baseline [11] where only manually tagged photos are used, with our approach after the face recognition. We also compare the recall and precision before and after retrieving the related phone contents, where in the first case initially selected seeds by the user are not extended. In this case the temporal metadata plays no role and the precision/recall is calculated only for the retrieved Facebook contents. In the second case our approach as a whole is evaluated as shown in Figures 5,6.

The baseline performs good only when enough manual tags exist in each album, and therefore our approach increases the recall considerably from an average of 59% in the baseline to 73% of the retrieved Facebook related contents. The recall of our system in retrieving the related contents in Facebook increases also after retrieving related photos on the mobile phone first. This is expected due to more recognized faces and thus more extracted tag-based and friendship-based features. The average recall increases to 81%. We evaluated the precision to see if the recall increase was at the cost of the precision. The precision slightly decreased from 61% in the baseline to 59% on average in our approach without extending the seeds on the phone. This slight decrease is tolerable considering the high gain of the recall. In our system we favour the recall slightly over the precision, because it is easier for the user to ignore and unselect the non-related photos and frames, while it is fairly hard to manually find the distributed contents between several social contacts. In some of the collections the precision is even slightly increased with the increase of the recall, because the new extracted tag-based features increase the accuracy. After extending the related seeds on the phone the precision of retrieving the related Facebook photos decreases slightly to an average of 56% which is an acceptable 8% decrease when comparing to 37% increase of the recall.

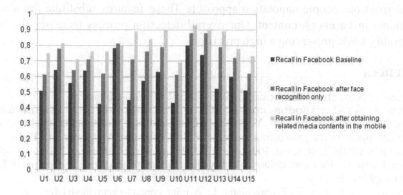

Fig. 5. A comparison in the recall of getting the related photos in Facebook

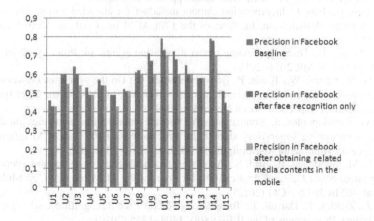

Fig. 6. A comparison in the precision of getting the related photos in Facebook

5 Conclusion

We introduced an approach to detect the distributed media content of an event where the metadata is preserved on the content of the smart phone while missing on the shared content in social networks. We exploit a face recognition approach that takes the social context of the user into consideration which achieves 40% improvement over the baseline depending on the number of people in the photo. The user starts with selecting as few as one video or photo on the smart phone and the distributed media contents of the same event are automatically detected. First, representative frames are extracted from the selected videos. Second, the related media contents are detected from both the mobile phone and Facebook. In this phase we introduced an offline novel face recognition approach where we use both manually added Facebook tags and the social context of the user to annotate the non-tagged photos with the respective people. In the last step we employ a probabilistic fusion approach with visual and people related features resulted from our people annotation approach. These features substitute the absence of metadata in Facebook content. Our overall detection process increases the recall considerably while preserving a high precision.

References

1. http://mobithinking.com/guide-mobile-web-Germany
2. http://royal.pingdom.com/2011/01/12/internet-2010-in-numbers/
3. Chatzichristofis, S.A., Boutalis, Y.S.: CEDD: Color and edge directivity descriptor: A compact descriptor for image indexing and retrieval. In: Gasteratos, A., Vincze, M., Tsotsos, J.K. (eds.) ICVS 2008. LNCS, vol. 5008, pp. 312–322. Springer, Heidelberg (2008)
4. Chen, L., Hu, B.: Face annotation for family photo album management. Int. Journal of Image and Graphics (2003)
5. Gkalelis, N., Mezaris, V., Kompatsiaris, I.: A joint content-event model for event-centric multimedia indexing. In: ICSC, pp. 79–84. IEEE (2010)
6. Ke, Y., Sukthankar, R., Huston, L., Ke, Y., Sukthankar, R.: Efficient near-duplicate detection and sub-image retrieval. In: ACM Multimedia, pp. 869–876 (2004)
7. Li, Z., Luo, H., Fan, J.: Incorporating camera metadata for attended region detection and consumer photo classification. In: Proc. of the 17th ACM Int. Conf. on Multimedia, MM 2009 (2009)
8. Liu, X., Troncy, R., Huet, B.: Finding media illustrating events. In: Proc. Int. Conf. Multimedia Retrieval, ICMR 2011 (2011)
9. Mavridis, N., Kazmi, W., Toulis, P., Ben-Abdelkader, C.: On the synergies between online social networking, face recognition and interactive robotics. In: Proc. of the 2009 Int. Conf. on Computational Aspects of Social Networks (2009)
10. Petkos, G., Papadopoulos, S., Kompatsiaris, Y.: Social event detection using multimodal clustering and integrating supervisory signals. In: Proceedings of the 2nd ACM International Conference on Multimedia Retrieval, ICMR 2012, pp. 23:1–23:8 (2012)
11. Rabbath, M., Sandhaus, P., Boll, S.: Analysing facebook features to support event detection for photo-based facebook applications. In: Proc. of the ACM Int. Conf. on Multimedia Retrieval, ICMR 2012. ACM (2012)
12. Stone, Z., Zickler, T., Darrell, T.: Toward Large-Scale Face Recognition Using Social network Context. Proceedings of the IEEE 98(8), 1408–1415 (2010)
13. Troncy, R., Malocha, B., Fialho, A.T.S.: Linking events with media. In: Proc. of the 6th Int. Conf. on Semantic Systems, I-SEMANTICS (2010)
14. Wang, G., Gallagher, A., Luo, J., Forsyth, D.: Seeing people in social context: Recognizing people and social relationships. In: Daniilidis, K., Maragos, P., Paragios, N. (eds.) ECCV 2010, Part V. LNCS, vol. 6315, pp. 169–182. Springer, Heidelberg (2010)
15. Wu, Z., Ke, Q., Sun, J., Shum, H.-Y.: Scalable face image retrieval with identity-based quantization and multireference reranking. IEEE Trans. Pattern Anal. Mach. Intell.
16. Zeng, X., Xie, X., Wang, K.: Instant video summarization during shooting with mobile phone. In: Proc. of the 1st ACM Int. Conf. on Multimedia Retrieval, ICMR 2011 (2011)
17. Zhang, L., Chen, L., Li, M., Zhang, H.: Automated annotation of human faces in family albums. In: Proc. of the Eleventh ACM Int. Conf. on Multimedia, MULTIMEDIA 2003 (2003)

Summarised Presentation of Personal Photo Sets

Nuno Datia[1,2], João Moura-Pires[2], and Nuno Correia[3]

[1] ISEL, Instituto Politécnico de Lisboa
[2] CENTRIA, Departamento de Informática, Universidade Nova de Lisboa
[3] CITI and DI, Faculdade de Ciências e Tecnologia, Universidade Nova de Lisboa

Abstract. People produce an increasing amount of digital photos to document events. Searching for a specific event can result in more photos than people can handle, making difficult judging their relevance. This paper presents a new algorithm, that summarises a set of photos described by attributes at different concept levels. It addresses the well-known human weakness to deal with large collections of distinct items, by presenting a low cardinality partition set. Each group yields a compact, yet distinct, description. The evaluation, including user tests, shows the algorithm outperforms others in context separation and informative power about the set being summarised.

Keywords: Multimedia summarisation, Human factors, Attribute induction, Clustering algorithm.

1 Introduction

Taking a photo is now a common activity, driven by technological evolutions in digital devices. They transformed common mobile phones into small computers with multimedia capture and production features. Such increasing number of photos cries for a different use of the metadata inserted during their creation, like the timestamp and latitude/longitude coordinates. The features of a personal photo collection pose challenges in the archiving and retrieval, that are different from general-purpose databases [1]. For instance, there is a strong relation between the semantics a user wants and its personal and social context. The actions taken on the collections include querying, sharing and creating derivative artefacts (e.g. albums) [2]. Querying a collection of personal photos implies remembering the spatio-temporal context of a specific event. Remembering is the result of linking, contextualising, and interpreting temporal, spatial, and social cues. Since the context is usually expressed by *when, where, who* and *what* (4Ws), a query is a specification of cues for the 4Ws. When a user wants to access a specific set of photos G, he queries the personal collection repository. Performing the query produces a set A, that can be different from the expected. The intersection between G and A can be empty; A can be a subset of G; G can be a subset of A; or A can partially overlap G. There are several reasons contributing to this. The user may enter wrong temporal cues. It is known [3] that a person may say that an event happened earlier than it did (backward

C. Gurrin et al. (Eds.): MMM 2014, Part I, LNCS 8325, pp. 195–206, 2014.

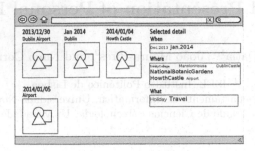

Fig. 1. Example of MSS in a user interface sketch

telescoping), or that it happened more recently (forward telescoping). The user may enter wrong or ambiguous spatial descriptors. For example, Paris can be a city in France or in the USA. The cardinality of A can be large and judging the relationship between G and A can be difficult, as it collides with our capacity to handle large volumes of information [4]. Understanding the context of A is important to reformulate the query, so the next query execution produces a set closer to the expected one. Thus, it is important how A is shown to users.

This work focused in presenting the user a query response of temporal and spatial referenced photos in a compact way, using a novel approach called *Multimedia Short Summary* (MSS). It comprises: *(i)* a partition over a set of photos, *(ii)* a small textual description for each element of the partition, and *(iii)* a selection of an auxiliary grain of detail. By reducing the displayed data, but keeping the necessary information, a user can judge the adequacy of the result without traversing it. Fig. 1 sketches a possible user interface using MSS, for displaying a set of photos documenting a visit to Ireland. On the left, the computed partition over the set of photos is presented, as a 2×3 matrix, where each group has a short, but informative, description. On the right, there is a selected detail for each non empty 4Ws, giving the user a notion of the underlying context of the set. In the example, the selected detail was the month, the place description and the annotated activities in the set. The interface can show hints for the *when*, *where*, and *what* that covers more photos, using a tag cloud approach. The set up follows the principles stated by Shneiderman [5] — *Overview first, zoom and filter, details on demand*. The overview is the partition with a proper description for each group. The selected detail can be used to filter the groups for each available dimension, and is the starting point to drill down and up the context' details.

The rest of the paper is organised as follows. In the next section, we formally define the problem and introduce some notation. In Section 3, the MSS algorithm is presented. Section 4 overview relevant related work. Section 5 presents the evaluation of the algorithm. Finally, we conclude and point out some future work in Section 6.

2   Problem Definition

Let $P = \{p_1, \ldots, p_m\}$ be a set of photos representing personal memories, described by a set of attributes. The attributes are organised in dimensions, denoted by W_i, where $W_1 = when$, $W_2 = where$, $W_3 = who$ and $W_4 = what$. The set of attributes of a dimension W_i is $A_i = \{A_{i1}, \ldots, A_{iN_i}\}$, where N_i is the number of attributes of W_i and A_{ij} represents the attribute j of dimension W_i. Let $D(A_{ij})$ denote the domain of A_{ij} and $\#D(A_{ij})$ its cardinality. Let $D(A_{ij}|P)$ denote the domain of A_{ij} observed in P and $\#D(A_{ij}|P)$ the number of distinct values of A_{ij} in P. The attribute values are textual, human-understandable descriptions of W_i at different conceptual levels. Examples of attributes are *Time-of-day* and *Season-in-year* for W_1, *Country* and *City* for W_2, *Name* and *Kinship* for W_3, and *Activity-type* for W_4. We assume the attributes for W_1 and W_2 are always present. This is a viable assumption as the photo's timestamp is ubiquitous in digital cameras, and the localization data is becoming more reliable in recent smartphones [6]. Attributes for W_1 are generated using known temporal algorithms [7]. Using reverse geocoding facilities from web services[1], it is possible to derive attributes used in W_2 from *WGS84* coordinates.

Let $<_s$ denote the *less specific* relation, where $A_{ij} <_s A_{ik}$ states the semantic of A_{ij} is less specific than A_{ik}. For example, *Month* is less specific describing an instant in time than *Day*, thus, *Month* $<_s$ *Day*. A relation $<_s$ is transitive and defines a partial order over attributes of the same dimension. The information about $<_s$ is stored in a knowledge base (KB). When two attributes have the same specificity, we denote the relation by $=_s$. This happens, for example, when the two are a translation of the same concept in two different languages, like *Year* and *Année*. It is possible that two attributes are incomparable. We denote that relation by inc_s, where

$$A_{ij} \; inc_s \; A_{ik} \Rightarrow \neg(A_{ij} <_s A_{ik}) \wedge \neg(A_{ik} <_s A_{ij}) \wedge \neg(A_{ij} =_s A_{ik})$$

For instance, *Quarter* inc_s *Season*. Given a partial order $(<_s, =_s, inc_s)$ based on the attribute specificity, we can define a compatible total order $(<_s^r, =_s^r)$, named *relaxed specificity*, obeying

$$\begin{cases} X <_s^r Y \; \text{ if } \; X <_s Y \; \vee \; (X \; inc_s \; Y \wedge \#D(X) < \#D(Y)) \\ X =_s^r Y \; \text{ if } \; X =_s Y \; \vee \; (X \; inc_s \; Y \wedge \#D(X) = \#D(Y)) \end{cases} \tag{1}$$

where X and Y are attributes of A_i. For summarisation purposes, we use (1) over a subset of A_i, to produce a non-decreasing ordered sequence governed by the attributes' specificity, denoted by A_i^r, where

$$\forall A_{ij}^r, A_{ik}^r \in A_i^r : A_{ij}^r \leq_s^r A_{ik}^r \Leftrightarrow j < k, i \in [1..4]$$

We define a matrix M_i for each dimension W_i, where the rows represent the photos in P, and the columns are given by A_i^r. Thus, the first column of M_i is

---

[1] e.g. http://www.geonames.org/

a less specific attribute and the last column is one most specific attribute. M_i does not have missing values.

The problem is stated as follows. Given a set of photos P, represented by the set $M = \{M_1, \ldots, M_4\}$, we want to reduce the data presented to the user, keeping the necessary information to understand the context. The summary comprises: *(i)* a partition over a set of photos, *(ii)* a small textual description for each element of the partition, and *(iii)* a selection of an auxiliary grain of detail. Fig. 1 sketches the output of the algorithm.

3   Multimedia Short Summary

In this section we will describe the three components of the MSS algorithm: *(i)* a partition $Q = \{Q_1, \ldots, Q_k\}$ over P, *(ii)* a description of each Q_i using one value for a selected column of each M_i, and *(iii)* a concise detail for P using, for each M_i, all values of a chosen column. From this point forward, and without loss of generality, we consider only the temporal and spatial dimensions. The generation of descriptors for identities and activities is out of scope of this paper. Nevertheless, MSS extends to the 4 dimensions, as long the information is complete for every photo.

3.1   Clustering Multimedia Objects

The clustering algorithm is used to present a summary of P. The summary is a partition Q, limited in size, as the capacity of working memory is a well known bottleneck in human information processing [8]. We consider 16 clusters as the largest accepted value, that enables a succinct presentation of P at most in a 4×4 matrix. When the temporal order of the photos is kept intra- and inter-cluster, the algorithm produces a *temporal dominated* cluster. If the places with equal descriptions are kept together, then Q has *spatial dominance*. The clustering algorithm has three inputs, namely, the limit for the number of clusters (κ), the set M and the dominance. For a better explanation of the algorithm we introduce the *less specific discriminant attribute*, denoted by $A_i^{LSD(\kappa)}$, with at most κ distinct values. It represents the first left most A_{ij} that guarantees $1 < \#D_s(A_{ij}^r|P$, or A_{i1} otherwise. The clustering algorithm starts selecting the $A_i^{LSD(\kappa)}$ for each matrix. The elements from P that share the value for the $A_i^{LSD(\kappa)}$ attribute in the dominant matrix are grouped. Then, for each group, we perform recursively the same action in the other matrix, using its own $A_i^{LSD(\kappa)}$. This creates a two level tree (considering only W_1 and W_2), where each cluster in the top level has one or more clusters in the bottom level. Do notice the order by which matrices are used defines the algorithm dominance. The cluster solution is found in the lowest level where the number of clusters do not exceed κ.

3.2   Descriptors for a Cluster

To achieve an effective summary, it is necessary to provide a cluster description that is: *(i)* short, *(ii)* textual and *(iii)* useful. The description is based on attribute values, since they are textual. To keep a short and useful description, each group is described with just one value of one attribute of each matrix. For each dimension, the chosen value (and consequently, the chosen attribute) has to be as representative as possible and simultaneously, as specific as possible. We decided to sacrifice the correction, keeping the information useful for the users. As an example, consider a set of 200 photos, taken on two consecutive days in 2013. 20 were taken in the Summer and 180 in the Spring. Describing the set as *"Photos from 2013"* is correct. But since its is to vague, it lacks informative power. If we use the description *"Photos from Spring, 2013"*, we correctly describe 90% of the photos, and produce a more specific cue to show the temporal location of the set.

For each cluster, the attributes used to describe it are picked among the most specific one in each matrix, as long as: *(i)* the most common value covers a minimum percentage of the photos in the group (*coverage*), and *(ii)* the most common value is *ratio* times greater than the second most common value. If no attribute satisfies both conditions, we choose the A_{i1}^r. In the experimental results, described in section 5, we used *coverage* = 75% and *ratio* = 3.

3.3   Selection of a Proper Level of Detail

As described in the two previous sections, we partition the original set into a reasonably small number of subsets (≤ 16). The spatio-temporal context is distinct among subsets, so we can briefly describe each one using a textual value from a spatial attribute and another value from a temporal attribute.

In this section we will describe how we find a set of textual descriptions for each dimension, allowing a globally description of P at a proper level of detail. Using a global selected detail has two major advantages: *(i)* shows a brief summary, starting at a given conceptual level, that complements the description of the clusters, and *(ii)* provides a way to filter on demand each dimension separately. Thus, we want a level where the information is detailed and limited, but sufficient to disclose the underlying context. For each matrix M_i, we select the most specific attribute that verifies $\#D_s(A_{ij}^r|P) \leq L_i$, being L_i a threshold for each dimension W_i. The value for L_i considers the nature of each dimension, enabling MSS to choose any attribute as a level of detail.

In M_1, the attributes we use[2] have values related to a specific date. For example, *08AM* is represented differently in two different days. We found that using $L_1 = 12$ adapts MSS to the temporal dispersion of the sets. For example, for one year sets there can be 12 different months. This means the selected attribute is between the *Hour* and *Month* levels.

---

[2] Year, Season, Month, WeekDayName, Day, PeriodOfDay, and Hour.

For M_2, the attributes used[3] show great cardinality differences between the two most specific attributes. The analysis of several geotagged personal set of photos confirmed that they include many places. If we use a small value, e.g. $L_2 = L_1$, MSS commonly choose the *City* as a level of detail. This lowers the detail relevance, since many sets have one city. Thus, L_2 depends on the number of unique values in the most specific attributes. In the experimental results, described in section 5, we set $L_2 = 40$.

3.4 Final Remark

Given a set P with n photos, the MSS output is: *(i)* a partition for P; *(ii)* a short textual description for each group in the partition; and *(iii)* a set of textual descriptions used to globally describe the context of P. This output depends, mainly, on the spatio-temporal context of the photos, not on their number. The MSS time complexity is, in worst case, $\mathcal{O}(n \log(n))$, provided the number of attributes used to describe the photos is much lower than the number of photos of P. Setting the partition is the major contributor to this complexity. MSS is computed fast, even for sets with thousands of photos, making it usable to take part in a user interface.

4 Related Work

Researchers tackled the summarisation of a set of multimedia objects using many approaches, for different purposes. Obrador et al. [9] presents a photo collection summarisation system that uses the one' social context, learnt from its online photo albums. The system uses principles of dramaturgy and cinematography, and helps creating new photo albums for online sharing. In our approach, the knowledge base enables MSS to deliver a summary even when there are no prior knowledge about the user interacting the system, contrary to [9]. Besides, changing the knowledge base adapts MSS to different setups and multiple cultures. Sinha et al. [10], uses temporal, spatial and content based features to select a semantically coherent subset of photos from a larger collection. The query made by the users serves to settle some parameters needed for the selection, namely, the weights to give to features present in the photos. The subset is tailored to summarise three fixed type of events. The MSS is independent of the type of event and, unlike [10], can summarise any multi-event set. Jaffe *et al.* [11], summarise sets of photos by ranking them and selecting the most important. Each photo has a spatial location and a tag entered by the user in Flickr. The first is used to cluster the collection and the latter to rank the photos. They place the collections on a map using representative textual tags on relevant locations in the viewed region. We built MSS to handle personal collection of photos, and its intrinsic characteristics. Thus, it uses features that do not rely on a full tagged collection, as happens in [11]. Adrienko *et al.* [12], suggest a spatio-temporal

---

[3] Continent, Country, City, and Place.

aggregation of photos from different users, based on a regular division of space and on temporal descriptors like day, week or year. Their goal is to analyse large sets of information using summarised data. Their division of time into cycles is similar to what we did in the attributes we use in MSS. However, in [12], the descriptions are omitted. But they are essential to inform the user about the temporal and spatial context of the groups. There is story in people's life that is lost when we omit them. To unveil the underlying context, Dork *et al.* [13], use a set of visualisation widgets. The widgets show the underlying context in a compact way, but also work as a filtering tool. They apply these visualisation widgets to three types of information, including photos. We use the proper level of detail in a similar way.

5   Experiments and Results

We test the sensitivity of MSS to changes in the dominant dimension and in κ, and compare the partitions with other clustering approaches. The user based tests assess the effectiveness of the MSS as a summarisation procedure.

5.1   Evaluating the Cluster Step

The photo sets commonly used in multimedia summary studies lack a strong personal context. We use 39 photo sets collected from personal collections of holiday photos, most of them available at Picasa Web Albums[4]. They average 188 photos, ranging from 10 to 583. The temporal context varies from 1 to 41 days, while the spatial context ranges from 1 to 65 different cities.

Sensitivity to κ and Dominance. We change the value of κ between [2..16]. The results show that MSS produces partitions that vary according to the sets being summarised. For higher values of κ, the partition seems to stabilise. Fig. 2 shows the aggregated results for the 39 datasets. Each individual set follows the tendency depicted. The number of clusters increases in steps, as κ increases. This is an expected behaviour, as κ limits the number of groups. The number of groups found seems to stabilise for $\kappa \geq 12$, independently of the dominance used. We found that MSS with spatial dominance produces less groups.

Comparison with Other Clustering Algorithms. The MSS was compared with other two clustering approaches: *(i)* one that uses attribute induction, identified by AOI [14], and an *(ii)* hierarchical clustering algorithm, named AGNES [15]. We choose them because MSS is both hierarchical and attribute induction based. For comparison purposes, we settle $\kappa = 16$, the maximum value admitted in the work.

AGNES produces 16 clusters for every set[5]. This results in similar descriptions for different clusters, which reduces their value as a summary. AOI creates

---

[4] The dataset can be downloaded from http://goo.gl/JQEVR1
[5] The only exception was one set which has only 10 photos.

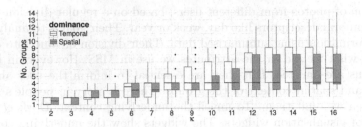

Fig. 2. Number of groups found varying κ

clusters that changed with the spatio-temporal context of P. On average, it produces 10 clusters for single day photo sets and 6 clusters for multiples day data sets. For single day photo sets the number of clusters are over-detailed for a summary, as it often uses the most specific attribute to discriminate the set. It is important to notice that AOI and AGNES do not guarantee temporal order in the partitions. This causes discontinuities in time between clusters, forcing more verbose descriptions.

We compare the inter- and intra-cluster distances for the clustering produced by each algorithm. The dissimilarity measure used was based in the *Gower's General Similarity Coefficient* [15]. For the inter-cluster distance, the dissimilarity is calculated using the medoid of the cluster. Table 1 shows the average dissimilarity for the partitions produced for each algorithm. The inter-cluster dissimilarity is very similar, with slightly advantage for MSS. The higher values for intra-cluster dissimilarity in MSS is a consequence of design, as we choose between less specific attributes the partition cuts. We prefer to have better separated groups than homogeneous ones, as they need to be described differently. The results shows the MSS found groups with similar performance as the clusters found by the other algorithms.

Table 1. Intra- and inter-cluster average dissimilarity

	Intra-cluster	Inter-cluster
MSS	0.1379	0.576
AOI	0.1044	0.572
AGNES	0.0855	0.522

5.2   User Testing

The MSS was tested by a set of volunteers. We wanted to know how long they take to answer a questionnaire about the spatio-temporal context of a set and how good they perform on the answers. We recruited 20 volunteers, mainly computer science students, both graduate and under-graduate. Their ages ranges from 21 and 38, with 40% of women. We choose 12 comparable photo sets based on their context. They average 203 photos, ranging from 18 to 566. The test used the following summarisation strategies: **G-Day** — group by day; **G-City** — group by city; **MSS-T** — MSS with *temporal* dominance; **MSS-S** — MSS with *spatial* dominance. In *G-Day*, the spatial description is the most common city (defaults to the city of the first photo). The temporal description in *G-City* is either a day or a range of days, if there is more than one day in the group. *G-Day* and *G-City* are relevant to stablish a baseline because they use important concepts to people. Currently, people make most of their activities in cities. According to [16], there is a strong relation between memory and city, where the last produces a dense network of encounters, that helps one to map memories to space and time. The notion of day, despite being a natural separator of activities, it is also the common denominator *"social cycle"* for events [17]. Also, users are aware of similar grouping procedures, since they are widely used in commercial programs, like Apple' iPhoto.

Test Design. Each test had 12 steps. In each one, the volunteers saw a photo set[6] summarised by one of the four strategics described above, guaranteeing a different summary in each group of 4 screens. The photo sets in each group shared similarities, namely: *(i)* 4 sets have less than 10 days and 10 cities; *(ii)* 4 sets have multiple months and more than 10 days; *(iii)* 4 sets have more than 6 cities in multiple continents. The interface used was similar to the one sketched in Fig. 1. With *G-Day* and *G-City*, the right side was absent and each group was labelled with the No. of photos, the temporal (mm/dd/yyyy) range and the city (the most common). There is no repetition of sets with different summarisations strategies.

The volunteers were divided in two groups. The first group watched the sets in a sequence, that was inverted for the second group. This would help us to check if the order changes the results. In the beginning of the test, each volunteer saw a small video explaining what we expected them to do.

The questionnaire had 4 questions about time and 5 about space: $Q1$ — "how many days are included in the set?" $Q2$ — "how many months the set has?" $Q3$ — "what is the time frame with more photos?" $Q4$ — "what is the text that best describes *when* the photos were taken?" $Q5$ — "how many countries the set has?" $Q6$ — "how many cities the set has?" $Q7$ — "how many places the set has?" $Q8$ — "what is the place with more photos?" $Q9$ — "what is the text that best describes *where* the photos were taken?" With this questions we can detect if the context is properly captured by the users.

Response Time Analysis. The order of screen presentation does not change the average response time in each screen ($p < 0.01$). Fig. 3 shows the response time variation for each screen, grouped by their type. The y-axis shows the response time in seconds. The x-axis tick' labels are pairs. Each one has the number of days and the number of cities in the dataset summarized in the screen. The results show the volunteers answer quicker when the summary is MSS. In average, they spent 54% more time answering at type *G-City* than at

---

[6] Visually represented by the first photo in the group.

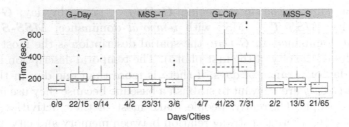

Fig. 3. Response times for each step, grouped by type of summary

type *MSS-S*, and 14% more time for type *G-Day* comparing with type *MSS-T*. It is also noticeable that two tests stand out with higher response times, both of type *G-City*. They share a higher number of cities, which causes an increase in the response times. Since the number of days in those sets is 41 and 7, we exclude the higher number of days as the cause of this increase. The screen with the highest number of cities, 65, uses type *MSS-S*. As we can see, the response time in that case is no different when compared with others with less cities.

The results confirm that limiting the number of groups presented to the user has advantages. Fig. 4 shows the response time for each summary, separating the sets with the number of cities above 16 from the others. We can see the response time with MMS is similar in both dominances. However, type *G-City* shows more often higher response times with higher number of cities ($p < 0.05$).

Response Accuracy Analysis. The users were able to answer to the questionnaire using MSS and, most of the time, the accuracy is the best among the four types. Notice the number of clusters with MSS is, on average, $\frac{1}{3}$ and $\frac{1}{5}$ less than using *G-Day* and *G-City*, respectively. *MSS-S* is always better than *G-City*. *MSS-T* is better than *G-Day* except in two cases. One is question $Q1$ (*how many days the set has*). We foresee this result, because the information available to users in screens using MSS was, most of the times, at a higher conceptual level (e.g. the Month). The other exception is question $Q7$ (*how many places the set has*). What should be reported as a "place" was left for the user to decide. In average, 50% of the volunteers choose the *City* as the conceptual level

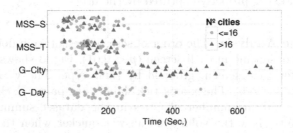

Fig. 4. Response time comparison for each type, separating the screens with more than 16 cities from the others

for "place", except for *MSS-T* screens, where only 25% choose the same level. Fig. 5 shows the adequacy of descriptions for the temporal and spatial context. As we can see, the descriptions are more often correct when the set of photos is displayed using MSS. We assume as acceptable, a term or phrase that covers most of the photos, and is a proper label for the holiday. For example, *"United States"* is a proper label for the spatial context of an holiday in the USA, even tough there is one day spent in Canada, during a visit to the Niagara Falls.

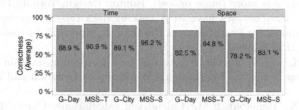

Fig. 5. Proper context description

6   Conclusions and Future Work

In this paper we described the *Multimedia Short Summary* algorithm, that uses concept generalisation to summarise a set of temporal and spatial referenced photos.

The results show the MSS has a proper inter-cluster context separation. It displays descriptions with enough information to give the users cues they need to identify correctly the spatio-temporal context of a set. They take less time understanding the context, comparing to other summarisation algorithms. We show that limiting the groups has advantages, at least for the spatial part of the context. The threshold of 16 groups proved its adequacy in the user test, for producing faster response times. This work relies on cluster's dominance to guaranty coherence in the summary. Selecting the proper dominance need further studies on how to decide the conditions where one is better than other to understand the context.

For future work, the algorithm should handle noisy data in *where* and incomplete information to support *who* and *what*. Since their source is, mostly, user annotations, they can be missing for some photos. We will apply attribute induction of social relation hierarchies (for identities), and rely on domain specific ontologies to hierarchically organise activities. The depth of the *what* hierarchy and the importance of some branches depends on the activities most important to the user. Future research is needed to decide on the attributes to form the matrix.

References

1. Sarvas, R., Turpeinen, M.: Social metadata for personal photography (2006)
2. Hardman, L.: Canonical processes of media production. In: Proceedings of the ACM Workshop on Multimedia for Human Communication: From Capture to Convey, pp. 1–6. ACM (2005)
3. Johnson, E., Schultz, L.: Forward telescoping bias in reported age of onset: an example from cigarette smoking. International Journal of Methods in Psychiatric Research 14(3), 119–129 (2005)
4. Buxton, W.: Less is more (more or less). Buxton Design, Toronto Ontario (2001)
5. Shneiderman, B.: The eyes have it: A task by data type taxonomy for information visualizations. In: Proceedings of the IEEE Symposium on Visual Languages, pp. 336–343. IEEE (1996)
6. von Watzdorf, S., Michahelles, F.: Accuracy of positioning data on smartphones. In: Proceedings of the 3rd International Workshop on Location and the Web, p. 2. ACM (2010)
7. Dershowitz, N., Reingold, E.M.: Calendrical calculations. Cambridge University Press (2007)
8. Baddeley, A.D., Hitch, G.: Working memory. Psychology of Learning and Motivation, vol. 8, pp. 47–89. Academic Press (1974)
9. Obrador, P., de Oliveira, R., Oliver, N.: Supporting personal photo storytelling for social albums. In: Proceedings of the International Conference on Multimedia, pp. 561–570. ACM (2010)
10. Sinha, P., Pirsiavash, H., Jain, R.: Personal photo album summarization. In: Proceedings of the 17th ACM International Conference on Multimedia, pp. 1131–1132. ACM (2009)
11. Jaffe, A., Naaman, M., Tassa, T., Davis, M.: Generating summaries and visualization for large collections of geo-referenced photographs. In: Proceedings of the 8th ACM International Workshop on Multimedia Information Retrieval, MIR 2006, pp. 89–98. ACM, New York (2006)
12. Andrienko, G., Andrienko, N., Bak, P., Kisilevich, S., Keim, D.: Analysis of community-contributed space-and time-referenced data (example of panoramio photos). In: Proceedings of the 17th ACM SIGSPATIAL International Conference on Advances in Geographic Information Systems, GIS 2009, pp. 540–541. ACM, New York (2009)
13. Dork, M., Williamson, C., Carpendale, S.: Towards visual web search: Interactive query formulation and search result visualization. In: WSSP, Madrid, Spain (2009)
14. Han, J., Fu, Y.: Attribute-oriented induction in data mining. In: Advances in Knowledge Discovery and Data Mining, pp. 399–421. American Association for Artificial Intelligence (1996)
15. Kaufman, L., Rousseeuw, P., et al.: Finding groups in data: an introduction to cluster analysis, vol. 39. Wiley Online Library (1990)
16. Pile, S.: Memory and the city. Temporalities, Autobiography and Everyday Life, 111–127 (2002)
17. Zerubavel, E.: Hidden Rhythms: Schedules and Calendars in Social Life. University of California Press (1985)

MOSRO: Enabling Mobile Sensing for Real-Scene Objects with Grid Based Structured Output Learning

Heng-Yu Chi[1,2], Wen-Huang Cheng[2],
Ming-Syan Chen[1,2], and Arvin Wen Tsui[3]

[1] Dept. of Electronic Engineering, National Taiwan University, Taiwan, R.O.C.
[2] Research Center for IT Innovation, Academia Sinica, Taiwan, R.O.C.
[3] Industrial Technology Research Institute, Taiwan, R.O.C.
{hengyuchi,whcheng,mschen}@citi.sinica.edu.tw, arvin@itri.org.tw

Abstract. Visual objects in mobile photos are usually captured in uncontrolled conditions, such as various viewpoints, positions, scales, and background clutter. In this paper, therefore, we developed a MObile Sensing framework for robust Real-scene Object recognition and localization (MOSRO). By extending the conventional structured output learning with the proposed grid based representation as the output structure, MOSRO is not only able to locate the visual objects precisely but also achieve real-time performances. The experimental results showed that the proposed framework outperforms the state-of-the-art methods on public real-scene image datasets. Further, to demonstrate its effectiveness for practical applications, the proposed MOSRO framework was implemented on Android mobile platforms as a prototype system for sensing various business signs on the street and instantly retrieving relevant information of the recognized businesses.

Keywords: Mobile Sensing, Object Recognition and Localization, Grid based Structured Output Learning, Composite Bounding Region.

1 Introduction

With the popularization and advancement of mobile devices, it has become a natural way for people to record life's moments by taking photos and videos using the built-in mobile cameras [1]. In augmented reality (AR) scenarios, users only need to move and point their mobile cameras to something real, say a fashion store in streets, and the information directly relevant to what appearing in front of the cameras can be retrieved and augmented on the mobile screen instantly [2], without requiring the users to step into the store personally. Obviously, the success of above multimedia applications would heavily rely on a key technique, effective object discovery in images of real-world scenes [3].

In particular, the discovery of visual objects in real-world scenes is a challenging mobile vision task. Early literatures mainly work on images with limited viewing angles and little cluttered background [4], as shown in Fig. 1(a) and 1(b).

C. Gurrin et al. (Eds.): MMM 2014, Part I, LNCS 8325, pp. 207–218, 2014.
© Springer International Publishing Switzerland 2014

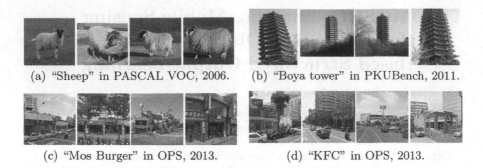

(a) "Sheep" in PASCAL VOC, 2006.   (b) "Boya tower" in PKUBench, 2011.

(c) "Mos Burger" in OPS, 2013.   (d) "KFC" in OPS, 2013.

Fig. 1. Mobile object sensing is a trendy but challenging research topic as reflected by the visual objects becoming diverse in public image benchmarks by years. The images in (a) VOC 2006 [5] and (b) PKUBench [6] usually consist of a salient object and little cluttered background. Relatively, the images in (c)(d) OPS dataset [7] tend to be captured in various viewpoints, positions, scales, and background clutter.

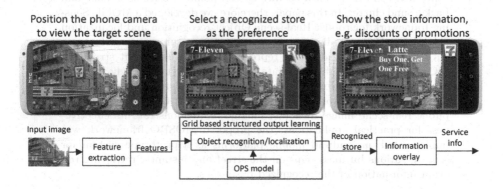

Fig. 2. The illustrative flowchart of the proposed MOSRO framework

Observing the fact that visual objects in mobile photos tend to be captured in various viewpoints, positions, scales, and background clutter, recent researches pay much attention to address such real data in real conditions [6][8][9], e.g., various life objects observed in street-view scenes [7], as shown in Fig. 1(c) and 1(d). In the literatures, sliding window search methods [10][11] are common techniques adopted to find the possible locations of a recognizable object in images. However, the obtained rectangular subwindow often cannot cover an object tightly. The free-shape based window [12][13] thus became more desirable in recent years for locating the visual object with a more precise bounding region. Although the free-shape based methods can produce a more precise window for the target object, they suffer from the inefficiency in computational complexity and would not be suitable for real-time applications.

In responding to the above challenges, in this work, we developed a MObile Sensing framework for Real-scene Object recognition and localization (MOSRO), as shown in Fig. 2, in which we extend the conventional *structured output*

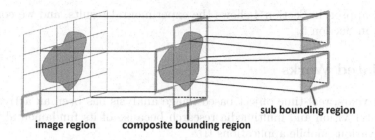

Fig. 3. Illustrations of the proposed grid based image representation, with the definitions for the concepts: "image region", "sub bounding region", and "composite bounding region" (See Section 3 for details)

learning [14] by the proposed grid based image representation as the output structure to learn a prediction of the object's presence and its extent in images directly (cf. Section 3). That is, we formulate the scene images as the input space $\mathcal{X}$ and the corresponding image areas that contain the desired objects as the output space $\mathcal{Y}$. In our approach, each image is represented by a set of non-overlapped image regions, as shown in Fig. 3, and the output space is thus a binary vector of the same size as the number of regions to indicate which image regions constitute the object parts, as detailed in Section 3. As a result, our approach is not only able to precisely localize a recognizable object but it is also computationally efficient because the solution to our tasks is a direct output of the proposed learning framework without extra optimization and decision heuristics. These properties are favored by computation-sensitive mobile vision applications. Our main contributions are summarized as follows.

- We presented a grid based representation as the output structure for object recognition and localization using structured output learning. Going beyond rectangular window based localization, our approach can locate the desired objects with a more precise and flexible composite bounding region. Also, it can efficiently recognize multiple objects at a time and even if a target object is broken into incomplete and disconnected image parts because of foreground obstacles.
- We solved the optimization for grid based structured output learning using branch-and-bound methods as detailed in Section 3.3. Significantly, the proposed approach can efficiently recognize objects in $O(1)$ by Equation(9).
- The experimental results show that the proposed MOSRO framework can achieve better performances than the state-of-the-art methods [13][15][16] on public benchmarks. In particular, our approach can achieve a 212.5% relative improvement compared with F-pLSA [15] according to average F-measure on OPS [7].
- We designed and implemented the proposed MOSRO framework on Android platforms as a real mobile application (cf. Section 4.4).

The rest of this paper is organized as follows. In Section 2, we briefly survey the related literatures, and Section 3 describes the detailed formulation of the proposed approach. Section 4 shows the experimental results, and we conclude our work in Section 5.

2   Related Works

In recent years, real-time object based image analysis has been an active topic in computer vision and multimedia research because of its fundamental role in creating various mobile applications [1].

In particular, sliding window search methods are commonly used in object localization for years [4][10][11]. Authors of [10] proposed an efficient subwindow search (ESS) method for object localization. The set of possible subwindows within a rectangular window is regarded as a node, each of them with a scoring function according to the bounding function. Rather than uniformly sampling for subwindows in ESS, authors of [11] developed a more efficient data-dependent sampling strategy for ESS. The proposed sampling strategy can eliminate most redundant subwindows by using the location of feature points as a prior for subwindow sampling.

Observing the fact that rectangular subwindows may not cover an object tightly, authors of [12] proposed an approach to search a free-shape subwindow for object localization. The feature points in images are classified by the learnt models to determine whether they are positive or not initially. Then the problem is formulated to find an optimal contour which can cover as many positive-score features and as few negative-score features as possible. Authors in [13] presented a branch-and-cut method for efficient region-based object localization. They formulated the problem as a maximum-weight connected subgraph problem by regarding the extracted superpixels as the nodes and their neighbors as the corresponding edges. The bounding area can be arbitrary in contrast to the previous rectangular window [10][11] based methods.

Although the free-shape based methods [12][13] can retrieve a more precise window for the objects, they suffer from the inefficiency in computation complexity. Taking both precise window and real-time efficiency into consideration, in this work, we developed a mobile sensing framework for real-scene object localization based on composite bounding regions with grid based structured output learning. The proposed approach is not only able to locate a visual object precisely but also computationally efficient, which is suitable for real-time mobile applications.

3   Grid Based Structured Output Learning

Structured output learning [4][14] is able to learn a predict function $f : \mathcal{X} \to \mathcal{Y}$ with a given set of input images $\{x_1, x_2, ..., x_n\} \subset \mathcal{X}$ and the corresponding complex output labels $\{y_1, y_2, ..., y_n\} \subset \mathcal{Y}$. In order to define appropriate output labels for our tasks, we first adopted a grid based representation to represent an image as a set of non-overlapped image regions, as illustrated in Fig. 3. For an image, each of the image regions that constitutes a recognizable object is called a sub bounding region (SBR), and the collection of all the SBRs is a composite bounding region (CBR). Without loss of generality, an image region can be as large as being a whole image, or as small as a single pixel, depending on the required precise level.

In this work, therefore, the defined output space is composed of a binary vector that can represent the CBR of a target object as: $\mathcal{Y} \equiv \{(w_1, w_2, ..., w_k) | w \in \{+1, -1\}\}$, each of w_i indicates whether the object is present or not in the corresponding image region b_i. The predict function f can be learnt by a discriminant function $F : \mathcal{X} \times \mathcal{Y} \to \mathbb{R}$ as follows.

$$f = \arg\max_{y \in \mathcal{Y}} F(x, y) \tag{1}$$

$F(x, y)$ is a measurement for the (x, y) pairs and will give a high score to those well-matched pairs.

The discriminant function F can be learnt by minimizing the convex quadratic function in the equation as follows.

$$\min_w \frac{1}{2} \|w\|^2 + C \sum_{i=1}^{n} \xi_i \tag{2}$$

$$s.t. \forall i : \xi_i > 0; \forall i, \forall y \in \mathcal{Y} \backslash y_i : \langle w, \phi(x_i, y_i) - \phi(x_i, y) \rangle \geq \Delta(y_i, y) - \xi_i$$

where $\phi(x_i, y)$ is a joint kernel map, and $\Delta(y_i, y)$ is a loss function regarded as a distance function between the possible output y and the ground truth output y_i. Previous work [14] has shown how to efficiently learn the F using constraint generation method. Initially, a small subset of constraints are used to learn the initial w. At each iteration, the new constraints will be added by finding the most violated constraint y in the equation as follows.

$$\arg\max_{y \in \mathcal{Y} \backslash y_i} \Delta(y_i, y) - (\langle w, \phi(x_i, y_i) \rangle - \langle w, \phi(x_i, y) \rangle) \tag{3}$$

Therefore, given an appropriate joint kernel map and a loss function, the optimization of the discriminant function F is equivalent to the optimization of Equation (3) and thus can be solved efficiently. In Section 3.1 and Section 3.2, we will propose an appropriate joint kernel map and loss function for object sensing respectively. The optimization procedures are presented in Section 3.3.

3.1   The Joint Kernel Map

Given an input image x_i and the corresponding object label y, the joint kernel map, $\phi(x_i, y)$, is defined to be a description of y in x_i. We adopted the bag-of-features scheme [18] for representing $\phi(x_i, y)$ as a regional histogram, cf. Section 4.1. Since the CBR of an object is constituted by SBRs, we thus can obtain $\phi(x_i, y)$ efficiently by fusing (e.g., the linear combination) of the associated SBR histograms if the regional histograms for each image region are precomputed. This property is significant and will be utilized in the optimization procedures in Section 3.3.

Algorithm 1. Branch-and-Bound Optimization

Input: k image regions, bounding function $\hat{g}(\hat{\mathcal{Y}})$, s.t. $\hat{g}(\hat{\mathcal{Y}}) \geq \arg\max_{y \in \hat{\mathcal{Y}}} g(y)$

Output: $y = \{(w_1, w_2, ..., w_k)|w_i \in \{-1, +1\}\}$, $y \in \hat{\mathcal{Y}}$

$\quad \hat{\mathcal{Y}} \leftarrow \mathbf{0} \in \mathbb{R}^{1 \times k}$, for representing a set of possible y

$\quad Q \leftarrow$ empty priority queue

$\quad$**while** $\hat{\mathcal{Y}}$ contains more than one element **do**

$\quad\quad$// *Branch process*

$\quad\quad$select an index i, where $\hat{\mathcal{Y}}(i) = 0$

$\quad\quad \hat{\mathcal{Y}}_1 \leftarrow \hat{\mathcal{Y}}$, $\hat{\mathcal{Y}}_1(i) \leftarrow 1$

$\quad\quad \hat{\mathcal{Y}}_2 \leftarrow \hat{\mathcal{Y}}$, $\hat{\mathcal{Y}}_2(i) \leftarrow -1$

$\quad\quad$// *Bound process*

$\quad\quad push(Q, (\hat{\mathcal{Y}}_1, \hat{g}(\hat{\mathcal{Y}}_1))$

$\quad\quad push(Q, (\hat{\mathcal{Y}}_2, \hat{g}(\hat{\mathcal{Y}}_2)), \hat{\mathcal{Y}} \leftarrow pop(Q)$

$\quad$**end while**

$\quad$**return** $y \leftarrow \hat{\mathcal{Y}}$

3.2 The Loss Function

Given two CBRs, y and $\bar{y}$, the loss function, $\Delta(y, \bar{y})$, is defined to reflect the mismatch between these two regions. The loss function needs to satisfy two criteria. The first one is $\Delta(y, \bar{y}) = 0$ iff $y = \bar{y}$. The second one is that $\Delta(y, \bar{y})$ increases towards 1 (totally mismatched) as y and $\bar{y}$ become more different and decreases towards 0 (perfectly matched) as y and $\bar{y}$ become more similar. Based on the above criteria, we utilized the area overlap [5] to construct the loss function:

$$\Delta(y, \bar{y}) = 1 - \frac{Area(y \cap \bar{y})}{Area(y \cup \bar{y})} \quad (4)$$

where $Area(y \cap \bar{y})$ represents the intersection area between y and $\bar{y}$ and $Area(y \cup \bar{y})$ represents the union area between them. The proposed loss function satisfies the aforementioned two criteria and has a desirable property that it can reflect the compatibility of partial detections during the learning process. Furthermore, the maximum margin learning is suitable for object sensing by the loss function. For those images containing no desired objects, no regions can reach a low loss since $Area(y \cap \bar{y})$ is always 0. For those images containing the desired objects, the correct CBRs will have the lowest loss.

3.3 Optimization Procedures

By applying the constraint generation method during the learning stage, the maximization in Equation (3) will be estimated iteratively for finding new constraints. Since the term $\langle w, \phi(x_i, y_i) \rangle$ does not vary with the most violated y and can be fixed as a constant, we thus rewrite the equation as follows.

$$\arg\max_{y \in \mathcal{Y} \backslash y_i} \Delta(y_i, y) + \langle w, \phi(x_i, y) \rangle \quad (5)$$

As motivated by [10], we apply a branch-and-bound method to solve the maximization in Equation (5). Assuming that an image is subdivided into k image regions, the search space $\mathcal{Y}$ is composed of all possible composite bounding regions y, with each of them has a k-dimensional labeling vector to indicate which image regions constitute the object. Meanwhile, we maintain a priority queue Q, which is comprised of sets of possible composite bounding regions $\hat{\mathcal{Y}} \subset \mathcal{Y}$ and they are ordered according to a corresponding value bounded by the given bounding function $\hat{g}(\hat{\mathcal{Y}})$, as defined below. At each iteration, the set $\hat{\mathcal{Y}}$ with the highest score is popped and split into two subsets $\hat{\mathcal{Y}}_1$ and $\hat{\mathcal{Y}}_2$. The iterative loop will be terminated once the popped set contains only one candidate y, which can maximize Equation (5). The detailed branch-and-bound method is summarized in Algorithm 1.

The upper bound of Equation (5) can be bounded by the summation of the upper bound of the loss function and the upper bound of the joint kernel map. The loss function $\Delta(y, \bar{y})$ can be bounded by finding a minimization of the intersection area and also a maximization of the union area as follows.

$$\hat{g}(\hat{\mathcal{Y}})_\Delta = \arg\max_{y \in \hat{y}} \Delta(y_i, y) = 1 - \frac{\arg\min_{y \in \hat{y}} Area(y_i \cap y)}{\arg\max_{y \in \hat{y}} Area(y_i \cup y)} \tag{6}$$

Since we adopted the bag-of-features method for representing the joint kernel map, $\phi(x_i, y)$ (cf. Section 3.1), we have $\phi(x_i, y) \geq 0$. The joint kernel map can thus be bounded by

$$\hat{g}(\hat{\mathcal{Y}})_F = \arg\max_{y \subset \hat{y}} \langle w, \phi(x_i, y) \rangle = \langle w^+, \phi(x_i, y_{max}) \rangle \tag{7}$$

where w^+ contains only positive values of w and y_{max} is the largest region in $\hat{\mathcal{Y}}$. Therefore, the bounding function $\hat{g}(\hat{\mathcal{Y}})$ can be solved by the summation of the upper bound of the loss function and the upper bound of the joint kernel map:

$$\hat{g}(\hat{\mathcal{Y}}) = \hat{g}(\hat{\mathcal{Y}})_\Delta + \hat{g}(\hat{\mathcal{Y}})_F$$
$$= 1 - \frac{\arg\min_{y \in \hat{y}} Area(y_i \cap y)}{\arg\max_{y \in \hat{y}} Area(y_i \cup y)} + \langle w^+, \phi(x_i, y_{max}) \rangle \tag{8}$$

In the object sensing stage, we need to find an appropriate composite bounding region y that can maximize f in Equation (1). Equation (1) is equivalent to the maximization of a joint kernel map, which is similar to the formulation of $\hat{g}(\hat{\mathcal{Y}})_F$. The maximization can thus be bounded in O(1) by

$$f = \arg\max_{y \in \mathcal{Y}} F(x, y) = \hat{g}(\hat{\mathcal{Y}})_F = \langle w^+, \phi(x_i, y_{max}) \rangle \tag{9}$$

4   Experimental Results

In this section, we validate the proposed framework, MOSRO, by using two public benchmarks, OPS [7] and PASCAL VOC 2008 [17].

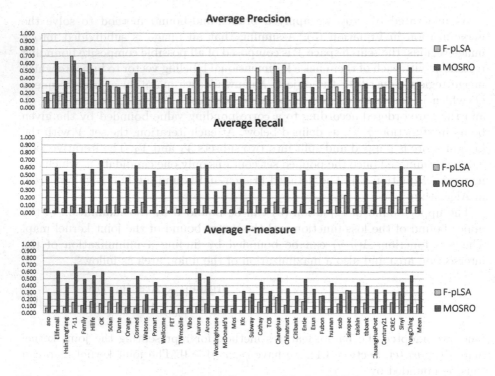

Fig. 4. Performances on pixel-level localization with the OPS data set, in terms of the average precision (AP), average recall (AR), and average F-measure (AF) on OPS

4.1   Experimental Setup

We first adopted a color based descriptor, i.e., the OpponentSIFT [19], to extract visual features in training images since we observed that both colors and local structures are important for real-scene objects. Then the extracted visual features were used to generate a codebook [18] for each object category. Next, each image region in the training data can be represented as a regional histogram. With the determined joint kernel map (cf. Section 3.1) and loss function (cf. Section 3.2), we exploited the structured output learning to obtain the learnt OPS models. We solved the optimization of finding the most violated constraint iteratively by the branch-and-bound algorithm in Section 3.3.

The Benchmarks. In addition to the PASCAL VOC 2008 [17], we utilize a newly developed and more challenging public real-scene image benchmark, the OPS data set [7], as shown in Fig. 1(c) and 1(d), for our evaluation. The OPS data set consists of 4,649 street view images, each of which contains one or multiple business signs and the corresponding labels, i.e., pixel-level masks. In the experiments, there are totally 43 object categories of the business signs included.

Fig. 5. Sample results of the recognized business signs in OPS data set

4.2   Performances on Pixel-Level Localization

In this section, we evaluate the localization performance of our method in pixel-level precision and recall on the real-scene OPS data set. We also compare our method with the state-of-the-art pLSA models [15]. As reported in [15], many of the pLSA variant methods were proposed and evaluated; however, there is no absolute winner. Therefore, we choose the Fergus *et al.*'s pLSA models [15] (abbreviated as F-pLSA hereafter) to be our comparing method, where the number of latent topics for all OPS models is set to eight as suggested in [15].

The localization performances of our method and F-pLSA are evaluated in pixel level, using precision, recall, and F-measure metrics. The precision and recall are respectively defined as follows.

$$\text{Precision} = \frac{|Area(R_{GT} \cap R_{RC})|}{Area(R_{RC})}, \text{Recall} = \frac{|Area(R_{GT} \cap R_{RC})|}{Area(R_{GT})}, \quad (10)$$

where R_{GT} is the ground truth area and R_{RC} is the recognized image portions. As shown in Fig. 4, the experimental results show the average precision (AP), the average recall (AR), and the average F-measure (AF) for each OPS category over its associated testing images. Note that since there is no explicit scheme for picking OPS models from the latent topics for F-pLSA, we chose to select from the eight latent topics of each OPS category one at a time as its OPS model and report the best recognition performance. Therefore, F-pLSA results in Fig. 4 can be viewed as the **upper bound** of the recognition performance that F-pLSA can achieve for each OPS category.

As shown in Fig. 4, the experimental results show that the average ARs of the proposed MOSRO and F-pLSA are 0.493 and 0.080, respectively. Our method can achieve a 514.8% relative improvement. Moreover, the MOSRO outperforms the F-pLSA for all object categories according to the individual AR, and the relative improvement is from a minimum of 115.43% to a maximum of over 2,000%. On the other hand, our method and F-pLSA are competitive according to the average APs, i.e., 0.339 versus 0.327. The result implies that in the same average AP level, the proposed MOSRO can be more powerful to recognize the existence of real-scene objects than F-pLSA. Overall, in terms of the AF,

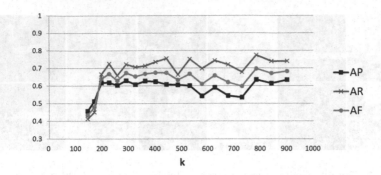

Fig. 6. Performances on varying size (k) of image regions among the selected OPS object categories

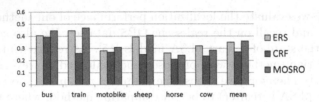

Fig. 7. Performances of our approach on the selected categories of PASCAL VOC 2008, in comparison with two state-of-the-art methods, i.e., ERS [13] and CRF [16]

the results show that the proposed MOSRO outperforms F-pLSA for all object categories and demonstrate the effectiveness of our approach.

Fig. 5 shows sample examples of recognized business signs in the OPS data set. The results show that our method can efficiently recognize multiple objects at a time. Also, our method can recognize the correct extent of a business sign even if the target business sign is broken into incomplete and disconnected image parts because of foreground obstacles, such as trees and poles.

Varying Size of Image Regions. In our approach, there is an adjustable parameter k, i.e., the number of divided image regions, during the structured output learning. Theoretically, the larger k is, the more precise the resulting composite bounding region for an object can be. However, empirically, as shown in Fig. 6, it can be found that both AP and AR are saturated with k over 200. Consequently, k is set to be 200 as the default value in the experiments.

4.3 Evaluations on PASCAL VOC 2008

We also evaluated our approach on another public benchmark, the segmentation data set in PASCAL VOC 2008 [17]. We compared our method against two state-of-the-arts algorithms, i.e., ERS method [13] and CRF method [16]. ERS method is proposed especially on the detection for objects with arbitrarily

(a) Take a photo of the target scene.   (b) Obtain the retrieved result.

Fig. 8. The proposed MOSRO framework deployed on an Android smart phone

non-rectangular shapes. For ERS, we followed the standard implementation in [13]. CRF method detects objects using a global connectivity potential within a standard CRF. We used the same features in [13] and followed the experimental setup in [16]. The PASCAL overlap criterion in [17] is used for the evaluations.

Overall, as illustrated in Fig. 7, the results show that the proposed MOSRO achieves the best performance, especially on the categories of artificial objects, such as bus, train, and motobike. Although ERS method is competitive with our approach on some categories, such as sheep, horse, and cow, the recognition time of our MOSRO is at most 0.1 seconds for a single image, which is much faster in comparison with ERS method that needs to be done in 0.29 seconds in average.

4.4 Mobile Application Demonstration

As shown in Fig. 8, we also developed a running prototype of the proposed MOSRO framework as a mobile sensing application on Android platforms. By using our application, users can move their mobile camera to capture a store-included target scene (cf. Fig. 8(a)) and then obtain the retrieved information of the interested stores (cf. Fig. 8(b)). As shown in Fig. 8(b), our application can currently provide the basic information, such as the store name and the logo, as enclosed within the red rectangle. If users click on the displayed logo, our application will give more detailed information, such as special offers, by connecting to the official business website.

5   Conclusions and Future Work

Observing the fact that visual objects in mobile photos are often captured in various viewpoints, positions, scales, and background clutter, in this work, we developed a mobile sensing framework, MOSRO, for robust real-scene object recognition and localization. Based on the proposed grid based representation, we can thus model the direct prediction from $\mathcal{X}$ to $\mathcal{Y}$ using the structured output learning. The experimental results showed that our approach is not only able to locate the target objects precisely but can also achieve real-time computational performances. In the future, we would like to further investigate along several research directions for improving our approach, such as extensions with online learning to incrementally update the learnt object models at run time.

References

1. Girod, B., Chandrasekhar, V., Grzeszczuk, R., Reznik, Y.A.: Mobile visual search: Architectures, technologies, and the emerging mpeg standard. In: IEEE MultiMedia (2011)
2. You, C.-W., Cheng, W.-H., Wen Tsui, A., Tsai, T.-H., Campbell, A.: Mobilequeue: an image-based queue card management system through augmented reality phones. In: UbiComp (2012)
3. Girod, B., Chandrasekhar, V., Chen, D.M., Cheung, N.-M., Grzeszczuk, R., Reznik, Y., Takacs, G., Tsai, S.S., Vedantham, R.: Mobile visual search. IEEE Signal Processing Magazine (2011)
4. Blaschko, M.B., Lampert, C.H.: Learning to localize objects with structured output regression. In: Forsyth, D., Torr, P., Zisserman, A. (eds.) ECCV 2008, Part I. LNCS, vol. 5302, pp. 2–15. Springer, Heidelberg (2008)
5. Everingham, M., Zisserman, A., Williams, C.K.I., Van Gool, L.: The PASCAL Visual Object Classes Challenge 2006 (VOC 2006) Results (2006), http://www.pascal-network.org/challenges/VOC/voc2006/results.pdf
6. Ji, R., Duan, L.-Y., Chen, J., Yang, S., Huang, T., Yao, H., Gao, W.: Pkubench: A context rich mobile visual search benchmark. In: ICIP (2011)
7. OPS data set, http://mclab.citi.sinica.edu.tw/dataset/ops62/ops62.html
8. Yu, F.X., Ji, R., Chang, S.-F.: Active query sensing for mobile location search. In: ACM Multimedia (2011)
9. Kuo, Y.-H., Lee, W.-Y., Hsu, W.H., Cheng, W.-H.: Augmenting mobile city-view image retrieval with context-rich user-contributed photos. In: ACM Multimedia (2011)
10. Lampert, C.H., Blaschko, M.B., Hofmann, T.: Beyond sliding windows: Object localization by efficient subwindow search. In: CVPR (2008)
11. Yeh, T., Lee, J.J., Darrell, T.: Fast concurrent object localization and recognition. In: CVPR (2009)
12. Zhang, Z., Cao, Y., Salvi, D., Oliver, K., Waggoner, J., Wang, S.: Free-shape subwindow search for object localization. In: CVPR (2010)
13. Vijayanarasimhan, S., Grauman, K.: Efficient region search for object detection. In: CVPR (2011)
14. Tsochantaridis, I., Joachims, T., Hofmann, T., Altun, Y.: Large margin methods for structured and interdependent output variables. J. Mach. Learn. Res. (2005)
15. Fergus, R., Li, F.-F., Perona, P., Zisserman, A.: Learning object categories from internet image searches. Proceedings of the IEEE (2010)
16. Nowozin, S., Lampert, C.H.: Global connectivity potentials for random field models. In: CVPR (2009)
17. Everingham, M., Van Gool, L., Williams, C.K.I., Winn, J., Zisserman, A.: The PASCAL Visual Object Classes Challenge 2008 (VOC 2008) Results (2008), http://www.pascal-network.org/challenges/VOC/voc2008/workshop/index.html
18. Nister, D., Stewenius, H.: Scalable recognition with a vocabulary tree. In: CVPR (2006)
19. van de Sande, K., Gevers, T., Snoek, C.: Evaluating color descriptors for object and scene recognition. IEEE Trans. Pattern Anal. Mach. Intell. (2010)

TravelBuddy: Interactive Travel Route Recommendation with a Visual Scene Interface

Cheng-Yao Fu[1], Min-Chun Hu[2], Jui-Hsin Lai[1], Hsuan Wang[3], and Ja-Ling Wu[1]

[1] Dept. of CSIE, National Taiwan University
[2] Dept. of CSIE, National Cheng Kung University
[3] Dept. of Civil Engineering, National Central University

Abstract. In this work, we propose a convenient system for trip planning and aim to change the behavior of trip planners from exhaustively searching information to receiving useful travel recommendations. Given the essential and optional user inputs, our system automatically recommends a route that suits the traveler based on a real-time route planning algorithm and allows the user to make adjustment according to their preferences. We construct a traveling database by collecting photos taken around famous attractions and analyzing these photos to extract each attraction's travel information including popularity, typical stay time, available visiting time in a day, and visual scenes of different time. All the extracted travel information are presented to the user to help him/her efficiently know more about different attractions so that he/she can modify the inputs to obtain a more favorable travel route. The experimental results show that our system can effectively help the user to plan the journey.

1 Introduction

Besides becoming a millionaire, traveling around the world is the most popular answer when people are asked about their fantasy. Traveling can always relax our mind, increase our knowledge and experience, widen our perspective, and create invaluable memories. With limited money and time, people would like to choose escorted tours planned by the travel agent and arranged to satisfy the average customer demand. Escorted tours are convenient but usually more expensive and with less flexibility than independent tours. With the popularity of internet and world wide web, people tend to search traveling information and plan suitable schedule on their own. However, this task has become more and more difficult due to internet information overflow. Trip planning websites/forums are then established for people to share their traveling experience so that one can find more reliable information and reorganize a new trip plan based on others' experiences. Unfortunately, the discussions on these websites/forums are usually disordered and without proper summarization.

The proliferation of social media websites and mobile devices motivate us to change the behavior of trip planners from exhaustively searching information to receiving useful recommendation. In this work, we propose a convenient system

C. Gurrin et al. (Eds.): MMM 2014, Part I, LNCS 8325, pp. 219–230, 2014.
© Springer International Publishing Switzerland 2014

with user friendly interface for trip planning. The proposed system framework is illustrated in Figure 1. We first construct a traveling attraction database containing famous attractions of each given city (e.g., Seville Cathedral in Barcelona and Eiffel Tower in Paris). These attractions are collected from TripAdvisor and Yahoo!Travel, which are the most two popular travel websites with helpful traveling information shared by experienced tourists [1]. Note that the GPS location of each attraction is also stored in the database. Moreover, photos of each attraction are retrieved from social media websites (e.g. Flickr) based on the attraction name and GPS information. We then analyze these photos to extract each attraction's travel information including popularity, typical stay time, available visiting time in a day, and visual scenes of different time.

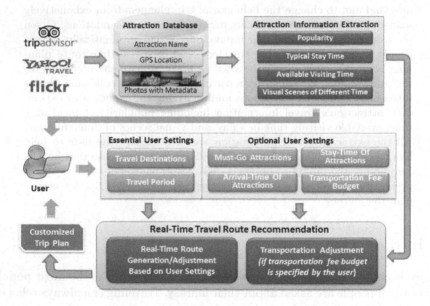

Fig. 1. The proposed system framework

When the user wants to plan a trip, he/she could simply input concise traveling constraints which are essential or optional. The essential user setting includes a sequence of city destinations (e.g. Barcelona→Paris→Roma) and the travel period (defined by the start time and the end time of the trip), while the optional user setting involves must-go attractions, stay-time of specific attractions, arrival-time of specific attractions, and transportation fee budget. The system presents attraction information of the input destinations to the user with a designed visual scene interface[1] so that he/she could further select must-go

---

[1] Visual Scene Interface: Each attraction may look different at different time in different seasons, and we design an interface to show representative photos of various time period for each attraction based on Flickr photos.

Fig. 2. Visual scenes of Kiyomizu Temple (Kyoto, Japan) at different time

(or must-not-go) attractions and arrival/stay-time of specific attractions according to his/her preference. The travel route recommendation component then customizes a more customized and favorable trip route strictly including (or excluding) these selected attractions in real time with the transportation information between two adjacent attractions. For example, after browsing visual scenes of Kiyomizu Temple taken at different time (as shown in Figure 2), the user would specify to go to Kiyomizu Temple in the afternoon if his/her travel is scheduled in November or December because he/she is attracted by the visual scenes of Kiyomizu Temple taken in the afternoon. Besides recommending a suitable travel route after the user first inputs essential settings, our system also allows the user to interactively adjust the recommended route according to their preference on each suggested attraction or transportation between two adjacent attractions. The system will re-customize the route according to the new feedback (i.e. the optional user settings) until the user is satisfied.

The main contributions of this work are threefold. 1) To the best of our knowledge, our system is the first travel recommendation system that considers the available visiting time of attractions and transportation budget for trip planning. To be more precise, we tackle the query of finding a best trip route such that the traveler will arrive at some attractions at some specified time subject to that the total travel time and the total transportation budget is constrained. This kind of query is very common in real-world trip planning scenarios, but none of the existing travel recommendation system is designed to solve this problem. 2) We designed a convenient visual scene interface to present travel information of each attraction, so that the user could quickly have an impression on each attraction. 3) Our system deals with the problem of effectively re-adjusting the travel route and the transportation way according to the users feedback.

The remainder of this paper is organized as follows: Section 2 expounds how we extract travel information of each attraction and Section 3 introduces the proposed real-time travel route recommendation algorithm. The experimental results and discussions are shown in Section 4, and conclusions are given in Section 5.

2   Attraction Information Extraction

As mentioned in Section 1, we construct the traveling database by searching famous attractions of each city on TripAdvisor and Yahoo!Travel and collecting Flickr photos taken around these attractions. All the collected Flickr photos (including the EXIF information) are used for extracting each attraction's travel information including popularity, typical stay time, available visiting time in a day, and visual scenes of different time. The extracted travel information will be presented to the user with the designed visual scene interface to help him/her efficiently know more about different attractions.

2.1   Estimation of Popularity

To estimate the popularity of each attraction, we assume that the more people had been to the attraction, the more popular the attraction is. This assumption is widely used by previous works [2, 3]. For each attraction A, we take the number of people who had taken photos at that attraction (i.e. the number of Flickr users who had uploaded photos taken at attraction A) as the corresponding popularity (popular score), denoted as $PS(A)$.

2.2   Estimation of Typical Stay Time

The estimation of typical stay time can be realized by the internal path discovering (IPD) method proposed by Lu et al. [2]. In general, one can utilize the information of photo taken time (embedded in the EXIF) to compute the time length of a photo sequence. For each photo sequence s_i composed of photos taken at attraction A by the same user on the same date, we set the corresponding time length, denoted as $TL(s_i)$, as the stay time of attraction A based on the sequence s_i. The estimated typical stay time of attraction A can be then defined by the average of all $TL(s_i)'s$. Yet photo sequences uploaded by users are usually sparse and incomplete. Hence the time length of photo sequence would often be much shorter than the real stay time of the attraction. IPD can concatenate multiple incomplete photo sequences to a complete one by discovering the internal path, and therefore the estimated typical stay time would be much more precise. However, for outdoor attractions, there is usually no specific visiting path, and therefore IPD may not successfully concatenate incomplete photo sequences. In this case, we apply the method proposed by Xie et al. [4] instead. That is, the typical stay time of an attraction A would be ($\omega * area\ size\ of\ A$), where ω depends on the popularity of the attraction. For instance, popular attractions are with higher ω than unpopular ones since people would like to spend more time in more popular attraction per unit area.

2.3   Estimation of Available Visiting Time

In order to automatically retrieve the available visiting time in a day of an attraction A, we exploit the photo taken time to gather statistics about when this attraction is available for visiting. If there are enough amount of photos taken around attraction A at certain time, then this time should belong to the available visiting time of attraction A. Practically, we quantize a day into 48 time intervals (i.e. each interval equals 30 mins) and calculate the amount of photos taken around attraction A during each time interval instead of at a specific time. If a time interval has few photos, it implies that A's open time may not include that time interval.

2.4 Visual Scenes Generation

We modify the representative photo extraction method proposed in [3] to generate visual scenes at different time for each given attraction. The flowchart of our method is illustrated in Figure 3. All photos of an attraction are clustered into N_D Date Clusters by their taken date to distinguish different scenes in a year. Photos in each Date Cluster are further clustered into N_T Time Clusters to distinguish different scenes in a day. Finally, for each Time Cluster we use the visual features to cluster all involved photos into N_V Visual Clusters [3]. Therefore, photos in the same Visual Cluster would look similar and photos in different Visual Clusters would look more different. We select the centroid photo from each Visual Cluster as the representative scene of the attraction at a certain time. In our implementation, N_D is set to 24, which means photos within the same half month interval would be clustered together. N_T is set to 3 since there are usually three quite different scenes (morning, afternoon, and evening) in a day of an attraction. N_v is set to $\sqrt{n/2}$ as suggested in [3], where n denotes the number of photos being clustered.

3 Real-Time Travel Route Recommendation

After gathering travel information of all attractions and the user inputs, we can compute and generate a suitable travel route for the user. In this section, we define the route recommendation problem and introduce how to solve it in real-time.

3.1 Problem Statement

The target travel route should fulfill the following criteria:

- **Plentiful.** The travel route should contain as many popular attractions as possible since in general tourists would like to visit as many popular and representative attractions as they can [5, 6]. The total popular score of a travel route R can be defined as:

$$TPS(R) = \sum_{i=1}^{RouteSize-1} PS(i), \qquad (1)$$

  where i denotes the ith attraction on the trip route, $i = 0$ means the start location, and $i = 1$ means the first visited attraction.

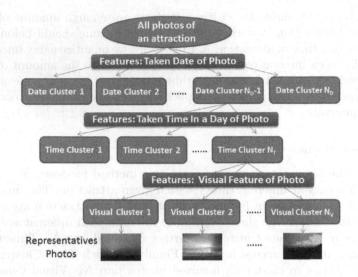

Fig. 3. Flowchart of Visual Scene Generation

- **Smooth.** The trip route should be smooth to avoid back and forth from one location to another. The non-smoothness of a trip route R is defined by

$$NS(R) = \sum_{i=1}^{RouteSize-1} [dist(i-1,i) + dist(i,i+1) - dist(i-1,i+1)], \quad (2)$$

where $dist(i-1,i)$ denotes the distance between the $i-1$th attraction and the ith attraction calculated based on the GPS locations. Besides, this criterion also implies that users would spend more time on staying in attractions instead of driving between attractions [5, 6].

- **Feasible.** Every attraction in the generated route should be visited or arrived at its available visiting time.

- **Customized.** Every users requirement should be satisfied, e.g. "the trip route must consist of attraction A", "arrives at attraction B in the evening of day 2", and "stay at attraction C for 3 hours".

Based on the criteria described above, the problem of generating a good trip route is to maximize $TPS(R)$ and minimize $NS(R)$ subject to the constraint that every attraction should be visited at its available visiting time interval and all user settings should be satisfied. The problem can be transformed to the Vehicle Routing Problem with Time Window (VRPTW) [7], which had been proven to be an NP-hard problem. However, in our application, to compute the travel route in real-time is essential. Inspired by [8], we propose an effective heuristic algorithm which can be executed in real-time but also yield good results.

3.2   The Proposed Heuristic Algorithm

As proposed in [8], in the beginning the travel route only consists of the start location and the end location, then the algorithm starts to insert one attraction at a time to the travel route until no more attraction can be inserted. To decide which attraction to insert, we first examine every of the rest attractions and determine which position in the current trip route is most proper to insert the examined attraction. The non-smoothness of inserting attraction A to position i is defined by

$$NS(A) = \min_i(dist(i, A) + dist(A, i + 1) - dist(i, i + 1)), \qquad (3)$$

where $i = 0, ..., CurrentRouteSize - 1$. Hence, the most proper inserting position of an attraction A can be determined by

$$MPIP(A) = \arg\min_i(dist(i, A) + dist(A, i + 1) - dist(i, i + 1)), \qquad (4)$$

where $i = 0, ..., CurrentRouteSize - 1$. After calculating the most proper inserting position for each of the rest attractions, we decide which attraction is selected to be inserted first. Since our goal is to maximize $TPS(R)$ and minimize $NS(R)$, we define the inserting priority score of an attraction A as

$$IPS(A) = \alpha * PS(A) - \beta * NS(A), \qquad (5)$$

where $\alpha, \beta \geq 0$. Each time, the attraction with the highest IPS is inserted to the original trip route if it is appropriate to be inserted into its most proper inserting position in the current route. An attraction is said to be appropriate if the new trip route still fulfills all the constraints after inserting it. The attraction insertion process will be terminated until no appropriate attractions is left, and then we obtain the final travel route. To customize the travel route, user feedback should be taken into account. With the proposed heuristic algorithm, user feedback can be easily integrated by inserting the user-specified must-go attractions prior to other attractions into appropriate position of the travel route to fullfill user-specified arrival-time (if any). Considering the real-time computation issue, we further apply the local search algorithm [7] to improve the original solution, as shown in Algorithm 1.

3.3   Transportation Budget Control

There usually exist multiple transportation choices from one attraction to another attraction, and each transportation choice may differ in the transportation duration and transportation fee. In most cases, the shorter the transportation duration is, the more expensive the transportation fee is. In other words, when users travel time and transportation budget are both limited, then the generated trip route should compromise between taking rapid but expensive transportation and taking slow but cheaper transportation. The problem becomes how to

---

Algorithm 1. Real-time Travel Route Recommendation Based on Local Search

---

Input: OTR(Original Travel Route), k(number of attractions being exchanged at a time), RA(set of the rest attractions).
Output: UTR(Updated Travel Route)
1: Initialize $UTR = OTR$
2: **do**
3:　　$OTR = UTR$;
4:　　$UTR = ExchangeWithinTripRoute(OTR, k)$;
5:　　$ExchangeBetweenTripRouteAndRestAttractions(UTR, k, RAset)$;
6: **while** $IsImproved(OTR, UTR)$
7: **function** EXCHANGEWITHINTRAVELROUTE(R_o, k, R_u)
8:　　$R_u = R_o$;
9:　　$R_{tmp} = R_o$;
10:　　**for** *every* $k - combination$ *of attractions in* R_o **do**
11:　　　　**for** *every possible exchanges among them* **do**
12:　　　　　　$R_{tmp} = travel\ route\ after\ exchanging$
13:　　　　　　**if** $IsFeasible(R_{tmp})$ *and* $IsImproved(R_o, R_{tmp})$ **then**
14:　　　　　　　　$R_u = R_{tmp}$;
15:　　　　　　　　$R_o = R_u$;
16:　　　　　　**end if**
17:　　　　　　**break**;
18:　　　　**end for**
19:　　**end for**
20:　　**return** R_u;
21: **end function**
22: **function** EXCHANGE($R_o, k, RAset, R_u$)
23:　　$R_u = R_o$;
24:　　$R_{tmp} = R_o$;
25:　　**for** *every* $k - combination$ *of attractions in* R_o *and* $RAset$ **do**
26:　　　　**for** *every possible exchanges among them* **do**
27:　　　　　　$R_{tmp} = travel\ route\ after\ exchanging$
28:　　　　　　**if** $IsFeasible(R_{tmp})$ *and* $IsImproved(R_o, R_{tmp})$ **then**
29:　　　　　　　　$R_u = R_{tmp}$;
30:　　　　　　　　$R_o = R_u$;
31:　　　　　　**end if**
32:　　　　　　**break**;
33:　　　　**end for**
34:　　**end for**
35:　　**return** R_u;
36: **end function**
37: **function** ISFEASIBLE(R)
38:　　**if** *all attractions in* R *satisfy all the constrains* **then**
39:　　　　**return** *true*;
40:　　**else**
41:　　　　**return** *false*;
42:　　**end if**
43: **end function**
44: **function** ISIMPROVED(R_o, R_u)
45:　　**if** $(TPS(R_u) > TPS(R_o)$ *and* $NS(R_u) \leq NS(R_o))$ *or* $(TPS(R_u) \geq TPS(R_o)$ *and* $NS(R_u) < NS(R_o))$ **then**
46:　　　　**return** *true*;
47:　　**else**
48:　　　　**return** *false*;
49:　　**end if**
50: **end function**

Table 1. Execution time for a 150-hours-long trip plan with different number of candidate attractions

Number of Candidate Attractions	50	100	150	200	250
Average Execution Time (secs)	0.273	0.655	1.039	1.589	2.319

choose every transportation way between every adjacent attractions pair in the trip route properly so that the generated trip route can be better?

To address the problem stated above, we propose the following algorithm. Each time we insert a new attraction into current trip route, we have to determine which transportation to take. However, at this time it is hard to decide which can yield better result in the end. Therefore, our algorithm would always choose the shortest duration but most expensive one, and keep inserting attractions regardless of whether user-specified transportation budget is exceeded until no more attractions can be inserted. Next, we start to reduce the total transportation fee to meet the transportation budget specified by the user. There are two ways to reduce the total transportation fee: (1) Replace one of the transportation way of the current trip route with another transportation choice which has longer duration but is cheaper. (2) Remove the attraction with lowest IPS from the current trip route. Our algorithm applies the first way. However, if there is no solution that meets the budget constraint, the second way will be utilized.

4 Experimental Results

In this section, we present and discuss the experiment results including the execution time of the proposed trip route planning algorithm, effectiveness of the proposed trip route planning algorithm.

4.1 Execution Time of the Proposed Trip Route Planning Algorithm

In order to evaluate whether the proposed trip route planning algorithm can compute a good trip route in real-time, we test our algorithm on the PC with Intel Dual Core i5-3210M CPU of 2.50 GHz. To simulate the user settings, we randomly generate a number of possible user queries. Then we average all the queries results to examine the average execution time. As Table 1 shows, although lots of constraints should be satisfied in order to achieve a suitable trip route, the proposed algorithm is still able to be accomplished in real-time even if the total traveling time is 150 hours long (a 7-days trip plan) and the number of attractions is 200 large.

4.2 Effectiveness of Proposed Trip Route Planning Algorithm

We crawled photos of attractions in popular travel area such as Eastern Area of Taiwan, Southern Europe and Kansai Area of Japan from Flickr to extract

Table 2. Results of trip route planning evaluated by objective criteria

	NAV	ETR	STR	TRPSR
Popular Insert First	17.47	0.98	66.16	0.73
Proposed Algorithm	22.22	0.97	77.67	0.78
Improvements Rate	1.27	0.99	1.17	1.06

travel information of attractions. Then we facilitate Google Maps API to fetch the distance and the duration of different transportation between attractions in the same travel area. We conduct subjective test to evaluate the quality of a planned trip route. However, there are some criteria which can objectively measure the quality of a trip plan, including Number of Attractions Visited (NAV), Elapsed Time Ratio (ETR, Total Travel Duration/Users Travel Time Budget), Stay Time Ratio (STR, Time Spent on Attraction/ Total Travel Duration), and Trip Route Popular Score Ratio (TRPSR, Total Popular Score of Attractions being Visited/Maximum Total Popular Score of All Generated Trip Route) [6]. When the other three criteria are the same for the two trip routes, the trip route with one of the criteria higher than the other trip routes would be considered better.

Because there doesnt exist trip plan generating algorithm which also takes the available visiting time of attractions into account to compare with, here we implement a baseline algorithm called PopularInsertFirst which at each round would select the attraction with highest popular score to insert into the trip route. We randomly generate a number of different user settings to simulate possible users queries and then average the results. The results in Table 2 show that proposed algorithm on average performs better than Baseline Algorithm except for the ETR. However, the difference between the two algorithms is quite small and both algorithms ETR are high enough. Proposed algorithm improved the most in NAV and its very reasonable because when smoothness is taken into account, the total travel time would be reduced and therefore more attractions can be added to the trip route. For the STR, the proposed algorithm also outperforms a lot. This is because that a smooth trip route also implies that there would be less redundant path and hence the transportation duration would be shorter. It is interesting that proposed algorithm also surpassed in IDR despite that at each round the attraction chosen by PopularInsertFirst has higher popular score than the one selected by proposed algorithm. The reason is that the improvements of NAV and STR are high enough to compensate for the slightly lower popular score of each rounds selection.

Except for the objective evaluation, we also recruit 20 people who had travel experience or knowledge in the travel destination to help evaluate the quality of the trip route generated by proposed algorithm. In this part, we also want to verify the importance of taking available visiting time into consideration. Therefore we ask users to compare the algorithm with previous works that does not consider available visiting time (denoted as NCA, NotConsiderAvailability). Besides, we also compare with the RT (RandomTrip) algorithm which randomly choose attractions and paths to generate a trip route. The questions we ask including:

1) Are the attractions being visited representative, interesting or popular enough?
2) Is the trip route smooth enough? 3) Is the stay time ratio reasonable? 4) Is
the arrival time of attractions being visited reasonable? 5) Overall, how do you
think the quality of the trip route? Users are asked to rate 1- 5 score to the
above questions. For questions 1 to 4, score 1 means not agree at all and score 5
means absolutely agree. For the fifth question, score 1 means the quality is very
bad, and score 5 means very good. The results are shown in Figure 4.

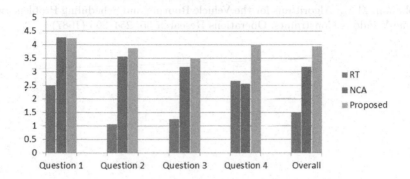

Fig. 4. Results of trip route planning evaluated by subjective questions

5 Conclusions

A convenient system for trip planning is proposed to change the behavior of trip
planners from exhaustively searching information to receiving useful travel recommendations. A traveling database is constructed by collecting photos taken
around famous attractions and analyzing these photos to extract each attractions
travel information including popularity, typical stay time, available visiting time
in a day, and visual scenes of different time. All the extracted travel information
are presented to the user to help him/her efficiently know more about different attractions in terms of visual scenes. Given the start location, destinations,
traveling dates, and transportation budget information as the input, the system
automatically recommends a route that suits the traveler based on a real-time
routing planning algorithm and allows the user to make adjustment according
to their preferences.

References

[1] Top 15 Most Popular Travel Websites (March 2013),
     http://www.ebizmba.com/articles/travel-websites
[2] Lu, X., et al.: Photo2Trip: Generating Travel Routes from Geo-Tagged Photos for
     Trip Planning. In: ACM MM 2010 (2010)
[3] Jiang, et al.: ContextRank: Personalized Tourism Recommendation by Exploiting
     Context Information of Geotagged Web Photos. In: IEEE ICIG 2011 (2011)

[4] Xie, et al.: CompRec-Trip: a Composite Recommendation System for Travel Planning. In: ICDE 2011 (2011)

[5] Yoon, H., Zheng, Y., Xie, X., Woo, W.: Smart Itinerary Recommendation Based on User-Generated GPS Trajectories. In: Yu, Z., Liscano, R., Chen, G., Zhang, D., Zhou, X. (eds.) UIC 2010. LNCS, vol. 6406, pp. 19–34. Springer, Heidelberg (2010)

[6] Yoon, H., et al.: Social itinerary recommendation from user-generated digital trails. Journal of Personal and Ubiquitous Computing (2012)

[7] Bräysy, O., Gendreau, M.: Vehicle Routing Problem with Time Windows, Part I: Route Construction and Local Search Algorithms. Transportation Science (2005)

[8] Solomon, M.M.: Algorithms for the Vehicle Routing and Scheduling Problems with Time Windows Constraints. Operations Research 35, 254–265 (1987)

Who's the Best Charades Player?
Mining Iconic Movement of Semantic Concepts

Yung-Huan Hsieh[1], Shintami C. Hidayati[1,2], Wen-Huang Cheng[1],
Min-Chun Hu[3], and Kai-Lung Hua[2]

[1] Research Center for Information Technology Innovation,
Academia Sinica, Taipei, Taiwan (R.O.C)
{yhhsieh,whcheng}@citi.sinica.edu.tw
[2] Dept. of Computer Science and Information Engineering,
National Taiwan University of Science and Technology, Taipei, Taiwan (R.O.C)
{d10115802,hua}@mail.ntust.edu.tw
[3] Dept. of Computer Science and Information Engineering,
National Cheng Kung University, Tainan, Taiwan (R.O.C)
anita_hu@mail.ncku.edu.tw

Abstract. Charades is a guessing game with the idea for one player to
act out a semantic concept (i.e. a word or phrase) for the other players
to guess. An observation from playing charades is that people's cognition on the iconic movements associated with a semantic concept would
be often inconsistent, and this fact has long been ignored in the multimedia research. Therefore, the novelty of this work is to propose an
automation for mining the most representative videos for each semantic
concept as its iconic movements from a large set of related videos containing various human actions. The discovered iconic movements can be
further employed to benefit a broad range of tasks, such as human action
recognition and retrieval. For our purpose, a new video benchmark is also
presented and the experiments demonstrated our approach potential to
human action based applications.

Keywords: video analysis, human actions, semantic concepts.

1 Introduction

How to define human actions? In daily life, it can be observed that when mentioning a specific kind of human action in words, most people would have had
a stereotype image of this action in their minds, from simple, classic actions
like "jump" and "walk", to more complex actions like "wearing glasses", or even
signature moves of celebrities such as "the dance moves of Michael Jackson" [1].
The individual's cognition on the stereotype image of each human action might
be similar but often in different degrees. It makes the interesting part for people
to play charades [15], an acting game with the idea for one player to use physical
(e.g. gesture or action) rather than verbal language to act out a *semantic concept*
(i.e. a word or phrase as the charades topic) for the other players to guess.

C. Gurrin et al. (Eds.): MMM 2014, Part I, LNCS 8325, pp. 231–241, 2014.
© Springer International Publishing Switzerland 2014

Fig. 1. Action variations in semantic concepts. Three users are asked to perform the iconic movements of "using a machine gun" in their minds.

In a charades game, different semantic concepts are known to vary in difficult level, i.e. some are easier to be guessed while some are not. This implies that semantic concepts are associated with certain *iconic movements* (i.e. representative action patterns), although the perceived stereotypes for easy concepts are more consistent among people and there are more variants over the ones for difficult concepts. The variations in defining iconic movements for a semantic concept have long been ignored in the multimedia literature [1,14], cf. Fig. 1. For example, in most of the existing human action video databases, such as the Weizmann and KTH datasets [1], the collected video clips for a same semantic concept (e.g. "running") tend to look identical in terms of the action patterns. The bias would limit the development of advanced human action based applications.

In this work, therefore, we take an alternative view to investigate the human action issues by considering a practical and challenging question: given a semantic concept (in words), how to find out the iconic movements which best match people's common stereotypes on this concept? Technically, different from the previous works which recognize and discriminate predefined basic actions (e.g. walking and waving) [1,2,4], the novelty of our work is to automatically mine the most representative videos for each semantic concept as its iconic movements from a large set of video clips containing human actions of the concept in various action patterns. The discovered iconic movements are important for understanding human actions, and can be regarded as mid-level features for benefiting human action recognition, retrieval, and other related tasks. For the experiments, we have also built a video dataset with 8 users acting out 32 semantic concepts, including the actions of imitating a celebrity's signature moves, playing a sport, and operating a physical object item, in which every video is rated by 10 other participants based on the subjective iconic judgement.

The rest of this paper is organized as follows. In Section 3, we present the proposed framework for analyzing the iconic movements of semantic concepts. We describe the details of our collected database and evaluations in Section 4. In Section 5 we conclude our work and give the directions for our future research.

2   Related Work

Our issue is about the relationship between semantic concepts and human action. In this section, we review previous studies that are relevant to our work.

To emerge the semantic meaning in multimedia, the TRECVid Multimedia Event Detection (TRECVid MED) [12] presented a video benchmark and evaluations of multimedia detection technologies, as revealed by [10]. The pre-specified events defined in TRECVid MED tasks (namely "making a cake ", "assembling a shelter", and "batting a run") are appeared in videos captured in unconstrained environments and settings. These events, exploited to extract the high-level features, facilitate user to manage and retrieve multimedia contents. Different from multimedia events detection, our purpose is to mine the action patterns between semantic concepts and human actions in an automatic manner.

There are a large number of works corresponding to human action analysis. Some surveys review and classify their approaches based on the recognition methodologies and the complexity of targeting human actions and activities [14,1]. In comparison with the reviewed human recognition approaches, our study discusses the different fields of action types. Rather than general purpose actions, our action videos are people acting out a semantic concept, e.g. "a celebrity/famous character", "a sport/action event", and "a daily object". Hence, the difference between our technique and the previous human recognition techniques can be listed as two-fold: *recognition algorithm* and *presented action dataset*.

Aggarwal and Ryo [1] classified human recognition algorithm according to model complexity. In their works, the type of action and activity events are limited to perform regularly and intuitively. These actions and activities are easily to figure out the action pattern of people. Furthermore, most of the defined simple actions (e.g. *waving* and *walking*) are periodic and the action patterns are temporary. The complex events and interactions (e.g. *fighting* and *picking up*) are predictable with grammatical structures and rules. However, in the human computing interface, recognizing the user's specific style of actions as a more semantic term, referred to concepts, are useful and challenging. The user tends to perform this concept as query to interact with the computer and to manipulate the multimedia content. It is interesting to discuss the human performing a semantic "concept" and the issues of the relationship between semantic concepts and human actions.

In this work, we are interested in exploiting consumer depth camera that facilitates human segmentation and forms human silhouette with depth information. In the most recent work, Sung *et al.* [13] and Li *et al.* [8] announced their datasets consisting of some simple actions. Sung *et al.* [13] proposed human activity detection approach based on a hierarchical maximum entropy Markov model (MEMM) that considers a person's activity as composed of a set of sub-activities, whilst Li *et al.* [8] proposed human activity detection approach based on action graphical model and depth image features. Different with their works,

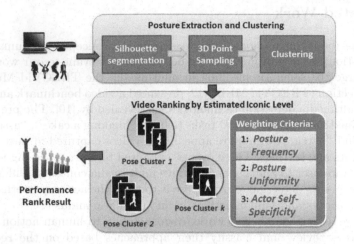

Fig. 2. The proposed framework

our work exploits depth image information of human action and discuss their usage to discover the specific and iconic movement in variant performances related to one concept.

3 The Methodology

The framework of our approach is shown in Fig. 2. First of all, we regard an action as a sequence of poses. The extracted silhouette of the pose is then exploited to constructed the sampling features [2,3,4]. Consequently, each pose of an action is represented by their sampling features. Next, following the similar ideas of bag-of-words models, we treat each pose as multiple points in the high dimensional space [8]. Similar poses from each semantic concept are then clustered into the same group by affinity propagation [6]. Finally, we propose the use of three criteria to evaluate the iconic level of a query action video to the corresponding semantic concept. Detailed procedures for mining the iconic movement will be explained below.

3.1 Posture Extraction and Clustering

We captured human actions using the Microsoft Kinect sensor and then applied the standard Application Programming Interface (API) of the Kinect Software Development Kit (SDK) to extract the human silhouettes. Thus, each action video consists of a sequence of silhouettes.

Given the sequences of extracted silhouettes with depth information, we obtain the 3D sampling points that represent the postures by [8,9]. The sampling process to select the representative 3D points from a depth map comprises three stages: projecting the depth map onto three orthogonal Cartesian planes, sampling the image contour, and retrieving the 3D points that are nearest to the sampled 2D points. For the sake of computation, in this work we sample the image contour on each plane to a specified number of points, i.e. totally 80 points with 36, 24, and 20 points to the xy-, zy-, zx-projections, respectively. Then, p number of sampling points of silhouette on the m-th frame of a video can be represented as $f_m = \{x_1, x_2, \ldots, x_p\}$.

Let a semantic concept denoted as α consists of K video action and the action video consists of M frames, the sampling points of silhouettes from all action videos in α can be written as $S^\alpha = \{f_1, f_2, \ldots, f_{KM}\}$. Here, the number of element in S^α, $|S^\alpha|$, is $K \times M$. Afterwards, as in [9], we build a similarity matrix of feature points in α, defined as

$$Sim = [sim_{f_i f_j}]_{f_i, f_j = 1}^{|S^\alpha|}, \tag{1}$$

where Sim is a $|S^\alpha| \times |S^\alpha|$ symmetric matrix, $sim_{f_i f_j}$ is similarity between f_i and f_j, and $f_i, f_j \in S^\alpha$. Hausdorff distance is used to compute $sim_{f_i f_j}$, that is

$$sim_{i,j} = -\Psi(f_i, f_j). \tag{2}$$

Given a similarity matrix Sim of semantic concept α, we employ affinity propagation [5] to cluster the similar poses of video actions in α into the same group. Different from [4] that clusters all frames from every predefined concepts in order to extract key poses, in this work we cluster each of the defined concepts separately and we mark the clusters as the salient postures for that concept. Consequently, T clusters considered as salient postures of semantic concept α can be written as $P^\alpha = \{p_1, p_2, \ldots, p_T\}$.

3.2   Video Ranking by Estimated Iconic Level

The subsequent ranking process exploits the extracted salient postures to rank the action videos in a semantic concept based on the estimated iconic level. To build our ranking model, as in [7], each video sequence that performs one kind of action is treated as a document with a set of words. Such document-word model is known as the Vector Space Model (VSM) [11], where a document is represented as a vector of its word-counts, in the space of possible words.

In our model, a document (i.e. a video sequence of action) is represented as a vector of term weight of salient postures. Having K action videos in semantic concept α, the weight of those videos can be represented as

$$W^\alpha = \{W_{v_1}, W_{v_2}, \ldots, W_{v_K}\}, \tag{3}$$

where W_{v_K} is a weight vector for video v_K. Let $\omega_{t,k}$ be the weight of posture $p_t \in P^\alpha$ in video v_k, the weight vector of video v_k is

$$W_{v_k} = \{\omega_{1,k} n_{1,k}, \omega_{2,k} n_{2,k}, \ldots, \omega_{T,k} n_{T,k}\}, \tag{4}$$

where $n_{t,k}$ is the number of posture that occurs in video v_k, $k \in \{1, 2, \ldots, |W^\alpha|\}$, and $t \in \{1, 2, \ldots, |P^\alpha|\}$.

To calculate the weight of posture in a video, we analyze the distribution of frames in each posture p_t, where $p_t \in P^\alpha$. Because a reasonable cluster of postures should include the silhouettes performed by multiple users, the postures with silhouettes performed by only one user will be filtered out. After the filtering step, weight of the filtered postures are calculated based on the distribution (histogram) of silhouette occurrences.

In our model, we apply two kinds of schemes to compute the weight vector of posture p_t for video v_k, denoted by $\omega_{t,k}$: posture frequency among performed action, $\omega_{t,k}^{freq}$, and posture uniformity among performed actions, $\omega_{t,k}^{uni}$.

Schemes 1: Posture Frequency among Performed Actions. This scheme calculates the ratio of each posture with respect to all postures in a semantic concept. If the ratio of the posture is high, it means that the posture occurs more frequently than the other postures. Let n_{p_t} denotes the number of silhouettes belongs to posture p_t and $p_t \in P^\alpha$, we formulate this scheme as

$$\omega_{t,k}^{freq} = \frac{n_{p_t}}{\sum_{t=1}^{|P^\alpha|} n_{p_t}}, \tag{5}$$

Schemes 2: Posture Uniformity among Performed Actions. The uniformity of a posture, as the entropy of the distribution, is the occurrence of that posture in all videos of a semantic concept. The entropy of a posture to video distribution indicates whether that posture appearing equally in all videos. If the entropy value is high, it means most players act out that posture as they perform the concept. Let n_{p_t,v_k} denotes the number of posture p_t that occurs in video v_k, we formulate this scheme as

$$\omega_{t,k}^{uni} = \sum_{k=1}^{|W^\alpha|} -prob_k(p_t) \log prob_k(p_t), \tag{6}$$

where $prob_k(p_t) = \frac{n_{p_t,v_k}}{\sum_{k=1}^{|W^\alpha|} n_{p_t,v_k}}$.

According to the above posture weight values as defined in Equations 5 and 6, there are two kinds of weight vectors related to one video. These vectors in the VSM form are rewritten as

$$W_{v_k}^{freq} = \{\omega_{1,k}^{freq} n_{1,k}, \omega_{2,k}^{freq} n_{2,k}, \ldots, \omega_{T,k}^{freq} n_{T,k}\}, \tag{7}$$

$$W_{v_k}^{uni} = \{\omega_{1,k}^{uni} n_{1,k}, \omega_{2,k}^{uni} n_{2,k}, \ldots, \omega_{T,k}^{uni} n_{T,k}\}. \tag{8}$$

In this work, three different criteria are then applied to rank the iconic level of an action video. The proposed criteria are calculated as follows.

Criterion 1: Ranking with the Posture Frequency. Based on the posture frequency in Equation 5, the video sequences in semantic concept α are ordered according to each frequency vector's norm (i.e. the sum of the vector $W_{v_k}^{freq}$). Hence, the ranking order using our first criterion is

$$rank_{freq}(v_i) < rank_{freq}(v_j) \quad \text{if} \quad \|W_{v_i}^{freq}\|_1 \leq \|W_{v_j}^{freq}\|_1. \tag{9}$$

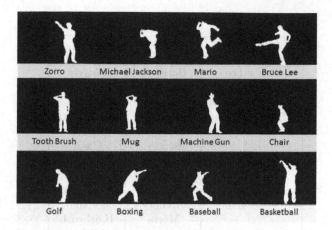

Fig. 3. Example silhouettes of some selected semantic concepts

Criterion 2: Ranking with the Posture Uniformity. Based on the posture uniformity in Equation 6, the video sequences in semantic concept α are ordered according to each uniformity vector's norm (i.e. the sum of the vector $W_{v_k}^{uni}$). Hence, the ranking order using our second criterion is

$$rank_{uni}(v_i) < rank_{uni}(v_j) \quad \text{if} \quad \|W_{v_i}^{uni}\|_1 \leq \|W_{v_j}^{uni}\|_1. \tag{10}$$

Criterion 3: Ranking with the Actor Self-specificity. Actor self-specificity indicates the degree of how many postures were appeared in a video. The more distinct postures of a video, the more specific video it will be. For a video action, if the value of this criterion is high, it means the actor performs more different postures to act out the semantic concept and his/her acting is more dramatically. To calculate the self-specificity of an actor in video v_k, the distribution entropy that represents the occurrence of posture p_t in v_k is exploited. Let n_{p_t,v_k} denotes the number of posture p_t in video v_k, the actor self-specificity value of video v_k is computed as

$$\omega_{v_k}^{spec} = \sum_{t=1}^{|P^{\alpha}|} -prob_t(v_k) \log prob_t(v_k), \tag{11}$$

where $prob_t(v_k) = \frac{n_{p_t,v_k}}{\sum_{t=1}^{|P^{\alpha}|} n_{p_t,v_k}}$. Thus, based on Equation 11, the ranking order using our third criterion is then written as

$$rank_{spec}(v_i) < rank_{spec}(v_j) \quad \text{if} \quad \|W_{v_i}^{spec}\|_1 \leq \|W_{v_j}^{spec}\|_1. \tag{12}$$

4 Experimental Results

4.1 Dataset and Experimental Settings

Due to the absence of public human action benchmarking datasets which supply the sequences of depth maps with labeled iconic levels, we collected a dataset that

Table 1. The selected semantic concepts

Celebrity/ Famous Character	Sport/ Action Event	Daily Object
Bruce Lee	Baseball	Bicycle
David Beckham	Basketball	Toothbrush
Michael Jackson	Boxing	Glasses
Michael Jordan	Taekwondo	Electric Drill
Super Mario	Karate	Watch
E.T	Soccer	Chair
Zorro	Golf	Mug
Superman	Swimming	Machine Gun
Spiderman	Shooting	Keyboard
	Archery	Cap
	Morra	Hairbrush
		Mirror

contains 32 unique semantic concepts. Figure 3 shows examples of silhouettes of some semantic concepts in our work. As listed in Table 1, we classified the 32 concepts into three sets, i.e. celebrity/famous character, sport/action event, and daily object. Each of these categories requests the participants to imitate other people, perform a sport/action event, or operate an object.

In our dataset, eight people participated as the action performers and all of them acted out once for each of the 32 concepts. We recorded each performing action using the Microsoft Kinect sensors, which outputs an RGB image together with aligned depths at each pixel at a frame rate of 30 Hz. The size of recorded depth image is 640×480 pixels with a range of 1.2 meters to 3.5 meters. On average, the length of our action videos is about 2.7 seconds.

We conducted a user study to obtain the human ratings for each action video in the dataset. The measurement rating is obtained using the typical Mean Opinion Score (MOS) test and it is based on whether the performing action is clear and representative enough to the corresponding concept. In this test, we set the score in the range 1 to 5, where 1 is the lowest perceived iconic level and 5 is the highest perceived iconic level measurement. We invited ten participants to score all of action videos in our dataset. For each action video, the average score of the ten participants are used as the ground truth iconic level.

4.2 Statistical Analysis

In contrast to human action recognition, the goal of our study is to evaluate the iconic level of a performing action video, i.e. whether it is vivid and expressive to describe the corresponding concept. To validate our automatic ranking model, we used Kendall's tau coefficient [11] to measure the similarity between our ranking results and the ground truth data. The coefficient ranges are from -1 to 1, where 1 indicates perfect agreement and -1 indicates perfect disagreement of the two rankings.

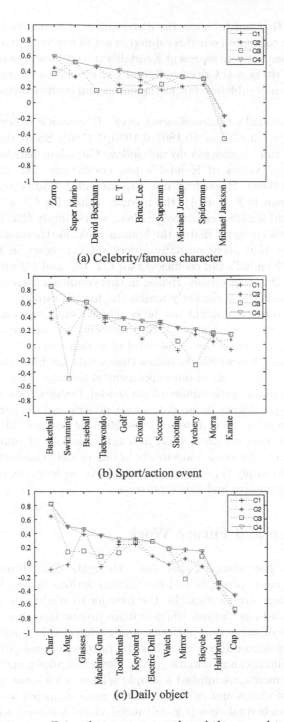

(a) Celebrity/famous character

(b) Sport/action event

(c) Daily object

Fig. 4. Kendall's tau coefficient between our rank and the ground truth rank in three different concept sets

Figures 4 (a), (b), and (c) show the distribution of Kendall's tau coefficients for each semantic concept in our three different set of concepts. In each subfigure, $C1$, $C2$, and $C3$ respectively represent Kendall's tau coefficient of our experiment when we apply Criterion 1, Criterion 2, and Criterion 3 as the ranking criteria. Moreover, $C4$ in each subfigure marks the maximum coefficient value of $C1$, $C2$, and $C3$.

On average, Kendall's tau coefficients over all semantic concepts defined in our work are $(C1, C2, C3) = (0.1807, 0.1930, 0.1580)$. Since the average values of Kendall's tau coefficients do not follow Gaussian distribution, we also report the median values of Kendall's tau coefficients, i.e. $(C1, C2, C3) = (0.2265, 0.2344, 0.1603)$. Regarding our experimental results on sport/action event set (as shown in Fig. 4(b)), the median values for $C1$, $C2$, and $C3$ are 0.3200, 0.2381, and 0.2308, respectively. These scores imply that around 70% of our rank order lists correctly match the human rank. Furthermore, even though for the worst case that occur on daily object set (as shown in Fig. 4(c)), the median values of Kendall's tau coefficients for $C1$, $C2$, and $C3$ are still high, i.e. 0.013, 0.1757, 0.1282, respectively. Hence, in this condition there are still around 60% of our rank order lists correctly match the human rank.

The results of our study might not be conclusive but quite encouraging. According to our observations, there is high probability for our approach to match the human's judgements on the iconic level of action videos for most semantic concepts. Especially, it is worthy to notice that we do not limit the participant's performing styles or thoughts in our experimental settings. This freedom causes some degradation on the performance of our model. Besides, some problems arise because few concepts defined in our dataset are generic and the participant's performing actions tend to be diverse, such as *Michael Jackson*. Moreover, action videos for concept *caps* are often short in length and lack of sufficient postures used to discriminate the iconic movements in our model. These cases with weak performance could imply that these concepts have no iconic movement due to lacking the discriminative and iconic postures.

5 Conclusion and Future Work

Human action is more complex than one's thoughts. The subjective relationship between semantic concepts and the human actions in people's perception is worthy of further investigation. Be the frontier to study this research topic, we collected a brand-new dataset which is more diverse than any existing action datasets in terms of the action patterns, and each video is annotated with the iconic level to the semantic concepts. For the purpose of evaluating our model, we attempt to build up an automatic judger like a charades game. For matching the human's judgments, we applied a simple scheme with some criteria to rank the iconic level of videos and our rank results are convincing and plausible. In the future, rich contextual cues (e.g. advanced visual features) will be explored in our model and the prevailing action recognition models could be exploited, such as graphical model and context-free grammar.

Acknowledgments. This study is conducted under the "On the Exploration and Design of Intelligent Living and Community Space Project" of the Institute for Information Industry which is subsidized by the Ministry of Economy Affairs of the Republic of China. This work was supported by National Science Council of Taiwan via NSC 102-2221-E-001-028.

References

1. Aggarwal, J.K., Ryoo, M.S.: Human activity analysis: A review. ACM Comput. Surv. 43, 16:1–16:43 (2011)
2. Baysal, S., Kurt, M.C., Duygulu, P.: Recognizing Human Actions Using Key Poses. In: International Conference on Pattern Recognition, pp. 1727–1730 (2010)
3. Brendel, W., Todorovic, S.: Activities as time series of human postures. In: Daniilidis, K., Maragos, P., Paragios, N. (eds.) ECCV 2010, Part II. LNCS, vol. 6312, pp. 721–734. Springer, Heidelberg (2010)
4. Cheema, S., Eweiwi, A., Thurau, C., Bauckhage, C.: Action recognition by learning discriminative key poses. In: IEEE International Conference on Computer Vision Workshops, pp. 1302–1309 (2011)
5. Chen, Y., Wu, Q., He, X., Du, C., Yang, J.: Extracting key postures in a human action video sequence. In: 10th IEEE Workshop on Multimedia Signal Processing, pp. 569–573 (2008)
6. Frey, B.J., Dueck, D.: Clustering by passing messages between data points. Science Journal 315, 972–977 (2007)
7. Hamid, R., Johnson, A., Batta, S., Bobick, A., Isbell, C., Coleman, G.: Detection and explanation of anomalous activities: representing activities as bags of event n-grams. In: IEEE International Conference on Computer Vision and Pattern Recognition, vol. 1, pp. 1031–1038 (2005)
8. Li, W., Zhang, Z., Liu, Z.: Action recognition based on a Bag of 3D Points. In: IEEE Computer Vision and Pattern Recognition Workshops, pp. 9–14 (2010)
9. Li, W., Zhang, Z., Liu, Z.: Expandable data-driven graphical modeling of human actions based on salient postures. IEEE Transactions on Circuits and Systems for Video Technology, 1499–1510 (2008)
10. Merler, M., Huang, B., Xie, L., Hua, G., Natsev, A.: Semantic Model Vectors for complex video event recognition. IEEE Transactions on Multimedia 14(1), 88–101 (2012)
11. Salton, G., Wong, A., Yang, C.S.: A vector space model for automatic indexing. ACM Communication Journal 18(11), 613–620 (1975)
12. Smeaton, A.F., Over, P., Kraaij, W.: Evaluation campaigns and TRECVid. In: 8th ACM International Workshop on Multimedia Information Retrieval, pp. 321–330 (2006)
13. Sung, J., Colin, P., Bart, S., Ashutosh, S.: Unstructured human activity detection from RGBD images. In: IEEE International Conference on Robotics and Automation, pp. 842–849 (2012)
14. Turaga, P., Chellappa, R., Subrahmanian, V.S., Udrea, O.: Machine recognition of human activities: A survey. IEEE Transactions on Circuits and Systems for Video Technology 18(11), 1473–1488 (2008)
15. Wikipedia: Charades - Wikipedia, The Free Encyclopedia, http://en.wikipedia.org/wiki/Charades (accessed March 1, 2012)

Tell Me about TV Commercials of This Product

Cai-Zhi Zhu[1], Siriwat Kasamwattanarote[1],
Xiaomeng Wu[2], and Shin'ichi Satoh[1]

[1] National Institute of Informatics,
2-1-2 Hitotsubashi, Chiyoda-ku, Tokyo 101-8430, Japan
[2] Media Information Laboratory, NTT Communication Laboratories,
3-1 Morinosato Wakamiya, Atsugi-shi, Kanagawa 243-0198, Japan

Abstract. TV commercial archives, once recorded the fashion and technology of our society, contain large amount of information deserved for deep analysis, for instance, discovery of hot products, exploration of the relationship between the air times and market sales of a product, analysis and prediction of the market trends, and so on. Levering a new text-to-features transformation and integrating many state-of-the-art video search techniques, we have built an interactive system on top of video retrieval in a large collection of three-year five-channel TV commercial videos. To the best of our knowledge, this is the largest commercial data set used for retrieval so far. To interact with the system, users can either use a keyboard to type keywords or use their mobile devices to snap a picture to describe their interested products, and the system will return relevant commercials in real time. Users are further able to browse videos and access their air patterns, such as air time and air frequency. This pattern usually reflects social behavior of viewers, *i.e.* which social groups (young or adult, male or female) are the targets of the product, when is the peak time for viewers to watch this commercial category according to the air pattern.

Keywords: TV Commercials, Video Retrieval, Text-to-features, Text-to-image, Video Browsing, Pattern Visualization.

1 Introduction

TV commercials record the technological advance in all aspects of our lives, such as food, finance, entertainment, education, travel and so on. Analysis on these valuable resources is believed to be beneficial to advertisers, manufacturers, retailers and consumers: advertisers are interested in the relationship between the market sales and air times of commercials, so as to smartly invest in advertisements; manufacturers are curious about hot sellers aired on TV in order to plan their product lines; retailers/consumers are willing to sell/buy hot products. Unlike other TV programs such as news, TV commercials usually have no closed captions associated in many countries. For example TV commercials in Japan lack meta data, thus are not directly searchable, which leads to those huge potentially valuable information being inaccessible.

C. Gurrin et al. (Eds.): MMM 2014, Part I, LNCS 8325, pp. 242–253, 2014.
© Springer International Publishing Switzerland 2014

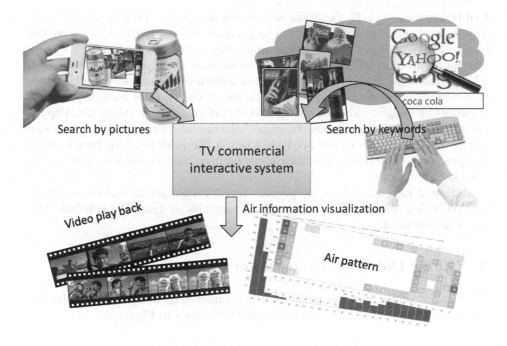

Search by pictures

TV commercial interactive system

Search by keywords

coca cola

Video play back

Air information visualization

Air pattern

Fig. 1. The TV commercial interactive system

Levering a new text-to-features transformation and integrating most state-of-the-art video search techniques, we have built an interactive system to reuse large archives of TV commercial videos, as shown in Figure 1. The interaction between users and the system can be fulfilled on PCs or mobile devices. Users are allowed to type query keywords, *e.g.*, the name of a logo, or snap and search products at hand with their mobile devices, and the system will return relevant commercials in real time. Afterward, commercial videos can be played back and related broadcasting information, such as air time and air frequency, is also accessible. We also keep recording the search log of users for future analysis on users' search behavior.

Related Work. Most work on TV commercials are about detection rather than retrieval [10]. A notable related work on this aspect was done by Del Bimbo and his team [4]: they proposed to retrieve commercials by their salient semantic features like colors, editing effects, motion, *etc.*, which is different from ours. There are similar works [5,12] on the application scenario of retrieving digital videos with photos snapped by mobile devices (note the type of video data is totally different), while the method we used is based on the state-of-the-art Bag-of-Words (BoW) framework [9] and the average pooling approach, for database videos, which leads the TrecVid Instance Search task [7,13]. Our method is more robust than the segmentation based method [12], and more scalable and faster than the method [5] of approximate nearest-neighbor search.

Contributions. (1) To facilitate text based search, we propose a new text-to-features transformation algorithm to generate representative features as the query. (2) We built a TV commercial interactive system on the largest TV commercial archives so far, which were extracted from three-year and five-channel TV programs. (3) Many state-of-the-art video retrieval techniques are integrated, as a result, our system exhibits multiple functions, such as querying by text keywords or images, video browsing, and air information visualization. The air pattern visualized usually reflects social behavior of viewers, *i.e.* which social groups are the targets of the product, when is the peak time for broadcasting the product. Advertisers, manufacturers, retailers and consumers will benefit from such a system.

Section 2 introduces how we prepare the TV commercial database offline. Section 3 describes the TV commercial interactive system. We conclude this paper in Section 4.

2   Offline Database Preparation

We first introduce how we detected TV commercials, and then describe the offline indexing process. The whole flowchart is shown in Figure 2.

2.1   TV Commercial Collection

We have currently archived three and a half years of full-day TV programs from seven Japanese TV channels. From which we picked three-year (starting from 17-Aug-2009 till 11-Aug-2012) and five-channel (TBS, TV Tokyo, NET, FUJI and NTV) commercials into consideration. First the dual-stage temporal recurrence hashing algorithm proposed by Wu and Satoh [10] was applied to detect commercials and clustered those detected commercial reruns into groups. In our experiment it took around 150 hours to analyze total fifteen years of full-day TV programs, from which we got around 4.5 million, three years' duration commercials (note in Japan commercials last 15 or 30 seconds). After clustering reruns together we had 34,709 groups. Finally we selected one representative video from each group to form our database, and got around 250 hours of TV commercials.

2.2   Offline Indexing

Since each TV commercial may consist of multiple shots corresponding to different scenes, we first detected shot boundary and segmented each commercial into shots. For each shot, multiple frames were sampled at a rate of 1 fps, and then for each frame, Hessian affine Root SIFT [1,6] features were extracted. In total, we got 356,019 shots, around 10 million frames, and more than 2 billion Root SIFTs. After that, we sampled a subset of 100 million Root SIFTs to train a 1 million vocabulary with the approximate k-means [8]. After the quantization of SIFT features each frame was represented by a BoW vector. Note if we take

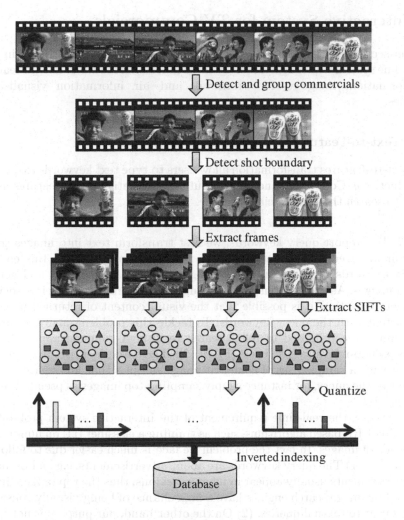

Detect and group commercials

Detect shot boundary

Extract frames

Extract SIFTs

Quantize

Inverted indexing

Database

Fig. 2. Offline commercial database preparation

frame as a basic unit, the database would be rather large. Since in this application navigation to a specific frame is not necessary, we took each shot as an unit in the database instead. To this end, multiple BoW vectors of frames contained in each shot were aggregated into one BoW vector by the average pooling algorithm [13], which won the TRECVID Instance Search 2011 [7]. In this case, taking the average pooling method is beneficial not only to the effectiveness but also to the efficiency, as the database size was significantly reduced from around 10 million to 356,019. Finally we built inverted indexing for BoW vectors of all shots for later online search.

3   Interactive System for TV Commercials

In this section we introduce the online interactive system, as shown in Figure 1. The system was concatenated by three functional modules: text-to-features transformation, image-to-video retrieval, and air information visualization modules.

3.1   Text-to-Features Transformation

The text-to-features transformation allow users to type text keywords to describe a product, *e.g.* Coca Cola, and will output representative visual features as the query to search the commercial database.

Why? To support query keywords, we first transform text into images (recall that our archived TV commercials have no closed captions). To this end, we submit keywords typed by users to Google image search engine, and get top ranked images. As existing image search engines rank images by their associated text metadata, thus it is possible that the visual content of returned images is inconsistent with the query keywords. To tackle this problem, many re-ranking algorithms were proposed to refine search results. A common approach is to utilize pseudo-relevance feedback (PRF) [2]. The basic idea is to perform re-ranking with a sampled set of confident image samples that are regarded to be relevant to the query, for instance simply sampling top images as pseudo-positive examples [11].

In our case, the real-time requirement of the interactive system prohibits us from using PRF based algorithms, such as training a classifier [11] on hundreds or thousands of images. In fact the problem we face is much easier due to following two reasons: (1) The query keywords are about advertisements, e.g., a logo name. As advertisements usually appear in many occasions, thus the top ranked images returned by image search engines have a high chance of being visually consistent and relevant to textual queries. (2) On the other hand, our purpose is not to get an entire re-ranked list, but to select representative query images to the keywords for later video retrieval. In fact, since our system is built on top of local features and the BoW framework, the query can be represented by a set of features that ideally correspond to relevant parts of images. We name it text-to-features transformation in this paper.

How? We propose an algorithm to transform query keywords into visual features, and representative images can be deduced as well, as stated in Algorithm 1. Given a text query, the algorithm incrementally crawls top-ranked images from the Google image search engine. Each time a set of SIFT features of a newly downloaded image will be matched with those of previous images by the *matching* function, and the number of matched features will be accumulated for each image. Finally the image having the largest number of matched features will be regarded as the representative image to the query keywords, accordingly

Algorithm 1. Text-to-features transformation

---

input : Text queries Q
output: Representative image I_{rep} and features F_{rep}
Search Google with Q and get urls of top N images $[I_0, \cdots, I_N]$;
$F(i) \leftarrow \emptyset, i \in [1, N], abort \leftarrow false$;
for $i \leftarrow 1$ **to** N **do**
    Download and resize image I_i;
    Extract SIFTs S_i from image I_i;
    for $j \leftarrow 1$ **to** $i - 1$ **do**
        $[good\_i\_idx, good\_j\_idx] \leftarrow matching(S_i, S_j)$;
        $F(i) \leftarrow unique(F(i), S_i(good\_i\_idx))$;
        $F(j) \leftarrow unique(F(j), S_j(good\_j\_idx))$;
        if $length(F(i)) > T$ **or** $length(F(j)) > T$ **then**
           | $abort \leftarrow true$;
        end
    end
    if $abort$ is $true$ **then**
        | break;
    end
end
$best\_i \leftarrow$arg $\max(length(F(i)))$;
$I_{rep} \leftarrow I_{best\_i}$;
$F_{rep} \leftarrow F_{best\_i}$;

---

those accumulated matched features will be outputted as representative features. For the sake of speed, the process can abort early once enough number of images are processed or enough matched features are found, N and T are the threshold, respectively. In our case we set $N = 5$ and leave T a big number.

Two options of the *matching* function are given in Algorithm 2 and Algorithm 3. The former one is relatively slow but accurate, in which mutual nearest neighbor pairs from two input sets of SIFT features are first found, and then inliers detected by a RANSAC based geometric verification algorithm are regards as good matches. In contrast the latter one is faster, as SIFT features quantized to the same codeword are simply regarded as good matches. Since the quantization of features is also needed in the subsequent search process, Algorithm 3 actually costs no much extra computation. In our experiments, we found both of them yielded good enough results, so Algorithm 3 is used for the consideration of speed.

An example of querying with the keyword "Asahi" (a Japanese brand of beer) is shown in Figure 3. Top 5 images returned by Google image search engine are shown in the second row from left to right in order, with the numbers of accumulated matched features shown beneath. The 5th image will be outputted as the representative image for the largest number (say, 96 in Figure 3) of matched features detected. The close-up of the representative image is shown in the first

---

Algorithm 2. Geometry based matching

input : Two sets of SIFT features S_i, S_j, homography estimation function
H_func
output: Corresponding index of inliers $good\_i\_idx$, $good\_j\_idx$
$[S_i^{nn}, S_j^{nn}] \leftarrow$ mutual_nn_search(S_i, S_j);
$[good\_i\_idx, good\_j\_idx] \leftarrow$ RANSAC$(S_i^{nn}, S_j^{nn}, H\_func)$;

Algorithm 3. BoW based matching

input : Two sets of SIFT features S_i, S_j, trained *dictionary*
output: Corresponding index of matching SIFTs $good\_i\_idx$, $good\_j\_idx$
$assign_i \leftarrow quantize(S_i, dictionary)$;
$assign_j \leftarrow quantize(S_j, dictionary)$;
$good\_i\_idx \leftarrow ismember(assign_i, assign_j)$;
$good\_j\_idx \leftarrow ismember(assign_j, assign_i)$;

| 71 | 44 | 49 | 42 | 96 |

Fig. 3. Representative image/features of "Asahi"

row with part of matched features overlaid in red. We can see that those features can really capture the salient regions, *i.e.* two logos of "Asahi" in this example.

3.2   Image-to-Video Search

Users are allowed to submit query images to the system from PC browser clients or iPhone/iPad mobile devices. The pipeline of online search is shown in Figure 4, where the query can be the representative image generated from the above transformation module, or images input by users, for instance, drag-and-drop an image from a local PC or snap a picture with a mobile phone. A similar process to the offline indexing will then be applied to get a BoW representation

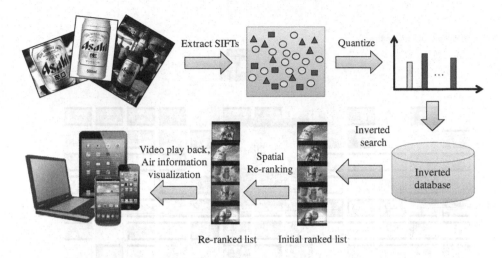

Extract SIFTs

Quantize

Inverted
search

Inverted
database

Video play back,
Air information
visualization

Spatial
Re-ranking

Re-ranked list   Initial ranked list

Fig. 4. The online image-to-video search pipeline

for the query. Following two points are worth mentioning: (1) If there are multiple query images, average pooling will be applied to get one BoW vector. In this way the system will be queried only once, and rank aggregation steps are completely avoided. The effectiveness of this method was proved in Zhu and Satoh's work [13]. (2) Only representative features will be took into account for the generated representative image.

The search process will be greatly sped up by the inverted indexing: normally the initial ranked results will be ready in milliseconds if only considering the search time. Then the initial results can be further refined by a Local Optimized RANSAC (LO-RANSAC) [3] based spatial re-ranking, which is fulfilled in parallel on 20 cores in our system. Thanks to the accurate spatial verification, our system is able to output a Boolean answer about whether the query product has appeared in the database or not.

The retrieval results by querying with "Asahi" are shown in Figure 5. More initial search results without spatial re-ranking are shown in Figure 6, we can see that our system performs very good and gets only one incorrect result in the 5th position of the Pizza example. After LO-RANSAC re-ranking all returns of those three examples are correct.

3.3 Air Information Visualization

With the air information visualization module, users are able to play back those returned commercials and access their broadcast information, such as air time and air frequency in different temporal granularities (hours, days, weeks and months). The air pattern of "Asahi" beer is shown in Figure 7, which includes (a) Hour-Dow (day of week) and (b) Hour-Frequency two graphs. The Hour-Dow pattern is interesting since it reflects Japanese policy on the air time of beer commercials: beer commercials are not allowed to be broadcasted from 2:00

Fig. 5. Querying with the text keyword "Asahi". By clicking one commercial thumbnail a float window will pop up, in which the commercial video will be streamed and related air information (TV channel, broadcast times, *etc.*) will be displayed beneath. The graphic air pattern can be accessed by the yellow text link at the bottom of the window.

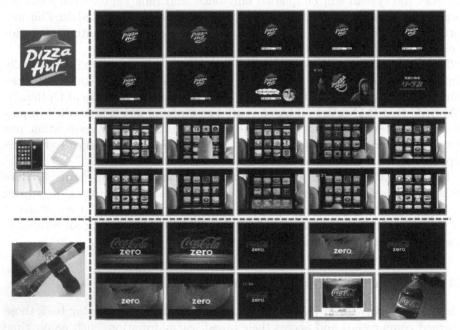

Fig. 6. Search results by querying with the Pizza Hut logo, a set of iPhone images, and a snapped pet bottle of Coca Cola Zero, respectively. Query objects are on the left side. On the right side, the top 10 thumbnails of videos are returned, in which positive (negative) returns are marked with green (red) bounding boxes. Note the representative image and features, detected by our method, of iPhone queries are highlighted.

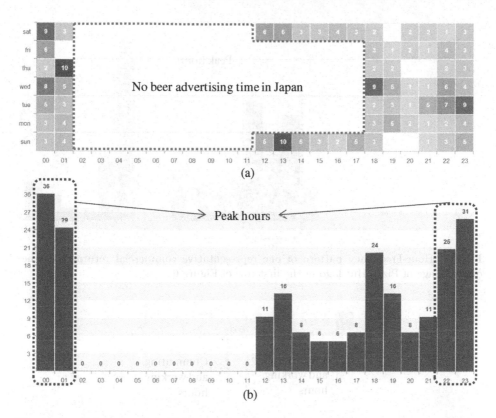

Fig. 7. Graphic air pattern of one representative commercial "Asahi" returned in Figure 5: (a) Hour (X-axis)-Dow (Y-axis) pattern, where the replay frequency per hour is shown in numbers; (b) Hour (X-axis)-Frequency (Y-axis) pattern

AM to 5:00 PM on weekdays and 2:00 AM to 11:00 AM on weekends. The Hour-Frequency pattern in Figure 7 (b) shows that the "Asahi" commercial is frequently broadcasted from 6:00 PM to 1:00 AM on average, and the peak hours, which are consistent with Japanese drinking time, are from 10:00 PM to 1:00 AM. We give other two interesting patterns in Figure 8 and 9 by querying with the Pizza Hut logo and the keyword "car", respectively. Figure 8 shows that pizza commercials are most frequently broadcasted right before the lunch (11:00 AM) and dinner (6:00 PM) time so as to remind viewers (mainly Japanese housewives) to order. Figure 9 shows that car advertisers try to skip commuting and working hours and focus on specific o'clocks when people have time to sit down and watch TV, such as lunch time or staying at home, and the peak hour is around 6:00 AM when people prepare to go to work. In this case the main target group could be office workers. This kind of air pattern is specially meaningful to some experts, such as advertisers: they are good at discovering and telling the difference behind broadcasting pattern among different products, thus to plan their marketing strategy accordingly.

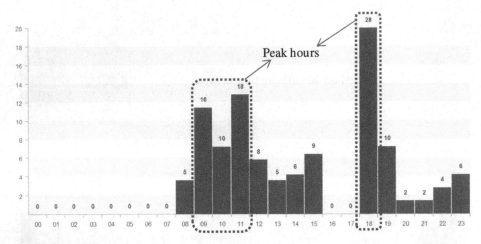

Fig. 8. Hour-Frequency pattern of one representative commercial returned by the query image of Pizza Hut logo in the first row of Figure 6

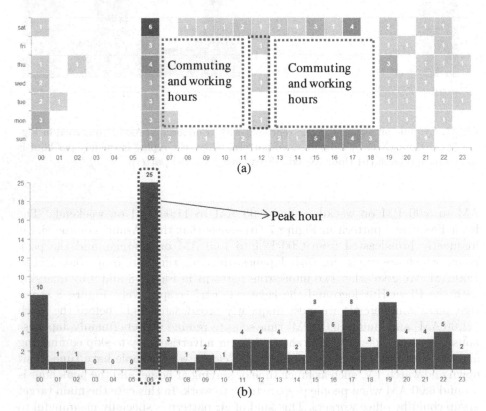

Fig. 9. Air pattern of one representative commercial returned by the query keyword "car": (a) Hour-Dow pattern; (b) Hour-Frequency pattern

4   Conclusions

In this paper, we proposed an interactive system to reuse large archives of TV commercial videos. To support querying by keywords, we designed an efficient text-to-features transformation algorithm, which takes advantage of an existing image search engine. To interact with this system, users can type text keywords, upload query images in hand, or simply snap and search with their mobile phones, and the system will return relevant TV commercials. Users can choose to browse returned commercials or access their air information. This system serves as a platform that can be expanded to various applications in the future.

References

1. Arandjelović, R., Zisserman, A.: Three things everyone should know to improve object retrieval. In: CVPR (2012)
2. Carbonell, J., Yang, Y., Frederking, R., Brown, R., Geng, Y., Lee, D.: Translingual information retrieval. In: IJCAI (1997)
3. Chum, O., Matas, J., Kittler, J.: Locally optimized RANSAC. In: Michaelis, B., Krell, G. (eds.) DAGM 2003. LNCS, vol. 2781, pp. 236–243. Springer, Heidelberg (2003)
4. Colombo, C., Bimbo, A.D., Pala, P.: Retrieval of commercials by semantic content: The semiotic perspective. Multimedia Tools Appl. 13(1), 93–118 (2001)
5. Henze, N., Boll, S.: Snap and share your photobooks. In: ACM Multimedia (2008)
6. Lowe, D.: Distinctive image features from scale-invariant key points. IJCV 60, 91–110 (2004)
7. Paul, O., Awad, G., Michel, M., Fiscus, J., Kraaij, W., Smeaton, A.F., Quéenot, G.: Trecvid 2011 - an overview of the goals, tasks, data, evaluation mechanisms and metrics. In: TRECVID (2011)
8. Philbin, J., Chum, O., Isard, M., Sivic, J., Zisserman, A.: Object retrieval with large vocabularies and fast spatial matching. In: CVPR (2007)
9. Sivic, J., Zisserman, A.: Video google: A text retrieval approach to object matching in videos. In: ICCV (2003)
10. Wu, X., Satoh, S.: Ultrahigh-speed tv commercial detection, extraction, and matching. IEEE Trans. Circuits Syst. Video Tech. 23(6), 1054–1069 (2013)
11. Yan, R., Hauptmann, A., Jin, R.: Multimedia search with pseudo-relevance feedback. In: Bakker, E.M., Lew, M., Huang, T.S., Sebe, N., Zhou, X.S. (eds.) CIVR 2003. LNCS, vol. 2728, Springer, Heidelberg (2003)
12. Zhu, C.-Z., Mei, T., Hua, X.-S.: Natural video browsing. In: ACM Multimedia (2005)
13. Zhu, C.-Z., Satoh, S.: Large vocabulary quantization for searching instances from videos. In: ICMR (2012)

A Data-Driven Personalized Digital Ink
for Chinese Characters

Tianyang Yi, Zhouhui Lian*, Yingmin Tang, and Jianguo Xiao

Institute of Computer Science and Technology, Peking University,
No. 128 Zhongguancun North Street, Haidian District, Beijing 100080, China
{yitianyang,lianzhouhui,tangyingmin,xiaojianguo}@pku.edu.cn

Abstract. In this paper, we propose a novel data-driven digital ink generating method for Chinese characters. When users write on tablets, they can select a specific rendering style from the predefined style database to generate character images in the selected style. Our method is able to learn personalized style information from a small set of Chinese characters in a given style and then generate the stroke style database using a specially-designed stroke segmentation model. The model divides a stroke into three kinds of segments: head, corner and middle segments. For each type of stroke segments, we employ the corresponding method to process the style information they contain. Experimental results demonstrate that our method not only works well for Chinese characters but is also effective for other kinds of shapes.

Keywords: digital ink, data-driven, stroke segmentation, style database.

1 Introduction

Chinese calligraphy has a long history throughout thousands of years. Ancient and modern people have created plentiful calligraphic styles using various techniques. Today, although many primary and middle schools in China have courses training students' handwriting skills, most people are not able to write as well as skillful calligraphers.

Along with the development of tablet PCs and other touch screen devices, there has been an increasing interest in exploiting screen touching input technology. Typically, users first create the trajectory of a character on a device, then the system helps to generate and beautify the character image and show it on the screen. A good rendering algorithm can generate artistic character effects even with simple input trajectories. The Digital Ink technique [1] has been proposed to do such work. However, we need to design it more carefully for Chinese characters compared to other general shapes.

So far, some commercial touching screen software with handwriting rendering function and related algorithms proposed by other researchers have exploited many ways to beautify users' handwriting inputs. However, their methods focus

---

* Corresponding author.

C. Gurrin et al. (Eds.): MMM 2014, Part I, LNCS 8325, pp. 254–265, 2014.

on beautifying handwritings in a somehow monotonous manner, which results in output character images of a fixed style. Users may be pleased to see different rendering styles for a same input trajectory, and they can choose their favorite style as the final "rendering rule".

In this paper, we propose a novel data-driven digital ink algorithm. The method can "learn" writing styles from a small set of training character images at stroke level, based on the fact that user's input character trajectory is composed of different stroke trajectories. To establish a style database, first we apply a semi-automatic method to extract strokes from training character images to get training stroke images. Afterwards, we utilize a novel stroke segmentation model to record strokes' style information. According to the Chinese handwriting custom, our algorithm relies on an intuition that the tip and corner parts of a stroke may contain more style information than middle parts (see Fig. 2).

For each writing style, we only need to learn it once to establish its own style database. For instance, when users write on a touch screen, styles that the system has learned will be selectable for them. We define a stroke trajectory as the move of a stylus between touching and leaving the touch screen. With the stroke trajectory, our method will choose the most suitable stroke information in the style database to generate the stroke image (see Fig. 1).

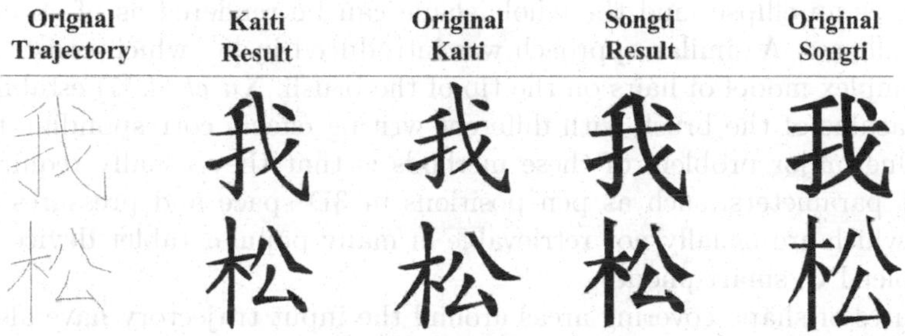

Orignal Trajectory	Kaiti Result	Original Kaiti	Songti Result	Original Songti

Fig. 1. Two examples of the results obtained by our method. Given a writing trajectory, it can be rendered in different styles selected by users (e.g., column 2 in Kaiti style and column 4 in Songti style). Comparing results generated using our method with original characters of the selected style, we can see that their writing styles are very similar but our results can still preserve the characteristics of original trajectories.

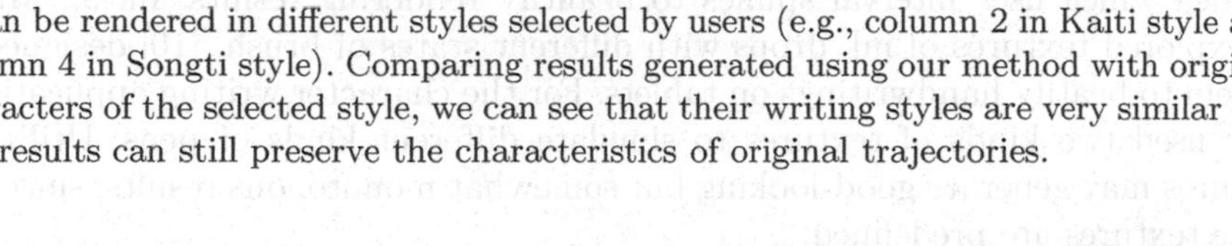

Fig. 2. Same characters with different writing styles. We can see obvious differences between corresponding strokes in different styles. Tip and corner positions have different shapes, while middle positions only differ in direction and width.

2 Related Work

Up to now, numerous methods have been developed to improve the handwriting quality of characters or drawings. Due to their intrinsic properties, those methods all have their own advantages and disadvantages.

Only a few studies have been carried out aiming at the similar goal as this paper: beautifying input Chinese character trajectories captured by styluses and tablet devices. In [2], they predefined some shapes in Kaiti style as skins, and the final result obtained is a simple combination of some chosen skins around the input trajectory bone. [3] and [4] share the same idea of generating the beautified character shape by interpolating the input trajectory and a existing character. Furthermore, in [4] they applied a learning process to get better results. Our method is motivated by some key ideas presented in above-mentioned papers: machine learning, skins around bone and shape interpolation. Furthermore, we also employ a novel stroke segmentation model, which makes our method work better than existing approaches for emerging different writing styles and generating beautiful results.

There also exist many algorithms that aim to render high-quality handwritings based on the physically simulating writing process of a real pen or brush. Wong and Ip [5] proposed a method simulating brush status when moving on paper. They assumed that the interactive area between the brush and paper can be modeled as an ellipse, and the whole shape can be rendered as of an array of moving ellipses. A similar approach was introduced in [6], which constructed a more complex model of hairs on the tip of the brush. Xu et al. [7] established a state machine of the brush with different writing effects corresponding to each state. One major problem of these methods is that they usually require a lot of input parameters, such as pen positions in 3D space and pressures on the screen, which are usually not retrievable in many popular tablet devices today (e.g. tablet PC, smart phone).

Textures or shape covering areas around the input trajectory have also been used to accomplish handwriting shape rendering. Su et al. [8] came up with an idea which uses interval splines to beautify rendering results. In [9], Mi et al. explored textures of ink drops with different states of brush. [10] designed a system to beatify handwritings on tablets. For the character writing application, they used two kinds of textures to simulate different kinds of pens. Utilizing textures may generate good-looking but somewhat monotonous results, since all these textures are predefined.

Another way to beatify handwritings is to adjust the input trajectory to achieve some specific purposes. Su et al. [11] matched hinting information with standard TrueType and input trajectories, making rendering results of different sizes to be good-looking. [12] extended this idea to drawings. Zhu and Jin [13] aimed at fusing the standard character shape with the scratchy input trajectory to make the result easier to read. More recently, Zitnick [14] proposed a novel approach which calculates the average of repeated inputs, to make the follow-up rendering result of the same content look better. Most of these methods adopt the simple way of rendering, which is not able to generate artistic handwriting effects.

To solve these problems and get better handwriting beautification results, we combine the ideas discussed above with a novel stroke segmentation model. In this way, our approach can generate various kinds of stylish characters with smooth contours and beautified appearances.

3   Approach

In this section, we give a detailed description of our method that generates stylish and beautified rendering results. Fig. 3 shows the pipeline of our approach. First, we introduce the four-level constraints for Chinese character stroke rendering, which indicate reusability relations between strokes in characters, and selecting training characters based on a chosen constraint. Then, with the images of training characters in a given style, we analyze how to extract stroke segment style information using a semi-automatic method. Finally, we apply a rendering algorithm to select the best suited information from the style database and generate the final character image from an input trajectory.

3.1   Select Training Characters

In this section, we analyze stroke reusability relations among different Chinese characters, and then select training characters based on the analysis. Basically, Chinese characters are hieroglyphs which consist of radicals and their radicals can be further decomposed into strokes. As shown in Fig. 4 (b), same kinds of strokes and radicals have reusable relations in different characters. Therefore, we want to set up reusability relations among strokes in characters.

Typically strokes in Chinese characters can be classified into 32 categories (see Fig. 4 (a)), and strokes in the same category may be written or designed in quite dissimilar ways.(see Fig. 4 (c)). So we need to subdivide strokes in the same kind into more categories. More specifically, we first classify all strokes

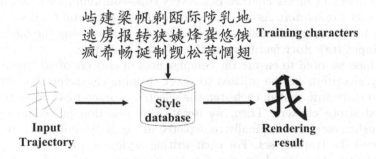

Fig. 3. Pipeline of our approach

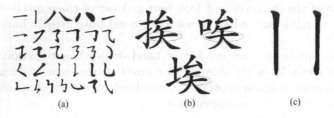

Fig. 4. 32 stroke kinds in Chinese characters (a), stroke reusable relations (b), Xuan Zhen shu and Chui Lu shu (c), which have different shapes at the end of strokes

into 32 categories, and then we select some representative strokes in the same category as cluster centers. Finally, a simple clustering algorithm is adopted to choose the best-suited cluster for a target stroke. After finer classification and clustering, we can claim that strokes in the same cluster are reusable.

When users write Chinese characters on tablets, assume that a separated input trajectory corresponds to one stroke, we can define four different levels of constraints which determine our strategy about how to select information from a given style database:

Level 0 constraint: No character related information is provided, we select information from the database simply based on trajectory features.

Level 1 constraint: Suppose the kinds of input strokes are already known, we select information from the database based on both the stroke kind and trajectory features.

Level 2 constraint: Suppose the kinds and cluster indexes of input strokes are already known, we make selections according to the stroke kind, cluster index and trajectory features.

Level 3 constraint: Suppose the kinds and cluster indexes of input strokes and which kind of radical the input stroke belongs to are already known, we make selections based on the radical kind, stroke kind, cluster index and trajectory features.

As we focus on rendering at stroke level and obtaining the radical classification information of all Chinese characters is a very time-consuming task, our method adopts *Level 2 constraint*. As we can see from our experimental results described below, it is already good enough to render stroke images using the information of both input trajectory features and the selected font style.

Therefore, we need to ensure the completeness of clusters of all stroke kinds. A greedy algorithm can be utilized to select training characters from the font library to make sure that in each step the selected character can cover the most uncovered stroke clusters. Then, we iterate this selection procedure until all stroke clusters are covered. Finally, as depicted in Fig. 3, 30 characters are chosen to compose the training set. For each writing style, we just need to provide images of these training characters for our system to learn and establish the style database.

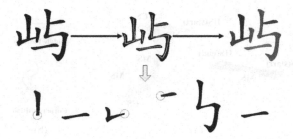

Fig. 5. Illustration of stroke extraction. We adopt a semi-automatic method by marking the rough skeleton for each stroke. Red circles in stroke images indicate regions where style information is lost.

3.2   Establish Style Database

In this section, we describe how to establish the style database from training character images. First, we need to extract raw stroke images from each character image. Afterwards, we adopt a novel stroke segmentation model to extract stroke segments, and then for each kind of segments we use the corresponding method to extract their features.

Extract Raw Stroke. As mentioned above, we first need to extract strokes from character images. To the best of our knowledge, existing automatic stroke extraction algorithm still cannot guarantee 100% correct extrations. Also, considering that the number of our training character images is small, we employ a semi-automatic stroke extraction method, which can precisely and completely segment all strokes out for every Chinese character.

There are three steps in this procedure as shown in Fig. 5. Firstly, we mark an inaccurate skeleton on a iPad for every stroke in the character image manually with a specially-designed IOS app. Secondly, our algorithm intends to find out regions roughly occupied by every stroke. Basically, we go through every valid pixel in the character image to seek for its nearest skeleton, and add the pixel to the corresponding stroke area. Finally, we implement some shape modifications to get final stroke images. Since most strokes in the Chinese character may intersect with other strokes, we need to automatically mark those stroke intersection regions. If a marked region is near the tip or corner part of stroke, the style information of this region may be incomplete. While, if the marked region is in the middle part, we will complete the stroke shape here by modifying contour between corresponding points on two separated contours.

Stroke Segmentation. After we have obtained raw stroke images, stroke segmentation that further decomposes a stroke into more segments should be carried out. To understand what features each kind of stroke segment contains, here we first analyze how calligraphers write Chinese characters. Calligraphers usually apply non-flat moving of pen or brush at stroke heads or stroke corners to make

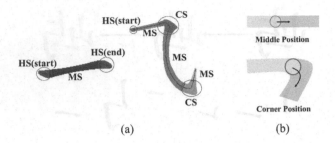

Fig. 6. Stroke Segmentation Model (a), Max circle rolling algorithm (b)

a nice appearance of the stroke, which results in various shapes at those positions. On the other hand, when writing at the middle position of the stroke, style information often appear as slightly-varied skeleton direction and stroke width. To catch the above-mentioned stroke features, we design a stroke segmentation model which works effectively for this task.

As shown in Fig. 6(a), we define the following three kinds of stroke segments:

Middle segment (MS): Stroke segments whose skeleton direction and stroke width are relatively stable.

Corner segment (CS): Stroke segments that are used to connect middle segments.

Head segment (HS): Start and end parts of the stroke.

Thus, the stroke can be represented as,

$$Stroke = (start\ HS) + MS + \{CS + MS\} + (end\ HS),$$

where {x} means x may not appear or repeat several times, (x) means x may not appear or repeat once. Note that if there is no obvious stylish appearance in a tip area, we should treat it as MS.

Since it is difficult to directly calculate the direction and width for stroke segments on the raw stroke image, we use the Max Circle Rolling Algorithm to first get the rough skeleton of the stroke. Our algorithm pretends that there is a circle rolling inside the valid stroke region, and the trajectory of rolling circles can roughly represent the real writing trajectory. The method can simulate the writing trajectory of stroke, thus facilitates the calculation of direction and width for strokes.

The Max Circle Rolling Algorithm consists of the following two steps: corner detection and circle rolling. In the first step, we detect all contour points to find *Inner Corner Contour Points(ICCPs)*, which determine the inner contours of corner parts. The second step is to roll circles through the whole character. As depicted in Fig. 6 (b), circle rolling should obey the following two rules: *Rule 1*: When rolling in middle or head segment areas, choose the nearby circle that covers most uncovered valid pixels to be the next rolling circle. *Rule 2*: When rolling in corner segment areas, in addition to follow the Rule 1, the circle must also be tangent to inner contour, namely covering at least one *ICCP*.

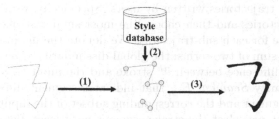

Fig. 7. Illustration of character rendering. (1) Segmentation of trajectory; (2) Select best suited stroke segments; (3) Rendering and smoothing.

With the rolling circle trajectory of the stroke, we then decompose the stroke into segments. To be specific, corner segments can be detected by rolling circles generated based on *Rule 2*. Since stroke shapes may vary in different styles, we first adopt a simple dividing method, and then apply manual checking and adjustment to get the accurate dividing index of circle trajectory for each stroke segment. Finally, according to dividing points and direction data, we set up "separating walls"(cutting lines) to decompose the stroke into stroke segments (see Fig. 6).

Segment Formatting. Due to the fact that different kinds of stroke segments contain quite different information, we design the corresponding special representation format to describe each kind of segments. Such design can not only save large amounts of storage spaces, but also make character rendering more convenient.

Middle segments: The middle segment is represented as a set of rolling circles. Because the values of stroke width and direction are approximately stable in middle segments, circles are selected evenly from circle trajectory, and their radius and direction information are recorded.

Head segments: Xia and Jin [2] predefined three Bezier curves to represent shape features, here we utilize a similar method, which implements a Bezier fitting method over contour points of the segment, producing an array of Bezier curves as segment features. The information stored in the style database includes entrance circle center, entrance width, entrance direction, and control points of Bezier curves.

Corner segments: Corner segments are represented in the similar way as head segments. Unlike head segments, corner segments have an entrance, an exit, inner and outer parts of contour points. The information stored in the style database includes entrance circle center, entrance width, entrance direction, exit circle center, exit width, exit direction, and control points of Bezier curves.

3.3   Rendering

As shown in Fig. 7, after extracting the style data from given training images, we need to choose appropriate information from the style database to generate stroke shapes for trajectories written by users. Specifically, we should first segment input trajectories and then choose the most similar stroke segment from the style database for each sub-trajectory. We denote the dissimilarity measure $Total\_E_{i,j}$ as the sum of two terms: the global dissimilarity $Global\_E_i$ that indicates the global difference between i^{th} stroke and the input trajectory, and the segment dissimilarity $Segment\_E_{i,j}$ that indicates segment difference between i^{th} stroke's j^{th} segment and the corresponding subset of the input trajectory. At *Level 0 constraint*, we select the stroke segment with least $Total\_E_{i,j}$ from all stroke segments in the style database, while at *Level 2 constraint*, we select the stroke segment with least $Total\_E_{i,j}$ from stroke segments with the same stroke kind and cluster type.

To generate the final rendering result, we first separately render the shape of each stroke segment, and then a smoothing method is adopted to adjust contour points within a stroke segment and connect contour points between nearby segments. Middle segments are simply rendered as rolling circles with trajectory points as centers and corresponding width values of the selected stroke segment as their radii. Shapes of head segments and corner segments need to be adjusted to ensure that subparts of two stroke segments in the joint region have almost the same width and direction. Afterwards, we apply Quadratic Bezier Curve fitting for each segment contour. For the joint region, we combine two Quadratic Bezier Curves into one Cubic Bezier Curve, keeping tip direction unchanged and generating smoothed joint parts.

4   Experiments

In this section, we show and discuss experimental results of our personalized digital ink. We develop an IOS app to obtain real handwriting trajectories on an iPad as inputs. Two typical styles of Chinese characters, i.e., Kaiti and Songti, are tested in our experiments. The size of a style stroke database is less than 1MB and the average time to render a stroke is less than 1 second.

Rendering Training Characters: In our first experiment, we test our method on training characters. From the comparison shown in the first two rows of Fig. 8, we can see that rendering result obtained using our method not only have the similar style as training data but also preserve the features of input trajectories.

Rendering Other Characters: In the second experiment, we render characters which are not contained in the training character set. As we can see from the last three rows in Fig. 8, where we put original character images after our rendering results, our method can effectively detect stroke segments and generate satisfactory results.

Comparison with Other Methods: We also carry out experiments on some characters tested by other researchers [3]. As we can see from Fig. 9, our results can appear in different styles selected by users, while rendering algorithms introduced in previous work can only generate relatively monotonous results.

Orignal Trajectory	Kaiti Result	Original Kaiti	Songti Result	Original Songti

Fig. 8. Experimental results of training characters and other characters

Orignal Trajectory	Kaiti Result	Original Kaiti	Songti Result	Original Songti	Wei' s Method	Li' s Method

				Youdao Cloud Note	Evernote

Fig. 9. Experimental results of characters compared with other methods and sotfware

Comparison with Commercial Software: Today, there exist some popular software for note writing on tablets. They have similar functions for Chinese handwriting beautification. We select some rendering results of two commonly-used software: Evernote and Youdao Cloud Notes. From results shown in Fig. 9, we can see that our results contains more style information, which is often desirable for many users.

Rendering Other Kinds of Shapes: Furthermore, if we follow *Level 0 constraint* instead of *Level 2 constraint* in our method, we can deal with any kind of shapes that can be decomposed into contour segments. Fig. 10 shows some

Fig. 10. Rendering results of an object and characters of other language

interesting results of characters in other languages and a cute object. As we can see, our method integrates Chinese writing styles into these results, and thus generating dramatic effects.

5   Discussion and Conclusion

In this paper, we propose a novel data-driven digital ink generating method for Chinese characters. Given training character images in a certain style, our system can generate the specific stroke style database. When users write on tablets, they can select a specified rendering style, and our system will generate stroke shapes around input trajectories with selected style. We propose a specially-designed stroke segmentation model to extract different style information carried by each kind of segments, which divides a stroke into three kinds of segments: head, corner and middle segments. Experimental results demonstrate the effectiveness and innovativeness of the proposed method.

Currently, our approach can only process offline inputs since according to *Level 2 constraint* we need to know what characters users are writing. If we extend our algorithm to support online processing, real-time rendering will become a key issue. In future work, we will optimize our approach using the delay-deforming method as mentioned in [14]. Specifically, we can use *Level 0 constraint* at first, to generate original stroke shape right after the user has finished stroke trajectory writing. When the user has finished writing the whole character, we can adopt an existing OCR method to recognize the writing character, and then we turn to *Level 2 constraint*to make adjustments for each stroke.

Since users may have thousands of input trajectories for a same character, we cannot ensure the aesthetics of whole character because our approach focuses on rendering at stroke level. Also, our method is not able to handle joined-up handwritings since we assume that every trajectory corresponds well to a stroke. Hence, adjusting positional relations between strokes and recognizing joined-up writing trajectories will be another key issue in our future work.

Acknowledgements. This work was supported by National Natural Science Foundation of China (Grant No.: 61202230) and China Postdoctoral Science Foundation (Grant No.: 2013T60038).

References

1. Hse, H.H., Newton, R.A.: Recognition and beautification of multi-stroke symbols in digital ink. Computers & Graphics 29(4), 533–546 (2005)
2. Xia, W., Jin, L.: A Kai style calligraphic beautification method for handwriting Chinese character. In: ICDAR 2009, pp. 798–802. IEEE (2009)
3. Li, H., Liu, P., Xu, S., Lin, S.: Calligraphy beautification method for chinese handwritings. In: ICDH 2012, pp. 122–127. IEEE (2012)
4. Xu, S., Jin, T., Jiang, H., Lau, F.C.: Automatic generation of personal chinese handwriting by capturing the characteristics of personal handwriting. In: IAAI, pp. 191–196 (2009)
5. Wong, H.T., Ip, H.H.: Virtual brush: a model-based synthesis of chinese calligraphy. Computers & Graphics 24(1), 99–113 (2000)
6. Chu, N.H., Tai, C.L.: An efficient brush model for physically-based 3d painting. In: Proceedings of the 10th Pacific Conference on Computer Graphics and Applications, pp. 413–421. IEEE (2002)
7. Xu, S., Lau, F., Tang, F., Pan, Y.: Advanced design for a realistic virtual brush. In: Computer Graphics Forum, vol. 22, pp. 533–542. Wiley Online Library (2003)
8. Su, S.L., Xu, Y.Q., Shum, H.Y., Chen, F.: Simulating artistic brushstrokes using interval splines. In: Proceedings of the 5th IASTED International Conference on Computer Graphics and Imaging, pp. 85–90 (2002)
9. Mi, X.F., Tang, M., Dong, J.X.: Droplet: a virtual brush model to simulate chinese calligraphy and painting. Journal of Computer Science and Technology 19(3), 393–404 (2004)
10. Hu, R.: Portable implementation of digital ink: collaboration and calligraphy. PhD thesis, The University of Western Ontario (2009)
11. Su, S.L., Wu, C., Xu, Y.Q.: A hybrid approach to rendering handwritten characters (2004)
12. Hsu, S.C., Lee, I.H.: Drawing and animation using skeletal strokes. In: Proceedings of the 21st Annual Conference on Computer Graphics and Interactive Techniques, pp. 109–118. ACM (1994)
13. Zhu, X., Jin, L.: Calligraphic beautification of handwritten chinese characters: a patternized approach to handwriting transfiguration. In: Proceedings of ICFH 2008, pp. 135–140 (2008)
14. Zitnick, C.L.: Handwriting beautification using token means. To appear in Proceedings of SIGGRAPH (2013)

Local Segmentation for Pedestrian Tracking in Dense Crowds

Clement Creusot

Toshiba RDC, Kawasaki, Japan
clementcreusot@gmail.com
http://clementcreusot.com/pedestrian

Abstract. People tracking in dense crowds is challenging due to the high levels of inter-pedestrian occlusions occurring continuously. After each successive occlusion, the surface of the tracked object that has never been hidden reduces. If not corrected, this shrinking problem eventually causes the system to stop as the area to track become too small. In this paper we investigate how hidden parts of one target object can be recovered after occlusions and propose challenging data to evaluate such segmentation-tracking technique in dense crowds. The segmentation/tracking problem is particularly difficult to solve for non-rigid objects. Here, we focus on pedestrians whose limbs and lower body parts often get occluded in crowded scene. We first investigate the unmet challenges of pedestrian tracking in crowds and propose a challenging video to evaluate segmentation-tracking robustness to inter-pedestrian occlusions. We then detail a fast segmentation-based method to overcome some aspects of the tracking-under-occlusion problem. We finally compare our results with two existing tracking methods.

1 Introduction

Tracking a single object/person in a video is a trivial task for any human operator. Machine vision systems are able, to some extent, to execute such a task. One of the most successful approaches used by many researchers has been to combine frequent object-detection with object-tracking algorithms. While these systems are not as robust as human perception, they now achieve very high results on video of sparse pedestrian scene (*e.g.* [1–3]). The success of these approaches is mainly due to the frequent human body or head detections which are very fast and fairly robust in non-occluded scene. The main limit of such techniques occurs when the detector is no-longer able to seed an initial solution to the tracking algorithm. This happens when pedestrians start to be occluded by other objects, which most of the time are also pedestrians. At this point, at least two research directions are possible: First, to develop even more sophisticated detectors that will reliably detect body parts under occlusion (*e.g.* Poselets [4]). This is a sound idea as humans are able to segment pedestrians' body parts from still images. Second, to develop trackers that do not rely as strongly on detection.

C. Gurrin et al. (Eds.): MMM 2014, Part I, LNCS 8325, pp. 266–277, 2014.

In this paper we focus on the second approach. We believe that robust tracking cannot be done while simply considering rectangular tracking windows. Indeed, a large proportion of the pixel in a tracking window will not belong to the tracked object (See Figure 1a). An efficient tracking system should know for every frame the set of pixels belonging to the object it is tracking. This segmentation problem is particularly difficult to solve in crowded scenes. One problem with the tracking/segmentation approach is that the area to track has a tendency to shrink over time as show in Figure 1b. In this paper we investigate simple prior-free methods to try to recover these parts.

(a)   (b)   (c)

Fig. 1. (a) Rectangle Window vs. Segmentation: here the pedestrian represents only a quarter of the rectangle selection area. (b) Point-tracking "shrinking" problem cause by occlusions. (c) Example of post-occlusion lower limbs recovery using our system.

The target application for this work is automatic or assisted surveillance. Requirements for automatic surveillance include people detection, tracking, re-identification, action recognition and ultimately suspicious behavior detection. While we focus on tracking in this paper, the automatic segmentation of pedestrians can be beneficial to all these different tasks. In non-automatic scenarios the people to track (suspicious individuals) are selected by surveillance personal. In practice the number of suspicious people will be far lower than the number of people in the scene. Tracking everybody approximately seems therefore less important than following one person in a robust way. Following a single individual also allows for computation to be local and therefore more efficient.

This paper is at the intersection of video segmentation and people tracking. The main difference with other tracking papers is that we focus on different performance measures (time before failure without detection and segmentation errors) as well as different challenges (heavily occluded pedestrians in crowds). The technical contributions of this paper are a novel heuristic method using segment sampling to re-grow missing regions after occlusions; a dense ground-truth dataset which is ideal for occlusion-robustness evaluations of pedestrian tracking method; and a fine-grain evaluation of segmentation errors for segmentation-tracking.

In the following sections, a brief literature review is given before presenting our proposed evaluation benchmark. The proposed method is detailed in section 4.

The results are presented in the following section. A discussion about how to approach pedestrian tracking in crowds in given in section 6.

2   Related Work

Most of the venues that require automatic assistance for video surveillance are also the places that show the highest level of pedestrian traffic and dense crowds (train stations, airports, department stores, stadiums and so on). Therefore, systems that can not deal with dense crowds are likely to be unusable in the very places they are needed the most. This has been acknowledged many times in the literature, but solutions for such problem are still scarce. Indeed a majority of pedestrian tracking methods only show results on videos of sparse set of people presenting limited and sparse cases of inter-pedestrian occlusions. In [5], first approaches have been tested to extend pedestrian detection-based tracking to crowds. In [6], a Bayesian clustering approach is used to group moving keypoints into individual pedestrian clusters. This model-free method is mainly based on motion information and show interesting results. However the density of point detection is usually low and, when apply to crowds, the system ends up tracking mainly the heads of people. Recently, [7] proposed a dense segmentation method to increase the density of tracking points. They cluster points motion into small segments that are then merged into larger segments using a motion invariant (geodesic) distance between segments. The final label attribution is done via an offline global optimization. In [8], a method to conciliate pedestrian detections and point trajectories was proposed. The tracking hypothesis based on detection windows (based on poselets [4]) and local optical flow motions are combined to get a more robust overall tracking.

In terms of data, there is no obvious benchmark to evaluate robustness to occlusion in crowds. Often crowds video present uniform motions. Indeed, corridors, sidewalk and zebra crossing scenes are defined along a single axis which can be used unintentionally by the researchers to smooth the trajectories. We consider that scene showing mutli-directional free motions are more appropriate to evaluate tracking algorithms. In [9], a dataset of crowded scenes was proposed. While most of the scene show crowds captured from long range which are not suitable for individual tracking, some of the videos show intermediate camera views with complex non-uniform crowd motion that are ideal to challenge tracking systems. In this paper, the density of the crowd refers to the number of people per surface area on the scene floor, not the number of people per pixel area in the image.

3   Problem Analysis

When tracking people in dense crowd the main problem is often to determine what pixels or patches of the input to track. A good description of this problem is given in [10]. If a rectangular window patch is tracked, either too few or too much data is used depending on the window size. A window that encompass the whole pedestrian is indeed polluted by background, occludees and occluders pixels. In order to track only the part of that rectangle corresponding to the object it is necessary to segment the target and track only features associated with these pixels.

In this paper, we assume the subject detection at the initial frame to be given as input to the system. For real applications this information can be provided by an automatic pedestrian detector (before any severe occlusion happens) or in another scenario by a human operator first selecting a suspicious individual to track on a security screen. The semi-supervise tracking scenario might sound counter-intuitive: if you have an operator to detect suspicious individuals why not ask him to do the tracking? The reason is simple: the operator would not be able to perform any other tasks (see unintentional blindness [11]) whether it is tracking someone else, answer the phone or detect other suspicious people. Besides, tracking is a very tedious task that requires constant attention and would grow tired any operator very quickly. Monitoring several automatically-tracked people is more easy than tracking one person manually. Our particular objective in this paper is to track a single target individual for as long as possible without intermediary automatic detections.

a   b   c

Fig. 2. Proposed benchmark data for segmentation-tracking evaluation in dense crowd: (a) Crowd scene video (879-38_1) from the UCF Crowd Dataset. (b) Example of manual segmentation for 10 pedestrians at frame 100. (c) 3D visualization of our temporally dense segmentation ground-truth for 300 frames.

3.1 Data

The main objective of our system is to improve robustness to occlusions observed in dense crowds. We choose one very challenging video (879-38_1.mov) from the UCF Crowds dataset [9] for our main evaluation. This video scene (see Figure 2a) is a scripted set up of numerous extras coming in and out of the screen in a dense crossing area. The average number of pedestrians per frame is around 30. The total length of the original video is 1285 frames (51.4 seconds). Sidewalks, corridors and two-ways zebra crossing scenes are limiting in that they only show two main directions for pedestrian motion. Here the actors cross each other in all possible directions making it an ideal benchmark to test against body occlusions.

Most of the actors wear suits (often dark) which are difficult to segment from each other when overlapping. The video is relatively low resolution (480×360) which make texture-based method less efficient. Despite being directed, the scene is a good representation of difficult cases that can occur in busy train stations or airports halls. The video present lot of the challenges observed in real life situation which makes it a good support for evaluation.

We consider pedestrians within the first 300 frames (12 seconds) of the videos. Since it was not possible to segment all pedestrians, we selected 10 for which the ground-truth has been manually defined for every frame (see Figure 2b). The manual ground truth we have generated is freely available for download on the author webpage[1]. To our knowledge it is the first temporally dense real video segmentation ground-truth available for a crowded scene.

4   Proposed Approach

Our first intuition to solve this problem is that very little information is required to track pedestrians.

When looking at a video of edge-segmented frames, a human can track the pedestrians without difficulty despite the lack of texture. The minimal amount of information for pedestrian motion detection/tracking have been studied for a long time for skeletons [12] and more recently for blob-like patterns in "emerging" images [13]. These studies strongly indicate that very little input information is required for human to track moving shapes. It is possible that the human brain use some sort of shape prior to perform such task. What is important to note here is that a video presenting only edges and motion contains enough information to solve the pedestrian segmentation/tracking problem. The question remaining is what general rules are needed to infer the solution from such input: 3D scene logic, continuity of motion and inertia, potential shape priors?

For this experiment we deliberately discard the direct use of texture information. The input images are used solely to generate segmentation and edge data, as well as to compute pixel displacement between frames.

4.1   Workflow

The input of our system is made of a video of a scene and an initial segmentation of one pedestrian. Our aim is to output for each frame of the video the propagated mask of this pedestrian despite all the occurring occlusions. Our process (see Figure 3) follows this following steps.

Mask Propagation (Push) - A first approximation of the pedestrian segmentation is computed by propagating points belonging to the previous frame mask to the current frame. The perceived motion of small local neighborhoods around these points is used as displacement. The mask at $t - 1$ is sampled using a

---

[1] http://www.clementcreusot.com/pedestrian/

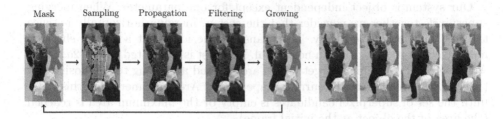

Fig. 3. Process workflow

sparse grid of 3x3 pixels. These points are pushed to the current frame using the standard Lucas-Kanade pyramidal optical flow [14]. We eliminate less robust elements by performing a back propagation of the keypoints using the reverse optical flow. We select only the initial keypoints which verify the following (unit in pixels):

$$Dist_{Eucl}(k, \tau^{-1}(\tau(k))) < 10 \qquad (1)$$

where τ is the optical flow translation function between frame $t-1$ and t and τ^{-1} the translation function between frame t and $t-1$.

Point Filtering - The optical flow is likely to give motion in different direction for the same object. In order to eliminate as much noise as possible we filter the points obtained using their associated translation direction. The mean direction for the last 5 frames are kept in a queue and used to filter incoherent displacements. The directions are clustered using their angle differences to the past mean and inliers within 0.5 radian of the main cluster centroid are used to compute the new mean. Points that are selected with this method are relatively sparse, especially in region presenting uniform texture. However they show coherent motion direction (See Figure 3).

Superpixel Expansion (Grow) - The propagated points provided by the previous method are sparse. If only those points were tracked, the system will fail after just a few frames. In this paper we investigate two different segmentation approaches: a simple Superpixel approach [15] and a hybrid method between edge-detection and superpixels. The segmentation is grown by merging the interior of all segments/superpixels which contain more than 5% of keypoints. If the superpixel usually follow the edges of objects, it is not always the case and this segmentation will have a tendency to leak outside the boundary of the object. Besides, in some case the occluder and occludee have exactly the same texture where the occlusion happens (*E.g.* sportwears during marathon, business-men dark suits, and so on). The masks retrieved by this method can also present some holes corresponding to regions where very few points were pushed.

Our system is object independent except for one parameter. When merging superpixels together we consider that the cost of merging horizontally is twice the cost of merging vertically. This translate the idea that a person shape is more likely to be vertical than horizontal. The cost is computed as $d = 2dx + dy$. All superpixel containing target points are ordered according to the position of their centroid relative to the current shape center. Areas are merged in this order until the set of superpixel candidates is empty or the maximum area is reached (the area of the object at the initial frame).

4.2   Implementation Details

The program is written in C++ using the OpenCV library and runs in around 0.09 second per pedestrian per frame. The hybrid segmentation is performed by detecting canny edges from which a first segmentation is produced using a ball filling of 7 pixels to close gaps between edges. Regions not detected by this method are segmented with a maximum of 50 superpixels [15]. In this article we only consider fast segmentation techniques that can be used for almost real-time applications. Level-set and graph-cut segmentation techniques are not discussed in this paper. Flood-fill segmentation methods have been tested but showed very low performances and are not presented here.

5   Experiments

Here we evaluate how this approach compares with two existing tracking systems and analyze the segmentation errors using our manual ground-truth.

5.1   Results

Qualitative results of our method are best seen in video. The videos presented on the author webpage[2] show the tracking results for the three tested methods as well as the segmentation error over time.

Time before failure -   One of the most important criteria for this method is to be able to follow the object for the longest period of time without any intermediary pedestrian detection. For each pedestrian we give the number of frames successfully tracked before the method fails. We consider that the tracking has failed if the recovered area represents less than 10% of the ground-truth mask for two consecutive frames or if the label changes.

In Figure 4a we see the influence of the segmentation method used. The techniques marked as SP-X are using superpixel segmentations with at most X cells in the local tracking area. The hybrid method is the one using edges (see Section 4.2). In Figure 4b we compare with [8] (Two-Granularity Tracking) and [16] (Compressive Tracking). Most of the tracking techniques we have tried completely failed on this crowd video as they rely heavily on detection.

---

[2] http://www.clementcreusot.com/pedestrian/

The best achieving existing tracking paper we have found on this dataset is [8] which was designed to compensate for the lack of detection and be more robust to occlusions. Here we plot the results from the first detection of the person by their system until the tracking is lost or the label changed. Among the 10 people we are monitoring, the longest period before failure with [8] is 70 frames (for id 5) and this case occur before large occlusions (see Figure 5). Most of the time the tracking is lost or the label is switched. To their defense, we do not believe that such dense crowd was considered when designing the method. Tracking people in dense crowd seems to raise a different set of problems altogether.

The results presented here differ a lot from what is usually presented in tracking papers. What is the point of a tracking that fails after 100 frames? This is because we evaluate visual tracking *methods/modules* rather than tracking *systems*. The aim here is to track for the longest possible time under occlusion *without* intermediary detections. By forbidding intermediary detections we can observe more closely the flaws of the visual tracking techniques.

It has to be noted that the compressive tracking method [16] is extremely sensitive to the initial detection window. If the initial bounding box is too large or too small the system fails after just a few frames. We therefore consider this method to be less robust than [8]. Note that in Figure 4, the results given for [16] are the best (max) results obtained over several runs introducing random noise on the initial window (not the mean).

In Figure 5 the level of occlusion of each pedestrian during the sequence is plotted. As expected the tracking failures are often correlated with pedestrian occlusions (peaks in the graphs). The person the most difficult to track in this experiment is pedestrian number 4. One possible explanation is that this person is already heavily occluded at initialization (50% occluded at frame 0).

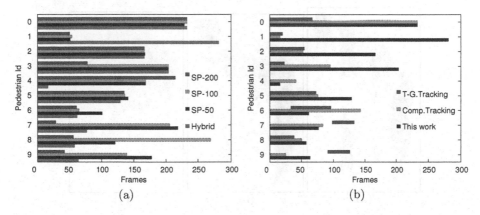

Fig. 4. (a) Results using different segmentation methods. (b) Comparison of the proposed method with [8] and [16]. Note that the detector used in [8] does not always give candidates for the first frame of the sequence. We plot the results from the first detection to the first failure.

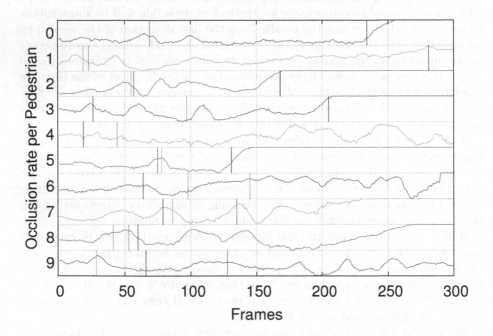

Fig. 5. Curves representing the ground-truth percentage of occlusions for each pedestrian in the sequence. A plateau at 100% of occlusion is reached when the pedestrian is off screen. The vertical lines represent the first-failure of the Two-Granularity tracking (red), Compressive tracking (green) and our segmentation tracking (blue).

Fig. 6. Qualitative results for pedestrian 1: (a) Ground-truth time-lapse mask. (b) Computed time-lapse mask. (c) Differences, missing area are shown in green (false negative), extra area in red (false positive).

Segmentation Errors - For each pedestrian and for each frame the difference between the recovered mask M_t and the ground-truth G_t can be computed in terms of false positive pixels (extra pixel that were not present in the ground-truth) and false negative pixels (missing pixels). We normalize these values using the total number of pixels in the ground truth mask in that frame.

$$FP_t = \frac{|M_t \setminus G_t|}{|G_t|} \quad \text{and} \quad FN_t = \frac{|G_t \setminus M_t|}{|G_t|} \tag{2}$$

In Figure 7, the hybrid segmentation is compared to standard superpixel approaches. An interesting point is that the pedestrian who is tracked for the longest period (pedestrian 1) is also the one that shows the lowest segmentation errors. However the correlation between the two metrics (segmentation error and time of tracking) is not straight-forward. While the hybrid segmentation gives better accuracy, it is in average less robust in time than simple superpixels (at least for pedestrian above id 4, that are usually further away from the camera and more occluded on the starting frame).

Please note that these results are raw data. We did not perform any temporal smoothing to detect and correct segmentation errors. Indeed this would make it more difficult to compare the underlying segmentation techniques. Visualization of the segmentation errors can be viewed for one pedestrian as a time-lapse in Figure 6 and for all pedestrians as a video in the supplemental materials.

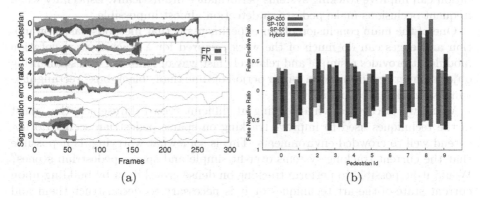

(a)   (b)

Fig. 7. (a) False Positive (FP) and False Negative (FN) errors over time for the hybrid segmentation. (b) Averaged FP and FN ratios for the different segmentations used in the growing stage.

5.2 Limitations

The proposed approach is very simple in its structure and relatively fast (0.09s per pedestrian per frame). However it focuses only on one segmentation/tracking issue. This process is thought to become essential when pedestrian detectors start to fail due to large occlusions and when visual tracking is required for many consecutive frames without re-detection. Our current approach is not using any object related prior (except for the merging cost indicating that the shape is mostly vertical). Hence, we can not segment touching object that move in the same direction for a long time. If the segmentation of the first object leak to the second object, the second object will be considered a part of the first object. Simple approaches can help deal with this situation. Co-segmenting objects within a close neighborhood can help resolve that kind of conflict where several objects claim a common set of pixels.

6   Discussion

In this article we presented a very challenging video and ground-truth for pedestrian tracking and investigate how previously hidden body parts can be recovered implicitly using local segmentation techniques. We analyzed these techniques after noticing that all conventional tracking methods fail on dense crowd videos, and posed the hypothesis that edge and motion are sufficient to perform pedestrian tracking.

It goes without saying that pedestrian tracking by visual tracking (including tracking by segmentation) and by detection are complementary. Any real-world applications will necessarily use a combination of those within a system and might have to deal explicitly with non-moving objects, long-term total occlusions and re-identification. We believe that more robust visual tracking and segmentation can improve tracking systems performance dramatically. Especially when frequent occlusions make pedestrian detection almost impossible.

One of the main conclusion of our experiments is that methods based on motion and edges can do much of the work required for a tracking-segmentation module. It provides a simple and relatively fast way of following and segmenting objects under occlusions for a longer period of time than "black-box" rectangular-window tracking techniques.

What we discovered while looking at difficult crowd datasets is that most of the techniques used to improve tracking on sparse pedestrian scenes do not extend well to crowded environments. This is somewhat worrying. Is it possible that the current tracking systems over-fit simple and sparse pedestrian scenes? Would it be possible to perform tracking on dense crowd data by building upon current state-of-the-art techniques or it is necessary to deconstruct them and start over with this new type of data in mind?

Our opinion is that a smooth transition from the techniques working well on sparse scenes to techniques working well on crowded scenes is possible if and only if fast occlusion-robust pedestrian detectors can be designed (e.g. fast poselets). If they cannot, we believe that the foundation for real-time pedestrian tracking in dense crowd will be very different from the current state-of-the-art tracking methods and might rely more on video segmentation techniques. Future work will investigate the use of priors (pedestrian shape statistical model, e.g. [17]) as well as spatio-temporal segmentation (e.g. [18]) to constrain the growing phase of the algorithm.

References

1. Benfold, B., Reid, I.: Stable multi-target tracking in real-time surveillance video. In: CVPR, pp. 3457–3464 (June 2011)
2. Yang, B., Nevatia, R.: Multi-target tracking by online learning of non-linear motion patterns and robust appearance models. In: CVPR, pp. 1918–1925 (2012)
3. Andriyenko, A., Schindler, K., Roth, S.: Discrete-continuous optimization for multi-target tracking. In: CVPR (2012)
4. Bourdev, L., Maji, S., Brox, T., Malik, J.: Detecting people using mutually consistent poselet activations. In: Daniilidis, K., Maragos, P., Paragios, N. (eds.) ECCV 2010, Part VI. LNCS, vol. 6316, pp. 168–181. Springer, Heidelberg (2010)
5. Zhao, T., Nevatia, R.: Tracking multiple humans in crowded environment. In: Proceedings of the 2004 IEEE Computer Society Conference on Computer Vision and Pattern Recognition, CVPR 2004, vol. 2, pp. 406–413 (2004)
6. Brostow, G.J., Cipolla, R.: Unsupervised bayesian detection of independent motion in crowds. In: IEEE Computer Vision and Pattern Recognition, pp. 594–601 (2006)
7. Iwasaki, M., Komoto, A., Nobori, K.: Dense motion segmentation of articulated objects in crowds. In: ICPR, pp. 861–865 (2012)
8. Fragkiadaki, K., Zhang, W., Zhang, G., Shi, J.: Two-granularity tracking: Mediating trajectory and detection graphs for tracking under occlusions. In: Fitzgibbon, A., Lazebnik, S., Perona, P., Sato, Y., Schmid, C. (eds.) ECCV 2012, Part V. LNCS, vol. 7576, pp. 552–565. Springer, Heidelberg (2012)
9. Ali, S., Shah, M.: A lagrangian particle dynamics approach for crowd flow segmentation and stability analysis. In: IEEE Conference on Computer Vision and Pattern Recognition, CVPR 2007, pp. 1–6 (2007)
10. Aeschliman, C., Park, J., Kak, A.: A probabilistic framework for joint segmentation and tracking. In: 2010 IEEE Conference on Computer Vision and Pattern Recognition (CVPR), pp. 1371–1378 (2010)
11. Simons, D.J., Chabris, C.F.: Gorillas in our midst: sustained inattentional blindness for dynamic events. Perception 28, 1059–1074 (1999)
12. Johansson, G.: Visual perception of biological motion and a model for its analysis. Perception & Psychophysics 14(2), 201–211 (1973)
13. Mitra, N.J., Chu, H.K., Lee, T.Y., Wolf, L., Yeshurun, H., Cohen-Or, D.: Emerging images. ACM Transactions on Graphics 28(5), 163:1–163:8(2009)
14. Lucas, B.D., Kanade, T.: An iterative image registration technique with an application to stereo vision. In: Proc. of 7th International Joint Conference on Artificial Intelligence (IJCAI), pp. 674–679 (1981)
15. Ren, C.Y., Reid, I.: gslic: a real-time implementation of slic superpixel segmentation. Technical report, University of Oxford, Department of Engineering Science (2011)
16. Zhang, K., Zhang, L., Yang, M.-H.: Real-time compressive tracking. In: Fitzgibbon, A., Lazebnik, S., Perona, P., Sato, Y., Schmid, C. (eds.) ECCV 2012, Part III. LNCS, vol. 7574, pp. 864–877. Springer, Heidelberg (2012)
17. Baumberg, A., Hogg, D.: Generating spatiotemporal models from examples. Image and Vision Computing 14(8), 525–532 (1996)
18. Apostoloff, N., Fitzgibbon, A.W.: Automatic video segmentation using spatiotemporal t-junctions. In: BMVC, pp. 1089–1098 (2006)

An Optimization Model for Aesthetic Two-Dimensional Barcodes*

Chengfang Fang, Chunwang Zhang**, and Ee-Chien Chang

School of Computing, National University of Singapore
13 Computing Drive, S117417, Republic of Singapore
c.fang@nus.edu.sg, {chunwang,changec}@comp.nus.edu.sg

Abstract. Given a message m and a logo image L, we want to generate an image that is visually similar to L and yet carries m in the payload with respect to a 2D barcode reader. This problem is similar to digital image watermarking, except that there is a requirement of using the specific barcode readers and applications, which are pre-installed in the end-users' devices. We formulate the generation as an optimization problem that considers operations carried out by the barcode readers, in particular, the sampling process and error correction. We propose a two-phase algorithm that solves the optimization problem. We adapt the algorithm to QR code and made a few observations to further enhance its performance.

Keywords: 2D barcodes, Aesthetic barcodes, Information Hiding.

1 Introduction

Two-dimensional barcodes can serve as a convenient unidirectional communication channel from printed materials or display panels to a mobile computing device equipped with a camera. They are designed for effective decoding by computing devices, and thus visually resemble random images that do not make sense to the naked eye. To encourage usages, many aesthetic barcodes (also known as artistic barcodes) are manually designed. These barcodes are visually interesting and yet can be read by the barcode readers.

A clear advantage of automation is the reduction in cost, as the aesthetic barcode is expensive to be manually designed. This is especially so when multiple barcodes embedding different messages but with the similar visual design are to be generated. In addition, automation allows fast generation, which is impossible to be supported by fully manual process. Examples of such applications could be delivering different promotion codes in advertisements or tracking printed documents [8]. There are also suggestions of using secure and automatically blended "visual cue" to detect malicious tampering [7]. In this paper, we propose a principled approach of aesthetic barcodes generation. Fig. 1 shows some examples of aesthetic barcodes generated by our proposed approach.

* This work is partially supported by the Singapore NRF under its IRC@SG Funding Initiative and administered by the IDMPO.
** Zhang is partially supported by the research grant R-252-000-514-112.

We consider a communication model shown in Fig. 2(a). The embedder, taking a message m and a logo image L as input, generates an aesthetic 2D barcode A. Now, A is to be viewed by a user, and to be read by a given barcode reader. The channel to the barcode reader is noisy, as the barcode A has to be printed out or rendered in a display panel, and to be captured by a camera. For a specific barcode reader, we want to design an embedder that minimizes the visual distance between A and L, and yet m can be successfully decoded with high probability.

The above model is closely related to digital image watermarking [6,4], where the encoder takes as input a message m' and a host image H, and generates a watermarked image I that is visually similar to the host image (Fig. 2(b)). The decoder, on input of a noisy watermarked image I', outputs the message with high probability.

However, there is a crucial difference between the two models. Note that the barcode reader in our model is "given" and is not part of the solution, in other words, there is a constraint of using a specific given barcode reader in the system. This constraint is due to the practical requirement of using well-accepted barcode

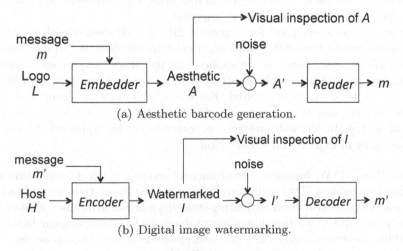

Fig. 1. Examples of aesthetic barcodes generated by our approach

(a) Aesthetic barcode generation.

(b) Digital image watermarking.

Fig. 2. Comparison of two models

standards and pre-installed applications in the end-users' devices. In contrast, there is not such requirement in the watermarking model, where both encoder and decoder can be modified to achieve higher performance.

We formulate the generation of aesthetic barcode as an optimization problem: on input L and m, find the barcode A that minimizes the visual distance from the logo L, and yet satisfies a condition whereby the message m can be successfully extracted with high probability. We adopt a weighted Euclidean distance to quantify the visual difference of L and A, where the saliency weights can be determined using known techniques or manually specified. Since saliency detection is not the focus of this work, we employ a simple edge-based weighting. To formulate the condition of successful extraction, we have to investigate steps carried out by a typical reader. There are two important steps. First, based on some sampling function, a value is sampled from each module, which is then rounded to binary value 0 or 1, giving a binary sequence. Next, the message m is extracted from the binary sequence based on some error correcting codes. We propose a distance function that estimates the expected number of additional bit-errors incurred by A compared to a default barcode. When the error is lower than a predetermined threshold, m can be extracted with high probability.

We give an efficient two-phase algorithm that solves the optimization problem. Essentially, the first phase determines the bits that can be "flipped" so as to minimize the visual distance, and the targeted sampled value for each module. This phase can be formulated as a knapsack problem with bounded weight, which is known to have polynomial time algorithm. The next phase finds the sub-image minimizing visual distance for each module, with the constraint that the sampled value is the same as the value determined by the first phase. The second phase can be formulated as a semidefinite quadratic programming problem, which is also known to have polynomial time algorithm. It is not difficult to prove that the overall two-phase algorithm gives the optimal solution. Although we only claim optimality for the choice of our distance functions, the two-phase approach can be applied for more refined distance functions.

Our algorithm is designed for a generic 2D barcode design with some simplified assumptions that differ from practical barcode designs. To demonstrate the feasibility of our approach, we applied our algorithm to QR code where the payload are web addresses. Interestingly, many applications inject additional redundancies that can be exploited. For example, note that domain name is case insensitive, and thus a message wWw.NuS.eDu.Sg would be corrected to www.nus.edu.sg by the web browser, i.e. correction at the application layer. We propose a way to exploit this observation.

Contributions. (1) We formulate the problem of aesthetic barcode generation as an optimization problem. Two main components in the formulation are a visual distance function, and a function estimating expected additional number of bit-errors. We also point out the relationship of our problem with the well-known digital watermarking communication model. (2) We propose an efficient two-phase algorithm that solves the optimization problem. (3) We adapt the algorithm to QR code and made a few observations to further enhance its performance.

2   Related Work

There are a number of automated or semi-automated methods, for example, replacing the center portion of the barcode by the intended logo (Fig. 3(a)), replacing each block (also known as the module) by an interesting logo (Fig. 3(b)), replacing each module by the corresponding region in the intended logo but with the total intensity shifted [2] (Fig. 3(c)), and watermarking technique [7] (Fig. 3(d)). There are many 2D barcode designs, for example, QR code [1] and the High Capacity Color Barcode (HCCB) [9] that uses colored triangles. Many barcodes are designed to encode data in printed copies. There are also proposals that use other types of sources. Collomosse et al. proposed "Screen codes" [3] for transferring data from a display to a camera-equipped mobile device, where the data are encoded as a grid of luminosity fluctuation within an arbitrary image. There are also other types of barcodes where the recognition is based on the topology, rather than geometry, of the codes [5].

    (a)   (b)   (c)   (d)

Fig. 3. Examples of automated generations

There are relatively few works in embedding images to barcodes. Fang et al. [7] propose a method to bind 2D barcodes to visual images similar to fragile watermarking. The idea is to choose among different representations of embedding the bits based on the intensity of a target logo. Zachi et al. [2] propose a method to change the intensity of each module towards the target logo. The output of their method is equivalent to a variant of our method that only carries out the second phase, and all the saliency and sampling weights are uniform. Fig. 3(c) gives an example generated using this variant.

3   Formulation and Notations

For clarity in exposition, we study the problem on a toy barcode design, and consider grayscale logo image. Section 5 applies our solution to popular barcodes, in particular QR code, and considers other practical issues.

The input of the embedder is the logo image L and a message m, and the output is a barcode image A. Each A and L is a $n \times n$ pixels grayscale image, where the intensity of each pixel ranges from 0 (black) to 1 (white). We treat the image A and L as two-dimensional arrays where each entry is the intensity

Table 1. Summary of Notations

$L, \ell_{i,j}$	: The $n \times n$-pixel logo image, and its pixel value at (i,j).
m	: The input message.
$B = \langle b_1, \ldots, b_t \rangle$	: The encoded message after ECC encoding being applied to m.
$A, a_{i,j}$	: The $n \times n$-pixel aesthetic barcode image and its pixel value at (i,j)
$W, w_{i,j}$	: The $n \times n$ saliency weights derived from L, and the weight at (i,j).
$M^1, \ldots, M^t$	: The modules in A. Each module is a $n' \times n'$ image.
$G, g_{i,j}$	: The $n' \times n'$ sampling weights of a module, and the weight at (i,j).
$S = \langle s_1, \ldots, s_t \rangle$	: The sampled intensity of the t modules in A.
$\mathcal{D}_W(A, L)$	: The visual distance function on A and L based on saliency weights W.
$\mathcal{F}(A, m)$	: The expected additional bit-errors incurred by A compared to the default barcode that carries m.
ϵ	: The amount of expected additional bit-errors allowed.
$\widetilde{B} = \langle \widetilde{b}_1, \ldots, \widetilde{b}_t \rangle$	: The rounded binary values of S.
$R_{0,\ell}, R_{0,u}$	: The lower/upper thresholds for module i where $b_i = 0$ (i.e. black).
$R_{1,\ell}, R_{1,u}$	: The lower/upper thresholds for module i where $b_i = 1$ (i.e. white).

of an pixel, and denote $a_{i,j}$ and $\ell_{i,j}$ to be respective value at index (i,j). The actual representation of the message m is not important in our formulation.

In the following sections, we first describe the toy barcode design, followed by the formulation of the two distance functions, and finally the optimization problem. Table 1 gives a summary of our notations.

3.1 A Toy Barcode Design

A barcode in the Toy Barcode is represented as a $n \times n$ pixels grayscale image. The barcode is divided into t equal-sized square images. Hence $\sqrt{t}$ divides n and each subimage is a $(n/\sqrt{t}) \times (n/\sqrt{t})$ pixels grayscale image. Following the terminologies of QR code, let us call each subimage a *module* and denote the sequence of modules in the barcode as $\langle M^1, M^2, \ldots, M^t \rangle$ where the modules are arranged in the two-dimensional grid under some serialization.

Default Encoder. Given a message m, a *default encoder* generates a barcode in the following steps.

1. Using some binary error correcting code, encodes m as a t-bit string. Let $B = \langle b_1, \ldots, b_t \rangle$ be such bits string.
2. Generates a barcode M where each module M^i contains only pixels of intensity b_i, i.e. it is either a black square image (when $b_i = 0$) or white square image (when $b_i = 1$).

Let us call the barcode generated by the default encoder the *default barcode*. Note that most 2D barcode designs include "control points" to aid correction of geometric distortion, for example, the prominent squares in QR code. Let us first omit these control points in the Toy Barcode.

Default Reader. A barcode reader on input an image, outputs the extracted message m, or an error message if it fails. Given a captured image $\widetilde{A}$, let $\langle M^1, \ldots, M^t \rangle$ be the modules in $\widetilde{A}$. The *default reader* carries out the following steps:

1. Computes the intensity of each module in $\langle M^1, \ldots, M^t \rangle$, and produces a sequence $\langle s_1, \ldots, s_t \rangle$. The intensity of a module is a weighted average of the pixel values in the module. The weights is represented as a $n' \times n'$ two-dimensional array G where $\sum_{i,j} g_{i,j} = 1$, $g_{i,j} \geq 0$ for all i, j's and $n' = n/\sqrt{t}$. Let us call G the *sampling weights*. For a module M^k, its intensity is computed as $s_k = \sum_{i,j} M_{i,j}^k \, g_{i,j}$, where $M_{i,j}^k$ is the pixel value at (i, j) within the module M^k.
2. Rounds the intensity $\langle s_1, \ldots, s_t \rangle$ to a binary seqeunce $\langle \widetilde{b}_1, \ldots, \widetilde{b}_t \rangle$. Specifically, $\widetilde{b}_i = 0$ iff $s_i < 0.5$.
3. Performs the ECC decoding on $\langle \widetilde{b}_1, \ldots, \widetilde{b}_t \rangle$ to extract the message m.

Note that the above reader is parameterized by the sampling weights G. If the weights are equal, then s_i becomes the average intensity value in the module. For robustness against noise and geometric distortion, it is desired to adopt a sampling function that emphasizes the region closer to the center.

We call the above the "default reader" as the actual implementation of a reader can vary in different mobile devices, leading to different outcomes on the same captured image. For instance, two different readers may performs different image enhancements, or employ different algorithms to correct geometric distortion. Nevertheless, the above are some common steps carried out in typical readers.

3.2   Visual Distance Function

There are extensive research efforts on formulating visual similarity between images. We adopt the weighted Euclidean distance to measure the visual distance between the logo L and the barcode image A. Let $w_{i,j}$ be the weight for the pixel at (i, j), and denote W as the array of weights. The visual distance is:

$$\mathcal{D}_W(A, L) = \sum_{i,j} w_{i,j}(\ell_{i,j} - a_{i,j})^2 \tag{1}$$

The weight W can be determined from L using some saliency detection algorithm, and let us call W the saliency weight. Since saliency detections are not the focus of this paper, we treat W as an input provided together with L. For the examples presented in this paper, we use a saliency weight with emphasis on regions near detected edges in the logo L.

3.3   Expected Additional Bit-Errors

Note that the sampled intensity of a module in the aesthetic barcode could differ from the corresponding intensity in the default barcode, and thus has a higher

chance of being wrongly read. We want to estimate how many more modules are
wrongly read.

The noise is a composition of many sources and difficult to formulate. In this
paper, we use the following simple noise model. Consider a white module (with
sampled intensity 1) in the default barcode. Suppose s_i is the sampled intensity
of the i-th module in the aesthetic barcode, let $\mathbf{E}_i = 0$ be the event that the
module is correctly read, and $\mathbf{E}_i = 1$ be the event that it is wrongly read. We
assume that when its intensity is more than a threshold $R_{1,u}$, then the module
will be correctly read with probability 1. If the intensity is in between $R_{1,\ell}$ and
$R_{1,u}$, then the module will be correctly read with probability 0.5. Finally, if the
intensity is lower than $R_{1,\ell}$, it is certain that it will be wrongly read. Hence, the
expected $\mathbf{E}_i$ is:

$$Exp\left(\mathbf{E}_i \mid i\text{-th module is white}\right) = \begin{cases} 0, & \text{if } s_i \geq R_{1,u}, \\ 0.5, & \text{if } R_{1,\ell} < s_i < R_{1,u}, \\ 1, & \text{otherwise.} \end{cases} \quad (2)$$

A reasonable and conservative choice of the parameters would be $R_{1,u} = 0.8$
and $R_{1,\ell} = 0.5$. We can similarly define the noise model and parameters $R_{0,u}$,
$R_{0,\ell}$ for black modules. By linearity of expectations, we can determine the total
number of expected errors incurred by the aesthetic barcode, which is an upper
bound on the additional errors compared to the default barcode. Let us take
this upper bound to estimate the additional bit-errors, and denote $\mathcal{F}(A, m)$ to
be the expected bit-errors incurred by A compared to the default barcode for
the message m. We want to keep $\mathcal{F}(A, m)$ within a threshold ϵ.

3.4   The Optimization Problem

Given the message m, the logo L, a threshold ϵ and saliency weight W, we want
to find an aesthetic barcode A s.t. the additional bit-errors incurred is no greater
than ϵ while minimizing the visual distance from the logo. Formally,

Problem 1. Aesthetic barcode generation

Given:	m, L, W
Find:	A
Minimize:	$\mathcal{D}_W(A, L)$
Subject to:	$\mathcal{F}(A, m) \leq \epsilon$

4   Proposed Solution

The above optimization problem greatly depends on the definition of $\mathcal{D}_W(\cdot, \cdot)$
and $\mathcal{F}(\cdot, \cdot)$. For our choice of functions, the problem can be efficiently solved using
a two-phase algorithm. Let A be the optimal solution. The first phase optimally
find $S = \langle s_1, \ldots, s_t \rangle$ which are the sampled intensities of the t modules in A.
After S is found, the value of each pixel in A is still not yet determined. The
second phase completes the process by solving a subproblem.

4.1   First Phase

Let us treat the logo image L as a barcode and partition it into modules, and let $\hat{s}_i$ to be the sampled intensity of the i-th module.

Let us consider a module where $b_i = 0$. Note that a smaller distance between s_i and $\hat{s}_i$ will reduce the visual distance, whereas a smaller s_i will not incur more bit-errors. Hence, the optimal s_i cannot be larger than $\hat{s}_i$. In addition, there are a few possibilities for s_i. (1) Case 1: $s_i \in [0, R_{0,\ell}]$; (2) Case 2: $s_i \in (R_{0,\ell}, R_{0,u})$; or (3) Case 3: $s_i \in [R_{0,u}, 1]$.

By our simple noise model, within each case, the incurred bit-error remains the same, regardless of the actual value of s_i. Let us consider case 2. If $\hat{s} > R_{0,u}$, then to minimize the visual difference, we must have $s_i = R_{0,u}$. If $\hat{s} \in (R_{0,\ell}, R_{0,u})$ then we must have $s = \hat{s}$. Similar argument holds for case 1 & 3. In other words, each module must be associated to one of the 3 cases, and we can determine the visual differences and bit-errors contributed by the module in each of the cases.

Now, the problem can be formalized as a variant of knapsack problem with bounded weights. Recall that a knapsack problem is defined by the values and weights of a set of items, and the goal is to select a subset of items that maximizes the sum of values, and yet the total weight is smaller than a threshold. In our problem, an item is a module in each of the 3 cases, its value is the reduction in visual difference, and its weight is the incurred bit-errors. Since there are only three possible weights for each item, there is a polynomial time dynamic programming algorithm that finds the optimal solution [10].

For our choice of distance functions, there is a more efficient algorithm. Due to space constraint, we will not describe the algorithm in this paper.

4.2   Second Phase

After the intensity $S = \langle s_1, \ldots, s_t \rangle$ of the modules in the optimal solution A is found, the second phase determines the pixel values in each module. Hence, the second phase essentially solves Problem 2, where M^k is the k-th module in A.

Problem 2. Second Phase

Given:	S, L, W
Find:	A
Minimize:	$\displaystyle\sum_{i,j} w_{i,j}(a_{i,j} - \ell_{i,j})^2$
Subject to:	$\displaystyle\sum_{i,j} M_{i,j}^k g_{i,j} = s_k, \quad \text{for } k = 1, \ldots, t.$

Since the constraint contains more unknowns than equations, the solution A is the (weighted) least square solution to an underdetermined linear system, which can be efficiently found. In general, even if a more sophisticated form of visual difference is adopted, as long as the objective function of the optimization problem remains semi-definite, there are efficient solvers [11].

4.3   Optimality

Although each of the above two phases is optimal, there is a question of whether the combined algorithm is optimal. We can show that it is indeed optimal by contradiction. Assume that an alternative solution A' is better than A. With equation (2), we have $\mathcal{F}(A, m) = \epsilon$, and therefore $\mathcal{F}(A', m) \leq \mathcal{F}(A, m)$.

Let us first consider the case where $\mathcal{F}(A', m) < \mathcal{F}(A, m)$. This implies that there is still "space" for the knapsack problem. Therefore, we can further "flip" some of the blocks in A' and yet the distance $\tilde{\mathcal{F}}(S', B)$ is still smaller than ϵ. This will contradict with the assumption that A' is optimal. On the other hand, $\mathcal{F}(A', m) = \mathcal{F}(A, m)$ would mean that there is another solution to problem 2 for some module M^i better than the least square fits. This also contradicts with Gauss-Markov theorem. Thus A is indeed the optimal solution.

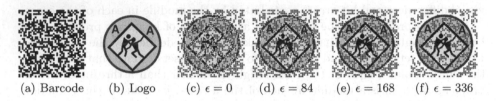

| (a) Barcode | (b) Logo | (c) $\epsilon = 0$ | (d) $\epsilon = 84$ | (e) $\epsilon = 168$ | (f) $\epsilon = 336$ |

Fig. 4. Experimental results on the Toy Barcode with different values of ϵ

5   Experiment

5.1   Aesthetic Barcodes on the Toy Barcode

We first conduct experiments on the Toy Barcode to examine the effect of the threshold ϵ on the visual difference. Fig. 4 shows the aesthetic barcodes for a logo image on different values of ϵ. Note that toy barcodes in Fig. 4 cannot be decoded by existing barcode readers. The Toy Barcode is composed of 41×41 modules and is a 410×410 pixels image (i.e. $t = 41^2$ and $n = 410$). The rounding thresholds $R_{0,\ell}, R_{0,u}, R_{1,\ell}, R_{1,u}$ are set to 0.3, 0.5, 0.5 and 0.8 respectively. The sampling function G gives higher weights to the central region and lower weights to the surrounding pixels. The saliency weight W is derived from the logo image by performing edge detection followed by a Gaussian low-pass filtering.

Fig. 5(a) shows the visual difference between the aesthetic barcode and the logo verse the threshold ϵ. As expected, with a larger ϵ, we can flip more "bits" and thus the barcode is closer to the logo. Note the white circular region surrounding the logo in Fig. 4(f), which is caused by higher saliency weights in that region. Fig. 5(b) shows the "rounded" binary barcode of Fig. 4(f), where a black (or white) module indicates that the corresponding module in Fig. 4(f) is read as "0" (or "1"). Fig. 5(c) shows the modules that are "flipped" in Fig. 4(f) as compared to the input toy barcode (Fig. 4(a)), where a black block indicates a bit-flip introducing 1 error and a gray block indicates a bit-flip introducing 0.5 error. Note that the modules flipped are concentrated around edges in the logo.

(a)   (b)   (c)

Fig. 5. (a) The visual difference verse ϵ; (b) the rounded binary barcode corresponding to Fig. 4(f); (c) the bits flipped in Fig. 4(f) as compared to toy barcode in Fig. 4(a)

5.2   Aesthetic Barcodes on QR Codes

Background on QR code. QR code is a popular 2D barcode. Given a message m, a default QR encoder first applies Reed-Solomon error correcting code on m. There are four levels of error correction capability, where the amount of codewords that can be restored varies from approximately 7% to 30%. After the encoded message B is acquired, the encoder produces black/white modules to encode B. Beside the modules that embed the message, there are also control modules in the QR code for version information, format information and registration information. These control modules are more important than the other modules.

Exploiting application layer error correction. The message we embedded to each QR code is $m = \mathtt{http://www.nus.edu.sg}$. As mentioned in the introduction, the application (i.e., web browser) on top of the barcode reader performs another layer of error correcting: convert upper case characters in the URLs to lower case. To further reduce the visual difference, we search for a combination of capitalizations that attains the lowest difference. Fig. 6 shows the effect of application level error correction and a weighted sampling function.

Implementation issues. In this section, the logos are colour images. We use a simple method to blend colour: first, generate a grey aesthetic barcode based on the intensity component of the logo image; the color components of the logo are then copied to the grey barcode. In QR code, the message is divided into a few sub-messages, whereby the Reed-Solomon code is independently applied. Hence, there is a structure of the bit-errors that can be exploited. We ignore this possibility. As the control points are important modules, we set the saliency weights in the control-point regions to 0 so that they will not be flipped.

Experimental setup. All QR codes generated in this experiment are in version 3 with 29×29 modules and have an error correction level of high (H). Each module is a block of 15×15 pixels. The rounding thresholds $R_{0,\ell}, R_{0,u}, R_{1,\ell}, R_{1,u}$ are set to 0.35, 0.5, 0.5 and 0.65 respectively. The sampling and saliency weights are same as before. We test barcode readers on 4 different devices, including "QR Barcode Scanner"[1] on an Android phone and an Android tablet, and "QR Code

---

[1] $\mathtt{http://www.qrbarcodescanner.com}$

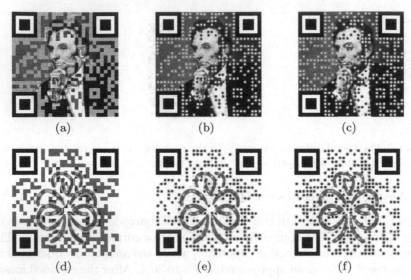

(a)   (b)   (c)

(d)   (e)   (f)

Fig. 6. Effect of application level correction and sampling function. Proposed method (center) compared to one with uniform sampling function (left column) and to one without application layer error correction (right column).

Scanner"[2] on an iPhone and an iPad. The experiment is carried out in an in-door environment with normal illumination, and on both a 20" LCD monitor (ACER P205H) and printed paper with a laser printer (LEXMARK T642). The aesthetic barcodes are displayed in 435×435 pixels (around $12cm \times 12cm$) on the monitor and $4cm \times 4cm$ in size on the printed paper, both with white background. In each trial, we take the following steps to ensure a fair test. We first position the camera near enough the surface to ensure that the aesthetic barcode completely fills the live preview window of the barcode reader, so that only the aesthetic barcode is being interpreted. Second, we make sure that the barcode is in focus by letting the autofocus stabilize and then hold the device still for 3 seconds. We say that an aesthetic barcode can be decoded if the barcode reader gives the correct URL address within 3 seconds, or cannot be decoded if it fails. We experiment on 6 different logo images and repeat the above decoding process 3 times for each aesthetic barcode and barcode reader.

Results and tests on robustness. Fig. 1, 6 and 7(a) give examples of the produced aesthetic barcodes on QR code. All of them can be decoded.

To test the robustness against geometric distortion, we randomly remove x columns of pixels from each of the aesthetic barcodes where x ranges from 5 to 45 and check whether the distorted barcode can be decoded. The result is shown in Table 2. Fig. 7 gives examples of one logo with different distortion levels.

To test noise resilience, we add Gaussian noise with zero mean and variance between 0.1 and 0.4 to the aesthetic barcodes. The result is shown in Table 2. Fig. 8 gives examples of one logo image with different noise levels.

---

[2] http://empowereddesignapps.com

Table 2. Successful decoding rate verse different distortion or noise levels

Distortion with x columns removed				Gaussian noise with variance δ			
$x = 5$	$x = 15$	$x = 30$	$x = 45$	$\delta = 0.1$	$\delta = 0.2$	$\delta = 0.3$	$\delta = 0.4$
100%	100%	35% (25/72)	8% (6/72)	100%	100%	43% (31/72)	11% (8/72)

(a) Original barcode   (b) $x = 15$ (c) $x = 30$

Fig. 7. Randomly remove x columns of pixels from barcodes

(a) Original barcode   (b) $\delta = 0.1$ (c) $\delta = 0.3$

Fig. 8. Add Gaussian noise with different variance δ

6 Conclusion

An aesthetic 2D barcode can be treated as a carrier over two channels, concurrently sending a logo image and a message to an end-user and a barcode reader respectively. As the human visual system and the barcode reader are sensitive to different properties of the barcode image, for example, the barcode reader effectively has "lower" resolution of modules, and the human visual system is more sensitive to edges, potentially such differences can be exploited. We formulate an optimization problem that captures such differences, and give an efficient two-phase algorithm. We also apply our formulation and algorithm to the popular QR codes, which produces promising results. It would be interesting to further exploit the coding structure in QR code, and take aesthetic measure into consideration to enhance effectiveness of the barcodes.

References

1. Q.C. International Organization for Standarization: Information Technology-Automatic Identification and Data Capture Techniques-Bar Code Symbology-QR Code (2000)

2. Baharav, Z., Kakarala, R.: Visually Significant QR Codes: Image Blending and Statistical Analysis. To be appeared in ICME (2013)
3. Collomosse, J., Kindberg, T.: Screen codes: visual hyperlinks for displays. In: Workshop on Mobile Computing Systems and Applications, pp. 86–90 (2008)
4. Costa, M.: Writing on dirty paper. IEEE Transactions on Information Theory, 439–441 (1983)
5. Costanza, E., Robinson, J.: A region adjacency tree approach to the detection and design of fiducials. In: Vision, Video and Graphics, pp. 63–70 (2003)
6. Cox, I.J., Miller, M.L., McKellips, A.L.: Watermarking as communications with side information. Proceedings of the IEEE, 1127–1141 (1999)
7. Fang, C., Chang, E.-C.: Securing interactive sessions using mobile device through visual channel and visual inspection. In: Annual Computer Security Application Conference, pp. 69–78 (2010)
8. Ninesigma. Improving Appearance of QR Codes. Request for proposal #67911 (2011), https://www.myninesigma.com/_layouts/RFPs/NineSigma_RFP_67911.pdf
9. Parikh, D., Jancke, G.: Localization and segmentation of a 2d high capacity color barcode. In: IEEE Workshop on Applications of Comp. Vision, pp. 1–6 (2008)
10. Pisinger, D.: Linear time algorithms for knapsack problems with bounded weights. Journal of Algorithms, 1–14 (1999)
11. Toh, K., Todd, M., Tütüncü, R.: Sdpt3a matlab software package for semidefinite programming, v1.3. In: Optimization Methods and Software, pp. 545–581 (1999)

Live Key Frame Extraction
in User Generated Content Scenarios
for Embedded Mobile Platforms

Alexandro Sentinelli and Luca Celetto

STMicroelectronics, Advanced System Technology,
Via della Vignolina 2, 20041, Agrate Brianza, Italy
{alexandro.sentinelli,luca.celetto}@st.com
http://www.st.com

Abstract. In this work we investigated the suitability of Key Frame Extraction solutions in embedded architectures for mobile platforms. In particular, in our scenario the list of key frames is requested right after a shooting session ends. The most interesting outcome has been the evaluation of performances in User Generated Content scenarios through an extensive survey based on Opinion Scores, interviews and other minor metrics testing five different solutions. Results suggest that pursuing sophisticated algorithms doesn't necessarily enrich the end user experience. We hope that this work will contribute to stimulate the debate on the KFE also in realistic scenarios and to pay attention to the user feedbacks to drive the investigation.

Keywords: Key Frame Extraction, Video Summarization, MOS, Embedded Architecture, Mobile Applications.

1 Introduction

In the wide world of Multimedia Information Retrieval the browsing tools that help users to navigate in Multimedia archives have been proved to be very relevant for the usability of any retrieval system. A popular problem in this area is the Video Summarization or Key Frame Extraction (KFE): the aim is to provide a set of frames extracted from the video that represent the content of the video itself. It is proven that a sequence of images is more comfortable and efficient than a bunch of words to read or, an hyperlink to click, open, then stream through a player. Instead a few images can easily describe the content of a rarely tagged video also because, usually, the user who is browsing in the archive is also the author of the video. We point out that User Generated Content (UGC) are far diverse from TV/professional contents. UGC are unpredictable, short, taken by one camera and content is not so interesting to watch. All these aspects didn't make UGC scenarios appealing for the research community until a few years ago. In this work we addressed the study of the KFE with the aim to embed a solution in a camera chipset for Smartphones. Since it is important to minimize the computational cost for battery consumption and latency, we tried to exploit, when

C. Gurrin et al. (Eds.): MMM 2014, Part I, LNCS 8325, pp. 291–302, 2014.
© Springer International Publishing Switzerland 2014

possible, the available accelerated algorithms on hardware. We will refer to the output of the KFE algorithm as a sequence of keyframes or visual storyboard. The added value of our approach in respect to traditional methods relies on the live computation of the stream: at the end of the shooting session the user has already the list of the keyframes. The aleatory behavior of the algorithm and the lack of specific objective metrics suggest being cautious in choosing a definitive solution. However the user must perceive the value of the solution otherwise it is more convenient to apply an arithmetical temporal sampling that brings average performance but remains extremely cheap. We implemented five different architectures and performed experiments on 14 video sequences with the aim to embrace a variety of indoor/outdoor private live scenarios. Our evaluation survey is made by Mean Opinion Score (MOS), minor metrics, face to face interviews. Through this work we aimed to contribute to the debate of the KFE addressing the discussion to market scenarios where resources can be limited. Finally we observe the importance of the end user feedbacks to drive the development.

2   Related Works

Lots of different applications in the Multimedia world are linked somehow to the problem of video summarization. It happens that the same algorithms can be used to perform video scene segmentation, semantic understanding, classification, event detection or a mix of tasks that are either overlapping or, at least, correlated to video summarization. In our context, we intend as video summary a limited list of frames picked up from the video that represent the content of the video itself. Such list constitutes the output of the KFE algorithm and is referred as list of keyframes or visual storyboard. In the past many techniques have been investigated to extract a list of keyframes from a video sequence. In [1] KFE is performed mainly through the motion estimation: the greater the intensity in the center of the frame the more representative is the keyframe. In [2], instead, the authors point out the steadiness of the camera to measure the relevance of a pose. Other more traditional methods extract color histograms and other low level features. Then the keyframes are selected through distance maximization in the feature space [3], or other clustering methods [5]. Such approach has a long tradition in partitioning problems, thus many mathematical tools are available and may be efficaciously reused. However, we also need to take care of the computational cost constraints. In [6] Geraci et al show a cheap and fast clustering method that doesn't affect the performances, but tests are still conducted on TV content. Recently the software R&D community started to embrace UGC scenarios in their tests after the mobile manufacturers understood the great potentials in the consumer electronics industry. In [8] Nokia buffers semantic metadata and features to discriminate frame after frame while the camera is streaming. Still on UGC content Kodak [7] uses also face detection engine and audio features. In [9] Intel had focused on the importance of redundancy while choosing keyframes. We believe it be an interesting approach although their system was not tested in UGC scenarios.

3   Simulation Platform

Fig. 1 shows the reference scenario and the basic idea of our architecture. As we said, the purpose is to design a solution that, up to negligible computation delay, provides at the end of the shooting session (we don't when) the keyframes list. The other development guideline is the integration of the algorithm in the camera chipset. Each raw frame captured by the camera is carried to a decision module that will discriminate if it has to be kept or discarded from the story-board buffer. Such a decision is made only through analysis and comparison of histograms or statistic information that are immediately available from the hardware. Our simulation model is based on GStreamer open source platform [10]. In the main loop of the decoding process each yuv frame transits temporarily in a buffer just like if it is coming from the camera sensor. Then the frame is passed to the KFE algo, which is the core of our investigation. The majority of input paramctcrs have been decided balancing the computational cost and the performance. Others have been based on heuristics mainly to discriminate similarities in the Duplicate Removal module. The end of the loop coincides to the end of the shooting session. It comes evident that the essential brick of the system is the description associated to the frame as frames are compared only through thcir compact descriptions. Resource constraints obliged us to not test any of the local descriptor family (SIFT, SURF and relatives) but to exploit the descriptions that are hardware accelerated (come for free). As we need to limit the memory stored in the chipset frames are parsed only oncc, then analyzed

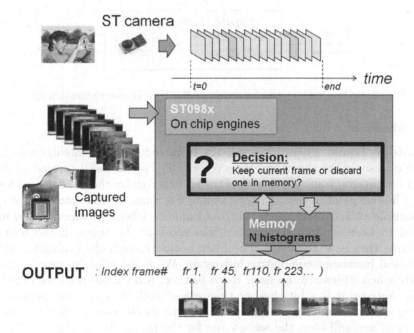

Fig. 1. System overview: the list of Key Frames is processed while shooting

and stored in the storyboard buffer through their available descriptions. This choice has pro and cons effects. On one side we save a lot of memory because the storyboard buffer do not store images neither raw nor compressed, but only a few integer arrays (a few kB are enough): the final output is represented, therefore, by a pure integer list indicating the absolute position of the keyframes in the video. On the other side, the information about the frame is lost so any subsequent analysis can be performed only through their compact description. In the rest of the paper, as they are referred to the same object, we'll refer either to frames or histograms equivalently. Then, in our simulation platform we have allocated additional memory for the keyframes also as images, of course, but in a separate buffer and only for evaluation purpose. In the next sub section we describe the main modules of the chain: a monotone frame filter, a buffer where the storyboard is updated, then a *Duplicate Removal* at the end of the shooting session (Fig. 2).

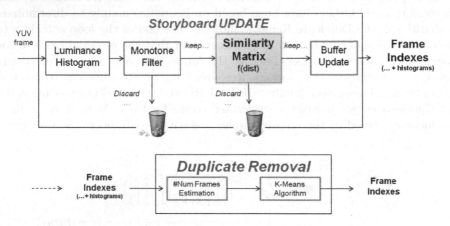

Fig. 2. System architecture overview of the Key Frame Extraction chain

3.1 Modules

Monotone Frame Filter. We will try to introduce more intelligence in the future releases of the algorithm. However, in UGC scenarios, it seemed easier to detect bad quality frames rather than trying to recognize the semantic relevance from a human point of view. Understanding if a frame is "too much monotone", for example a black frame or a fade in/out frame or, when the camera is obscured, seemed to be very simple and efficacious most of the times: if the frame is monotone then it is discarded. The filter works through the evaluation of the traditional luminance cumulative histogram. We found effective representation already when #bins=16; no need to go further. After some tests we opted for a *Zero Forcing*: it is a bit cheaper than other methods based on variance (no root squares to compute). It works with two thresholds, one to select a sufficient number of non null bins, the second one for the height. If the frame is "good", then is passed to the similarity matrix block.

Similarity Matrix. If the buffer stores N frames (alias N histograms) there will occur $(N)(N-1)/2$ distances computations among the corresponding histograms. All the possible distances are stored in a squared symmetrical matrix M_{diff}. We implemented two options: briefly, maximize the element's sum of the matrix or look for the minimum distance among histograms.

- Maximize Eq 1: it identifies the frame k that, when discarded, maximizes the sum of the elements of M_{diff}:

$$\arg\max_k \sum_{i,j} M_{diff}(i,j) \; \forall \; (i,j) \neq k \tag{1}$$

   We point out that, in this case, the M_{diff} has to compute all similarity values for each new incoming frame. That may be not a big deal for simulations performed by a PC desktop, but can be for the limited resources of an embedded architecture.
- min $M_{diff}(i,j)$: the algorithm looks for the minimum value in the whole M_{diff}. Then the frame i or j is discarded.

StoryBoard Buffer. When the frame f_k has been identified through one of the above options the storyboard buffer is updated by replacing f_k with the new incoming frame from the camera.

Duplicate Removal. The last module clusters the frame representations through the k-Means algorithm. As the latter needs the #clusters as input, we first need to estimate K which, in our case, becomes the very length of the storyboard. Histograms, or array in case of other frame representations, in order to keep a natural order of the events, are first ordered temporally and a simple distance among adjacent frames is computed. When the distance is above a certain threshold the #clusters K is incremented. Finally the initial set of K centroids is chosen by keeping them as equidistant as possible.

3.2 Distance and Similarity

Frames are compared through distances among their histogram representation, which are in fact same dimension vectors. We used the *Generalized Jaccard Distance*(GJD):

$$D(s, z) = 1 - \frac{\sum_i min(s_i, z_i)}{\sum_i max(s_i, z_i)} \tag{2}$$

where s and z are two generic vectors, min and max gets the minor and greater in (s_i, z_i). Eq 2, in respect to the more traditional Bachattarayya or Euclidean distance, saves around 5% of the computational time over the whole chain.

In the next sections we briefly describe the different architectures that we implemented and evaluated.

4   Methods Description

As we said we implemented 5 chains of KFE. We assumed that an average UGC video length is between ~15 seconds and ~2 minutes. Then we set the storyboard buffer dimension $B_{dim} \cong 30$ which is also *max* size of the storyboard. For each chain we will provide a brief description.

4.1   Cumulative Histogram

The *CumulHist* chain relies on a popular histogram representation adopted for images: the cumulative luminance histograms. After many trials we understood that the cumulative histogram, known also as Scalable Color Descriptor in the MPEG-7 standard [11], despite its simplicity and ease of computation cannot evaluate more than a macroscopic presence of global light/dark in the scene. The rest of the chain reflects the architecture as shown in Fig. 2. We tested the two options concerning the M_{diff} update but we didn't observe remarkable differences.

4.2   Zonal Statistics

The *ZnlStat* architecture is equal to the previous one: only the histogram representation is different. *ZnlStat* frame representation comes from our camera chipset and provides information about the spatial distribution of the color inside the image. The image is first segmented by a grid. Therefore statistical information values of red, green, blue, are collected for each area of the grid. The similarity matching seems to perform better; however, this chain doesn't make a notable difference over the whole dataset of video considered.

4.3   Temporal Sampling (fix)

This solution is nothing but a pure arithmetical division between the length of the video N (#frames) and the length L of the storyboard initially set by the user (Fig. 3).

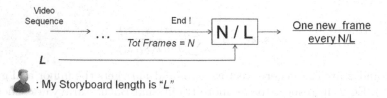

Fig. 3. Pure Temporal Sampling

After a first benchmarking of the algorithm, when asking to end users, we remarked that the natural temporal sampling of the video sequence had the best ever tradeoff between the output storyboard quality and the cost of the implementation. The only concern was about the redundancy of the keyframes. That's when we thought about a hybrid chain that mixes a pure temporal sampling until the end of the shooting, and then applies the DR to reduce the redundancy.

4.4   Pseudo Temp *ZnlStat*

As the ratio N/L is dependent by N, which is unknown while sampling, we need to progressively increase the interval among adjacent frames as shown in Fig. 4. In particular, by choosing the buffer dimension $L =$ a power of 2, and doubling ∇T every L samples, we keep a perfect equal distance among sampled frames. While sampling we also store the relative *ZnlStat* histograms in the storyboard buffer, though they are not used until the end of video shooting. Then the duplicate removal module is applied.

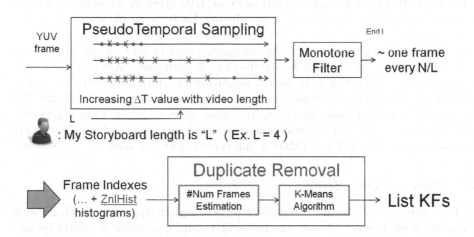

Fig. 4. Pseudo Temporal ZnlStat chain: architecture overview

4.5   Temporal Sampling (var)

The last solution is nothing but the variable version of the temporal sampling algorithm: the possibility to vary L depending by the length of the video through 4 fixed thresholds (Fig. 5).

For example the storyboard is made by 24 keyframes if the video last between 30 and 60 seconds. Although the approach is very simple the trade off cost/performance makes it one of the best KFE candidates for UGC content.

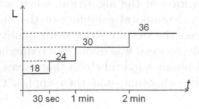

Fig. 5. Variable Temporal Sampling: storyboard size depends by the video length

5 Experiments and Results

We benchmarked these 5 solutions on a set of 14 videos taken with a Nokia N900 (res. 848x480). As we didn't find a dedicated dataset of UGC content we shot our own set of video sequences with the aim to cover a variety of indoor/outdoor scenes. We evaluated: Mean Opinion Score (MOS) on 24 users, #Bad Frames, #Redundant Frames, #Key Frames vs an estimated optimal length. We just recall that MOS is an averaged quality score in [1,10] given by a human subject to the algorithm output: in this case a storyboard. Each visual storyboard appears as a small image composite made by the keyframes thumbnails. Since it's a tool for browsing in video archives we encourage the user to look at the storyboard as a whole rather than looking for details in each keyframe. MOS proved to be a reasonable index to estimate so far the algorithm's performance. After watching the videos evaluators were asked to give a score for each storyboard (totally 70). Then, through one to one interviews we asked users to reveal their subjective perceptions in order to better understand the inner mechanism of the storyboard summarization. The full test (MOS + interview) lasts ~ 1 hour.

5.1 Mean Opinion Score

In Fig. 6(top) we note that a temporal pure sampling approach (both fix and variable) is often good to meet the needs of an end-user without efforts by the developer and hardware resources. That may depend by both the low quality of the UGC and the length of a typical video sequence shot through a camera phone (rarely more than 2 minutes). It is kind of intuitive that an arithmetical sampling should represent on average the different moments of the story. On the other side, if the story doesn't change keyframes can be unpleasantly redundant to a human eye. So the hybrid solution that samples temporally and removes the adjacent duplicates ("PsdSmpZnlStat") got the highest MOS. We can so far identify trends but not the ideal solution for all videos. For example, the sequence "Elevator_up" shows by far (Fig. 7) the best MOS with the cumulative histogram chain although the average results tend to point out to the pseudo temporal sampling approach. In Fig. 6 (bottom) we show also a measure of the standard deviation of MOS in respect to the videos. The best algorithm in terms of average MOS is not the most predictable (or stable). This is a non trivial

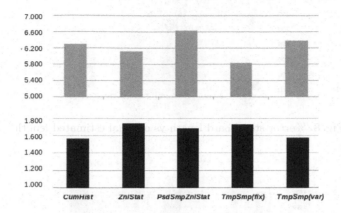

Fig. 6. Average MOS (top), Standard Deviation (bottom) per algorithm

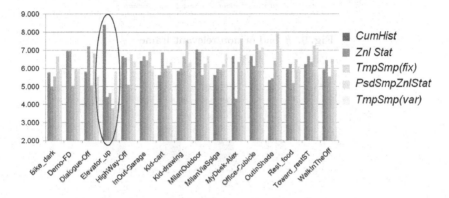

Fig. 7. Average MOS per video sequence

aspect when dealing with algorithms that need to be integrated into products for the consumer market. If performances are only good on average but not predictable it becomes tricky to consciously drive the development of future releases.

5.2 Secondary Metrics

In Fig. 8 we evaluated the error of the storyboard length (= #keyframe) for each solution in respect to a ground truth set of (manually) pre-defined storyboards. It is pretty expectable that, for this index, the last performing is the temporal sampling as it doesn't discriminate: any frame can be stored. The same effect is witnessed in Fig. 9 where frames like obscured camera shots are however retrieved. From Fig. 10, instead, we infer that redundancy raises as with the ratio storyboard size over the video length. In other words, if the algorithm samples less, keyframes have more chance to be different.

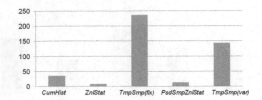

Fig. 8. %error storyboard length vs optimal estimated length

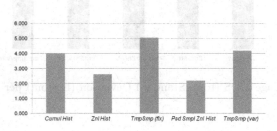

Fig. 9. # bad or non relevant frames

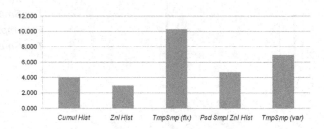

Fig. 10. # redundant frames

5.3   Issues and Remarks

As expected, we met more issues during the survey in respect to the development. First we had to find a set of videos that represent the content of a typical camera phone user. Second, the evaluation deals with aspects belonging to the variegated world of HMI. The storyboard, in fact, is evaluated through a HMI that can affect dramatically the survey experience. The behavior of the algorithm can be hidden or misunderstood by a wrong evaluation setting and future improvements may take a wrong direction. For example, if thumbnails used for the survey are too small, blurred shapes are less evident, thus any need of sharpness analysis is just ignored. Another aspect to consider is the visual symmetricity of the storyboard composit. If thumbnails are collated 4 per lines, some user may prefer a storyboard with 16 keyframes rather than a storyboard with 13 with less redundancy, just because symmetricity is pleasant to the eye. Finally we integrated the MOS collection with one to one interviews asking what they would like to put in the storyboard and what are the frames that can be considered semantically "important". We are confident in saying that CV colleagues were a bit or very much conditioned by their daily activity on images and video analysis, no matter if junior or senior scientists. Because of that the last set of 10 people has been selected from people outside the company or the CV field. Through this benchmark it has been possible to identify the general trends of the five solutions and some highlights for designers who consider to evaluate a KFE algorithm:

- Redundancy is bad, especially among adjacent frames;
- Monotone frames can be part of the story;
- Beauty is somehow important. A colorful and/or symmetrical storyboard is more pleasant than a representative summary;
- The storyboard length may depend by the redundancy and relevancy of the keyframes;
- Blurred shapes in key frames are perceived as "bad";
- Better a bit of redundancy, but keeping the story as it is;
- The "story" seems to be a key point to consider: the longer the video, the longer the story, the longer the storyboard;

Before leaving this paragraph we point out that KFE may serve different tasks: browse in a video, browse in a database, recall the content. The developer needs an overview of the whole system and the interaction with the user has to be understood, possibly, prior the development.

6   Conclusion

Although video summarization is a widely discussed topic in the Computer Vision community, the investigation in mobile use case scenarios did not receive much attention. In this work we focused on a runtime implementation performing the Key Frame Extraction (KFE) for User Generated Contents (UGC) shoot by cellular phones. The computational constraints forced us to look at low cost algorithms trying to exploit, when possible, the information available on the camera chipset. We developed and compared five solutions that have been benchmarked through an extensive evaluation in terms of Mean Opinion Score, interviews and some minor metrics. The most appreciated algorithm was far from embedding complex semantic operations but based on a natural temporal sampling approach with a few more computations to detect low level similarities. This work aims to stimulate the debate on KFE in real market scenarios and point out the importance of the end user feedback to drive the investigation to efficient solutions.

References

1. Narasimha, R., Savakis, A., Rao, R., Queiroz.: Key frame extraction using MPEG-7 motion descriptors. In: Proceedings of the Asilomar Conference on Signals, Systems, and Computers, Pacific Grove, CA

2. Wolf, W.: Keyframe selection by motion analysis. In: Proceedings of the ICASSP Conference, vol. 2, pp. 1228–1231 (1996)
3. Yeung, M.M., Liu, B.: Efficient matching and clustering of video shots. In: Proceedings of the International Conference on Image Processing, October 23-26, vol. 1, p. 338 (1995)
4. Zhang, H.J., Wu, J., Zhong, D., Smoliar, S.W.: An integrated system for content-based video retrieval and browsing. Pattern Recognition 30, 643–658 (1997)
5. Zhuang, Y., Rui, Y., Huang, T.S., Mehmtra, S.: Adaptive Key Frame Extraction Using Unsupervised Clustering. In: Proc. ICIP 1998, vol. I, pp. 866–870 (1998)
6. Furini, M., Geraci, F., Montangero, M., Pellegrini, M.: VISTO: visual storyboard for web video browsing. In: CIVR 2007: Proceedings of the 6th ACM International Conference on Image and Video Retrieval (2007)
7. Jiang, W., Cotton, C., Loui, A.: Automatic Consumer Video Summarization By Audio And Visual Analysis. In: ICME 2011, Barcelona, Spain (2011)
8. Zeng, X., Xie, X., Wang, K.: Instant Video Summarization during Shooting with Mobile Phone. In: ICMR 2011, Trento, Italy (2011)
9. Wang, T., Gao, Y., Wang, P.P., Li, E., Hu, W., Zhang, Y., Yong, J.: Video summarization by redundancy removing and content ranking. In: MM 2007: Proceedings of the 15th ACM Multimedia, Germany (2007)
10. http://gstreamer.freedesktop.org/ (accessed on May 23, 2012)
11. Manjunath, B.S., Ohm, J.-R., Vasudevan, V.V., Yamada, A.: Color and texture descriptors. IEEE Trans. Circ. Syst. Video Technol. 11(6), 703–715 (2001)

Understanding Affective Content of Music Videos through Learned Representations

Esra Acar, Frank Hopfgartner, and Sahin Albayrak

DAI Laboratory, Technische Universität Berlin,
Ernst-Reuter-Platz 7, TEL 14, 10587 Berlin, Germany
name.surname@tu-berlin.de

Abstract. In consideration of the ever-growing available multimedia data, annotating multimedia content automatically with feeling(s) expected to arise in users is a challenging problem. In order to solve this problem, the emerging research field of video affective analysis aims at exploiting human emotions. In this field where no dominant feature representation has emerged yet, choosing discriminative features for the effective representation of video segments is a key issue in designing video affective content analysis algorithms. Most existing affective content analysis methods either use low-level audio-visual features or generate hand-crafted higher level representations based on these low-level features. In this work, we propose to use deep learning methods, in particular convolutional neural networks (CNNs), in order to learn mid-level representations from automatically extracted low-level features. We exploit the audio and visual modality of videos by employing Mel-Frequency Cepstral Coefficients (MFCC) and color values in the RGB space in order to build higher level audio and visual representations. We use the learned representations for the affective classification of music video clips. We choose multi-class support vector machines (SVMs) for classifying video clips into four affective categories representing the four quadrants of the Valence Arousal (VA) space. Results on a subset of the DEAP dataset (on 76 music video clips) show that a significant improvement is obtained when higher level representations are used instead of low-level features, for video affective content analysis.

Keywords: Affect Analysis, Learning Feature Representations, Convolutional Neural Network, Support Vector Machine.

1 Introduction

Accessing online videos through services such as video on demand has become extremely easy thanks to equipments including DVB set top boxes (terrestrial, cable or satellite), Tablet PCs, high-speed Internet access or digital media-streaming devices. However, among the growing amount of multimedia, finding video content matching the current mood and needs of users is still a challenge. Within this context, video affective analysis is an emerging research field that targets this problem of affective content analysis of videos. The affective content of a video is defined as the intensity (i.e., arousal) and type (i.e., valence) of emotion (both are referred to as affect) that are expected to arise in the user while watching that video [7]. In order to deal with this

C. Gurrin et al. (Eds.): MMM 2014, Part I, LNCS 8325, pp. 303–314, 2014.

challenging affective analysis problem, machine learning methods are mainly used. The performance of machine learning methods is heavily dependent on the choice of data representation (or features) on which they are applied [2]. Therefore, one key issue in designing video affective content analysis algorithms is the representation of video content as in any pattern recognition task. The common approach for video content representation is either to use low-level audio-visual features or to build hand-crafted higher level representations based on the low-level ones (e.g., [5, 8, 13, 21]). Low-level features have the disadvantage of losing global relations or structure in data, whereas creating hand-crafted higher level representations is time consuming, problem-dependent, and requires domain knowledge. Besides, no dominant feature representation has emerged yet in the literature. In recent years, there has been a growing interest for learning features directly from raw data in the field of audio or video content analysis. Within this context, deep learning methods such as convolutional neural networks (CNNs) and deep belief networks are shown to provide promising results (e.g., [10, 15]). The advantages of deep learning architectures are: (1) *feature re-use*: constructing multiple levels of representation or learning a hierarchy of features, and growing ways to re-use different parts of a deep architecture by changing the depth of the architecture; (2) *abstraction and invariance*: more abstract concepts can often be constructed in terms of less abstract ones and have potentially greater predictive power (i.e., less sensitive to changes in input data) [2].

Inspired by the recent success of the deep learning methods in the field of audio-visual content analysis (e.g., [10, 12, 15]), we propose to directly learn feature representations from automatically extracted low-level audio-visual features by deep learning for the task of video affective content analysis. The aim of this work is to investigate the discriminative power of mid-level audio-visual representations which are learned from raw data by CNNs for modeling affective content of videos.

Our approach differs from the existing works (presented in Section 2) in the following aspects: (1) we learn both audio and visual feature representations from automatically extracted raw data by using a CNN and fuse these representations at decision-level for the affective classification of music video clips by SVM; (2) we show that the learned mid-level audio-visual representations are more discriminative and provide more precise results than low-level audio-visual ones.

The paper is organized as follows. Section 2 explores the recent developments and reviews methods which have been proposed for affective content analysis of video material with an emphasis on the feature representation of videos. In Section 3, we introduce our method for the affective classification of music video clips. We provide and discuss evaluation results on a subset of the DEAP dataset [11] in Section 4. Finally, we present concluding remarks and future directions to expand our method in Section 5.

2   Related Work

The most common type of affective content analysis approaches is to employ low-level audio-visual features as video representations. In [7], Hanjalic et al. utilize motion, color and audio features to represent arousal and valence. Soleymani et al. [16] address the affective representation of movie scenes based on the emotions that are actually felt by the audience, where audio-visual features as well as physiological responses of participants are employed to estimate the arousal and valence degree of scenes. Canini et al. [4] aim at defining the emotional identity of a movie in a multidimensional space (along natural, temporal and energetic dimensions) based on audio-visual features in order to retrieve emotionally similar movies based on their emotional identities. In [6], a method for mood-based classification of TV Programs on a large-scale dataset is presented, in which frame-level audio-visual features are used as video representations. Srivastava et al. [17] address the recognition of emotions of movie characters. Low-level visual features based on facial feature points are employed for the facial expression recognition part of the work, whereas lexical analysis of dialogs is performed in order to provide complementary information for the final decision. Cui et al. [5] address affective content analysis of music videos, where they adopt audio-visual features for the construction of arousal and valence models. Intended emotion tracking of movies is a subject addressed by Malandrakis et al. [13], where audio-visual features are extracted for the affective representation of movies. In [19], a combined analysis of low-level audio- and visual representations based on early feature fusion is presented for the facial emotion recognition in videos. Yazdani et al. [23] present a method which uses audio-visual features as representation for the affective analysis of music video clips.

The methods of the second category are based on mid-level or hierarchical representations of videos. These solutions construct mid-level representations based on low-level ones and employ these mid-level representations for the affective content analysis of videos. The work presented in [21] combines both low-level audio-visual representations with higher-level video representations. In a first step, movies of different genres are clustered into different arousal intensities (i.e., high, medium, low) with fuzzy c-means [3] using low-level audio-visual features of video shots. In a second step, the results from the first step (i.e., higher level video representations) are employed along together with low-level audio-visual features in order to perform emotional movie classification. Irie et al. [8] propose a latent topic driving model (LTDM) in order to address the issue of classifying movie scenes into affective categories at video shot level. For emotion modeling, the authors adopt Plutchik's eight basic emotions [14] and add a "neutral" category in order to reject movie scenes that arouse no emotion. Video shots are represented with a histogram of quantized audio-visual features and emotion topics are subsequently extracted via latent Dirichlet allocation. Emotions contained in a movie shot are determined based on the topics of the movie shot and predefined emotion transition weights based on the Plutchik's emotion theory. In an extended version of this work [9], Irie et al. propose to represent movie shots with so-called Bag-of-Affective Audio-visual Words and then apply the same LTDM architecture. Xu et al. [22] present a three-level affective content analysis framework, in which the purpose is to detect the affective content of videos (i.e., horror scenes for horror movies, laughable sections for sitcoms and emotional tagging of movies). They introduce mid-level representations which indicate dialog, audio emotional events (i.e., horror sound and laughter) and textual concepts (i.e., informative emotion keywords).

The abovementioned works represent videos with low- or mid-level hand-crafted features. However, in attempts to extend the applicability of methods, there is a growing interest for directly and automatically learning features from extracted low-level (i.e., raw) audio-visual features rather than representing audio or video data based on manually designed features. Schmidt et al. [15] address the feature representation issue for automatic detection of emotions in music by employing regression-based deep belief networks to directly learn features from magnitude spectra instead of manually designed feature representations. By taking into account the dynamic nature of music, they also investigate the effect of combining multiple timescales of aggregated magnitude spectra as a basis for feature learning. These learned features are then evaluated in the context of multiple linear regression. Li et al. [12] propose to perform feature learning for music genre classification and use CNNs for the extraction of musical pattern features. Ji et al. [10] address the automated recognition of human actions in surveillance videos and develop a novel 3D CNN model for action recognition. The proposed model extracts features from both spatial and temporal dimensions by performing 3D convolutions to capture motion information encoded in multiple adjacent frames. Information coming from multiple channels is combined into a final feature representation. They propose regularizing the outputs with high-level features and combining the predictions of a variety of different CNN models. The proposed method is tested on the TRECVID surveillance video dataset and has proven to achieve superior performance in comparison to baseline methods. Different from the aforementioned existing works, we learn both audio and visual feature representations by using a CNN and perform the affective classification of music video clips by fusing these representations at decision-level by a couple of SVMs. In this work, it is also experimentally shown that the learned audio-visual representations are more discriminative than low-level audio-visual ones.

3   The Video Affective Analysis Method

In this section, we present our approach for the affective classification of music video clips into the four quadrants of the VA-space. An overview of our method is illustrated in Figure 1. Music video clips are first segmented into pieces, each piece lasting 5 seconds, and subsequently MFCC and color values in the RGB space are extracted. After this feature extraction, the next step in the training phase is to build mid-level audio and visual representation generators using CNNs. Eventually, mid-level audio and visual representations are constructed and classifiers are generated using these representations. In the test phase, audio and visual representations are created by using the mid-level representation generators. These representations are used in order to classify a video segment of 5-second length into one of the four quadrants in the VA-space.

The audio and visual feature learning phases of our method are discussed in detail in Section 3.1 and 3.2, respectively. The generation of an affective analysis model is discussed in more detail in Section 3.3.

3.1   Learning Audio Representations

We extract MFCC values for each video segment. The resulting MFCC feature vectors are given as input to a CNN. The first layer (i.e., the input layer) of the CNN is a 125x13 map which contains the MFCC feature vectors from 125 frames of one music video segment. In Figure 2, the CNN architecture used to generate audio representations

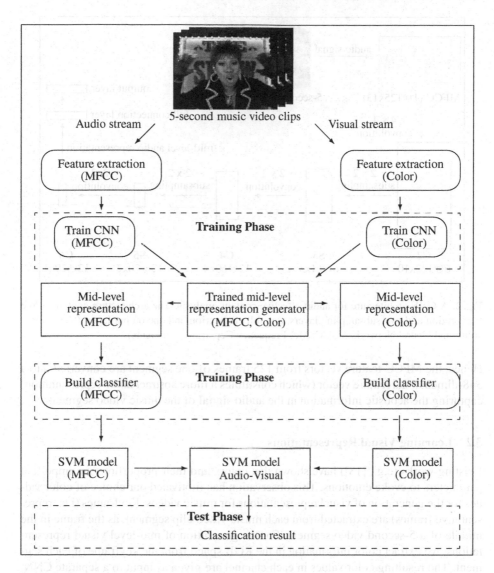

Fig. 1. A high-level overview of our method for video affective analysis. Final classification decisions are realized by a majority voting process. (CNN: Convolutional Neural Network, MFCC: Mel-Frequency Cepstral Coefficients, SVM: Support Vector Machine).

is presented. The CNN has three convolutional and two subsampling layers, one full connection and one output layer (this network size in terms of convolution and subsampling layers has experimentally given satisfactory results). The CNN is trained using the backpropagation algorithm. After training the CNN, the output of the last convolutional layer is used as the mid-level audio representation of corresponding video segment.

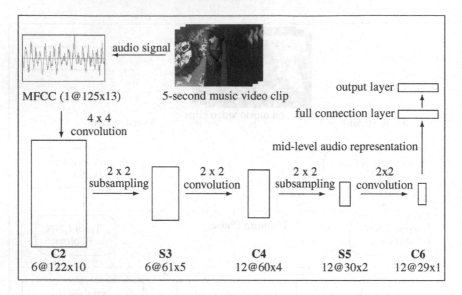

Fig. 2. A CNN architecture for audio affective content analysis. The architecture contains three convolution and two subsampling layers, one full connection and one output layer. (CNN: Convolutional Neural Network, MFCC: Mel-Frequency Cepstral Coefficients).

Hence, the MFCC feature vectors from 125 frames of one segment are converted into a 348-dimensional feature vector (which constitutes a more abstract audio representation) capturing the acoustic information in the audio signal of the music video segment.

3.2   Learning Visual Representations

Existing works (e.g., [18]) have shown that colors and their proportions are important parameters to evoke emotions. This observation has motivated our choice of color values for the generation of visual representations for music videos. Keyframes (i.e., representative frames) are extracted from each music video clip segment, as the frame in the middle of a 5-second video segment. For the generation of mid-level visual representations, we extract color information in the RGB space from the keyframe of each segment. The resulting color values in each channel are given as input to a separate CNN. In Figure 3, the CNN architecture used to generate visual representations is presented. The first layer (i.e., the input layer) of the CNN is a 160x120 map which contains the color values from one color channel of the keyframe. The CNN has three convolutional and two subsampling layers, one full connection and one output layer (this network size in terms of convolution and subsampling layers has also experimentally given satisfactory results). The training of the CNN is done similarly to the training of the CNN in the audio case. As a result, the color values in each color channel are converted into an 88-dimensional feature vector. The feature vectors generated for each of the three color channels are concatenated into a 264-dimensional feature vector which forms a more abstract visual representation capturing the color information in the keyframe of the music video segment.

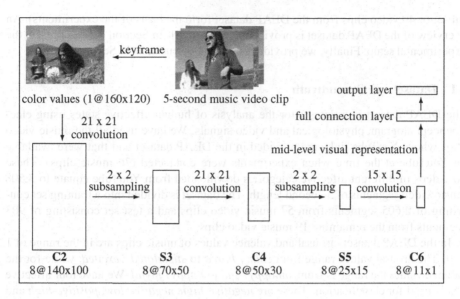

Fig. 3. A CNN architecture for visual affective content analysis. The architecture contains three convolution and two subsampling layers, one full connection and one output layer. (CNN: Convolutional Neural Network).

3.3 Generating the Affective Analysis Model

In order to generate an audio affective analysis model, mid-level audio representations are fed into a multi-class SVM. Similarly, a visual affective analysis model is also generated by feeding mid-level visual representations into a second multi-class SVM. The probability estimates of the two SVM models are subsequently fed into a third multi-class SVM to generate an audio-visual affective video analysis model. Normally, in a basic SVM, only class labels or scores are output. The class label results from thresholding the score, which is not a probability measure. The scores output by the SVM are converted into probability estimates using the method explained in [20].

In the test phase, mid-level audio and visual representations are created by using the corresponding CNN models for music video segments of 5-second length based on MFCC feature vectors and color values in the RGB color space. The music video segments are then classified by using the affective video analysis models. Final decisions for the classification of music video segments are realized by a majority voting process as in [10].

4 Performance Evaluation

The experiments presented in this section aim at comparing the discriminative power of mid-level audio-visual representations against low-level audio-visual features. A direct comparison with the methods (e.g., [23]) which are also tested on the DEAP dataset is limited due to the usage of different subsets of the DEAP dataset (e.g., in [23], only a subset of 40 video clips from the DEAP dataset form the basis of the experiments). An overview of the DEAP dataset is provided in Section 4.1. In Section 4.2, we present the experimental setup. Finally, we provide results and discussions in Section 4.3.

4.1 Dataset and Groundtruth

The DEAP dataset is a dataset for the analysis of human affective states using electroencephalogram, physiological and video signals. We have used all the music video clips whose YouTube links are provided in the DEAP dataset and that were available on YouTube at the time when experiments were conducted (76 music clips). These 76 videos of different affective categories downloaded from YouTube equate to 3,508 music video segments of 5-second length. The dataset is divided into a training set consisting of 2,605 segments from 57 music video clips and a test set consisting of 903 segments from the remaining 19 music video clips.

In the DEAP dataset, arousal and valence values of music clips are in the range of 1 to 9. The arousal values range from *calm / bored* to *stimulated / excited*, while for the valence values the range is from *unhappy / sad* to *happy / joyful*. We have four affective labels used for classification. These are *negative-high*, *negative-low*, *positive-high* and *positive-low* each representing one quadrant in the VA-space. The online ratings of music video clips provided within the DEAP dataset are used in order to determine the label of the music video clips. First of all, the average of the online ratings of the music video clips is computed both for the arousal and valence dimensions. If the average arousal value of a music video clip is above / below 5, then the music video clip is labeled as *high / low*. Similarly, a music video clip is labeled as *positive / negative*, when the average valence value of the music video clip is above / below 5. Table 1 summarizes the main characteristics of the dataset in more detail.

Table 1. The characteristics of training and test datasets (NH: *negative-high*, NL: *negative-low*, PH: *positive-high*, PL: *positive-low*)

Dataset	Clips	Segments	NH	NL	PH	PL
Train	57	2,605	12	14	12	19
Test	19	903	4	5	4	6
Whole	76	3,508	16	19	16	25

4.2 Experimental Setup

We employed the MIR Toolbox v1.4[1] to extract the 13-dimensional MFCC features. Frame sizes of 40 ms without overlap are used to temporally match with the 25-fps video frames. Mean and standard deviation for each dimension of the MFCC feature vectors are computed, which compose the low-level audio representations of music video segments. In order to generate the low-level visual features of music video segments, we constructed 256-bin color histograms for each color channel in the RGB

---

[1] https://www.jyu.fi/hum/laitokset/musiikki/en/research/coe/
materials/mirtoolbox

color space resulting in 768-dimensional low-level visual feature vectors. We used the Deep Learning toolbox[2] in order to generate mid-level audio and visual representations with a CNN. We trained the multi-class SVMs with an RBF kernel using libsvm[3] as the SVM implementation. Training was performed using audio and visual features extracted at the music video segment level. More specifically, we trained one SVM using the CNN-based mid-level audio features, one SVM using the CNN-based mid-level visual features and two other SVMs using the low-level audio and visual features as input, respectively. Fusion of audio and visual features is performed at decision-level using an SVM for both low-level and mid-level audio-visual representations. The SVM parameters were optimized by 5-fold cross-validation on the training data. Our approach was evaluated using a training-test split (75% training and 25% test).

4.3   Results and Discussions

Table 2 reports the classification accuracies of our method compared to the method which employs low-level audio-visual features on the DEAP dataset. We achieved 52.63% classification accuracy with mid-level audio-visual representations learned from raw data, while the classification accuracy is 36.84% for the method which uses low-level audio-visual features. When CNN-generated representations are used for training, trained classifiers are able to discriminate better between videos with varying affective content. This demonstrates the potential of our approach for video affective content analysis.

Table 2. Classification accuracies on the DEAP dataset (with audio-visual representations)

Method	Accuracy (%)
Our method (mid-level audio-visual)	**52.63**
The low-level audio-visual method	36.84

We present the classification accuracies in the case where only one modality is employed in Table 3. It gives an estimation of the influence in the performance of each modality (either audio or visual) in more detail.

Table 3. Classification accuracies on the DEAP dataset (with unimodal representations (audio- or visual-only))

Method	Accuracy (%)
Our method (mid-level audio)	**47.37**
Our method (mid-level visual)	36.84
The low-level audio method	36.84
The low-level visual method	31.58

---

[2] https://github.com/rasmusbergpalm/DeepLearnToolbox/
[3] http://www.csie.ntu.edu.tw/~cjlin/libsvm/

312   E. Acar, F. Hopfgartner, and S. Albayrak

One significant point which can be inferred from Table 3 is that audio representations are more discriminative than visual features for the affective analysis of music videos. This is true for the mid-level audio representation outperforming the mid-level visual one, and also for the low-level audio representation compared to the low-level visual representation. Additionally, we see that the improvement of using mid-level audio features (around 10% gain compared to low-level audio) compared to the improvement of using mid-level visual features (around 5% gain compared to low-level visual) is higher.

Another conclusion that can be derived from the results is that mid-level audio and visual representations outperform low-level audio and visual representations, respectively. The most significant point which can be deduced from the overall results is that the fusion of mid-level audio and visual representations further improves the performance, while the fusion of low-level audio and visual features provides no significant performance improvement over low-level audio representations. In other words, using color histograms provides a slight improvement (at the music segment level) for the affective analysis of music videos. In Figure 4, the confusion matrices of the classification results of our method (with audio-only and audio-visual representations) for the DEAP dataset are illustrated. The confusion matrix on the left (Figure 4(a)) represents the performance of our method with CNN-generated audio representations, while the confusion matrix on the right (Figure 4(b)) represents the performance of our method with CNN-generated audio and visual representations that are fused at decision-level using a multi-class SVM. The detailed definition of the labels presented in Figure 4 is given in Section 4.1. It is observed from these two matrices that more discriminative video representations are built by incorporating mid-level color information and that video clips which were wrongly classified as *positive-low* with audio-only representations are classified correctly as *negative-low*. As illustrated in Figure 4(b), *negative-low - positive-low* pairs are the most confused affective label pairs. Another point to mention is that our method shows difficulties in discriminating video segments of *positive-high* and classifies these segments mainly as *positive-low*. These results suggest that there is a need to incorporate additional audio and visual features for the representation of video segments.

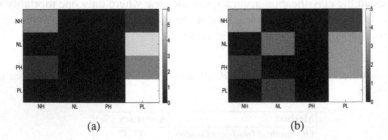

(a)   (b)

Fig. 4. Confusion matrix on the DEAP dataset with (a) mid-level audio and (b) mid-level audio-visual representations learned from raw data (Mean accuracy: 47.37 for audio-only and 52.63 for audio-visual). (NH: *negative-high*, NL: *negative-low*, PH: *positive-high*, PL: *positive-low*).

5   Conclusions

In this paper, we presented an approach for the affective labeling of music video clips, where higher level representations were learned from low-level audio-visual features using CNNs. More specifically, MFCC was employed as audio features, while color values in the RGB space formed the visual features. We utilized the learned audio-visual representations to classify each music video clip into one of the four quadrants of the VA-space using multi-class SVMs. Experimental results on a subset of the DEAP dataset support our belief that higher level audio-visual representations which are learned using CNNs are more discriminative than low-level ones. The current method only exploits MFCC and color features for the generation of video representations. An interesting research question would consist in determining whether augmenting the feature set by including an extensive set of audio and visual features (especially motion-related features such as optical flow) would provide a significant gain in performance. Hence, as future work, we plan to study the representation of videos with additional audio- and visual features. Exploring the architecture of the CNNs (i.e., optimizing the number of convolution layers and kernel size) is another direction for future research. Another direction in the context of learning features by deep learning is to use unsupervised deep learning methods (e.g., restricted Boltzmann machines or stacked auto-encoders [1]) for modeling the data before applying SVM. Comparing the performance of learned audio-visual representations to hand-crafted mid-level feature representations such as the bag-of-words approaches is another track for future work.

References

1. Bengio, Y.: Learning deep architectures for ai. Foundations and trends® in Machine Learning 2(1), 1–127 (2009)
2. Bengio, Y., Courville, A., Vincent, P.: Representation learning: A review and new perspectives (2013)
3. Bezdek, J.C.: Pattern recognition with fuzzy objective function algorithms. Kluwer Academic Publishers (1981)
4. Canini, L., Benini, S., Migliorati, P., Leonardi, R.: Emotional identity of movies. In: 2009 16th IEEE International Conference on Image Processing (ICIP), pp. 1821–1824. IEEE (2009)
5. Cui, Y., Jin, J.S., Zhang, S., Luo, S., Tian, Q.: Music video affective understanding using feature importance analysis. In: Proceedings of the ACM International Conference on Image and Video Retrieval, pp. 213–219. ACM (2010)
6. Eggink, J., Bland, D.: A large scale experiment for mood-based classification of tv programmes. In: 2012 IEEE International Conference onMultimedia and Expo (ICME), pp. 140–145. IEEE (2012)
7. Hanjalic, A., Xu, L.: Affective video content representation and modeling. IEEE Transactions on Multimedia, 143–154 (2005)
8. Irie, G., Hidaka, K., Satou, T., Kojima, A., Yamasaki, T., Aizawa, K.: Latent topic driving model for movie affective scene classification. In: Proceedings of the 17th ACM International Conference on Multimedia, pp. 565–568. ACM (2009)
9. Irie, G., Satou, T., Kojima, A., Yamasaki, T., Aizawa, K.: Affective audio-visual words and latent topic driving model for realizing movie affective scene classification. IEEE Transactions on Multimedia 12(6), 523–535 (2010)

10. Ji, S., Xu, W., Yang, M., Yu, K.: 3d convolutional neural networks for human action recognition. IEEE Transactions on Pattern Analysis and Machine Intelligence, 221–231 (2013)
11. Koelstra, S., Muhl, C., Soleymani, M., Lee, J.-S., Yazdani, A., Ebrahimi, T., Pun, T., Nijholt, A., Patras, I.: Deap: A database for emotion analysis; using physiological signals. IEEE Transactions on Affective Computing 3(1), 18–31 (2012)
12. Li, T., Chan, A.B., Chun, A.: Automatic musical pattern feature extraction using convolutional neural network. In: Proc. Int. Conf. Data Mining and Applications (2010)
13. Malandrakis, N., Potamianos, A., Evangelopoulos, G., Zlatintsi, A.: A supervised approach to movie emotion tracking. In: 2011 IEEE International Conference on Acoustics, Speech and Signal Processing (ICASSP), pp. 2376–2379. IEEE (2011)
14. Plutchik, R.: The nature of emotions human emotions have deep evolutionary roots, a fact that explain their complexity and provide tools for clinical practice. American Scientist 89(4), 344–350 (2001)
15. Schmidt, E.M., Scott, J., Kim, Y.E.: Feature learning in dynamic environments: Modeling the acoustic structure of musical emotion. In: International Society for Music Information Retrieval, pp. 325–330 (2012)
16. Soleymani, M., Chanel, G., Kierkels, J., Pun, T.: Affective characterization of movie scenes based on multimedia content analysis and user's physiological emotional responses. In: Tenth IEEE International Symposium on Multimedia, ISM 2008, pp. 228–235. IEEE (2008)
17. Srivastava, R., Yan, S., Sim, T., Roy, S.: Recognizing emotions of characters in movies. In: 2012 IEEE International Conference on Acoustics, Speech and Signal Processing (ICASSP), pp. 993–996. IEEE (2012)
18. Valdez, P., Mehrabian, A.: Effects of color on emotions. Journal of Experimental Psychology: General 123(4), 394 (1994)
19. Wimmer, M., Schuller, B., Arsic, D., Rigoll, G., Radig, B.: Low-level fusion of audio and video feature for multi-modal emotion recognition. In: 3rd International Conference on Computer Vision Theory and Applications, VISAPP, vol. 2, pp. 145–151 (2008)
20. Wu, T.-F., Lin, C.-J., Weng, R.C.: Probability estimates for multi-class classification by pairwise coupling. The Journal of Machine Learning Research 5, 975–1005 (2004)
21. Xu, M., Jin, J.S., Luo, S., Duan, L.: Hierarchical movie affective content analysis based on arousal and valence features. In: Proceedings of the 16th ACM International Conference on Multimedia, pp. 677–680. ACM (2008)
22. Xu, M., Wang, J., He, X., Jin, J.S., Luo, S., Lu, H.: A three-level framework for affective content analysis and its case studies. Multimedia Tools and Applications (2012)
23. Yazdani, A., Kappeler, K., Ebrahimi, T.: Affective content analysis of music video clips. In: Proceedings of the 1st International ACM Workshop on Music Information Retrieval with User-Centered and Multimodal Strategies, pp. 7–12. ACM (2011)

Robust Image Restoration via Reweighted Low-Rank Matrix Recovery*

Yigang Peng[1,2], Jinli Suo[1], Qionghai Dai[1], Wenli Xu[1], and Song Lu[2]

[1] TNLIST and the Department of Automation, Beijing 100084
[2] National Computer Network Emergency Response Technical Team Coordination
Center of China (CNCERT or CNCERT/CC), Beijing 100029
pengyigang@cert.org.cn, {jlsuo,qionghaidai,xuwl}@tsinghua.edu.cn

Abstract. In this paper, we propose a robust image restoration method via reweighted low-rank matrix recovery. In the literature, Principal Component Pursuit (PCP) solves low-rank matrix recovery problem via a convex program of mixed nuclear norm and ℓ_1 norm. Inspired by reweighted ℓ_1 minimization for sparsity enhancement, we propose reweighting singular values to enhance low rank of a matrix. An efficient iterative reweighting scheme is proposed for enhancing low rank and sparsity simultaneously and the performance of low-rank matrix recovery is prompted greatly. We demonstrate the utility of the proposed method on robust image restoration, including single image and hyperspectral image restoration. All of these experiments give appealing results on robust image restoration.

Keywords: robust image restoration, low-rank matrix recovery, iterative reweighting, non-uniform singular value thresholding.

1 Introduction

Low-rank matrix recovery is a highly developing research topic with close relationship to compressed sensing [1] and matrix rank minimization [2], [3] and has wide applications in computer vision and image processing [4], [5], [6], [7], signal processing [8], and many other fields. It considers recovering a low-rank matrix from gross sparse errors, also known as Robust Principle Component Analysis (RPCA) [4]. However, both minimizing the ℓ_0 norm and matrix rank function are not directly tractable due to the highly nonlinear, highly nonconvex properties of the objective functions. A surprising result given by Candès et al. [4] shows that under rather weak assumptions, solving the following convex optimization problem:

$$\min_{\mathbf{L},\mathbf{S}} \quad \|\mathbf{L}\|_* + \lambda\|\mathbf{S}\|_1, \quad \text{s.t.} \quad \mathbf{M} = \mathbf{L} + \mathbf{S}, \tag{1}$$

which is dubbed as Principal Component Pursuit (PCP), can exactly recover the groundtruth low-rank matrix $\mathbf{L}^0$ and the sparse matrix $\mathbf{S}^0$ (the observation

* This work was supported by the project of National Nature Science Foundation of China (NSFC) No. 61120106003, and No. 61171119, and National High-tech R&D Program of China (863 Program) No. 2011AA010601.

C. Gurrin et al. (Eds.): MMM 2014, Part I, LNCS 8325, pp. 315–326, 2014.
© Springer International Publishing Switzerland 2014

matrix $\mathbf{M}$ is generated by $\mathbf{M} = \mathbf{L}^0 + \mathbf{S}^0$), where $\|\mathbf{L}\|_*$ is the nuclear norm of matrix $\mathbf{L}$ (i.e. the sum of all its singular values), $\|\mathbf{S}\|_1$ is the ℓ_1 norm of matrix $\mathbf{S}$ (i.e. the sum of absolute value of all its entries), $\mathbf{M}, \mathbf{L}, \mathbf{S}, \mathbf{L}^0, \mathbf{S}^0 \in \mathbb{R}^{m \times n}$, and $\lambda > 0$ is a weighting parameter.

Due to the inherent sparse or/and low rank property in images and videos, such a low-rank matrix recovery model and its variants have been successfully applied in many image processing and computer vision problems. A patch based robust video restoration method, which groups similar patches in spatial-temporal domain and then formulates the restoration problem as joint sparse and low rank matrix approximation, is proposed by Ji et al. [9]. In [4], by modeling the stable background as low rank part and foreground variations as sparse part, PCP demonstrates its successful application in surveillance video background modeling. A robust alignment method by sparse and low rank decomposition (RASL) is proposed in [10], which simultaneously aligns a batch of linearly correlated images in spite of gross corruption. Low-rank structure and sparse modeling is harnessed for holistic symmetry detection and rectification by a method called transform invariant low-rank texture (TILT) [11]. These techniques have been extended for camera calibration [12], generalized cylindrical surfaces unwrap-pig [13], urban structures reconstruction [14], text detection [15], etc. In a recent work [16] on image completion, they correctly repair the global structure of a corrupted texture by harnessing both low-rank and sparse structures in regular or near regular textures. More applications also lie in dynamic MRI reconstruction [17] and many others.

A lot of practical algorithms are developed to solve ℓ_1 minimization or/and nuclear norm minimization, such as singular value thresholding (SVT) [18], accelerated proximal gradient (APG) algorithm [19], augmented lagrangian multiplier (ALM) method [20], [21] etc. In [4], they empirically show the algorithm's ability of matrix recovery with varying rank from errors of varying sparsity.

1.1   Motivation

This work is inspired by reweighted ℓ_1 minimization for enhancing sparsity [22]. By minimizing a sequence of weighted ℓ_1 norm, Candès et al. show significant performance improvement on sparse recovery or/and estimation.

Reweighted ℓ_1 norm is defined as $\|\mathbf{w} \odot \mathbf{x}\|_1$ for some appropriate weights $\mathbf{w}$, where $\mathbf{x} = \{x_i\} \in \mathbb{R}^n, \mathbf{w} = \{w_i\} \in \mathbb{R}^n_{++}$. Note that, if for each item x_i, the weights w_i are exactly inversely proportional to the absolute value of x_i, i.e., $w_i = \frac{1}{|x_i|}$ (for $x_i \neq 0$), we have $\|\mathbf{w} \odot \mathbf{x}\|_1 = \|\mathbf{x}\|_0$, where $\odot$ denotes the component-wise product (Hadamard product) of two variables, and $\|\cdot\|_0$ denotes the ℓ_0 norm (the number of nonzero entries). While, to minimize $\text{rank}(\mathbf{X})$, it is equivalent to minimize $\|\sigma(\mathbf{X})\|_0$, where $\sigma(\mathbf{X}) = \{\sigma_i(\mathbf{X})\}$ is the singular value vector constructed from $\mathbf{X}$. Inspired by reweighted ℓ_1 norm minimization, we propose reweighted nuclear norm which is defined as $\sum_i w_i \sigma_i(\mathbf{X})$ for

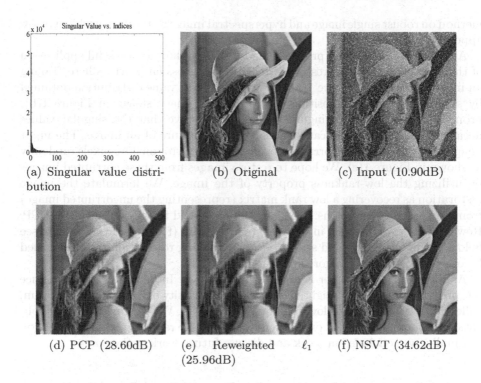

(a) Singular value distribution

(b) Original

(c) Input (10.90dB)

(d) PCP (28.60dB)

(e) Reweighted ℓ_1 (25.96dB)

(f) NSVT (34.62dB)

Fig. 1. Restoration of a single image 'Lena', 20% of whose pixels are corrupted by large noise. Figure(a) shows the singular value distribution of the original image shown in Figure(b). Figure(c) shows the corrupted image to be restored. Figure(d), (e), (f) are restored images obtained by PCP, Reweighted ℓ_1, and NSVT, respectively. Here, Peak Signal-to-Noise Ratio (PSNR) is given below each image.

some appropriate weights w_i. Moreover, if $w_i = \frac{1}{\sigma_i(\mathbf{X})}$ (for $\sigma_i(\mathbf{X}) \neq 0$), we have $\sum_i w_i \sigma_i(\mathbf{X}) = \text{rank}(\mathbf{X})$. In practice, we estimate the weights w_i^{k+1} from the previous iteration. For instance, in $(k+1)$-th iteration, $w_i^{k+1} = \frac{1}{\sigma_i(\mathbf{X}^k)+\epsilon}$ (where, $\epsilon > 0$ is a preset constant).

1.2 Methodology

In this paper, we propose a Non-uniform Singular Value Thresholding (NSVT) operator to enhance low rank for matrix rank minimization, which shares a similar philosophy with nonuniform soft-thresholding operator for reweighted ℓ_1 norm minimization for sparsity enhancement. We apply NSVT accompanying with reweighted ℓ_1 norm minimization to RPCA problem so as to promote the performance of low-rank matrix recovery. We show that by properly reweighting singular values for low-rank matrix and reweighting ℓ_1 norm for sparse matrix, better matrix recovery performance can be obtained. Applications of our proposed NSVT method on robust single image and hyperspectral image restoration problems show appealing results.

As the first glance of our proposed algorithm, we show a successful application of the NSVT algorithm in restoring images from gross large errors here. Taking an image as a matrix, Figure 1(a) shows the singular value distribution obtained by singular value decomposition of the original image shown in Figure 1(b). From the distribution of singular values, we observe that the singular values decay very fast, which indicates the low-rank structure of an image. The input images to be restored are corrupted by gross sparse error on 20% pixels randomly as shown in Figure 1(c). We hope to restore images from those corrupted images by utilizing the low-rankness property of the image. We formulate the image restoration as recovering a low rank matrix (representing the uncorrupted image) from corrupted observations. We show the recovered images obtained by PCP, Reweighted ℓ_1, and NSVT in Figure 1 (d), (e), and (f), respectively (please see below for details). We could see the visual appealing restoration result obtained by our proposed NSVT algorithm from Figure 1(f).

As a reminder, the paper is organized as follows. In Section 2, we introduce our proposed iterative reweighted low rank and sparsity enhancement algorithm, called NSVT for short, for low-rank matrix recovery. We demonstrate the applications of proposed method on different robust image restoration tasks in Section 3. Finally, we conclude our work and discuss future work in Section 4.

2 An Iterative Algorithm for Reweighted Low-Rank Matrix Recovery

2.1 Problem Description: Minimization of Mixed 'Weighted' Nuclear Norm and ℓ_1 Norm

The idea of reweighted ℓ_1 norm minimization is to use large weights so as to discourage nonzero entries and small weights so as to encourage nonzero entries. Our proposed NSVT method is motivated by such a similar idea for singular values. Notice that the rank of a matrix is actually the number of nonzero singular values of the matrix. Thus, similar to the philosophy of reweighted ℓ_1 norm minimization, we explore large weights to discourage nonzero singular values and small weights to encourage nonzero singular values, which leads us to consider the following 'weighted' nuclear norm and ℓ_1 norm mixed minimization problem:

$$\min_{\mathbf{L},\mathbf{S}} \sum_{j=1}^{n} w_{\mathbf{L},j} \cdot \sigma_j + \lambda \|\mathbf{W}_\mathbf{S} \odot \mathbf{S}\|_1 \quad \text{s.t.} \quad \mathbf{M} = \mathbf{L} + \mathbf{S}, \tag{2}$$

where $\mathbf{w}_\mathbf{L} = \{w_{\mathbf{L},j}\} \in \mathbb{R}^n_{++}$ and $\mathbf{W}_\mathbf{S} \in \mathbb{R}^{m \times n}$ are weights for singular values of $\mathbf{L}$ and entries of $\mathbf{S}$, respectively, and $\{\sigma_j\}$ are singular values of matrix $\mathbf{L}$. Without loss of generality, suppose $m \geq n$. Note that, the 'weighted' nuclear norm, i.e. weighted sum of singular values, isn't a convex function in general and thus it isn't a norm any more.

2.2   An Iterative Algorithm for Enhancing Low Rank and Sparsity

One immediate question is how to choose weights $\mathbf{w_L}$ and $\mathbf{W_S}$ wisely to improve matrix recovery. As suggested in [22], the weights $\mathbf{W_S}$ are inversely proportional to signal magnitude for reweighted ℓ_1 norm minimization. Similarly, the weights $\mathbf{w_L}$ are chosen to be inversely proportional to the magnitudes of singular values. In practice, the favorable weights $\mathbf{w_L}$ and $\mathbf{W_S}$ are updated based on estimations of $(\mathbf{L}^{(k)}, \mathbf{S}^{(k)})$ in the previous iteration according to (3).

Thus, we propose the following iterative low rank and sparsity enhancement scheme for matrix recovery by alternatively estimating $(\mathbf{L}, \mathbf{S})$ and redefining weights $(\mathbf{w_L}, \mathbf{W_S})$. The whole algorithm is summarized in Algorithm 1.

Algorithm 1. Enhancing low rank and sparsity for low rank matrix recovery

Step 1: Set the iteration counter $k = 0$, and $\mathbf{w_L}^{(0)} = \mathbf{1} \in \mathbb{R}^n_{++}$, $\mathbf{W_S}^{(0)} = \mathbf{1} \cdot \mathbf{1}^T \in \mathbb{R}^{m \times n}$ (supposing $m \geq n$).

Step 2: Enhance low rank and sparsity using **Algorithm 2** by weights $\mathbf{w_L}^{(k)}$ and $\mathbf{W_S}^{(k)}$, and get solutions $(\mathbf{L}^*, \mathbf{S}^*) \rightarrow (\mathbf{L}^{(k)}, \mathbf{S}^{(k)})$.

Step 3: Update weights: the weights for each $i = 1, ..., m$ and $j = 1, ..., n$ is updated by

$$w_{\mathbf{L},j}^{(k+1)} = \frac{1}{\sigma_j^{(k)} + \epsilon_L}, \quad w_{\mathbf{S},ij}^{(k+1)} = \frac{1}{|s_{ij}^{(k)}| + \epsilon_S}, \tag{3}$$

where ϵ_L and ϵ_S are predetermined positive constants, and the singular value matrix $\mathbf{\Sigma}^{(k)} = \mathrm{diag}\left(\left[\sigma_1^{(k)}, \cdots, \sigma_n^{(k)}\right]\right) \in \mathbb{R}^{n \times n}$ with

$$\left[\mathbf{U}^{(k)}, \mathbf{\Sigma}^{(k)}, \mathbf{V}^{(k)}\right] = \mathrm{svd}(\mathbf{L}^{(k)}), \tag{4}$$

where $\mathbf{U}^{(k)} \in \mathbb{R}^{m \times n}$, $\mathbf{V}^{(k)} \in \mathbb{R}^{n \times n}$.

Step 4: Terminate on convergence or when k attains a specified maximum number of iterations k_{max}, and output solutions $(\tilde{\mathbf{L}}, \tilde{\mathbf{S}})$. Otherwise, increase k and go to **Step 2**.

2.3   Solving Inner Optimization via an Inexact Augmented Lagrangian Multiplier Method

Next, we show how an inexact augmented Lagrangian multiplier (IALM) method [20] can be adopted to solve problem (2). The augmented Lagrangian function is defined as

$$l(\mathbf{L}, \mathbf{S}, \mathbf{Y}) = \sum_{j=1}^{n} w_{\mathbf{L},j} \cdot \sigma_j + \lambda \|\mathbf{W_S} \odot \mathbf{S}\|_1$$

$$+ \langle \mathbf{Y}, \mathbf{M} - \mathbf{L} - \mathbf{S} \rangle + \frac{\mu}{2} \|\mathbf{M} - \mathbf{L} - \mathbf{S}\|_F^2, \tag{5}$$

where $\langle , \rangle$ denotes the inner product of two matrices.

Algorithm 2. The IALM for solving the mixed 'weighted' nuclear norm and ℓ_1 norm minimization problem (2)

INPUT: Initialize $(\mathbf{L}_0, \mathbf{S}_0) \in \mathbb{R}^{m \times n} \times \mathbb{R}^{m \times n}$, $\mathbf{Y}_0 \in \mathbb{R}^{m \times n}$, $\mu_0 > 0$, $\rho > 1$, $t = 0$.
WHILE not converged **DO**
$\quad \mathbf{L}_{t+1} = \mathcal{D}_{\mu_t^{-1}\mathbf{w}_\mathbf{L}}\left[\mathbf{M} - \mathbf{S}_t + \mu_t^{-1}\mathbf{Y}_t\right]$;
$\quad \mathbf{S}_{t+1} = \mathcal{S}_{\lambda\mu_t^{-1}\mathbf{w}_\mathbf{S}}\left[\mathbf{M} - \mathbf{L}_{t+1} + \mu_t^{-1}\mathbf{Y}_t\right]$;
$\quad \mathbf{Y}_{t+1} = \mathbf{Y}_t + \mu_t\left(\mathbf{M} - \mathbf{L}_{t+1} - \mathbf{S}_{t+1}\right)$;
$\quad \mu_{t+1} = \rho\mu_t$; $\quad t \leftarrow t + 1$.
OUTPUT: solutions $(\mathbf{L}^*, \mathbf{S}^*)$.

The IALM takes advantage of the separable forms with respect to $\mathbf{L}$ and $\mathbf{S}$, and solves a sequence of programs alternatively between

$$\min_{\mathbf{L}} \; l(\mathbf{L}, \mathbf{S}, \mathbf{Y}), \quad \text{(fixing } \mathbf{S}\text{)}, \tag{6}$$

and

$$\min_{\mathbf{S}} \; l(\mathbf{L}, \mathbf{S}, \mathbf{Y}), \quad \text{(fixing } \mathbf{L}\text{)}, \tag{7}$$

and then updates the Lagrange multiplier matrix via

$$\mathbf{Y} \leftarrow \mathbf{Y} + \mu(\mathbf{M} - \mathbf{L} - \mathbf{S}). \tag{8}$$

It is not hard to proof that the optimal solution of (7) can be given by non-uniform soft thresholding operator, which is defined as follows. For a vector $\mathbf{x} \in \mathbb{R}^n$ and thresholding weights vector $\mathbf{w} \in \mathbb{R}^n_{++}$, the non-uniform soft thresholding operator $\mathcal{S}_\mathbf{w}$ is introduced as:

$$\mathcal{S}_\mathbf{w}[\mathbf{x}] = \{\text{sign}(x_i)(|x_i| - w_i)_+\}, \tag{9}$$

where $t_+ = \max(t, 0)$ is the positive part of t.

To solve (6), we introduce the Non-uniform Singular Value Thresholding (NSVT) operator as follows:

$$\mathcal{D}_\mathbf{w}[\mathbf{X}] = \mathbf{U}\mathcal{S}_\mathbf{w}[\boldsymbol{\Sigma}]\mathbf{V}^\mathbf{T}, \tag{10}$$

where $\mathcal{S}_\mathbf{w}[\boldsymbol{\Sigma}] = \text{diag}(\{(\sigma_i - w_i)_+\})$. Note that, for a matrix $\mathbf{X} \in \mathbb{R}^{m \times n}$ of rank r, we denote its Singular Value Decomposition (SVD) as

$$\mathbf{X} = \mathbf{U}\boldsymbol{\Sigma}\mathbf{V}^\mathbf{T}, \quad \boldsymbol{\Sigma} = \text{diag}(\{\sigma_i\}_{1 \leq i \leq r}), \tag{11}$$

where $\mathbf{U} \in \mathbb{R}^{m \times r}$ and $\mathbf{V} \in \mathbb{R}^{r \times n}$ are with orthonormal columns respectively, the singular values $\sigma_i > 0$, and $\boldsymbol{\Sigma} \in \mathbb{R}^{r \times r}$ is the diagonal matrix with $\sigma_1, \cdots, \sigma_r$ on the diagonal and zeros elsewhere.

In this way, the matrix $\mathbf{L}$ is updated by the non-uniform singular value thresholding (NSVT) operator

$$\mathbf{L} \leftarrow \mathcal{D}_{\mu^{-1}\mathbf{w}_\mathbf{L}}\left[\mathbf{M} - \mathbf{S} + \mu^{-1}\mathbf{Y}\right], \tag{12}$$

with the thresholding value $\mu^{-1}\mathbf{w_L}$, and the matrix $\mathbf{S}$ is updated by the non-uniform soft thresholding operator

$$\mathbf{S} \leftarrow \mathcal{S}_{\lambda\mu^{-1}\mathbf{W_S}}[\mathbf{M} - \mathbf{L} + \mu^{-1}\mathbf{Y}], \qquad (13)$$

with the thresholding value $\lambda\mu^{-1}\mathbf{W_S}$. Specially, such a method is also known as the alternating direction method of multipliers (ADMM), which is popular in and useful for optimizing problems with separable variables, also see [23], [21].

The IALM algorithm for solving the mixed 'weighted' nuclear norm and ℓ_1 norm minimization problem (2) is summarized in Algorithm 2. Note that the objective function in (2) is not a convex optimization problem and the convergence of the algorithm is still under investigation. However, the numerical simulation and robust image restoration experiments give empirical evidence on the performance of the proposed algorithm.

3   Experiments and Analysis

We show performance our proposed method on robust single image and hyperspectral image restoration problems. In the experiments, the weighting parameter λ is set to be $1/\sqrt{\max(m, n)}$ the same as in PCP problem [4]. In Algorithm 1, we fix the constant $\epsilon_\mathbf{L} = 0.01$ and $\epsilon_\mathbf{S} = 0.01$ empirically. In Algorithm 2, we set $(\mathbf{L_0}, \mathbf{S_0}) = (\mathbf{0}, \mathbf{0})$, $\mathbf{Y_0} = \mathbf{M}/\max(\|\mathbf{M}\|, \lambda^{-1}\|\mathbf{M}\|_\infty)$, where $\|\mathbf{M}\|$ denotes the spectral norm of matrix $\mathbf{M}$ and $\|\mathbf{M}\|_\infty$ is the maximum absolute value of entries in matrix $\mathbf{M}$, $\mu_0 = 1.25/\|\mathbf{M}\|$, $\rho = 1.5$, and the stopping criterion for the iteration is $\|\mathbf{M} - \mathbf{L}_t - \mathbf{S}_t\|_F/\|\mathbf{M}\|_F < 10^{-7}$.

We denote our proposed rank and sparsity enhancement scheme as NSVT for simplicity. We compare our proposed method with the original convex PCP problem and one scheme which only applies Reweighted ℓ_1 for sparsity enhancement in low-rank matrix recovery, i.e.,

$$\min_{\mathbf{L},\mathbf{S}} \quad \|\mathbf{L}\|_* + \lambda\|\mathbf{W_S} \odot \mathbf{S}\|_1 \quad \text{s.t.} \quad \mathbf{M} = \mathbf{L} + \mathbf{S}, \qquad (14)$$

where $\odot$ denotes the component-wise product of two variables. The problem (14) is also considered in [24], and we call it Reweighted ℓ_1 for short. For all the three problems of PCP, Reweighted ℓ_1 and NSVT, we use IALM to solve them[1].

3.1   Single Image Restoration

We show the application of proposed method on single image restoration in Figure 1 and 2. Most natural images are approximately low-rank, as can bee seen from Figure 1(a) and 2(a), for example. Note that, these two images contain different scenes: Figure 1(a) shows a portrait of a woman, and 2(a) shows a scene of bridge and lighthouse. Taking an image as a matrix, Figure 1(a)

---

[1] The code for solving PCP problem using IALM is downloaded from
http://perception.csl.illinois.edu/matrix-rank/home.html.

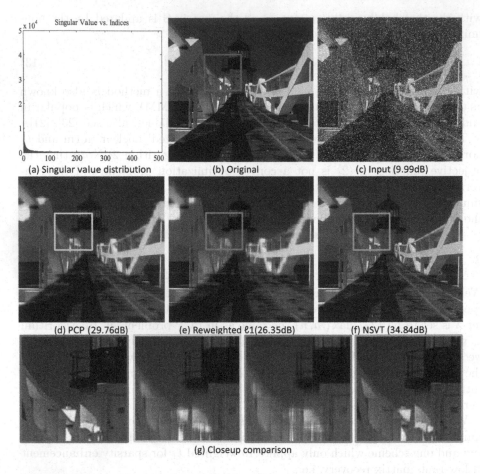

Fig. 2. Restoration of a single image, 20% of whose pixels are corrupted by large noise. Figure(a) shows the singular value distribution of the original image shown in Figure(b). Figure(c) shows the corrupted image to be restored. Figure(d), (e), (f) are restored images obtained by PCP, Reweighted ℓ_1, and NSVT, respectively. The PSNR is given below each image. Figure(g) shows the close-up patches of Figure(b), (d), (e), and (f), respectively.

and 2(a) show the singular value distribution obtained by SVD of the images, which are shown in Figure 1(b) and 2(b)[2], respectively. From the distribution of singular values, we observe that the singular values decay very fast, which indicates the low-rank structure of an image. The low-rank prior is used for restoring images from corruptions. In the experiment, we set $k_{max} = 1$ for both Reweighted ℓ_1 method and our proposed NSVT method. The original images of 515×512 pixels are shown in Figure 1(b) and 2(b). The input images to be restored are corrupted by gross sparse error on 20% pixels randomly as shown

---

[2] Original images are downloaded from: http://decsai.ugr.es/cvg/CG/base.htm.

in Figure 1(c) and 2(c). We hope to restore images from those corrupted images. Note that, this is different from low-rank matrix completion or tensor completion for restoring visual data from missing values [25], [26]. Therein, the locations of missing values are known. While, in our problem, the observation is corrupted by gross errors with unknown positions, which is generally an even more challenging problem. Since we don't know positions of the corrupted pixels in advance, we formulate the image restoration as recovering a low rank matrix (representing the uncorrupted image) from corrupted observations. We show the recovered images obtained by PCP, Reweighted ℓ_1, and NSVT in Figure 1 & 2 (d), (e), and (f), respectively. To have a close-up observation, we also show a 128×128 patch of Figure 2(b), (d), (e), and (f) in Figure 2(g). We can observe that much more appealing results are obtained by NSVT. This is because low rank and sparsity are enhanced simultaneously and in a balanced manner in our algorithm, which helps to capture data structure.

Note that, our proposed method is different from the video restoration method via joint sparse and low rank matrix approximation in [9] or the robust video denoising method using low-rank matrix completion in [27]. In their work, the low-rank structure of similar image patches is harnessed, while, our method utilizes the holistic low-rank structure of images.

3.2 Hyperspectral Image Restoration

Hyperspectral images are well-localized in wavelength and they represent the scene radiance much more faithfully comparing with traditional RGB images [28]. By stacking image of each frequency band into a vector and arranging all the image vectors into a matrix, a hyperspectral image is represented as a data matrix $\mathbf{H} \in \mathbb{R}^{m \times n}$ with m pixels and n frequency bands. Low-rank assumption of a hyperspectral image assumes that a hyperspectral image can be well explained by nonnegative linear combination of an endmember matrix $\mathbf{B} \in \mathbb{R}_{++}^{m \times k}$, where $k < \min(m, n)$, i.e., $\mathbf{H} = \mathbf{BC}$, where $\mathbf{C} \in \mathbb{R}_+^{k \times n}$ is a nonnegative abundance matrix [29]. For instance, Figure 3(a) shows the singular value distribution obtained by SVD of a hyperspectral image, one image of whose frequency bands is shown in Figure 3(b). The image contains a piece of clothes with rich texture. There are 31 frequency bands in total and an image of each frequency band is of 256×256 pixels, i.e., $\mathbf{H} \in \mathbb{R}^{65536 \times 31}$. One can clearly observe the low-rank structure as the first few singular values are much larger than the remains. We show in this subsection that how to restore hyperspectral image from corrupted observations using the low-rank prior. The 40% pixels of a hyperspectral image to be restored are corrupted by large noise randomly. Figure 3(c) shows such a corrupted image of its one frequency band, which is very difficult for a human to tell what's in the image! We use the low-rank prior to restore a hyperspectral image from severe corruptions. We show the recovered results obtained by PCP, Reweighted ℓ_1, and NSVT in Figure 3 (d), (e), and (f), respectively. We set $k_{max} = 1$ for both Reweighted ℓ_1 and our proposed NSVT as before. The restored image obtained by PCP still contains a lot of large noise (the white dots), as can be seen from Figure 3(d). Close-up 128×128 patches of Figure 2(b), (d), (e), and (f) are

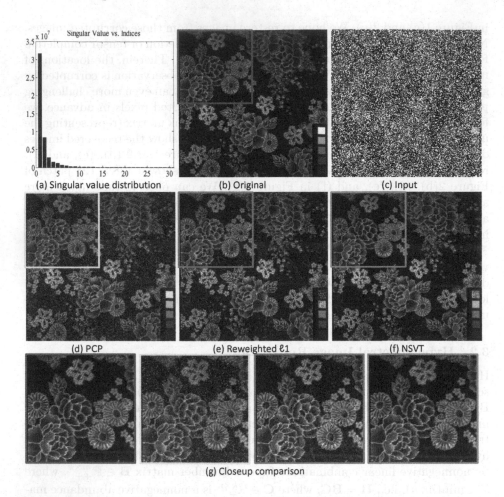

(a) Singular value distribution   (b) Original   (c) Input

(d) PCP   (e) Reweighted ℓ1   (f) NSVT

(g) Closeup comparison

Fig. 3. Restoration of a hyperspectral image 'Cloth', 40% of whose pixels are corrupted by large noise. Figure(a) shows the singular value distribution of the original hyperspectral image, an image from one of whose frequency bands is shown in Figure(b). Figure(c) shows the corresponding corrupted image to be restored. Figure(d), (e), (f) are restored images obtained by PCP, Reweighted ℓ_1, and NSVT, respectively. Figure(g) shows the close-up patches of Figure(b), (d), (e), and (f), respectively.

shown in Figure 3(g). Although Reweighted ℓ_1 relieves the noise in some part of the image comparing to PCP, it brings in annoying noise in other part. Again, the NSVT method gives the best restoration result among the three methods.

4   Conclusions and Future Work

In this paper, we propose a reweighted low-rank matrix recovery method and show its applications on different robust image restoration problems.

A non-uniform singular value thresholding (NSVT) operator is used for enhancing low rank property of a matrix. Significant improvements on recovery of corrupted images are observed, outperforming the original PCP optimization and Reweighted ℓ_1 minimization to a board extend. In light of promise of NSVT, deeper investigations of the algorithm remain: How to adaptively choose the parameters such as ϵ_L and ϵ_S to further improve the performance is considered as one of the future work; Theoretical analysis of the proposed method, such as the global convergence, is one of the important future research directions.

References

1. Candès, E., Romberg, J., Tao, T.: Robust uncertainty principles: Exact signal reconstruction from highly incomplete frequency information. IEEE Transaction on Information Theory 52(2), 489–509 (2006)
2. Candès, E., Recht, B.: Exact matrix completion via convex optimization. Foundations of Computational Mathematics 9(6), 717–772 (2009)
3. Recht, B., Fazel, M., Parrilo, P.: Guaranteed minimum-rank solutions of linear matrix equations via nuclear norm minimization. SIAM Review 52(3), 471–501 (2010)
4. Candès, E., Li, X., Ma, Y., Wright, J.: Robust principal component analysis? Journal of the ACM 58(3), 1–37 (2011)
5. Zhao, K., Zhang, Z.: Successively alternate least square for low-rank matrix factorization with bounded missing data. Computer Vision and Image Understanding 114(10), 1084–1096 (2010)
6. Li, K., Dai, Q., Xu, W., Yang, J., Jiang, J.: Three-dimensional motion estimation via matrix completion. IEEE Transactions on Systems, Man, and Cybernetics, Part B: Cybernetics 42(2), 539–551 (2012)
7. Lu, X., Gong, T., Yan, P., Yuan, Y., Li, X.: Robust alternative minimization for matrix completion. IEEE Transactions on Systems, Man, and Cybernetics, Part B: Cybernetics 42(3), 939–949 (2012)
8. Konishi, K., Furukawa, T.: A nuclear norm heuristic approach to fractionally spaced blind channel equalization. IEEE Signal Processing Letters 18(1), 59–63 (2011)
9. Ji, H., Huang, S., Shen, Z., Xu, Y.: Robust video restoration by joint sparse and low rank matrix approximation. SIAM Journal on Imaging Sciences 4(4), 1122–1142 (2011)
10. Peng, Y., Ganesh, A., Wright, J., Xu, W., Ma, Y.: RASL: Robust alignment by sparse and low-rank decomposition for linearly correlated images. IEEE Transactions on Pattern Analysis and Machine Intelligence (PAMI) 34(11), 2233–2246 (2011)
11. Zhang, Z., Ganesh, A., Liang, X., Ma, Y.: TILT: Transform-invariant low-rank textures. International Journal of Computer Vision 99(1), 1–24 (2012)
12. Zhang, Z., Matsushita, Y., Ma, Y.: Camera calibration with lens distortion from low-rank textures. In: Proceedings of IEEE Conference on Computer Vision and Pattern Recognition (2011)
13. Zhang, Z., Liang, X., Ma, Y.: Unwrapping low-rank textures on generalized cylindrical surfaces. In: Proceedingsof International Conference on Computer Vision (2011)

14. Mobahi, H., Zhou, Z., Yang, A.Y., Ma, Y.: Holistic reconstruction of urban structures from low-rank textures. In: Workshop on 3D Reconstruction and Recognition, International Conference on Computer Vision (ICCV) (2011)

15. Yao, C., Tu, Z., Ma, Y.: Detecting texts of arbitrary orientations in natural images. In: Proceedings of IEEE Conference on Computer Vision and Pattern Recognition (2012)

16. Liang, X., Ren, X., Zhang, Z., Ma, Y.: Repairing sparse low-rank texture. In: Fitzgibbon, A., Lazebnik, S., Perona, P., Sato, Y., Schmid, C. (eds.) ECCV 2012, Part V. LNCS, vol. 7576, pp. 482–495. Springer, Heidelberg (2012)

17. Lingala, S.G., Hu, Y., DiBella, E., Jacob, M.: Accelerated dynamic MRI exploiting sparsity and low rank structure: k-t SLR. IEEE Transactions on Medical Imaging 30, 1042–1054 (2011)

18. Cai, J., Candès, E., Shen, Z.: A singular value thresholding algorithm for matrix completion. SIAM Journal on Optimization 20(4), 1956–1982 (2010)

19. Toh, K.C., Yun, S.: An accelerated proximal gradient algorithm for nuclear norm regularized least squares problems. Pacific Journal of Optimization 6, 615–640 (2010)

20. Lin, Z., Chen, M., Wu, L., Ma, Y.: The augmented lagrange multiplier method for exact recovery of corrupted low-rank matrices. Submitted to Mathematical Programming (UIUC Technical Report UILU-ENG-09-2215) (2009)

21. Yuan, X., Yang, J.: Sparse and low-rank matrix decomposition via alternating direction methods. Pacific Journal of Optimization (2011) (forthcoming)

22. Candès, E., Wakin, M., Boyd, S.: Enhancing sparsity by reweighted ℓ_1 minimization. Journal of Fourier Analysis and Applications 14(5), 877–905 (2007)

23. Afonso, M., Bioucas-Dias, J., Figueiredo, M.: An augmented lagrangian approach to the constrained optimization formulation of imaging inverse problems. IEEE Transaction on Image Processing 19(9), 2345–2356 (2010)

24. Deng, Y., Dai, Q., Liu, R., Zhang, Z., Hu, S.: Low-rank structure learning via nonconvex heuristic recovery. IEEE Trans. on Neural Networks and Learning Systems 24(3), 383–396 (2013)

25. Liu, J., Musialski, P., Wonka, P., Ye, J.: Tensor completion for estimating missing values in visual data. In: Proceedings of IEEE International Conference on Computer Vision (2009)

26. Huang, J., Zhang, S., Li, H., Metaxas, D.: Composite splitting algorithms for convex optimization. Computer Vision and Image Understanding 115(12), 1610–1622 (2011)

27. Ji, H., Liu, C., Shen, Z., Xu, Y.: Robust video denoising using low rank matrix completion. In: Proceedings of IEEE International Conference on Computer Vision and Pattern Recognition (2010)

28. Yasuma, F., Mitsunaga, T., Iso, D., Nayar, S.K.: Generalized assorted pixel camera: Postcapture control of resolution, dynamic range, and spectrum. IEEE Transaction on Image Processing 19(9), 2241–2253 (2010)

29. Zymnis, A., Kim, S.J., Skaf, J., Parente, M., Boyd, S.: Hyperspectral image unmixing via alternating projected subgradients. In: Proceedings Asilomar Conference on Signals, Systems, and Computers, pp. 1164–1168 (2007)

Learning to Infer Public Emotions from Large-Scale Networked Voice Data

Zhu Ren[1,2], Jia Jia[1,2], Lianhong Cai[1,2], Kuo Zhang[3], and Jie Tang[1]

[1] Department of Computer Science and Technology, Tsinghua University, Beijing, China
[2] TNList and Key Laboratory of Pervasive Computing, Ministry of Education
bamboo.renzhu@gmail.com, {jjia,clh-dcs,jietang}@tsinghua.edu.cn
[3] Sogou Corporation, Beijing, China
zhangkuo@sogou-inc.com

Abstract. Emotions are increasingly and controversially central to our public life. Compared to text or image data, voice is the most natural and direct way to express ones' emotions in real-time. With the increasing adoption of smart phone voice dialogue applications (e.g., Siri and Sogou Voice Assistant), the large-scale networked voice data can help us better quantitatively understand the emotional world we live in. In this paper, we study the problem of inferring public emotions from large-scale networked voice data. In particular, we first investigate the primary emotions and the underlying emotion patterns in human-mobile voice communication. Then we propose a partially-labeled factor graph model (PFG) to incorporate both acoustic features (e.g., energy, f0, MFCC, LFPC) and correlation features (e.g., individual consistency, time associativity, environment similarity) to automatically infer emotions. We evaluate the proposed model on a real dataset from Sogou Voice Assistant application. The experimental results verify the effectiveness of the proposed model.

Keywords: public emotions, acoustic features, correlation features, factor graph model.

1 Introduction

It is an emotional world we live in. Emotions, which are associated with subjective feelings, cognitions, impulses to action and behavior [1], can be recognized from many different information sources, e.g., human voice [3], facial expression [15], physiological signal [16], or their multimodal combination [5]. Compared to individual emotions, **public emotions** pay attention to the major emotions of the public induced by social events. Previous studies have shown the success of using networked text [2] or image data [17] to infer public emotions. Nowadays, with the rapid development of smart phone voice dialogue applications (e.g., *Siri*[1] and *Sogou Voice Assistant*[2]), people can share voice messages to their friends or make requests to the voice

---

[1] http://www.apple.com/ios/siri/, an intelligent personal assistant and knowledge navigator which works as an application for Apple's iOS.
[2] http://yy.sogou.com, an smart phone voice dialogue application developed by Sogou (one of China's largest internet service providers).

C. Gurrin et al. (Eds.): MMM 2014, Part I, LNCS 8325, pp. 327–339, 2014.

assistant easily. Voice is the most direct way to express emotions. And emotions can be conveyed by not only linguistic information but also acoustic information. For example, a user who intends to share happiness with his friend on special days may send voice messages saying "Happy New Year" or "Happy birthday" with pleasant tone of higher pitch. Since people's voice data can be regarded as microscopic instantiations of emotions, the collection of all the public voice data uploaded over a given time period or around a special social event can unveil the trends of public emotions at a macroscopic scale.

Previous researches have been conducted for empirical analyses of emotion based on text or image data from social networks. Some of these analyses focus on public emotions around specific events [17], while others further analyze broader social and economic trends [7]. However, due to the lack of availability of large-scale networked voice data, few have been done in studying public emotions from voice signals. As voice is the fastest and the most natural method of communication [9], it can express people's emotional states in a much more vivid and efficient way. Therefore, using available large-scale networked voice data to perform public emotion analyses can significantly reduce the costs and efforts. Furthermore, it can benefit lots of fields, e.g., improve the user-friendly voice communication applications, or help companies formulate marketing strategies [2].

In this paper, employing a mobile voice assistant application as the basic of our experiments, we systematically study the problem of inferring public emotions from networked voice data. The problem is non-trivial and poses a set of unique challenges. First, the emotion patterns in human-mobile voice communication are quite different from that in human-human voice communication. It is unclear how to identify the underlying emotion patterns behind the human-mobile voice communication. Second, former studies have confirmed that acoustic features [3-5] can reflect the individual's emotions, while in different social environment, the acoustic features might be quite different. Third, technically, how to design a principled model to automatically infer public emotions by considering both the acoustic features and social environment?

To address the above challenges, we make our efforts and make contributions on three aspects:

- **Emotion Patterns.** Based on the observations of networked voice data, we investigate the primary emotions in human-mobile voice communication by combining the linguistic information with acoustic information. Furthermore, we identify two interesting emotion patterns behind the human-mobile voice communication.
- **Features.** Besides the selected acoustic features which can reflect emotions, we take into consideration three social correlation features (individual consistency, time associativity and environment similarity), which can be combined with acoustic features for performance improvement in different social environment.
- **Model.** We formulate our problem into a novel partially-labeled factor graph model (PFG) to infer public emotions by incorporating both acoustic features and correlation features.

Our experiments are based on a real networked voice dataset, which is from *Sogou Voice Assistant*. The experimental results demonstrate that the proposed model can achieve better performance than alternative method using SVM. Discussion and analysis of the experimental results rationally verify the contribution of combining the acoustic features with the correlation features to improve the performance.

The rest of this paper is organized as follows: Section 2 gives the basic formulation of our problem. Section 3 introduces the data observation and dataset setup. Section 4 presents the PFG model for inferring public emotions. Section 5 carries out the experiments employed to analyze the feature contribution and evaluate the performance of the proposed model. We'd like to show some interesting case studies in this section too. Finally, Section 6 summarizes this paper.

2   Problem Formulation

Fig. 1 gives an illustration of inferring public emotions from networked voice data. Each utterance in the voice dataset not only has its own acoustic features, but also correlations with other utterances, such as individual consistency (blue line), time associativity (pink line), and environment similarity (green line). Part of the utterances in the dataset are labeled with emotions, and our task is predicting the emotions of unlabeled utterances. For further clarification, in this section, we give some essential definitions and subsequently present the problem formulation.

The network of input utterances can be represented as $G = (V, E)$, where $V = \{u_1, ..., u_N\}$ is the set of $|V| = N$ utterances, $E \subset V \times V$ is the set of $|E| = K$ relationships between utterances. Each edge e_{ij} indicates u_i having a correlation with u_j (e.g. u_i and u_j recorded by the same speaker or recorded in the same city within a short time). We aim at learning a model that can effectively infer emotions from networked voice data. For this reason, we first define the speaker's emotions and the partially labeled network as follows.

Definition 1. **Emotions**: The emotion category of an utterance u_i is denoted as $y_i \subset A$, where A is the emotion space that contains the primary emotions in human-mobile communication. The investigation on primary emotions will be described in details in Section 3.

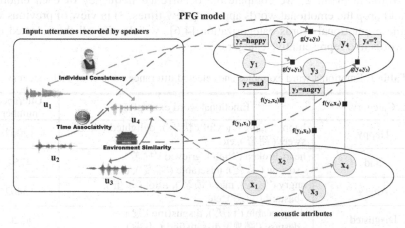

Fig. 1. The illustration of inferring public emotions from large-scale networked voice data using partially labeled factor graph model

Definition 2. **Partially labeled network:** The partially labeled network is denoted as $G = (V^L, V^U, E, \Gamma)$, where V^L and V^U are respectively the set of labeled and unlabeled utterances with $V^L \cup V^U = V$; E is the correlations between utterances; Γ is an attribute matrix associated with utterances in V with each row corresponding to an utterance, each column representing an attribute and an element x_{ij} denoting the value of the jth attribute of utterance u_i. The label of utterance u_i is denoted as $y_i \subset A$.

Problem. **Learning task:** Given a partially labeled network G, the objective is to infer the emotion categories of utterances by learning a predictive function

$$f: G = (V^L, V^U, E, \Gamma) \rightarrow \Lambda \qquad (1)$$

where $\Lambda = \{S_1, \dots, S_M\}$ is the set of inferred results, with each S_m belonging to one emotion category in A.

3   Data Preparation and Emotion Pattern Analysis

3.1   Data Collection and Observation

We collected a corpus of large-scale networked voice data from *Sogou Voice Assistant*. The raw dataset contains 6,891,298 utterances recorded in Chinese by 405,510 users during year 2013. Each utterance has some basic information (e.g. user ID, record time, geographical position) and the corresponding speech-to-text information provided by Sogou Corporation.

For training and evaluating the proposed model for inferring emotions, we firstly need to know the primary emotions of networked voice data, and establish an experimental dataset with emotional labels as ground truth. Due to the massive scale of our dataset, manually annotating the emotion category for each utterance is not practical. Considering that linguistic information contained in voice data can help us understand speaker's emotion, we conduct the investigation on primary emotions as following steps: 1) we screen all the utterances' speech-to-text information and find the emotional words in them; 2) we compute the occurrence frequency of each emotional word, and drop the emotional words appearing few times; 3) in view of previous work on Chinese emotional words categorization [4,6], we classify all the selected emotional words into representative categories.

Table 1. Emotional word examples and selected utterance number of each category

Category	Emotional word examples	Utterance number
Happy	happy ('高兴'), joyful ('快乐'), delighted ('开心'), sweet ('甜蜜'), etc.	12067
Sad	heart-broken ('伤心'), grieved ('痛苦'), sorrow ('悲哀'), miserable ('难受'), etc.	4754
Angry	angry ('生气'), rage ('愤怒'), idiot ('笨蛋'), bastard ('可恶'), etc.	13407
Disgusted	disagreeable ('讨厌'), disgusting ('恶心'), despise ('鄙视'), dissatisfied ('不满'), etc.	4320
Bored	bored ('无聊'), tired ('累'), toilsome ('辛苦'), etc.	13663

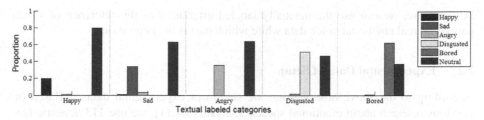

Fig. 2. The proportion of manually labeled emotions in each textual labeled category

The classification results show the emotional words finally cluster to five main categories: *Happy*, *Sad*, *Angry*, *Disgusted*, and *Bored*. We can see that these five primary emotions in human-mobile communication are different from Ekman's six basic emotions proposed for human-human communication. Specifically, *Fear* and *Surprise* in Ekman's six emotions are replaced by *Bored*. Following the above steps, from the raw dataset, we finally pick out 48,211 utterances whose speech-to-text information contains emotional words that only belong to one of the five primary emotion categories. We annotate these utterances with that specific category as their textual labels. Table 1 shows the emotional words and selected utterance number of each category.

In order to further explore how an utterance's textual label is consistent with its real emotion, we randomly selected 200 utterances from each textual labeled category (1,000 utterances in total). Then we invite three human labelers to annotate each utterance with an emotion category manually. Besides the above five emotions, we also allow the labelers to give *Neutral* annotation. When they have disagreement, they stop and discuss until they have final agreed views. The manually labeled results are regarded as the real emotions for these utterances.

We observe the distinctions between textual labeled categories and manually labeled emotions. The observation results are quite interesting. The proportion of manually labeled emotions in each textual labeled category is shown as Fig. 2. We can easily find two phenomena as follows:

- **Phenomenon I:** In some cases, the textual labels are not consistent with the real emotions. For example, some of the utterances with textual labels *Happy* are actually *Angry* emotion. It may happen when a user says "I'm really happy with what you said", but actually he is angry and what he says means an irony. This phenomenon indicates the **Emotion Pattern I**: There exist insincerities in human-mobile communication (this pattern exists in human-human communication too). It also means that we cannot use the textual labels as real emotion categories directly.
- **Phenomenon II:** A large part of the utterances, whose speech-to-text information contains emotional words, are actually neutral voice data. This phenomenon indicates the **Emotion Pattern II**: The expressions in human-mobile communication are more rational and implicit. People pay much attention to linguistic information rather than paralinguistic information to express their meanings. It leads us to find that each textual labeled category is approximately consisted by two parts, the real emotional voice data, and the neutral voice data.

Therefore, for each textual labeled category, we can further conduct a 2-cluster classification using acoustic features to separate it into two parts. Then we can establish an experimental dataset with emotional and neutral voices respectively.

Furthermore, we can use the manually labeled utterances as the reference of which part is the real emotional voice data while which part is the neutral one.

3.2 Experimental Dataset Setup

According to the above observations, we setup our experimental dataset. Based on previous research about emotional speech analysis [3,4,11], we use 113 acoustic features to conduct the 2-cluster classification on 48,211 utterances:

- Energy features (13): the energy envelop applied with 13 functionals (mean, std, max, min, range, quartile1/2/3, iqr1-2/2-3/1-3, skewness, kurtosis).
- F0 features (13): the fundamental frequency contour, which are extracted using a modified STRAIGHT procedure[10], applied with 13 functionals the same as Energy.

Algorithm 1. The feature selection algorithm

Input:
features[1..n][1..d]: acoustic features matrix with each row corresponding to an utterance, each column representing one kind of feature
labels[1..n]: the manual labels of n utterances from one text category
threshold: a const in {0.05, 0.01, 0.001, 0.0001, 0.00001, 0.000001}
Output:
The order numbers of the selected feature set
```
1:      dnum ← 0
2:      for j ← 1 to d do
3:          calculate the p-value for testing the hypothesis of no correlation between
                    features[1..n][j] and labels[1..n]
4:          if p-value < 0.05 then
5:              dnum ← dnum + 1
6:              fset[dnum].p ← p-value
7:              fset[dnum].ind ← j
8:          end if
9:      end for
10:     sort structure array fset[1..dnum] along its element p in ascending order
11:     initialize tag[1..dnum] = 0
12:     calculate a matrix P of p-values for array features[1..n][1..d]
13:     current ← 1
14:     tag[current] ← current
15:     repeat
16:         for j ← 1 to dnum do
17:             if tag[j] = 0 then
18:                 if P[fset[current].ind][fset[j].ind] < threshold then
19:                     tag[j] ← current
20:                 end if
21:             else
22:                 if P[fset[current].ind][fset[j].ind] < P[fset[tag[j]].ind][fset[j].ind] and tag[j] != j
                    then
23:                     tag[j] ← current
24:                 end if
25:             end if
26:         end for
27:         find the minimum value of k satisfying the equation tag[k] = 0
28:         current ← k
29:         tag[current] ← current
30:     until all the elements in tag[1..dnum] are not zero
31:     find the same values as in tag[1..dnum] without repetitions into tset[1..gnum]
32:     return fset[tset[1..gnum]].ind
```

Table 2. The results of feature selection and emotion classification

Category	Manual labeled sample numbers	Feature selection		Emotion Classification	
		Threshold	Selected feature set	F1-Measure	Sample numbers in final dataset
Happy	H: 102 NH: 161	0.001	f0_quartile3, lfpc8_mean, energy_range, sc_skewness, f0_kurtosis, lfpc4_std, mfcc11_mean, sr_max	70.78%	H: 5721 NH: 6346
Sad	S: 167 NS: 133	0.01	lfpc9_std, syldur_quartile3, mfcc3_mean	75.84%	S: 2562 NS: 2192
Angry	A: 100 NA: 129	10E-4	mfcc2_std, lfpc11_mean, lfpc10_std, mfcc9_mean, energy_skewness, mfcc8_mean, mfcc6_std, mfcc12_mean, mfcc2_mean, mfcc7_std	70.97%	A: 7892 NA: 5515
Disgusted	D: 103 ND: 97	0.01	lfpc7_mean, mfcc5_std, mfcc9_mean, lfpc11_std, syldur_iqr1-2	74.64%	D: 2328 ND: 1992
Bored	B: 125 NB: 75	10E-4	sr_std, f0_iqr1-2, mfcc5_std, mfcc4_mean, mfcc6_mean, mfcc3_mean, lfpc7_std, lfpc9_mean, energy_iqr2-3, mfcc6_std, energy_iqr1-2	79.20%	B: 6968 NB: 6695

- MFCC features (26): the mean and standard deviation of mel-frequency cepstral coefficients 1-13.
- LFPC features (24): the mean and standard deviation of log frequency power coefficients 1-12, which are extracted using the method in [11] with $\alpha=1.4$.
- Spectral Centroid (SC) features (13): the spectral centroid contour applied with 13 functionals the same as Energy.
- Spectral Roll-off (SR) features (13): the spectral roll-off contour applied with 13 functionals the same as Energy.
- Syllable Duration (SD) features (11): the syllable duration sequence, which is extracted using the method in [12], applied with 11 functionals (mean, std, max, min, range, quartile1/2/3, iqr1-2/2-3/1-3).

All the spectral features (MFCC, LFPC, SC, SR) are extracted from voiced segments of the utterances with the 20ms frame length and 10ms frame shift. Each kind of feature is normalized first, hence the mean is zero and the standard deviation is one.

We make use of the correlation coefficients and their significances between acoustic features and manual labels for the feature selection (Algorithm 1 gives the details). The manual labels of utterances are defined as X / NX, where X∈{H, S, A, D, B} and NX is *Neutral*. By changing the threshold in Algorithm 1, we can obtain diverse feature sets. As we pay more attention to clustering most of the positive samples into one class, we apply the F1-Measure of positive samples to evaluate the performance of clustering. For simplicity, we run the classic k-means clustering for each category, and compute the F1-Measure by using the Hungarian algorithm [8] to assign which class is corresponding to the positive samples. The best classification results and the selected feature sets they used are summarized in Table 2.

By the above method, we finally establish an experimental dataset with emotional labels as ground truth. The dataset contains 48,211 utterances consisted of five primary emotions as well as *Neutral*: *Happy* (5721), *Sad* (2562), *Angry* (7892), *Disgusted* (2328), *Bored* (6968), and *Neutral* (22740).

The emotional labels in our experimental dataset certainly still have some errors. But in the statistical level, it can be ignored in some ways such as using large-scale data. The most importance is the above method can help us avoid the impossible mission of manual annotation on large-scale networked voice data. Since our prime concern is the accuracy of public emotions inferring, which is quite different from the task of individual emotion recognition, we believe the above method provide the good balance between efficiency and performance.

4 Proposed Method

4.1 Prediction Model

As networked voice data are disposed in this paper, we take advantage of some social correlation information to improve the performance and identify three kinds of correlation features:

- Individual Consistency (IC): whether two utterances are recorded by the same speaker.
- Time Associativity (TA): whether two utterances are recorded within the same hour of one day.
- Environment Similarity (ES): whether two utterances are recorded in the same city.

Since the correlations between utterances are hard to be modeled by traditional classifiers such as SVM, we utilize a partially-labeled factor graph model (PFG) [13] to learn and infer public emotions. All the utterances recorded by ordinary speakers can be formalized as variables and observation factor functions in a factor graph, basing on the theory of FGM [14]. Each utterance u_i can be mainly described as one kind of primary emotion, which can be mapped as an emotional node n_i in the PFG model. The labels of emotional nodes are denoted as $Y = \{y_1, ..., y_N\}$, where y_i is a hidden variable associated with n_i. As the emotions in G are partially labeled, they can be divided into two subset Y^L and Y^U corresponding to the labeled and unlabeled emotions. For each emotional node n_i, we define the emotional attributes into a vector x_i, considering that the speaker-independent acoustic features may contain information about emotions. At the same time, we can also find the basic intuition that the relationships between the utterances (correlation features) can constitute the correlations between hidden variables in our model. Based on the above intuitions, we define the following two factors:

- **Attribute factor:** $f(y_i, x_i)$ represents the posterior probability of the emotion y_i given the attribute vector x_i.
- **Correlation factor:** $g(y_i, R(y_i))$ denotes the correlation among the emotions, where $R(y_i)$ is the set of correlated emotions to y_i.

Given a partially-labeled network $G = (V^L, V^U, E, \Gamma)$, we can define the joint distribution over Y as

$$P(Y|G) = \prod_i f(y_i, x_i) g(y_i, R(y_i)) \tag{2}$$

Since the two factors can be instantiated in different kinds of ways, we only give a general definition for them by using exponential-linear function. In particular, we define the attribute factor as

$$f(y_i, x_i) = \frac{1}{Z_\alpha} \exp\{\alpha^T \cdot x_i\} \tag{3}$$

where α is a weighting vector of Γ and Z_α is a normalization factor.

The correlation factor can be naturally modeled in a Markov random field. Thus, by the fundamental theorem of random fields, it can be defined as

$$g(y_i, R(y_i)) = \frac{1}{Z_\beta} \exp\left\{\sum_{y_j \in R(y_i)} \beta_{ij} \cdot h_{ij}(y_i, y_j)\right\} \tag{4}$$

where $h_{ij}(y_i, y_j)$ is a feature function that captures the correlation between emotional nodes n_i and n_j; β is the weighting of this function; Z_β is also a normalization factor.

Finally, the joint probability defined in (2) can be written as

$$P(Y|G) = \frac{1}{Z} \exp\left\{\sum_{y_i \in Y} \left[\alpha^T \cdot x_i + \sum_{y_j \in R(y_i)} \beta_{ij} \cdot h_{ij}(y_i, y_j)\right]\right\} \tag{5}$$

where $Z = Z_\alpha Z_\beta$ is a normalization factor.

Learning the predictive model is to estimate a parameters configuration $\varphi = (\{\alpha\}, \{\beta\})$ from the partially-labeled dataset, so that it can maximize the log-likelihood objective function $\Theta = \log P(Y|G)$, i.e. $\varphi^* = \arg\max \Theta(\varphi)$.

4.2   Model Learning

After learning the parameter values, we turn to address the problem of estimating the remaining free φ and inferring speakers' emotions. Specifically, we first calculate the gradient of each parameter with regard to the objective function:

$$\frac{\partial \Theta(\varphi)}{\partial \alpha_j} = E[f_j(y_{ij}, x_{ij})] - E_{P_{\alpha_j}(y_i|x_{ij}, G)}[f_j(y_{ij}, x_{ij})] \tag{6}$$

where $E[f_j(y_{ij}, x_{ij})]$ is the expectation of feature function $f_j(y_{ij}, x_{ij})$ given by the data distribution and $E_{P_{\alpha_j}(y_i|x_{ij}, G)}[f_j(y_{ij}, x_{ij})]$ is the expectation of feature function $f_j(y_{ij}, x_{ij})$ under the distribution $P_{\alpha_j}(y_i|x_{ij}, G)$ given by the estimated model. Similar gradients can be derived for parameter β. Then we update the parameters by $\varphi_j^{new} = \varphi_j^{old} + \gamma \cdot \frac{\partial \Theta(\varphi)}{\partial \varphi}$, where γ is the learning rate.

Given the observed value x and the learned parameters φ, the inference task is to find the most likely y, as follows

$$y^* = \arg\max_y p(y|x, \varphi) \tag{7}$$

Finally, we utilize the loopy belief propagation to compute the marginal probability of each emotional node and then predict the type of an unlabeled emotion as the label with largest marginal probability.

5   Experiments and Discussions

5.1   Experimental Setup

We use the dataset described in Subsection 3.2 in our experiments, which contains 48,211 utterances that carefully chosen from networked voice data recorded by 25,370 speakers and labeled with one of the six primary emotions. We perform five-fold cross validation and quantitatively evaluate the performance of inferring public emotions in term of *Accuracy* and *F1-Measure* computed as

$$\text{Accuracy} = \frac{\sum_i \text{correctly predicted sample number of category } i}{\sum_i \text{sample number of category } i} \tag{8}$$

$$\text{F1-Measure} = \frac{\sum_i \text{F1-Measure of category } i}{\text{category number}} \tag{9}$$

For the purpose that justifying whether the correlation information can help infer public emotions from utterances, we define a baseline method using the classic machine learning technique Support Vector Machine (SVM). We compare the performance achieved by PFG model utilizing both acoustic features and correlation features (*Individual Consistency (IC)*, *Time Associativity (TA)*, and *Environment Similarity (ES)*) with the baseline method.

5.2   Results and Discussions

Performance Comparison. Table 3 shows the *Accuracy* and *F1-Measure* for proposed PFG and SVM. The *Accuracy* of the proposed PFG model achieves 86.11%, while the SVM model achieves only 49.15%. For *F1-Measure* shown in Table 3, PFG also shows clearly the best performance and yields an 27.04-83.51% improvement compared with SVM. These results demonstrate the effectiveness of our proposed method on inferring emotions from networked voice data. Furthermore, the proposed PFG model utilizes both acoustic features and correlation features, while SVM model cannot express the relationships between utterances. So the experimental results also verify that the correlation information among utterances can compensate the deficiency of acoustic features and help infer public emotions in our problem.

Table 3. Performance of emotion prediction with different method

Model / Category	SVM	F1-Measure (%) PFG	
		Acoustic	All
Happy	7.83	56.27	**84.92**
Sad	6.50	49.65	**69.46**
Angry	8.56	69.34	**92.07**
Disgusted	8.47	26.07	**67.64**
Bored	5.04	51.24	**68.02**
Neutral	64.98	89.74	**92.02**
Average	16.90	57.05	**79.02**
Accuracy (%)	49.15	73.86	**86.11**

Feature Contribution Analysis. Fig. 3 shows the *F1-Measure* of each kind of acoustic or correlation feature when inferring emotions. Comparing the *F1-Measures* of different correlation features, we find that *TA* makes the improvement in the majority of the emotion categories, while *ES* benefits the prediction of several categories as well. Since the mean number of utterances per speaker is 1.9 and 62.03% of the utterances are recorded by the speakers who only have one utterance in the dataset, there are lesser *IC* correlations than *TA* and *ES*, which leads to its lower performance. The aforementioned results confirm that public emotions are closely related to time and environment. For acoustic features, the spectral features (*MFCC, LFPC*) make a great contribution to inferring most emotions, which is consist with the feature selection results described in Table 2.

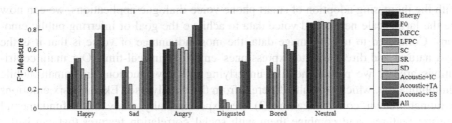

Fig. 3. Feature contribution analysis

Case Study. To further demonstrate the effectiveness of the proposed model, we would like to show two interesting case studies. Since Beijing, Shanghai and Shenzhen are the top 3 active cities with the most users in *Sogou Voice Assistant*, we use the utterances that respectively uploaded in these three cities as our case study data. The proposed PFG model is used to infer emotions from them. By analyzing the results, we find the trends of public emotions related to specific time period or social event. We take positive emotion *Happy* and negative emotion *Angry* as examples:

- We all know that Beijing suffered the severest fog and haze in history during the week of Jan 21 to 27, 2013. What was the major public emotion inferred from the voice data in that period in Beijing? The answer is people felt less happy and more agonizing day by day, shown in Fig.4a(1). Comparing with Shanghai and Shenzhen's results shown as Fig. 4a(2) and Fig. 4a(3), Beijing obviously had more people in bad moods due to the environmental problem. These results are rational and common in our daily life.
- A strong earthquake struck the southwestern Chinese province of Sichuan on April 20, 2013. Were the public emotions affected by the emergency? The answer is yes. Taking Beijing as an example, we can find an increasing of negative emotion from April 20 (Fig. 4b). After the earthquake, people were generally worried about the disaster and suffering from the tragedy of losing compatriots, so their anxieties caused the negative public emotions. These results also indicate the networked voice data can reflect the changes of public emotions on emergency in real-time.

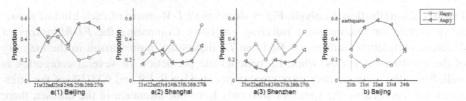

Fig. 4. Case study: a) The major emotional trends in three cities from Jan 21 to 27, 2013. b) The major emotional trends in Beijing from April 20 to 24, 2013.

6   Conclusions

With the increasing adoption of smart phone voice dialogue applications, we can now use the large-scale networked voice data to achieve the goal of inferring public emotions. Compared to text or image data, the most advantage of voice is that it is the most natural and direct way to express ones' emotions in real-time. Our main contributions are: 1) we reveal the five underlying primary emotions in human-mobile communication, which are quite different from the widely-used Ekman's six emotions in human-human communication; 2) we experimentally analyze the fundamental acoustic features, and combine them with social correlation features that can better reflect emotions in different social environment; 3) we formulate the problem into a PFG model for inferring public emotions from large-scale networked voice data, turning out good results.

For future works, we are planning to investigate and model more social phenomenon such as conformity in human-mobile communication for further improving the inferring accuracy.

Acknowledgements. This work is supported by the National Basic Research Program of China (2012CB316401), National Natural, and Science Foundation of China (61370023, 61003094). And this work is partially supported by 973 Program of China (2013CB329304). We also thank Microsoft Research Asia-Tsinghua Univertity Joint Laboratory for its support.

References

1. Plutchik, R.: Emotion: A Psychoevolutionary Synthesis. Harper & Row, New York (1980)
2. Tang, J., Zhang, Y., Sun, J., Rao, J., Yu, W., Chen, Y., Fong, A.: Quantitative study of individual emotional states in social networks. IEEE Transactions on Affective Computing 3(2), 132–144 (2012)
3. Wu, D., Parsons, T.D., Narayanan, S.: Acoustic feature analysis in speech emotion primitives estimation. In: Proc. of INTERSPEECH 2010, Makuhari, Japan, pp. 785–788 (2010)
4. Cui, D.: Analysis and Conversion for Affective Speech. Tsinghua University, Beijing (2007) (in Chinese) (doctoral dissertation)
5. Nicolaou, M.A., Gunes, H., Pantic, M.: Continuous prediction of spontaneous affect from multiple cues and modalities in valence-arousal space. IEEE Transactions on Affective Computing 2(2), 92–105 (2010)

6. Mei, J.: Tongyici Cilin (version 2). Shanghai Dictionary Press, Shanghai (1996) (in Chinese)
7. Bollen, J., Mao, H., Pepe, A.: Modeling Public Mood and Emotion: Twitter Sentiment and Socio-Economic Phenomena. In: Proc. AAAI 2011, San Francisco, California, USA, pp. 450–453 (2011)
8. Kuhn, H.W.: The Hungarian Method for the assignment problem. Naval Research Logistics Quarterly 2, 83–97 (1955)
9. Ayadi, M.E., Kamel, M.S., Karray, F.: Survey on speech emotion recognition: Features, classification schemes, and databases. Patter Recognition 44, 572–587 (2011)
10. Kawahara, H., Cheveigne, A., De, B.H., Takahashi, T., Irino, T.: Nearly Defect-free F0 Trajectory Extraction for Expressive Speech Modifications based on STRAIGHT. In: Proc. of INTERSPEECH 2005, Lisboa, pp. 537–540 (2005)
11. Nwe, T., Foo, S., Silva, L.D.: Speech emotion recognition using hidden Markov models. Speech Commun. 41, 603–623 (2003)
12. Wang, D., Narayanan, S.: An acoustic measure for word prominence in spontaneous speech. IEEE Transactions on Speech, Audio, and Language Processing 15(2), 690–701 (2007)
13. Tang, W., Zhuang, H., Tang, J.: Learning to infer social ties in large networks. In: Gunopulos, D., Hofmann, T., Malerba, D., Vazirgiannis, M. (eds.) ECML PKDD 2011, Part III. LNCS, vol. 6913, pp. 381–397. Springer, Heidelberg (2011)
14. Frey, B., Dueck, D.: Mixture modeling by affinity propagation. In: Weiss, Y., Schölkopf, B., Platt, J. (eds.) NIPS, pp. 379–386 (2006)
15. Fasel, B., Luettin, J.: Automatic facial expression analysis: a survey. Pattern Recognition 36(1), 259–275 (2003)
16. Fairclough, S.H.: Fundamentals of physiological computing. Interacting with Computers 21, 133–145 (2009)
17. Jia, J., Wu, S., Wang, X., Hu, P., Cai, L., Tang, J.: Can We Understand van Gogh's Mood? Learning to Infer Affects from Images in Social Networks. In: Proc. of ACM Multimedia, Nara, Japan (2012)

Joint People Recognition across Photo Collections Using Sparse Markov Random Fields

Markus Brenner and Ebroul Izquierdo

School of EECS, Queen Mary University of London, UK
{markus.brenner,ebroul.izquierdo}@eecs.qmul.ac.uk

Abstract. We show how to jointly recognize people across an entire photo collection while considering the specifies of personal photos that often depict multiple people. We devise and explore a sparse but efficient graph design based on a second-order Markov Random Field, and that utilizes a distance-based face description method. Experiments on two datasets demonstrate and validate the effectiveness of our probabilistic approach compared to traditional methods.

1 Introduction and Background

Faces, along with the people's identities behind them, are an effective element in organizing a personal collection of photos, as they represent *who* was involved. However, along with the fact that current face recognition approaches (that tell apart and identify people based on already detected face appearances) require a notable amount of user interaction for training a classification model, the accurate discrimination and thus recognition of faces itself is still very challenging. Particularly in the uncontrolled environments of consumer photos, faces are usually neither perfectly lit nor captured. Wide variations in pose, expression or makeup are common and difficult to handle as well.

One way to address this issue is to keep improving face recognition techniques; for example, by researching novel visual features. Another way is to incorporate context: to consider additional information aside from just faces within photos and across entire collections. As a wide variety of literature [1–7] shows, such contextual information might include time, location or scene. The people appearing in photos can also provide further information. For instance, a person's demographics such as age and gender are often easier to surmise than his or her identity [8]. While there are many works researching face recognition in general, only some works like [1, 2, 9, 6] focus on the task of recognizing people in Consumer Photo Collections. Compared to independently sourced photos, such collections usually provide rich contextual information.

Whereas early works are limited to generating candidate lists or clustering people (often based on holistic face recognition approaches), recent studies use more effective methods to more accurately resolve people's identities in photo collections. However, many studies often do this by separately predicting one face appearance at a time; for example, by using *traditional* supervised classification

C. Gurrin et al. (Eds.): MMM 2014, Part I, LNCS 8325, pp. 340–352, 2014.

methods like Support Vector Machines. Thus, such studies do not interconnect much context across multiple faces nor entire photo collections as we envision.

There are works like [4, 6, 10, 11] that incorporate contextual cues like clothing or social semantics, but few works research the more general question on how to best combine or fuse such contextual information, and in particular, in a jointly manner. Dealing with multiple information domains in a structured way is usually not straightforward. An apparent trend of recent studies [2, 5, 7] is the use of graphical models to interconnect information. However, most of these studies do not aim for a sparse and efficient graph representation, which is essential to keep computational complexity manageable even with larger photo collections. [2], for example, simply always establishes a fixed number of graph edges. Therefore, it is of interest to research approaches and graph designs that retain and preferably even improve people recognition performance while leading to a sparser and more efficient graph model.

2 Objective and Approach

In this work we present a framework to jointly recognize people across a Consumer Photo Collection, and which builds upon and extends our prior research [11, 10, 12, 8]. Compared to traditional approaches where a classifier is typically trained to recognize a single random face at a time, we intend to consider and recognize all faces within an entire dataset simultaneously in a semi-supervised fashion. We also wish to consider the specifies of personal photos which, for example, often depict multiple people within a single photo. Thus, one focus of this paper is on solving the task of jointly recognizing all people by finding the best overall solution. To accomplish this, we lay out a graphical model with a similarity or distance-based face description technique at its core. We show how to model such a graph and explore ways to improve the recognition performance by *optimizing* the graph's topology while aiming for a sparse and efficient graph design. Because of the probabilistic and semi-supervised nature of our proposed framework, we believe that it could serve as a basis for future studies that wish to incorporate and model additional contextual information for the task of people recognition.

The remainder of this paper is structured as follows: In the next section we set forth the techniques we employ to detect and differentiate faces. Then, we explain the basics of our graph-based approach and how a sparse but effective model can be achieved. We demonstrate the effectiveness of the overall framework with experiments thereafter and present our conclusions.

2.1 Face Detection and Basic Recognition

We choose to utilize the seminal work of Viola and Jones included in the OpenCV package to detect faces. Their detection framework builds upon Haar-like features and an AdaBoost-like learning technique. The face recognition technique we introduce next provides some leeway for minor misalignment. Thus, the only *normalization* we perform is scaling the patches identified as faces to a common

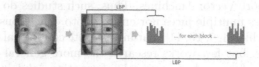

Fig. 1. We divide the face input image into blocks of a grid. For each block, we extract Local Binary Patterns features and compute a histogram.

size (in our case, 96 pixels in both dimensions) and converting them to a gray-scale representation. Instead of using holistic face recognition approaches that typically require training, we turn to a feature-based method using histograms of Local Binary Patterns [13] that allows us to directly compute face descriptors and subsequently compare them with each other based on a distance measure (e.g. utilizing χ^2 Statistics). First, we divide a face image into a grid (of 8×8 similarly sized patches); Figure 1 illustrates this with fewer patches. For each grid-patch, we then extract Local Binary Pattern features (8 neighbors at radius 2) and bin the gained features into histograms. We retain only uniform feature patterns (and account for all remaining patterns using an additional bin); thus, each of our histograms is composed of 59 elements. A common technique to further improve discrimination performance is to repeat the mentioned extraction steps for multiple resolution scales. We follow a pyramid-model, where we simply half the image resolution and quarter the amount of grid-patches for each smaller scale. To retrieve the distance between two faces, we compute the histogram distance for each block or histogram, and then compute the mean over all sub-results.

To actually recognize faces, the most straightforward approach is then nearest-neighbor matching against a set of known face descriptors. Doing so for all pairwise combinations is usually not fast, but it might be adequate for demonstration and because Consumer Photo Collections are typically only around a few thousand photos large. For larger collections, however, it is advisable to utilize more efficient KD- or Ball tree models as we do, or to incorporate hashing or sparse-coding retrieval techniques, as proposed in [14].

Although the employed face features and methods to extract and process them may not be *state-of-the-art* – for example, we do not perform face registration and alignment – they are sufficient to demonstrate the proposed graph-based recognition approach introduced next. Its main input are face similarity scores, which, for example, could also be confidence scores gained through a traditional classifier (e.g. class probabilities emitted by a Random Forest classifier).

2.2   Graph-Based Recognition

Inspired by previous works [2, 5, 7], we employ a graphical model to further improve recognition of people. Such models are factored representations of probability distributions that include Bayesian and Markov Networks, where nodes represent random variables and edges probabilistic relationships. Whereas

Bayesian Networks are directed graphical models, Markov Networks are undirected ones. The latter suit our purpose since we do not intend to model directed relationships.

Let $\mathbf{x} = x_1, \ldots, x_N$ be a set of N random variables, then a Markov Network may model probability distributions over these variables in the following form:

$$Pr(\mathbf{x}) = Z^{-1} \prod_{c=1}^{C} \phi_c [\mathbf{s}_c] \tag{1}$$

In the aforementioned Equation 1, ϕ represents potential functions (or short *potentials*) that operate on subsets of $\mathbf{s} \subset \mathbf{x}$, also called cliques (hence the subscript c). These potentials take non-negative values only, and although seemingly similar, they are not the same as (conditional) probabilities. Ideally, a partition function Z normalizes the product such that the outcome results in a valid probability distribution.

Ultimately, we wish to infer the states – as in a discrete model – of the random variables. An approach is to compute the individual marginal distributions, and then, as Equation 2 shows, the largest marginal values would reflect the *world states* that are individually most probable.

$$\hat{w}_n = \arg \max_{w_n} Pr(w_n | \mathbf{x}) \tag{2}$$

We can also find the maximum a posteriori (MAP) solution that is the most likely (overall) configuration and thus reflects the joint statistics:

$$\hat{\mathbf{w}} = \arg \max_{\mathbf{w}} Pr(\mathbf{w} | \mathbf{x}) \tag{3}$$

Note that both approaches can be computationally quite complex. Therefore, exact inference is often intractable – even if the models are small – but approximate techniques such as Belief Propagation can provide feasible solutions.

Markov Networks can be modeled in various ways; for instance, as chains or trees. More sophisticated are Markov Random Fields (MRFs) that are typically associated with a grid of undirected connections. In a MRF of second order (also called a pairwise MRF), observations are also conditionally dependent on the states, thus additionally incorporating pairwise potentials.

Finding the MAP solution of such a model can be expressed as the following equation, where u defines a unary and p a pairwise function term, and where the subscripts n and m each represent a person's appearance:

$$\hat{\mathbf{w}} = \arg \max_{\mathbf{w}} \sum_{n=1}^{N} u_n(w_n) + \sum_{n,m \in C} p_{nm}(w_n, w_m) \tag{4}$$

A MRF as introduced above is a generative model; for instance, it can be utilized to optimize for $Pr(\mathbf{x}, \mathbf{y})$, where both $\mathbf{x}$ and $\mathbf{y}$ define random data. MRFs can be more flexible in expressing dependencies, but in practice are also sometimes

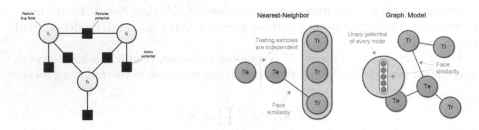

Fig. 2. Left: Exemplary factor graph modeling the identity of three face appearances. Unary potentials encode likelihoods of faces belonging to particular people (based on face similarities), and pairwise potentials encode a spatial smoothness as well as a uniqueness or exclusivity constraint. Right: Face and people recognition by traditional nearest-neighbor matching (each testing sample Te is independently matched against all training samples Tr) and our graph-based framework that considers all people's appearances within an entire dataset simultaneously (faces are represented by inter-connected nodes).

more difficult to handle. The discriminative equivalent to a MRF is a Conditional (Markov) Random Field, which generally shares the same principles. However, it models only the target variables (without their joint distribution); for example, $Pr(\mathbf{x}|\mathbf{y})$, where $\mathbf{x}$ is unobserved but $\mathbf{y}$ observed data.

People Recognition. We now explain how to use such an undirected graphical model for the task of people recognition. In our proposed approach, people's appearances (e.g. as represented by their faces) correspond to $\mathbf{x}$ and thus the nodes in a MRF. We set up one single graph $G = (V, E)$ with nodes (vertices) V relating to a set combined of a testing and a training set (signified by samples Tr and Te in Figure 2) of people's appearances. V is composed of a total of N nodes corresponding to the overall number of people's appearances across an entire photo collection.

Unary Potentials. We use the graph's unary node potentials to express how likely people's appearances belong to particular individuals (in the training set). In other words, the states of the nodes V correspond to $\mathbf{w}$ and reflect the people's identities. For simplicity, we constrain ourselves to at least one training sample per label class. As such, we are able to determine the number of distinct people and thus states when setting up the graph. We define the unary term as in Equation 5, where s signifies a face similarity function conveying the pairwise distances among faces (based on the histogram-based distance measure intro-duced in Section 2.1), and Z_s an optional normalization constant. In particular, we encode each state in s by computing the mean distance over the k nearest-neighbors corresponding to training samples. We test two variations: In the first and simpler case, k globally spans over all training samples; and in the second and finer-grained case, k separately spans over multiple sets of training samples, where each set is defined by its class label.

$$u\left(\mathbf{w}\right) = Z_s^{-1} s(\mathbf{w}) \tag{5}$$

Pairwise Potentials. As there is one unary potential for each node, there is one pairwise potential for each edge (connecting two nodes) in a pairwise MRF. They allow us to model the combinations of states two connected nodes can take, and thus, to encourage a spatial smoothness among neighboring nodes in terms of their states. Like in [2], they also allow us to enforce an exclusivity or *uniqueness* constraint that no individual person can appear more than once in any given photo i. The latter is a simple, yet presumably effective contextual cue: unlike with studio shots (e.g. passport photos), it is not uncommon for personal photos to depict multiple people. By incorporating such a constraint, we handle any number of detected faces in a single photo, even if they are very similar, as distinct and unique persons.

$$p\left(w_n, w_m\right) = \begin{cases} \tau, & \text{if } w_n = w_m \wedge i_n \neq i_m \\ 0, & \text{if } w_n = w_m \wedge i_n = i_m \\ 1, & \text{otherwise} \end{cases} \tag{6}$$

The larger the spatial smoothness τ in Equation 6, the more we encourage neighboring nodes to take on the same state, therefore leading to fewer but larger clusters. Apart from damping *noise*, we also use this effect to model the observation that photo collections are often organized into smaller groups of consecutive photos (*events*), which are usually *limited* to only a few individuals. Note that for the smoothness and exclusivity constraint to be effective, we need to establish edges E reflecting direct dependencies among nodes – in the latter case, simply among all nodes that share the same photos.

Topology. Recall that edges reflect direct dependencies among nodes, therefore influencing the performance a great deal. They also enable the aforementioned spatial smoothness and exclusivity constraint. There are many ways a graphical model can be connected in terms of its nodes. While densely or fully connected graphs are possible, they complicate inference and are often not more effective. The main reason is the increasing complexity arising from cycles (or loops) in the graphs. Although techniques such as Loopy Belief Propagation (LBP) [15], Tree-Reweighted Belief Propagation or Generalized Belief Propagation can deal with graph cycles, we shall generally strive for as few cycles as possible and as few edges as necessary.

For our aim of a sparse but effective graph representation, we only connect nodes representing the most similar face appearances of people with each other. In other words, we connect each node with its η closest matches among other nodes based on the feature-space distance or similarity of their associated faces. However, unlike [2] and other works, which simply use a globally fixed value for

η and thus treat all node connections the same way, we introduce a regularization based on rank and differences among similarities (as retrieved through a k nearest-neighbor search). In addition, we propose to apply this step in several variations and combinations thereof: among only the testing samples (denoted by subscript $_{te}$); among the testing and the training samples (denoted by subscript $_{tr}$); and globally among the combined testing and training samples. In each variation, we regularize and thus indirectly control η based upon a dynamic threshold such that $\check{s} - s_x < \varepsilon\check{s}$, where $\check{s}$ denotes the closest match (with the smallest feature-space distance and thus highest face similarity s), s_x the next x closest matches, and $\varepsilon \in [0, 1]$ the regularization constant. Additionally, it is possible to enforce that at least a certain number $\check{\gamma}$ of the most closest matches shall be connected, and similarly, cap the number of most closest matches by an upper bound $\hat{\gamma}$. Again, both $\check{\gamma}$ and $\hat{\gamma}$ can be controlled separately for each variation of the possible connection strategies. Lastly, we introduce a lower threshold $s > \mu\bar{s}$ to discard *all-too dissimilar* candidates that might otherwise still be considered as closest matches.

The proposed threshold compares the similarity s between a node and a candidate with the average $\bar{s}$ over all within-class similarities that we can compute based on known training samples. Note that we utilize three KD-tree models (global, testing-testing, testing-training) to substantially lower the complexity of all involved nearest-neighbor searches.

Joint-Recognition. As briefly mentioned in the beginning of this work, our main aim for employing a graph-based approach is to consider all people's appearances within an entire dataset simultaneously. As a result, we perform recognition over the entire dataset (and thus inference over the entire graph) and not for a singular appearance of a person. When adding new photos to a dataset, an incremental (perhaps approximated) solution might be possible, but for simplicity and lack of space we restrict ourselves to always performing inference over the entire graph. Note, however, that we can efficiently store and re-use all information (e.g. any extracted features) except the graph-structure.

3   Experiments

3.1   Datasets

Typical face datasets like *FERET* or *Labeled Faces in the Wild* are not suitable for our aim of recognition in Consumer Photo Collections because they usually only contain single faces stripped of their contexts. Instead, we are using two dedicated photo collections, each composed of personal photos centering around a single person and the person's family and friends.

Our first dataset is the public *Gallagher Dataset* [9]. It contains 589 family photos, all of which are shot in an uncontrolled environment with a typical consumer camera. Many of the photos show the main subjects, a couple and their children, in a broad variety of settings and scenes both indoors and outdoors

Fig. 3. Exemplary photos of the *Gallagher Dataset* (left) and our larger dataset (right)

(see Figure 3). The dataset depicts 32 different individual people over a period of roughly two years. In total, there are 931 face appearances. Infants and children appear quite often, some of whom seem to be rather indistinguishable upon initial viewing of only their faces.

To better validate and demonstrate our proposed approach (e.g. with respect to scalability), we additionally create a new, more challenging dataset that spans a significantly longer time period (thus presumably offering more face variations) and that contains notably more photos and face appearances. Similarly to the *Gallagher Dataset*, all photos are shot in an uncontrolled environment. The main subjects, a couple and their friends and family, are depicted throughout numerous changing settings, scenes and locations (e.g. living in multiple places with different circles of friends). In total, our dataset depicts 106 different individual people over a period of roughly ten years. Altogether, there are nearly 3500 face appearances spread over 2341 photos, many of which depict the few main subjects. Most individuals appear at least ten times in total. Note that infants and children appear quite seldomly; most people are adults. The number of distinct people appearing together throughout *events* (e.g. when we consider an *event* to last around one day) seems to be mostly well under ten, but is in few cases more than 15. Both ground truths specify the true eye/face-positions along a unique class label.

3.2   Implementation Details and Setup

Direct performance comparisons with related works are difficult - their specific implementations and their datasets are not public. Thus, our basis for comparison is recognition by traditional nearest-neighbor matching, by a supervised Random Forest classifier, and by evaluating against a *fixed* and denser graph design as used in [2]. To also evaluate the recognition performance without the influence (outcome and specifies) of the face detection method, we only consider correctly detected faces as verified against the ground truth from the overall dataset denoted by D; then: $D' \subseteq D$. Our primary scores are the accuracy (1-hit recognition rate) as well as the Normalized Mutual Information (NMI) metric. Both scores are in the range $[0, 1]$, where higher values indicate better results.

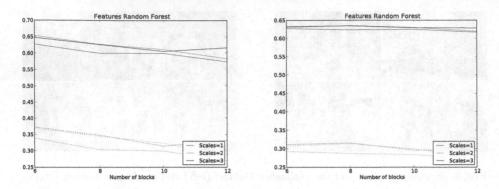

Fig. 4. Recognition accuracy (solid) and NMI (dotted) depending on the feature extraction configuration for the *Gallagher Dataset* (left) and our larger dataset (right)

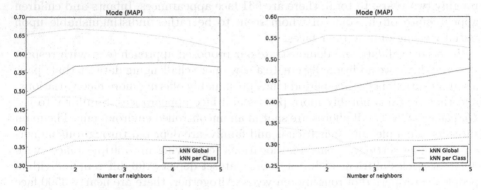

Fig. 5. Recognition accuracy (solid) and NMI (dotted) depending on how unary potentials are computed: either by finding k nearest-neighbors over all training classes, or over each training class separately. Left: *Gallagher Dataset*. Right: our larger dataset.

Our motivation for additionally using NMI compared to only the accuracy alone is the observation that most Consumer Photo Collections show a highly unbalanced distribution with respect to the amount of appearances by the individual people. For example, there might be 50 times as many photos depicting the main subjects than other people.

For all experiments, we split the dataset D' into a training set D'_{tr} and a testing set D'_{te}, such that D'_{tr} represents a small random but stratified subset of D' (with t being the percentage) with at least $\check{t}$ samples of each label class. We repeat all experiments 5 times and average the results. If not otherwise mentioned, we base all experiments on the following default configuration: MAP-based LBP graph inference with $k = 2$; unused Z_s; $\tau_{tr} = 1.0$; $\tau_{te} = 1.2$; $\hat{\gamma}_{tr} = 0$; $\hat{\gamma}_{te} = 2$; $\check{\gamma}_{tr} = \check{\gamma}_{te} = 0$; $\mu = 0.5$; and $\varepsilon_{tr} = \varepsilon_{te} = 0.925$. Lastly, we default to a rather small training set of only $t = 5\%$ and $\check{t} = 3$ samples.

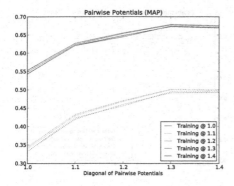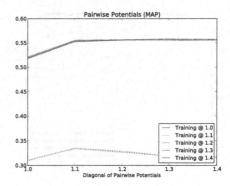

Fig. 6. Recognition accuracy (solid) and NMI (dotted) depending on how pairwise potentials are set up. Left: *Gallagher Dataset*. Right: our larger dataset.

3.3 Results

Face Detection First, we briefly evaluate the performance of detecting faces in photos. We compute a ratio IoU of intersection area over union area given detected face boundary and ground truth face boundary. We consider a face detected if $IoU > 0.5$. In case multiple detected faces match this criteria for a given ground truth face, we choose the detected face with the highest IoU ratio.

Given D, the Viola-Jones face detection routine we utilize (a cascade trained on frontal faces) finds 816 faces (*true positives*) out of the 931 faces specified in the ground truth of the *Gallagher Dataset*. This represents a rate of 87.6%. For our dataset, we correctly detect 3085 faces. Since personal photos often depict numerous people on the side or in the background (*bystanders*), it would be a time-consuming task to manually annotate ground truth data for virtually all depicted people within a photo collection. Thus, we are not able to compute the precision of our face detection results. However, the evaluation approach is sufficient for our purpose as it allows us to retain only correctly detected faces intersecting with D for all subsequent experiments.

Graph-Based People Recognition Figure 4 shows the recognition performance depending on the mode of face feature extraction. We see that too few or too many grid blocks tend to degrade performance. Dividing the face images into grids of 8×8 patches seems to lead to a good balance with respect to both performance and storage requirements of the computed histogram feature vectors. More importantly, however, is the observation that a multi-scale extraction approach improves performance by up to 5%. The gain of using three or more scales seems to be less significant in our experiments though.

In Figure 5, we evaluate the performance depending on how we define and compute the unary potentials of the graph nodes. Computing the mean over $k = 2$ or $k = 3$ nearest-neighbors, which however span over each label class separately, performs by a large margin (up to 30% gain) best for both datasets.

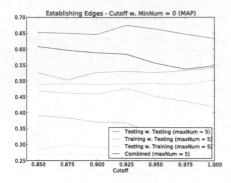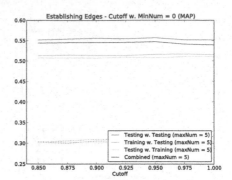

Fig. 7. Recognition accuracy (solid) and NMI (dotted) depending on threshold controlling the way edges among nodes are established. Left: *Gallagher Dataset*. Right: our larger dataset.

Figure 6 shows the recognition performance with respect to τ, which controls the spatial smoothness. We see that $\tau_{te} = 1.3$ leads to a better recognition performance than not encouraging any spatial smoothness among testing samples at all (that is when $\tau_{te} = 1$). As expected, we also see that increasing τ_{tr} has almost no effect; thus, encoding the training labels in only the unary potentials suffices (additional *clamping* seems unnecessary).

Next, Figure 7 depicts results depending on ε that implicitly regularizes the number of graph edges. We notice that values in the range of 0.85 to 0.925 tend to best improve performance; in particular for the smaller *Gallagher Dataset*. However and more importantly, the same figure also illustrates that establishing edges mostly only among the testing samples leads to notably better performance compared to all other connection strategies (e.g. testing-training or vice versa). This observation is also supported by Figure 8, which shows performance depending on the maximum number of considered candidates $\hat{\gamma}$. Here, $\hat{\gamma}_{te} = 1$ or $\hat{\gamma}_{te} = 2$ suggest that our topology-regularized framework already achieves a reasonably good recognition performance with only up to one or two connections per node. Therefore, the proposed regularization leads to fewer overall edges and thus a sparser graphical model.

To an end, we compare the recognition performance of our graph-based framework to a baseline nearest-neighbor approach as well as to a supervised Random Forest classifier. For the latter, we first concatenate all feature histograms of each face appearance into a combined feature vector. We see in Figure 9 that our framework clearly outperforms both other approaches on the smaller *Gallagher Dataset*, especially when looking at the NMI metric. Considering both metrics, our approach also shows a better overall performance on our larger dataset. Moreover, judging by the overall poorer NMI performance of the Random Forest classifier, it seems that our graph-based approach can better cope with unbalanced class label distributions that usually prevail in Consumer Photo Collections. Figure 9 also illustrates how performance changes when varying the

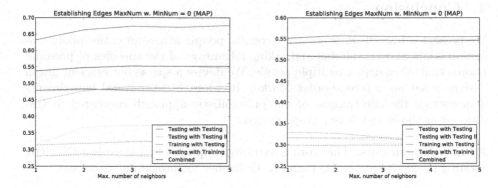

Fig. 8. Recognition accuracy (solid) and NMI (dotted) depending on the amount of maximum neighbors to consider when establishing edges. Left: *Gallagher Dataset*. Right: our larger dataset.

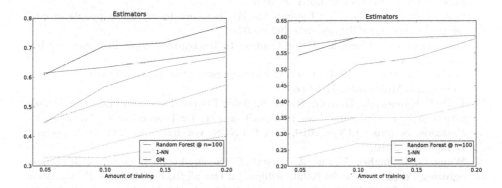

Fig. 9. Recognition accuracy (solid) and NMI (dotted) depending on the amount of training samples, and compared to other estimators: 1-NN and Random Forests. We keep the testing set constant at 0.8. Left: *Gallagher Dataset*. Right: our larger dataset.

amount of training samples. The results show that our graph-based framework generally performs better than both other approaches when the amount of training samples is low (which is preferred to keep user interactions for training at a minimum), and that its gain over both other approaches only gradually decreases with increasing training amounts.

Lastly, without detailing complexity, we list the following observations (on a AMD X6 2.6 GHz): Viola-Jones face detection takes about 0.2 seconds per photo; face feature extraction takes a few milliseconds per person; setting up the entire graph-structure takes below a second given pre-computed face distances; and graph inference over the entire dataset finishes between around one second and one minute depending on the configuration and the (size of the) dataset. Since our proposed approach is feature-based, all stages except similarity-search and graph-inference scale linearly in terms of computational complexity.

4   Conclusion

We present a framework to jointly recognize people across an entire photo collection while also considering and taking advantage of the specifies of personal photos that often depict multiple people. We devise a sparse but efficient graph design based on a second-order Markov Random Field. Several experiments demonstrate the effectiveness of our probabilistic approach compared to traditional methods and denser graph designs.

Acknowledgements. This work is partially supported by EU project CUbRIK (grant FP7-287704). We also thank A. Gallagher for providing his dataset.

References

1. Anguelov, D., Lee, K., Gokturk, S.: Contextual identity recognition in personal photo albums. In: CVPR, pp. 1–7 (2007)
2. Gallagher, A.C., Chen, T.: Using a Markov network to recognize people in consumer images. In: ICIP, pp. 489–492 (2007)
3. Gallagher, A.C., Chen, T.: Using Context to Recognize People in Consumer Images. TCVA 1 (2009)
4. O'Hare, N., Smeaton, A.F.: Context-aware person identification in personal photo collections. Multimedia 11 (2009)
5. Lin, D., Kapoor, A., Hua, G., Baker, S.: Joint People, Event, and Location Recognition in P. P. C. Using Cross-Domain Context. In: Daniilidis, K., Maragos, P., Paragios, N. (eds.) ECCV 2010, Part I. LNCS, vol. 6311, pp. 243–256. Springer, Heidelberg (2010)
6. Wang, G., Gallagher, A., Luo, J., Forsyth, D.: Seeing People in Social Context: Recognizing People and Social Relationships. In: Daniilidis, K., Maragos, P., Paragios, N. (eds.) ECCV 2010, Part V. LNCS, vol. 6315, pp. 169–182. Springer, Heidelberg (2010)
7. Kapoor, A., Lin, D., Baker, S.: How to Make Face Recognition Work: The Power of Modeling Context. AAAI Work (2012)
8. Brenner, M., Izquierdo, E.: Gender-aided People Recognition in Photo Collections. In: MMSP (2013)
9. Gallagher, A.C., Chen, T.: Clothing co-segmentation for recognizing people. In: CVPR (2008)
10. Brenner, M., Izquierdo, E.: Recognizing People by Face and Body in Photo Collections. In: FG (2013)
11. Brenner, M., Izquierdo, E.: Graph-based Recognition in Photo Collections using Social Semantics. In: MM SBNMA (2011)
12. Brenner, M., Izquierdo, E.: People Recognition in Ambiguously Labeled Photo Collections. In: ICME. IEEE (2013)
13. Ahonen, T., Hadid, A.: Face description with local binary patterns: Application to face recognition. PAMI 28(12) (2006)
14. Wang, D., Hoi, S., He, Y., Zhu, J.: Retrieval-based face annotation by weak label regularized local coordinate coding. In: MM (2011)
15. Koller, D., Friedman, N.: Probabilistic graphical models: principles and techniques. MIT Press (2009)

Event Detection by Velocity Pyramid

Zhuolin Liang, Nakamasa Inoue, and Koichi Shinoda

Department of Computer Science, Tokyo Institute of Technology
{zhuolin,inoue}@ks.cs.titech.ac.jp, shinoda@cs.titech.ac.jp

Abstract. In this paper, we propose velocity pyramid for multimedia event detection. Recently, spatial pyramid matching is proposed to introduce coarse geometric information into Bag of Features framework, and is effective for static image recognition and detection. In video, not only spatial information but also temporal information, which represents its dynamic nature, is important. In order to fully utilize it, we propose velocity pyramid where video frames are divided into motional sub-regions. Our method is effective for detecting events characterized by their temporal patterns. Experiment on the dataset of MED (Multimedia Event Detection) has shown 10% improvement of performance by velocity pyramid than without this method. Further, when combined with spatial pyramid, velocity pyramid provides an extra 3% gains to the detection result.

Keywords: Event detection, spatial pyramid, velocity pyramid, GMM supervectors.

1 Introduction

With the development of various web services, the amount of videos available on the Internet is growing exponentially. These open source videos usually have a large variety of contents composed by scenes, objects, motions and audio cues. How to search videos effectively becomes a heated issue. Event detection in large scale unconstrained videos is a research topic towards promoting understanding of video contents. Here, an event is defined as a complex activity occurring at a specific place and time which involves people interacting with other people and/or object(s) [1]. For example, the event of "Birthday party" may be indicated by the observation of the following aspects: a scene as indoor, an object like a birthday cake, activity of opening a gift, and even audio cues such as people cheering. Event detection is thus more challenging than object detection in a still image or activity detection in a video.

A typical flow to detect an event from a video is described as follows: 1) feature extraction, 2) feature encoding, 3) recognition. In order to capture different characteristics of an event, researchers tend to integrate multiple kinds of features into an event detection framework [4]. For example, GIST [5] features can be used for capturing global scene characteristics of an image; HOG, SIFT, HOG3D can capture object appearance information; STIP [15], Dense Trajectory [16],

C. Gurrin et al. (Eds.): MMM 2014, Part I, LNCS 8325, pp. 353–364, 2014.
© Springer International Publishing Switzerland 2014

MoSIFT [18] are able to encode temporal evolution. After feature extraction, features are often encoded into fixed-length histogram. Usually a standard Bag-of-Words approach is used for this encoding process. Soft encoding methods such as Fisher vector encoding [9], GMM supervector encoding [12] have also been applied. After obtaining the encoded vectors, classifies such as SVMs are used for detection with early or late fusion of features.

Spatial pyramid [8], which is originally proposed for scene recognition of static images, is often used as features in event detection. Inspired by this spatial pyramid, we propose velocity pyramid in this paper. Instead of utilizing pyramid structure which represents spatial information, we construct a pyramid structure which represents dynamic nature of video. We first divide video into several components using motion information. Then, for each motion component, we model the distribution of features, which is expected to follow a fixed pattern. For example, in the event *parade*, people are likely to move in the horizontal direction than the vertical direction. In the event *repairing_an_appliance*, people's hands tend to move in every direction.

We construct a framework for event detection with GMM supervectors. The GMM supervector is used for feature encoding, which has been applied to event detection and outperforms Bag-of-Words models [12]. We verify the system's effectiveness on the challenging dataset of Multimedia Event Detection (MED) task of TRECVID [1]. In this dataset, multiple conditions of scenes, objects, motion patterns exist. We will show that the velocity pyramid can capture the rough dynamic information of the video. Furthermore, when combined with spatial pyramid, the performance of the system is further improved.

The rest of this paper is organized as follows: Section 2 describes related works for event detection; Section 3 focuses on the proposed method of velocity pyramid; Section 4 introduces the steps for constructing the detection system; Section 5 describes the evaluation dataset, evaluation measures, and experiment results; Finally Section 6 gives a conclusion.

2   Related Work

As video has its own nature as a spatial and temporal sequence, various spatial-temporal features have been applied to event detection in video. STIP [15] selects spatial-temporal interest points by detecting 2D corners with rapid velocity change. MoSIFT [18] finds interest points that have both discriminative appearance and sufficient amount of motion. Dense Trajectory [16] represents videos as a set of trajectories obtained from tracked points. Motion Histogram [17] integrates appearance and motion by calculating a motion histogram for each visual word. Relative motion histogram is also computed for each visual word pair between every two frames. However, all these features explore deeply into the internal structure of an video volume by detecting local maximum or by tracking objects, thus have the problem of a heavy computation cost and face the storage problem especially when applied to large-scale unconstrained videos.

Our purpose is to construct a structure that can encode both spatial and dynamic information effectively and efficiently.

Spatial pyramid matching [8] is proposed to introduce coarse geometrical information into Bag-of-Words approach. Several methods to construct the spatial pyramid have been explored. In [13], both feature specific and event specific tiling has been examined. They show that for all kinds of features, including both appearance and motion, spatial pyramid improves the detection performance. Also in [14], 12 different Regions of Interest (ROI) are defined and their contributions to the detection cost are evaluated separately, In [10] soft tiling is proposed where a sample point is assigned to several tiles and gives significant improvement.

Feature encoding methods used in event detection include Bag-of-Words [2], GMM supervectors [12], Fisher Vectors [9] etc. GMM models a set of features as mixture of Gaussians with different means and covariances. Fisher vector consists of the first and second order differences to the cluster center. The latter two are based on generative probability models and use soft assignment to mitigate the influence of code word miss assignment.

Here we focus on how to integrate appearance and motion features effectively. Since a global motion histogram and appearance features from a certain frame may be independent with each other, it's better to calculate motion histogram for each object, e.g. tree, people, etc. From another view, appearance features with similar motion should belong to the same feature set characterized by the motion. This idea is similar to spatial pyramid representation in images, which models feature distribution in each spatial subregion. Instead of spatial information, we utilize the dynamic features of video. We build a velocity pyramid which captures coarse dynamic information of appearances. This method is also efficient compared to spatial-temporal interest point approaches because it does not need tracking or local maximum exploration.

3   Velocity Pyramid

An illustration of the procedure to construct the velocity pyramid subregions is given in Fig. 1. First, extract appearance features and motion vectors. Second, quantize motion vectors into motion bins. Third, calculate an encoded appearance histogram for each motion bin. Lastly, concatenate the encoded histograms to form an input vector for a classifier.

Appearance Features. The pyramid representation can be applied to any kind of low level visual features, including densely of sparsely sampled features, e.g. SIFT, HOG, STIP, etc. Here we use a single type of features for simplicity of explanation. For one frame which consists of n feature samples, the set of low level features can be represented as

$$X = \{\mathbf{x}_i\}_{i=1}^n, \tag{1}$$

where $\mathbf{x}_i$ is the ith sample.

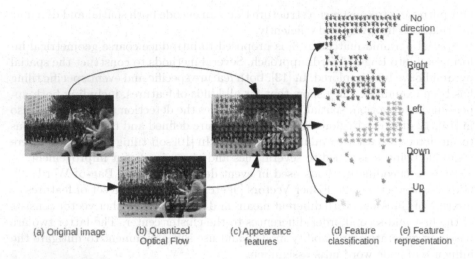

| | (a) Original image | (b) Quantized Optical Flow | (c) Appearance features | (d) Feature classification | (e) Feature representation |

Fig. 1. Subregion generation in velocity pyramid. (a) Original image which shows people who are parading right meet a car passing left. (b) Optical flows calculated from two adjacent frames. The flow vectors are quantized into 4 directions, and each color represents one direction. Gray dots have 0 motions. (c) Appearance features. In this figure, we use dense oriented gradient based features. (d) Partition result of (c) according to the quantization result in (b). (e) Feature representation, e.g. histograms from the Bag-of-Words model.

Motion Vectors. Motion information is captured by optical flow computed by *Farnebäck* algorithm [20]. We calculate velocity vectors for the same coordinates as in X. So the set of optical flows can be expressed as

$$V = \{\mathbf{v}_i\}_{i=1}^{n}, \tag{2}$$

where $\mathbf{v}_i$ is a velocity vector of the ith sample.

Motion Quantization. In order to relate appearance information with motion information, we assign each feature vector $\mathbf{x}$ to a certain motion orientation bin. For each non-zero velocity vector $\mathbf{v}$, let us introduce a P-dimensional orientation vector $\mathbf{o}$, in which each element is a binary variable. The appearance features that have motion are classified into P categories and in each category one of the dimensions $o(p)$ is equal to 1 and other dimensions are equal to 0. In other words, $o(p) \in \{0, 1\}$ and $\sum_p o(p) = 1$. The value of $\mathbf{o}$ is determined by quatizing a motion vector

$$o(p) = 1 \quad \text{if } \frac{2\pi}{P}p \leq \theta < \frac{2\pi}{P}(p+1), \ p \in \{0, ..., P-1\}, \tag{3}$$

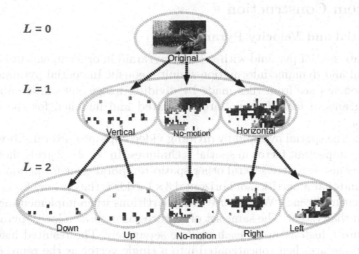

Fig. 2. Velocity Pyramid. In $L = 1$ and $L = 2$, the horizontal part is mainly comprised of a passing parade and a car; the vertical part mainly comes from people's legs; and no motion part mainly corresponds to backgrounds.

where θ is the orientation of optical flow vector $\mathbf{v}$ ranging from 0 to 2π. This motion bin vector is calculated for each sample. Consequently, for each frame we have a set of orientation vectors which can be explained as

$$O = \{\mathbf{o}_i\}_{i=1}^n, \qquad (4)$$

Feature Classification. Features are classified according to their orientation quantization result

$$X_p = \{\mathbf{x}_i | \mathbf{v}_i \neq (0,0) \text{ and } o_i(p) = 1\} \ (p \subset \{0, ..., P-1\}), \qquad (5)$$

$$X_P = \{\mathbf{x}_i | \mathbf{v}_i = (0,0)\}. \qquad (6)$$

where P is the number of quantized orientation bins, and X_p, $p \in \{0, 1, ..., P\}$ is a feature set. X_P represents features from an extra zero bin which is more likely to come from background features. The total number of pyramid components is $P + 1$, and the relationship among sets satisfies $X = X_0 \cup X_1 \cup ... \cup X_P$.

Feature Encoding. For each set of features X_p, $p \in \{0, 1, ..., P\}$, the corresponding histogram of encoded features is calculated by the encoding method in Subsection 4.2. Each histogram is a component of the velocity pyramid. And a final representation of the video is the concatenation of all the histograms.

Pyramid Level. Let us define L as the level of velocity pyramid, so the number of non-zero motion components is given by $P = 2^L$. $L = 0$ represents the original appearance feature; $L = 1$ divides these features into horizontal and vertical components; $L = 2$ includes 4 equally quantized orientations including left, right, up, down; etc. We illustrate velocity pyramid in Fig. 2.

4   System Construction

4.1   Spatial and Velocity Pyramid

We integrate spatial pyramid with velocity pyramid in order to capture a coarse geometrical and dynamic information simultaneously. In spatial pyramid, pyramid components are from tiles made by dividing images into finer subregions, and histograms of local features are calculated and matched for the resulted subregions.

We make the spatial pyramid by dividing video clips into several sub-volumes. Tiling is an important factor in spatial techniques. In [7], 2×2 grids shows good result. They also evaluate several other spatio-temporal grids for action recognition, and find that a partition of horizontal 3×1 grids is the optimal for capturing layout of natural scenes. We also use these partitions when implementing spatial pyramid in the system. The same set of features as used in velocity pyramid are encoded into a histogram for each volume separately. The resulted histograms for all volumes are then concatenated into a single vector as the representation of the video.

4.2   GMM Supervectors

Gaussian mixture model (GMM) and SVMs was proposed in the context of speaker verification, and applied successfully to multimedia event detection [12]. It outperforms Bag-of-Words, because it realizes soft assignment by considering covariance information.

Given a set of features $X = \{x_i\}_{i=1}^n$, the probability distribution function of X conditioned on a Gaussian Mixture Model is given by

$$p(x|\theta) = \sum_{k=1}^{K} w_k \mathcal{N}(x|\mu_k, \Sigma_k), \tag{7}$$

where x represents a d dimensional feature vector, K is number of Gaussian mixtures, $\theta = \{w_k, \mu_k, \Sigma_k\}_{k=1}^K$ are parameters for Gaussian functions, w_k, μ_k, Σ_k are the weight, mean, and covariance of the kth Gaussian probability distribution function $\mathcal{N}(\cdot|\mu_k, \Sigma_k)$.

It is difficult to precisely estimate the parameter θ for one video since its number of feature samples is quite small. In this case, Maximum a Posteriori (MAP) adaptation technique is utilized, because it performs well with a small amount of data [11]. In MAP adaptation the priori knowledge comes form an universal background model (UBM), a GMM whose parameters are estimated by EM algorithm from all training data. After obtaining the UBM, GMM parameters for each video are estimated by MAP adaptation in the following way

$$\hat{\mu}_k = \frac{\tau \mu_k^{(U)} + \sum_{i=1}^n c_{ik} x_i}{\tau + \sum_{i=1}^n c_{ik}}, \tag{8}$$

$$c_{ik} = \frac{w_k^{(U)} \mathcal{N}(x_i|\mu_k^{(U)}, \Sigma_k^{(U)})}{\sum_{k=1}^K w_k^{(U)} \mathcal{N}(x_i|\mu_k^{(U)}, \Sigma_k^{(U)})}, \tag{9}$$

where $\theta^{(U)} = \{w_k^{(U)}, \mu_k^{(U)}, \Sigma_k^{(U)}\}_{k=1}^K$ is a set of parameters for Gaussian components in UBM. c_{ik} is the confidence of Gaussian mixture k for observing feature point x_i. τ is a pre-defined hyper parameter. Note that here only mean vectors for each video are adapted. Next, the signature of a video clip is represented as a concatenation of the K adapted mean vectors

$$\Phi(X) = (\tilde{\mu}_1^T \tilde{\mu}_2^T ... \tilde{\mu}_K^T)^T, \tilde{\mu}_k = \sqrt{w_k^{(U)}}(\Sigma_k^{(U)})^{-\frac{1}{2}}\hat{\mu}_k. \tag{10}$$

In the formula above, each mean vector is normalized by the corresponding weight $w_k^{(U)}$ and covariance $\Sigma_k^{(U)}$.

4.3 SVM Detection

GMM supervectors are the inputs to Support Vector Machines (SVMs) with RBF kernel $k(X, X')$ for detection:

$$k(X, X') = \exp(-\gamma \sum_{p=0}^{P} \|\Phi(X_p) - \Phi(X_p')\|_2^2), \tag{11}$$

where γ is set to be the inverted averaged distance between GMM supervectors. The detection confidence of SVM is given by

$$f(X) = \sum_{l=1}^{L} a^{(l)}k(X, X^{(l)}) + b. \tag{12}$$

where $X^{(l)}$ is a set of features from a training video, L is the number of support vectors, $a^{(l)}$ and b are parameters obtained during SVM training.

4.4 Fusion of Spatial and Velocity Pyramid

Spatial and temporal information are integrated in a late fusion manner. Suppose $f_E^{(sp)}$ and $f_E^{(vp)}$ are detection scores of spatial and velocity pyramid, respectively. The final confidence of one video for event E is:

$$S_E = a * f_E^{(sp)}(X) + (1 - a) * f_E^{(vp)}(X). \tag{13}$$

where a ($0 \leq a \leq 1$), is the fusion weight, which is determined by cross validation for each E during training.

5 Experiment

In this part, we will first introduce the dataset and metric for evaluation, then show experiment result with respect to: effectiveness of spatial-velocity pyramid and the influence of pyramid levels.

5.1   Dataset

The dataset we used is a part of HAVIC data collected by Linguistic Data Consortium, including MED10 data set, MED11 development and test data set [1]. The videos are user-generated videos posted to various Internet video hosting sites. These datasets are for system development and evaluation in the Multimedia Event Detection (MED) task of TRECVID workshop, aiming at permitting users to define their own complex events and quickly and accurately searching large collection of multimedia clips. In MED, an event is defined in an **event kit** including event name, definition, explication, description, and example video clips.

The MED10 dataset is a collection of 3,468 videos (~115h), of which 1,744 videos are for training and 1,724 for testing. There are three events: *Assembling_shelter*, *Batting_in_run*, and *Making_cake*. Approximately 50 positive clips are provided for each event to train its event model. The MED11 dataset is a collection of 44,904 videos (~1406h) including 10 events, in which 13,083 videos are for training, and 31,821 for testing. Each event has 80-230 positive samples in its event kit. These datasets are challenging due to the following reasons: user-generated videos diverse in resolution, length, and quality; unconstrained videos usually have unavoidable camera motions (e.g. *Getting_a_vehicle_unstuck*), clustered background (e.g. *Parade*), various viewing angles, etc.

5.2   Evaluation Criterion

The evaluation criterion is Normalized Detection Cost (NDC) which is used in the TRECVID MED task. This criterion is the linear combination of two kinds of error rates: missed detection rate (P_{MD}) and false alarming rate (P_{FA}). When applying a certain threshold T to the detection scores, the calculation of NDC, P_{MD}, and P_{FA} are defined by the formulas below

$$NDC(T) = w_{MD}P_{MD}(T) + w_{FA}P_{FA}, \qquad (14)$$

$$P_{MD}(T) = N_{MD}(T)/N_{pos}, \qquad (15)$$

$$P_{FA}(T) = N_{FA}(T)/N_{neg}, \qquad (16)$$

where w_{MD} and w_{FA} are the weighting factors for the two error rates respectively. In MED task, $w_{MD} = 1.0$ and $w_{FA} = 12.4875$. N_{MD} is the number of videos that are real positives but have a confidence score lower than the detection threshold. Oppositely, N_{FA} is the number of videos that are real negatives, but assigned a higher confidence score than the detection threshold. N_{pos} and N_{neg} are total numbers of positive and negative videos in test set, respectively.

We use a posterior way to tune the detection threshold T to find the minimum NDC (MNDC). We will report MNDC for each event separately as well as the mean MNDC across all events,

$$MNDC = \min_{T} NDC(T). \qquad (17)$$

Fig. 3. Clip examples with significant improvements by velocity pyramid than without it. This figure lists 5 events in columns. For each column, the first row is original frame, the second row shows optical flow frame, and the third row is the result of flow quantization. In the third row, color indicates flow orientations, and saturation indicates flow magnitudes.

5.3 Performance of Velocity Pyramid

In this section, we compared the performance of the original representation without pyramid, spatial pyramid, and velocity pyramid. We use dense HOG feature [6] since dense features have shown better results than features from sparse interest points in several applications. They are sampled from 4×4 pixels grid, counted in a patch of 20×20 image pixels divided by a 2×2 window. For each window we generate an 8-bin histogram, summing up to a 32-dimensional HOG feature. PCA is applied to it without reducing its dimension. The spatial pyramid follows a pattern of *original*, 2×2, and 3×1, totally 8 spatial components. All pyramid techniques utilize the same set of HOG features.

Evaluation on MED10 Dataset. For MED10 data, we use a single level of velocity pyramid $L = 2$, totally 6 velocity components including the original one. The result is shown in Table 1. MNDC for spatial pyramid, 0.635, is better than the original HOG features, 0.661. Velocity pyramid outperforms spatial pyramid for 2 out of 3 events. Furthermore, the combination of spatial and velocity pyramid achieves the best MNDC, 0.607. For a motion intensive event *Batting_in_run*, velocity pyramid is especially effective. For a complex event *Making_cake*, which is comprised by multiple objects and activities, spatial pyramid and velocity pyramid collaborate well to obtain large gain from original HOG features.

Evaluation on MED11 Dataset. MED11 dataset containing 10 events is a more challenging dataset. Since we get a better result when combining $L = 1$ and $L = 2$ (see the next subsection), we apply this setting to the MED11 data.

Table 1. MNDC of HOG original, spatial pyramid, velocity pyramid for 3 events in MED10. SP = spatial pyramid, VP = velocity pyramid.

Event	HOG original	HOG SP	HOG VP	HOG SP&VP
Assembling_shelter	0.768	0.772	0.776	0.751
Batting_in_run	0.453	0.446	0.434	0.442
Making_cake	0.761	0.688	0.642	0.628
Mean	0.661	0.635	0.617	0.607

Table 2. MNDC of HOG original, spatial pyramid, velocity pyramid for 10 events in MED11. SP = spatial pyramid, VP = velocity pyramid.

Event	HOG original	HOG SP	HOG VP	HOG SP&VP
Birthday_party	0.860	0.749	0.762	0.739
Changing_a_vehicle_tire	0.698	0.600	0.598	0.573
Flash_mob_gathering	0.412	0.364	0.366	0.362
Getting_a_vehicle_unstuck	0.556	0.523	0.597	0.512
Grooming_an_animal	0.774	0.712	0.746	0.705
Making_a_sandwich	0.856	0.753	0.768	0.761
Parade	0.698	0.607	0.599	0.594
Parkour	0.574	0.484	0.498	0.486
Repairing_an_appliance	0.645	0.598	0.518	0.519
Working_on_a_sewing_project	0.810	0.783	0.752	0.746
Mean	0.688	0.617	0.620	0.600

The evaluation result is shown in Table 2. The spatial-velocity pyramid outperforms spatial pyramid in 8 out of 10 events, which further verifies the effectiveness of the method. The velocity pyramid method obtained a competitive result with spatial pyramid; for 4 out of 10 events it outperforms spatial pyramid. Specifically, we observed a significant improvement on *Repairing_an_appliance* and *Working_on_a_sewing_project*. In these two events, motions are clearer than the others. This may provide a large gain in the detection result. These two events also have targets whose motions are widely spread across one frame. In this case, velocity pyramid performs better than spatial pyramid. Meanwhile, velocity pyramid is not effective for some events. These include *Parkour* where the motion area is small, and *Getting_a_vehicle_unstuck* which has hand-held camera motions that hides real object motion. However, the overall performance is improved by utilizing spatial-temporal information. Fig. 3 shows some examples in which velocity pyramid has significant improvements than without motion information.

The best performance for MED task in TRECVID 2012 was achieved by AXES team. They reported a MNDC of 0.411 on MED11 dataset [3], while our best performance on the same dataset is a MNDC of 0.495. The higher performance of AXES team may owe to the robustness of Motion Boundary Histogram (MBH) against camera motion.

Table 3. Influence of pyramid levels on MNDC. $L = 0$ is the detection result by the original appearance features; $L = 1$ means horizontal, vertical and 0-bin; $L = 2$ means 4 equally quantized orientation plus 0-bin; $L = 1,2$ means an early fusion of supervectors from different levels. Note that in $L = 1$ and 2, supervector from $L = 0$ is also used. The values in bold are the best MNDC score for each event.

Event	$L = 0$	$L = 1$	$L = 2$	$L = 1,2$
Assembling_shelter	0.768	**0.695**	0.776	0.744
Batting_in_run	0.453	0.448	0.434	**0.410**
Making_cake	0.761	0.684	0.642	**0.628**
Mean	0.661	0.609	0.617	**0.594**

To reduce both computation and storage costs, we extract features every 60 frames (2 seconds). In our experiment, velocity pyramid's computation cost is 20% of STIP, whose cost is reported to be much less than Motion SIFT [19]. In addition, the metadata size of velocity pyramid is 15% of dense trajectory.

5.4   Influence of Pyramid Levels

The effects of different pyramid levels L are evaluated in this subsection. The result is shown in table 3. When $L = 1$, which means the components only include 0-bin, horizontal and vertical, the result is surprisingly good. By this division, we can have a rough separation of foreground and background. The combination of different pyramid levels of $L = 1,2$ is better than a single pyramid level $L = 1$ or $L = 2$.

6   Conclusion

In this paper, we propose velocity pyramid as an image representation for multimedia event detection. While spatial pyramid divides appearance in a 2D spatial domain, velocity pyramid models appearance in the motion. The resulted velocity pyramid together with a representation by Gaussian Mixture Models is applied to the challenging MED dataset and shows effectiveness for detecting events. In the case of MED11 dataset, MNDC of 0.620 is obtained by velocity pyramid. Further, the MNDC score is reduced to 0.600 when velocity pyramid is combined with spatial pyramid. Future work includes the cancelling of camera motions in unconstrained videos, such as using the camera motion canceled feature MBH instead of HOG. In addition, since currently only orientation information is taken into consideration, we plan to use not only orientation information, but also flow magnitude information in velocity pyramid.

Acknowledgments. Thanks for Canon Incorporation for providing with computation resources and technical supports.

References

1. 2013 TRECVID Multimedia Event Detection Track,
   http://www.nist.gov/itl/iad/mig/med13.cfm
2. Jiang, Y.G., Zeng, X., Ye, G., et al.: Columbia-UCF TRECVID2010 Multimedia
   Event Detection: Combining Multiple Modalities, Contextual Concepts, and Temporal Matching. In: Proc. of TRECVID Workshop (2010)
3. Aly, R., McGuinness, K., et al.: AXES at TRECVid 2012. In: Proc. of TRECVID
   Workshop (2012)
4. Jiang, L.: Alexander G. Hauptmann, G. Xiang: Leveraging High-level and Low-level
   Features for Multimedia Event Detection. ACM Multimedia 12, 449–458 (2012)
5. Torralba, A., Oliva, A.: Modeling the shape of the scene: a holistic representation
   of the spatial envelope. IJCV 42(3), 145–175 (2001)
6. Dalal, N., Triggs, B., Schmid, C.: Human Detection Using Oriented Histograms of
   Flow and Appearance. In: Leonardis, A., Bischof, H., Pinz, A. (eds.) ECCV 2006.
   LNCS, vol. 3952, pp. 428–441. Springer, Heidelberg (2006)
7. Laptev, I., Marszalek, M., Schmid, C., Rozenfeld, B.: Learning Realistic Human
   Actions from Movies. In: Proc. CVPR, pp. 1–8 (2008)
8. Lazebnik, S., Schmid, C., Ponce, J.: Beyond Bags of Features: Spatial Pyramid Matching for Recognizing Natural Scene Categories. In: Proc. CVPR,
   pp. 2169–2178 (2006)
9. Sun, C., Nevatia, R.: Large-scale Web Video Event Classification by use of Fisher
   Vectors. In: 2013 IEEE Workshop on Application of Computer Vision, pp. 15–22
   (2013)
10. Viitaniemi, V., Laaksonen, J.: Spatial extensions to bag of visual words. In: Proc.
    CIVR. ACM (2009)
11. Inoue, N., Shinoda, K.: A Fast and Accurate Video Semantic-Indexing System
    Using Fast MAP Adaptation and GMM Supervectors. IEEE Transactions on Mul-
    timedia 14(4-2), 1196–1205 (2012)
12. Kamishima, Y., Inoue, N., Shinoda, K., Sato, S.: Multimedia Event Detection Using
    GMM Supervectors and SVMs. In: Proc. ICIP, Florida, pp. 3089–3092 (2012)
13. Yu, S., Xu, Z., Ding, D., Sze, W.: Informedia E-Lamp@TRECVID 2012. In: Proc.
    of TRECVID Workshop (2012)
14. Cheng, H., Liu, J., Ali, S., Javed, O.: SRI-Sarnoff AURORA System at TRECVID
    2012 Multimedia Event Detection and Recounting. In: Proc. of TRECVID Work-
    shop (2012)
15. Laptev, I.: On space-time interest points. IJCV 64, 107–123 (2005)
16. Wang, H., Klser, A., Schmid, C., Liu, C.L.: Action recognition by dense trajectories.
    In: Proc. CVPR, pp. 3169–3176 (2011)
17. Wang, F., Jiang, Y.G., Ngo, C.W.: Video Event Detection Using Motion Relativity
    and Visual Relatedness. In: Proc. ACM Multimedia, pp. 239–248 (2008)
18. Chen, M., Hauptmann, A.: MoSIFT: Recognizing Human Actions in Surveillance
    Videos. CMU-CS-09-161, Carnegie Mellon University (2009)
19. Bermejo Nievas, E., Deniz Suarez, O., Bueno García, G., Sukthankar, R.: Violence
    Detection in Video using Computer Vision Techniques. In: Real, P., Diaz-Pernil, D.,
    Molina-Abril, H., Berciano, A., Kropatsch, W. (eds.) CAIP 2011, Part II. LNCS,
    vol. 6855, pp. 332–339. Springer, Heidelberg (2011)
20. Farnebäck, G.: Two-frame motion estimation based on polynomial expansion.
    In: Bigun, J., Gustavsson, T. (eds.) SCIA 2003. LNCS, vol. 2749, pp. 363–370.
    Springer, Heidelberg (2003)

Fusing Appearance and Spatio-temporal Features for Multiple Camera Tracking

Nam Trung Pham, Karianto Leman, Richard Chang,
Jie Zhang, and Hee Lin Wang

Institute for Infocomm Research, Singapore
{ntpham,karianto,rpchang,zhangj,hlwang}@i2r.a-star.edu.sg

Abstract. Multiple camera tracking is a challenging task for many surveillance systems. The objective of multiple camera tracking is to maintain trajectories of objects in the camera network. Due to ambiguities in appearance of objects, it is challenging to re-identify objects when they re-appear in other cameras. Most research works associate objects by using appearance features. In this work, we fuse appearance and spatio-temporal features for person re-identification. Our framework consists of two steps: preprocessing to reduce the number of association candidates and associating objects by using the probabilistic relative distance. We set up an experimental environment including 10 cameras and achieve a better performance than using appearance features only.

Keywords: Multiple camera tracking, feature fusion, metric learning.

1   Introduction

Maintaining trajectories of objects in a wide area is an interesting research issue for many surveillance systems. If the camera network in the system has a full coverage of the surveillance area, objects probably can be monitored in a wide area. Due to many issues such as cost of cameras, storage and communication bandwidth, it is not affordable to have a full coverage camera network in the surveillance area in most cases. Hence, tracking objects over non-overlapped cameras becomes important for surveillance systems. Although there have been significant improvements in the multiple non-overlapped camera tracking, this problem is still an open issue because of challenges such as ambiguities in appearance of objects, light changing between cameras, and changing of object poses in multiple cameras.

With the recent developments of tracking methods [1], [2], [3], performance of object tracking under occlusions improves significantly. These methods can handle a crowd up to 30-40 objects in a local camera. When objects go from a camera to another one, it is required to have methods to re-identify them and maintain their trajectories. If cameras are overlapped, some methods can be used such as [4], [5]. However, these methods are not suitable for multiple non-overlapped camera tracking due to assumptions on overlapped cameras.

C. Gurrin et al. (Eds.): MMM 2014, Part I, LNCS 8325, pp. 365–374, 2014.
© Springer International Publishing Switzerland 2014

The problem of multiple non-overlapped camera tracking can be summarized as follows: given tracks of objects which disappeared in cameras and tracks of objects just appeared in other cameras, we need to find the correspondences between disappeared tracks and new appeared tracks. Most methods for multiple non-overlapped camera tracking consists of two steps: feature extraction and object association.

Features in multiple non-overlapped camera tracking can be appearance features and spatio-temporal features. First, a brightness transfer function is introduced to compensate the lighting change between cameras [6]. Then, features are extracted for the object association. Appearance features include color features in [7], [8], [9], textures (Gabor features) [9], [10], covariance features [11], histogram of gradient orientation [12] and local descriptor features [13]. Features can also be color name [14]. They can be fused together to improve the performance. Using appearance features can find objects that have consistent appearance across multiple cameras. However, they also find difficulties when other objects are similar in appearance. To overcome this problem, some methods proposed to use the spatio-temporal features [15], [16], [17]. However, these features do not contain enough information when the direction of tracks changes when the objects move between cameras.

For object association, some methods can be applied such as Markov chain Monte Carlo [16], support vector machine (SVM) [9], probabilistic relative distance (PRD) [10], rankboost [14], Munkres assignment algorithms [15]. It is observed that the object association will be better if the number of candidates for the association is reduced.

In this paper, we propose a spatio-temporal feature to take care of changes of directions when objects move between cameras. Appearance features are obtained from color distributions by discriminative color representations [18]. Then, appearance and spatio-temporal features are fused in PRD [10]. Moreover, a pre-processing step is to reduce the number of track candidates by using time and movement directions is also proposed in the paper.

2   Problem Description

The problem can be described as follows. Let $\mathcal{T} = \{T_1, T_2, ..., T_N\}$ be tracks of objects moved out of cameras and $\mathcal{S} = \{S_1, S_2, ..., S_L\}$ be tracks just appeared (at least 5 frames) in cameras. $\mathcal{T}$ and $\mathcal{S}$ can be obtained from single camera tracking methods. For each track $S \in \mathcal{S}$, the problem is to find track $T \in \mathcal{T}$ that can be associated with S. If track T can be found, the ID of track S will be the same with the ID of track T and some features of track T can be transferred to track S. Otherwise, a new ID will be assigned to track S. When the number of tracks in $\mathcal{T}$ is large, the performance of the association can be reduced. It is observed that finding a person in a crowd is much more difficult than finding a person in a less crowded situation. Hence, in the next section, we will propose a preprocessing method to reduce the number of candidates for the object association. Sec. 4 will introduce the feature extraction and the object association method.

3 Preprocessing for Object Association

Let $S \in \mathcal{S}$ be a track. When the number of tracks in $\mathcal{T}$ is large, the possible of wrong associations with S will increase. In this section, we try to reduce possible candidates in $\mathcal{T}$ to associate with S. Let $T \in \mathcal{T}$ be another track. Track T will be a potential candidate of associating with S when it is satisfied three constraints: camera topology constraint, time constraint and direction constraint.

First, camera topology constraint allows objects move from one camera to defined neighbor cameras. The camera topology is described by graph $G^c = \{V^c, E^c\}$, where V^c is a set of cameras and E^c is a set of edges represented for movements between cameras. Weight w for an edge is 1 if the movement between cameras on this edge is allowed. Otherwise, this edge weight will be 0. The camera topology constraint is defined as

$$\text{TopologyConstraint}\,(S,T) = \begin{cases} 1, \text{ if } w\,(v_T, v_S) = 1 \\ 0, \qquad \text{otherwise} \end{cases} \tag{1}$$

where v_T, v_S are cameras capture S and T. The time constraint will be

$$\text{TimeConstraint}\,(S,T) = \begin{cases} 1, \text{ if } \alpha_1 < t_b\,(S) - t_e\,(T) < \alpha_2 \\ 0, \qquad\qquad\quad \text{otherwise} \end{cases} \tag{2}$$

where $t_e\,(\cdot)$ and $t_b\,(\cdot)$ are functions to extract the end time and the begin time of tracks. α_1 and α_2 are time thresholds for moving between cameras. The direction constraint will be

$$\text{DirectionConstraint}\,(S,T) = \begin{cases} 1, & \begin{aligned} &\text{if } dir\,(T) = D\,(v_T | E^c\,(v_T, v_S)) \\ &\wedge dir\,(S) = D\,(v_S | E^c\,(v_T, v_S)) \end{aligned} \\ 0, & \qquad\qquad \text{otherwise} \end{cases} \tag{3}$$

where v_T, v_S are camera nodes in G^c represented for cameras which capture T and S. $dir\,(\cdot)$ is the direction function of a track defined as in Fig. 1. $D\,(\cdot)$ is the direction function of movements between cameras that is trained by experimental data.

Fig. 1. Directions of tracks move between cameras

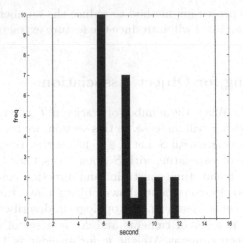

Fig. 2. Time thresholds for Time Constraint. Red lines are upper bound and lower bound for time constraint.

4   Object Association across Multiple Cameras

4.1   Spatio-temporal Feature

Let consider track $S \in \mathcal{S}$ and the track $T \in \mathcal{T}$. The spatio-temporal feature for associating T with S is defined as $f^s(S,T) = \{f_{a_e}, f_{a_s}\}$ where f_{a_e}, f_{a_s} are angle differences between track T and S with typical tracks. They can be obtained as follows. Let $\mathcal{K} = \{\{T_1, S_1\}, \{T_2, S_2\}, ..., \{T_P, S_P\}\}$ be a set of training tracks move from the camera capture T to the camera capture S. A set of directions of training tracks is obtained from $\mathcal{K}$, $\mathcal{D} = \{\{d_1^e, d_1^s\}, \{d_2^e, d_2^s\}, ..., \{d_P^e, d_P^s\}\}$ where d_i^e is the direction of T_i and d_i^s is the direction of S_i. The direction of a track is the vector from the start position to the end position of the track. After using K-Mean clustering on $\{d_1^e, d_2^e, ..., d_P^e\}$ and $\{d_1^s, d_2^s, ..., d_P^s\}$, we can have typical directions of moving between cameras $D^e = \{\bar{d}_1^e, \bar{d}_2^e, ..., \bar{d}_Q^e\}$ and $D^s = \{\bar{d}_1^s, \bar{d}_2^s, ..., \bar{d}_Q^s\}$ where Q is the number of clusters. An example of extracting representation vectors is shown in Fig. 3. Spatio-temporal feature f_{a_e}, f_{a_s} will be

$$f_{a_e} = \min_{\bar{d}^e \in D^e} \left(d_{\cos}\left(v\left(T\right), \bar{d}^e\right)\right) \tag{4}$$

$$f_{a_s} = \min_{\bar{d}^s \in D^s} \left(d_{\cos}\left(v\left(S\right), \bar{d}^s\right)\right) \tag{5}$$

where $d_{\cos}(\cdot)$ is the cosine similarity distance and $v(\cdot)$ is the function to extract the vector from the start position to end position of a track.

4.2   Appearance Feature

In this paper, the color distribution is applied to obtain the appearance feature. Due to illumination changes between cameras, color values of image patches of a

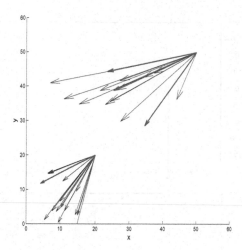

Fig. 3. An example of vector representations for movements between two cameras. Red vectors are representations, blue vectors are from the previous camera and black vectors are from the current camera.

person are also not consistent when this person moves across cameras. This cause many difficulties in using color distribution to re-identify persons. Fortunately, in [18], Khan etal. tried to cluster color values based on the Divisible Information Theoretic Clustering method [19]. Color clusters will be trained so that they have an optimum discriminative power of classification for the training data. This discriminative power is measured by the mutual information theory. These clusters are called discriminative color representations. Here, we use 25 color clusters from [18]. The color distribution for the track S is $h(S) = [h_1^S, ..., h_{N_H}^S]$, where $N_H = 25$, and color distribution for the track T is $h(T) = [h_1^T, ..., h_{N_H}^T]$. An example of the robustness of the color distribution by using the discriminative color representations is shown in Fig. 4. In this figure, although color changes due to different illuminations, densities of major colors in two color distributions are still similar. The Hellinger distance for these two color distributions is $\frac{1}{2}\sum_{i=1}^{N_H}\left(\sqrt{h_i^S} - \sqrt{h_i^T}\right)^2$. Hence, the appearance feature for associating tracks is defined as

$$f^c(S, T) = \left[\left|\sqrt{h_1^S} - \sqrt{h_1^T}\right|, ..., \left|\sqrt{h_{N_H}^S} - \sqrt{h_{N_H}^T}\right|\right] \qquad (6)$$

4.3 Probabilistic Relative Distance for Fusing Appearance and Spatio-temporal Features

Features for an association between the track $T \in \mathcal{T}$ and the track $S \in \mathcal{S}$ will be $f(S, T) = \{f^s(S, T), f^c(S, T)\}$, where $f^s(S, T)$ is the spatio-temporal feature in Sec. 4.1 and $f^c(S, T)$ is the appearance feature in Sec. 4.2. For simplification, $f(S, T) = f_T$. Let consider $T, T' \in \mathcal{T}$ where T is the correct match to S. In [10],

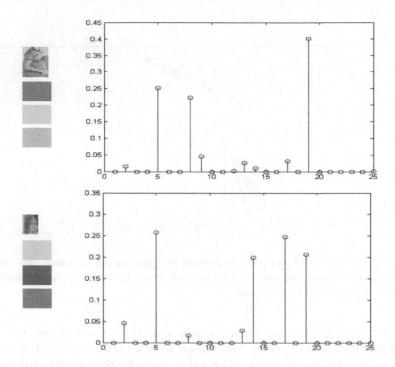

Fig. 4. Color distribution by using discriminative color representations when an object acrosses two cameras

a probabilistic relative distance function r is proposed so that $r(f_T) < r(f_{T'})$. r is defined as

$$r(f) = f^T \mathbf{M} f, \quad \mathbf{M} \succeq \mathbf{0} \tag{7}$$

where $\mathbf{M}$ is a semi-definite matrix. r will be trained so that the probability of $r(f_T) < r(f_{T'})$ needs to be maximized. The probability of $r(f_T) < r(f_{T'})$ is

$$p(r(f_T) < r(f_{T'})) = \frac{1}{1 + \exp\{r(f_T) - r(f_{T'})\}} \tag{8}$$

4.4 Object Association

For each track $S \in \mathcal{S}$ and track $T \in \mathcal{T}$, the feature for the association will be $f(S,T)$ as in Sec 4.1 and Sec. 4.2. The weight for the association is the probabilistic relative distance as in Sec. 4.3 and [10]. Then, the association problem becomes finding $\hat{\mathcal{X}} = \{X_{ST}, (S \in \mathcal{S}) \wedge (T \in \mathcal{T}), X_{ST} \in \{0,1\}\}$ so that

$$\hat{\mathcal{X}} = \arg\max_{\mathcal{X}} \sum_{S \in \mathcal{S}} \sum_{T \in \mathcal{T}} w(S,T) X_{ST} \tag{9}$$

$$\sum_{S \in \mathcal{S}} X_{ST} = 1, \sum_{T \in \mathcal{T}} X_{ST} = 1$$

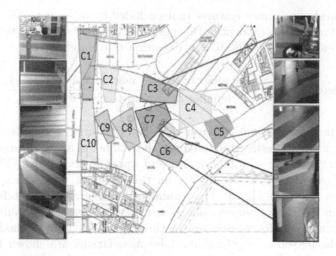

Fig. 5. Camera coverage of our experimental environment

Note that when $w(S,T) < \eta$, $w(S,T) = 0$. That means the association between S and T is not enough confidence. The problem in Eq. (9) can be solved by using the Munkres algorithm.

5   Experimental Results

To evaluate the performance of the proposed method, we set up a camera network including 10 cameras. The coverage of cameras is shown in Fig. 5. Some cameras are different in both zoom and angle, for examples camera 6 and camera 7. The Illumination different is large in some cameras such as camera 3 and camera 4. We collected and annotated 20 videos from 10 cameras at different time. Each video is about 20 minutes. Half data is used for training. For local camera tracking, we applied a method to fuse head detection with single object tracking results for multiple object tracking. The local tracking method can track up to 30 persons per camera view. When a person moves from one camera to another camera and

Cameras	PRD with color distribution	Our system
C5-C4	68.6%	88.6%
C2-C3	63.3%	83.3%
C7-C3	57.2%	78.2%
C4-C3	53.3%	71.43%

Fig. 6. Detection rate for person re-identification on four best camera pairs

has well-defined appearance features (not occluded by other persons), we apply the proposed method to re-identify this person. The results on detection rate for camera pairs are shown in Fig. 6. The detection rate is defined as

$$\text{Detection rate} = \frac{\text{number of correct association detections}}{\text{number of ground-truth associations}} \qquad (10)$$

The reason that our method is better than the PRD with color distribution is that we have a preprocessing step to reduce the number of association candidates and the fusing between spatio-temporal and appearance features. The overall accuracy of our method is about 71% on this data set. One example of the association is shown in Fig. 8. This example demonstrates the ability of our method when associating objects across multiple cameras given multiple. The number of candidates for association in this case is 59. Our method still can re-identify the person correctly. Some false associations are shown in Fig. 7. False associations can be caused by occlusions and ambiguous in appearance of persons whose have similar movements with training data. To improve the algorithm with these cases is a challenging task.

Fig. 7. Typical false alarm cases for association

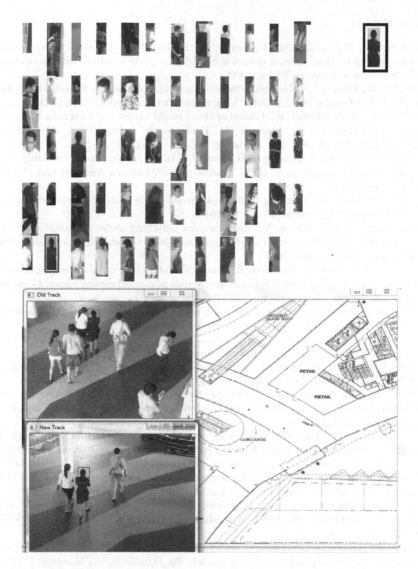

Fig. 8. An example of the object association in multiple camera tracking. Red track is a previous track and blue track is a new track.

6   Conclusions

In this work, we proposed a spatio-temporal feature and fuse it with appearance features for multiple non-overlapped camera tracking. The method consists of two stages: preprocessing to reduce the number of object association candidates and object association. Our method can be applied for real time surveillance applications. Results also showed that our method is better than using appearance feature only for multiple non-overlapped camera tracking.

References

1. Breitenstein, M.D., Reichlin, F., Leibe, B., Koller-Meier, E., Gool, L.V.: Robust tracking-by-detection using a detector confidence particle filter. In: International Conference on Computer Vision (2009)
2. Benfold, B., Reid, I.: Stable multi-target tracking in real-time surveillance video. In: IEEE Conference on Computer Vision and Pattern Recognition (2011)
3. Andriyenko, A., Schindler, K.: Global optimal multi-target tracking on a hexagonal lattice. In: Daniilidis, K., Maragos, P., Paragios, N. (eds.) ECCV 2010, Part I. LNCS, vol. 6311, pp. 466–479. Springer, Heidelberg (2010)
4. Fleuret, F., Berclaz, J., Lengagne, R., Fua, P.: Multi-camera people tracking with a probabilistic occupancy map. IEEE Transaction on Pattern Analysis and Machine Intelligence (2008)
5. Eshel, R., Moses, Y.: Tracking in a dense crowd using multiple cameras. International Journal of Computer Vision (2010)
6. Prosser, B., Gong, S., Xiang, T.: Multi-camera matching under illumination change over time. In: Workshop on Multi-Camera and Multi-modal Sensor Fusion Algorithms and Applications (2008)
7. Javed, O., Shafique, K., Rasheed, Z., Shah, M.: Modeling inter-camera space-time and appearance relationships for tracking across non-overlapping views. Computer Vision and Image Understanding (2008)
8. Porikli, F.: Inter-camera color calibration using cross-correlation model function. In: IEEE International Conference on Image Processing (2003)
9. Processer, B., Zheng, W.S., Gong, S., Xiang, T.: Person re-identification by support vector machine. In: British Machine Vision Conference (2010)
10. Zheng, W.S., Gong, S., Xiang, T.: Re-identification by relative distance comparison. IEEE Transaction on Pattern Analysis and Machine Intelligence (2013)
11. Corvee, E., Bak, S., Bremond, F.: People detection and re-identification for multi surveillance cameras. In: International Conference on Computer Vision Theory and Applications (2012)
12. Martinel, N., Micheloni, C.: Re-identify people in wide area camera network. In: IEEE Conference on Computer Vision and Pattern Recognition Workshops (2012)
13. Farenzena, M., Bazzani, L., Perina, A., Murino, V., Cristiani, M.: Person re-identification by symmetry-driven accumulation of local features. In: IEEE International Conference on Computer Vision and Pattern Recognition (2010)
14. Kuo, C.H., Khamis, S., Shet, V.: Person re-identification using semantic color names and rankboost. In: IEEE Workshop on Applications of Computer Vision (2013)
15. Kuo, C.-H., Huang, C., Nevatia, R.: Inter-camera association of multi-target tracks by on-line learned appearance affinity models. In: Daniilidis, K., Maragos, P., Paragios, N. (eds.) ECCV 2010, Part I. LNCS, vol. 6311, pp. 383–396. Springer, Heidelberg (2010)
16. Meden, B., Lerasle, F., Sayd, P.: MCMC supervision for people reidentification in nonoverlapping cameras. In: British Machine Vision Conference (2010)
17. Chen, K.W., Lai, C.C., Hung, Y.P., Chen, C.S.: An adaptive learning method for target tracking across multiple cameras. In: IEEE International Conference on Computer Vision and Pattern Recognition (2008)
18. Khan, R., Weijer, J.V.D., Khan, F.S., Muselet, D., Ducottet, C., Barat, C.: Discriminative color descriptors. In: IEEE International Conference on Computer Vision and Pattern Recognition (2013)
19. Dhillon, I., Madella, S., Kumar, R.: A divisive information theoretic feature clustering algorithm for text classification. Journal of Machine Learning Reserach (2003)

A Dense SURF and Triangulation
Based Spatio-temporal Feature for Action Recognition

Do Hang Nga and Keiji Yanai

Department of Informatics, The University of Electro-Communications
Chofu, Tokyo 182-8585 Japan
{dohang,yanai}@mm.cs.uec.ac.jp

Abstract. In this paper, we propose a novel method of extracting spatio-temporal features from videos. Given a video, we extract its features according to every set of N frames. The value of N is small enough to guarantee the temporal denseness of our features. For each frame set, we first extract dense SURF keypoints from its first frame. We then select points with the most likely dominant and reliable movements, and consider them as interest points. In the next step, we form triangles of interest points using Delaunay triangulation and track points within each triple through the frame set. We extract one spatio-temporal feature from each triangle based on its shape feature along with the visual features and optical flows of its points. This enables us to extract spatio-temporal features based on groups of related points and their trajectories. Hence the features can be expected to be robust and informative. We apply Fisher Vector encoding to represent videos using the proposed spatio-temporal features. We conduct experiments on several challenging benchmarks, and show the effectiveness of our proposed method.

1 Introduction

For action recognition in a video, both spatio and temporal features are needed to represent actions, while only spatio features such as SIFT and SURF are needed for object or scene recognition on a still image. There exist several well-known features to represent the movements of actors performing the actions such as Histogram of Oriented Optical Flow (HOOF) [1] or trajectory of interest points [2]. In addition to exploiting spatial or temporal features separately, using spatio-temporal (ST) features which integrate both visual and motion characteristics of actions is also preferred among approaches of action recognition in videos. Some ST features have been proposed so far as spatio-temporal extensions of local image descriptors, such as Cuboids [3,4], 3D-SIFT [5], HOG3D [6], and Local Trinary Patterns [7].

To extract local spatio-temporal features from videos, one of the most popular methods is based on Cuboids [3,4]. However, to decide the cuboid size is a tough task. Instead of detecting local Cuboids before extracting features from them, some recent methods tracked interest points in video sequence then leveraged the motion information from their trajectories [2,8]. These approaches obtained good results for action recognition. To track interest points, either tracker based technique or point matching based technique has been employed. Our method proposed in this paper is also based on trajectories of interest points. We apply LDOF [9] to estimate optical flows of all video frames.

C. Gurrin et al. (Eds.): MMM 2014, Part I, LNCS 8325, pp. 375–387, 2014.
© Springer International Publishing Switzerland 2014

In this paper, we propose a dense spatio-temporal feature that is comparable to state-of-the-arts. Empirical results have shown that dense features perform better for complex videos [2,10,11,12]. Our idea is inspired by the work of Noguchi et al. [13]. This method is the baseline we refer in this paper. They proposed to extract spatio-temporal features based on moving SURF keypoints. We address some problems of their method such as their failure in feature extraction of videos containing camera motion or holistic decision of motion threshold in the selection of interest points. We propose simple yet effective solutions to solve these problems. That means, similar to their method, our method is also based on SURF interest points with robust movements, nevertheless how we determine those points is different. Moreover, we propose to improve their feature by exploring more aspects of selected points and introducing several novel spatio-temporal descriptors. The experimental results show significant improvements of our method over the baseline.

According to the success of the BoV (Bag of Visual words) on the research field of image classification, it has became the most popular model for video representation. Noguchi et al also applied BoV model to encode videos. In this paper, we use Fisher Vector to represent videos instead. Fisher vector encoding technique was first applied to image classification task several years ago, shown to extend the BOV representation [14]. The advantage of this technique has been demonstrated that it is not limited to the number of occurrences of each visual word but it also encodes additional information about the distribution of the descriptors. The methodology is that, (1) extracting local features from images, (2) modelling the distribution of those features as mixtures of Gaussian (GMM), training a soft codebook, (3) applying Fisher kernels on obtained codebook to encode each image as a Fisher Vector. Recently, some works on action recognition have also employed this approach to encode videos and showed the effectiveness of Fisher Vector encoding over the traditional BoV [15]. In this paper, we apply Fisher encoding technique as described in [14].

We aim to recognize human actions in realistic videos where the background is complex. We conduct our experiments on large-scale benchmark datasets for action recognition such as UCF-50, HMDB51 adn UCF-101 to validate our proposed method. These datasets are not only large-scale but also challenging benchmarks for human action recognition in realistic and challenging settings such as large variations in camera motion, object appearance and pose, viewpoint and complicated background.

Our contribution is three-fold: first, a method of extracting spatio-temporal features which is comparable to the state of the arts, second, a simple yet efficient selection of interest points, and finally, novel descriptorization of spatio-temporal features. The reminder of this paper is organized as follows: Section 2 discusses more about some related works. In Section 3, we overview the proposed method, and In Section 4 the method to extract the ST proposed feature is described in detail. Section 5 explains about conducted experiments on three large-scale action recognition datasets, and presents the results. Conclusions are presented in Section 6.

2   Related Work

So far, in the research field of action recognition, local spatio-temporal features has been explored to represent videos. Many methods of extraction and descriptorization of spatio-temporal features [3,16,12,4] have been proposed over the past few years. To determine space-time regions (called as Cuboids) where features are extracted, Dollar et al. [3] proposed to apply 2-D Gaussian kernels to the spatial space and 1-D Gabor filters to the temporal direction. Laptev et al. [16] proposed an extended Harris detector to extract Cuboids. As another method other than using Cuboids, extracting local features based on trajectories of interest points also showed good results for action recognition [2,8,15]. Matikainen et al. [17] proposed to extract trajectories using a standard KLT tracker, cluster the trajectories, and compute an affine transformation matrix for each cluster center.

In this paper, we propose to improve method of extracting ST feature proposed by Noguchi et al. [13]. Following [13], we also extract features based on moving SURF points and use Delaunay triangulation to model the spatial relationships between interest points. We address some problems of their method such as the inability to handle camera motion or holistic decision of motion thresholds for selecting points which cause failure in extracting features of some videos. We propose to solve these problems by simple yet efficient methods of motion compensation and point selection.

As treatment for camera motion, Cinbis et al. [18] applied video stabilization using homography-based motion compensation approach. They estimated camera flow by calculating the homography between consecutive frames and compensate optical flow of points by removing estimated camera flow. Similarly, Jain et al. [8] also removed camera motion from original optical flow, nevertheless they consider affine motion as camera motion. Wu et al. [19] decomposed Lagrangian particle trajectories into camera-induced and object-induced components for videos acquired by a moving camera. In. [2], Heng Wang et al. did not compensate camera motion in advance but employed motion boundary histograms which already have constant motion removed. We also reduce the influence of any existed camera motion by canceling the constant motion. Our proposed method of motion compensation improves significantly performance of feature extraction over the baseline since it helps not only extract features in case that camera motion exists but also detect more robust interest points.

3 Overview of the Spatio-temporal Proposed Feature

Our spatio-temporal feature is based on dense SURF keypoints with dominant and reliable movements. Here dominant and reliable points are supposed to belong to the human who is performing the action. Our feature investigates the shape of triangles produced by applying Delaunay triangulation on those informative points along with the movements (flow vectors) of points within the same Delaunay triple. We show that concatenating these features with SURF features of interest points can form a powerful spatio-temporal feature.

Our proposed feature is dense in both space and time. We extract features with temporal step size of N frames. We set the value of N to be small so that the extracted features are temporally dense. We operate tracking of interest points through N frames. As trajectories may drift from their precise locations during the tracking process, limiting the tracking process within short duration like this is supposed to be able to overcome this problem. In our experiments, we fix N as 5. The process of extracting our proposed spatio-temporal feature from each frame set is summarized as follows:

1. Extract dense SURF keypoints of the first frame using Dense SURF [20].
2. Compute optical flows from k^{th} frame ($k = 1, 2, ..., N - 1$) to the next frame ($k + 1^{th}$ frame) using LDOF [9].
3. Estimate camera motion in each frame and compensate motion if camera motion detected (Section 3.2).
4. Select points which are expected being more informative than the others (Section 3.3) and from triangles of selected points using Delaunay triangulation.
5. Extract ST features from each triangle based on its shape along with motion features of its points through the frame set (Section 3.4).

The main improvements of our method over our baseline can be summarized as follows: (1) treatment of camera motion, (2) selection of interest points and, (3) enhancement on descriptorization of ST features. We explain in details these improvements in following subsections.

4   The Detail of the Proposed Method

4.1   Detection and Compensation of Camera Motion

In Noguchi et al.'s work [13], once camera motion has been detected in a frame set, obtained information would be considered as noise, thus no points would be selected. Consequently, no features are extracted if the whole video contains camera motion. We propose to solve this problem by following a simple 2-step technique:

1. Step 1: Confirm the existence of camera motion based on optical flows of SURF keypoints. If detecting camera motion, determine the direction and magnitude of camera motion before going to the next step.
2. Step 2: Compensate motion by canceling camera motion from original flows of SURF keypoints.

Detection of Camera Motion: At the first step, we aim to find out at each frame how the camera move in both horizontal direction (forward or backward) and vertical direction (up or down). This step is based on our assumption that if most points move toward the same direction, camera motion exists. Let denote P^{x^+} and P^{x^-} as number of points with positive and negative optical flows respectively, $P_m^{x^+}$ and $P_m^{x^-}$ as number of moving points which shift forward and backward respectively, so that we suppose that camera is moving forward if Eq.1 and Eq.2 are satisfied or backward if Eq.3 and Eq.4 are satisfied:

$$P_m^{x^+} \geq kP^{x^+} \tag{1}$$

$$P_m^{x^+} > P_m^{x^-} \tag{2}$$

$$P_m^{x^-} \geq kP^{x^-} \tag{3}$$

$$P_m^{x^-} > P_m^{x^+} \tag{4}$$

Here, k is a fraction threshold representing minimal required proportion of moving points over all points with the same direction. In our experiments, we set k as $\frac{2}{3}$.

Fig. 1. An example that shows efficiency of proposed method of reducing camera motion and selecting interest points. The first row presents a frame set of consecutive frames which contains camera motion. In this case, camera is moving (to the right), thus interest points are not detected according to the baseline. The most left figure of the second row shows optical flows of extracted dense SURF keypoints before the camera motion compensated. The middle figure and the right figure of the second row respectively present points determined as moving points by the baseline (with fixed motion threshold) and our method (with flexible threshold). Point selection of the both are performed after compensating camera motion. This example shows that our method is not only able to reduce the effect of camera motion but also to select more representative interest points than the baseline.

A point is considered as moving points if its absolute optical flow is larger than or equal to 1.The camera is supposed as horizontally stable if none of above condition is satisfied. If the camera is detected as being moved, camera motion is calculated as average of absolute optical flows of points which moved to the same direction as camera. Camera motion for vertical direction is estimated in the similar manner.

Compensation of Camera Motion: Flow of each SURF keypoint is compensated simply as follows:

$$f_i = f_i - df_{camera} \tag{5}$$

Here, f_i refers to flow of point i, f_{camera} refers to camera flow. d equals 1 if camera moved to positive direction or -1 if camera moved to negative direction. f_{camera} is measured separately for all considered directions (forward, backward, up and down) and compensation is operated in each of those directions. By our manner, camera motion can be compensated in most cases except for zooming. Handling this case of camera motion is one of our future works. See Fig. 1 for an example result of our motion compensation method.

4.2   Selection of Interest Points

According to the baseline, selection of interest points is based on their optical flows between the first frame and the middle frame of the frame set. A point is believed as an interest point if its flow is larger than the pre-defined motion threshold. As a result, in case of no significant movement than the threshold from the first to the middle frame, no feature can be extracted. Moreover, their motion threshold is determined in holistic manner and fixed for every video of every type of actions. However, due to camera motion, video resolution, movements of background objects, and especially the diversity of actions as well as actors, points selected based on a constant motion threshold may not always be representative. For example, even though that significant movements are expected to be caused mainly by the actor, in the background there may be objects which move dominantly at several frames. Hence, the points belong to these objects may be imprecisely taken as interest points. In addition, magnitude of movement may vary largely from action to action. For instance, sport activities such as jumping trampoline or swimming are supposed to cause large displacements. On the other hand, daily activities such as drinking or talking in general generate smaller optical flows. We demonstrate that in order to overcome these problems, motion threshold should be flexible.

We propose to determine motion threshold flexibly and select as reliable moving points as possible. The idea is that robustness of a point should be compared to its fellows at the same time rather than to a fixed threshold. In our method, motion threshold is estimated for every frame in all directions based on flows of its SURF points. The following equation represents how we calculate motion threshold for a frame in forward direction (x^+). Thresholds for the remaining directions are similarly calculated.

$$thresh_{f_{x+}} = aver_{f_{x+}} + \alpha(max_{f_{x+}} - aver_{f_{x+}})$$ (6)

Here, $thresh_{f_{x+}}$ means the motion threshold for frame f in x^+ direction. $aver_{f_{x+}}$ and $max_{f_{x+}}$ respectively refer to the average and the maximal flow magnitude at frame f in x^+ direction. The qualification that a point should satisfy to be considered as a moving point is that in at least one of four considered directions, its flow magnitude is somewhat greater than the average flow of that direction. The constant α controls that qualification. In our experiments, we set α as 0.5. Thus, the motion threshold is near to the median of the average and the max flows. However, in some case, at some frames, all objects including actor stay still, thus it is not necessary that there always must be moving points. We suppose that nothing in a frame moved if all of its thresholds are smaller than 1.

After determining which points are moving points through the frame set, instead of simply taking all points which ever moved like Noguchi et al, we aim to select as many representative points as possible. We postulate a hypothesis that points with more movements are more reliable and informative. For example, through the whole frame set, points moved 2 times are expected to be more reliable as well as representative than points moved only once. Based on this hypothesis, we propose to select points greedily based on number of times they moved through the frame set. Our algorithm of point selection is described as follows:

(Algorithm for selecting interest points.)
M = maximal number of movements ($M \leq N - 1$)
T = total number of moving points
GS = group of selected points (initialized as empty)
for $i = M$ to 1 **do**
    $GS = |GS$, points moved i times $|$
    if $|GS| \geq \beta T$ **then**
        break;
    end if
end for

Following the above algorithm, the group of selected points is only a proportion of moving points but expected to consist of most representative points. In our experiments, we set β as $\frac{1}{2}$. Fig.1 shows the effectiveness of our method of selecting interest points over the baseline.

4.3 Descriptorization of ST Features

After selecting interest points, following the baseline, we apply Delaunay triangulation to form triples of them. One ST feature can be obtained from each triple. Our proposed feature extracted from a triple is constructed based on following descriptors. We classify them to *spatial descriptor* which represents static visual features of points, *temporal descriptor* which presents movements of points through the frame set and *spatio-temporal descriptor* which characterizes trajectory-based visual features of points or group of points. Below we describe in detail each descriptor.

Spatial Descriptors. To form spatial descriptor, we combine SURF descriptors of three points of the triple at the first frame. Each SURF is extracted with subregions of 3 by 3 pixels, Haar filters of 4 by 4 pixels and 4 subregions for SURF. Thus we obtain 64-dimension SURF descriptor for each point [20]. However, concatenating three SURF descriptors of three points forms a descriptor which is high-dimensional ($3 \times 64 = 192$)but may consist of repeated information. Thus, we apply PCA on this descriptor to acquire a lower dimensional but more representative one, as well as reduce computational cost. We denote this dimension reduced descriptor as PSURF. In our experiments, PSURF is a 96-dimension vector.

Temporal Descriptors. We propose to extract following 2 temporal features:

(1) **A histogram of Optical Flow (HOOF).** $3(N-1)$ flow vectors of 3 points are binned to B_o-bin histogram. Following [1], each flow vector is binned according to its primary angle from the horizontal axis and weighted according to its magnitude. That means, a flow vector $v = [x, y]$ with its angle $\theta = tan^{-1}(y\frac{x}{})$ in the range:

$$-\frac{\pi}{2} + \pi\frac{b-1}{B_o} \leq \theta < -\frac{\pi}{2} + \pi\frac{b}{B_o} \qquad (7)$$

will contribute by $\sqrt{x^2 + y^2}$ to the sum in bin b. Finally, the histogram is normalized to sum up to 1.

(2) **A Histogram of Direction of Flows (HDF).** Following Noguchi et al., we binned flow vectors of 3 points within the triple according to their direction. However, in [noguchi], one histogram is calculated for each point, thus many bins become zeros. This makes the motion descriptor have small effectiveness on action discrimination. Here we propose to bin all flow vectors ($3(N-1)$ flows) into 4 bins: $x^+(|f_{x+}| \geq 0), x^-(f_{x-} \geq 0), y^+(f_{y+} \geq 0), y^-(f_{y-} \geq 0)$. Similarly to HOOF, this histogram is also weighted by flow magnitude and normalized to sum up to 1.

Spatio-temporal Descriptors. We propose to generate the following 3 descriptors. The first two represent visual characteristics of triangles through the frame set. The last one descriptorizes the shape of trajectories. The last two are newly introduced by us. Refer to Fig.2 for illustration of these proposed two features.

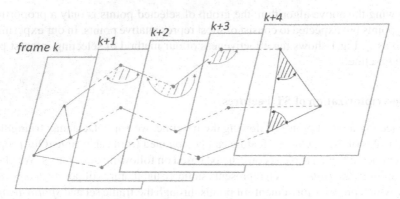

Fig. 2. Illustration of proposed spatio-temporal features. We additionally explore characteristics of interest points by exploiting angles of triangles formed by them (red ones) and angles shaped by consecutive trajectories of them (green ones). We show here an example of trajectories of grouped interest points in a frame set of 5 frames. 2×5 smallest angles of triangles are binned to obtain a HAT and 3×3 trajectory based angles are binned to obtain a HAF following proposed method described in Section 3.4.

(1) **Areas of Triangle (AT):** Following Noguchi et al.' work, the areas of the triangle at all frames are calculated then concatenated and normalized to form a N-dimension descriptor.

(2) **A Histogram of Angles of Triangle (HAT).** To better explore the shape characteristics of obtained triangles, we propose to investigate their angles by binning them based on their magnitude. Here, we consider only two angles since given the degrees of any two out of three angles, it is sufficient to characterize the shape of a triangle. Using two optional angles is not preferred here since they may be not representative for their triangle. Thus, one can consider use two largest or two smallest angles. However, two largest angles can range from 0° to 180° while two smallest angles range only from 0° to 90°. Hence, we select two smallest angles since binning them is expected to be more efficient and easier to define histogram bin. Moreover, it can not happen that both of two smallest angles are larger than 60°. Based on this observation, we set up histogram bin as follows: for $\theta > 60^o$, the histogram bin is of size 30, otherwise, the histogram bin is of size 15. In this manner, $2 \times N$ smallest angles are binned to 5 bins: [0-15], [15-30], [30-45], [45-60], [60-90]. Each angle is weighted by sum of magnitude of its two edges and normalized at the end.

A Histogram of Angles of Flows (HAF). To exploit trajectories of interest points for modelling the action, some work straightly employ them as descriptors [2]. However, this approach suffers from the problem that trajectories may vary largely due to the velocity of the actor. To reduce the effect of variety of velocity, we propose to extract features based on angles shaped by trajectories. These angles are supposed to be more informative than trajectories themselves(See Fig3). The angles are binned by the same method as shown in Equ. 7. Number of histogram bin for HAF is denoted as B_a.

Finally all above descriptors are concatenated to form our ST feature which has 96 (PSURF) + B_o (HOOF) + 4 (HDF) + N (AT) + 5 (HAT) + B_a (HAF) = 105+B_o+N+B_a

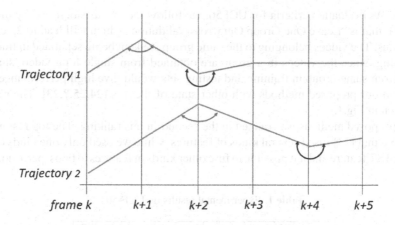

Fig. 3. An example that illustrates the effect of variety in velocity on action recognition and the efficiency of our proposed method. We show trajectories of points which belong to two actors performing the same action in 6 consecutive frames. We assume that the actors move in similar way but at different speed. As shown here, Trajectory 1 which corresponds to faster actor and Trajectory 2 which belongs to lower actor only match at first, thus trajectory based descriptors become nearly totally different. On the other hand, according to our method, exploiting angles shaped by trajectories help to find out more the similarity between these two trajectories. The similar angles (marked by same color) can be binned to the same bin, hence this angle based descriptor can be expected to reduce the effect of diversity in velocity.

dimension. In our experiment, we set $N = 5$, $B_o = 6$ and $B_a = 4$, thus we obtain a 120-dimension ST descriptor.

5   Experiments and Results

We made experiments on large-scale benchmark datasets for three large-scale action recognition datasets including UCF-50[1], HMDB51 [21] and UCF-101[2] to validate our proposed method. These datasets are not only large-scale but also challenging benchmarks for human action recognition in realistic and challenging settings such as large variations in camera motion, object appearance and pose, viewpoint and complicated background. The above three datasets consists of Web video clips collected from YouTube manually. Each of them has 50, 51 and 101 action categories, respectively.

To validate the enhancement of proposed feature over the baseline, we conduct experiments with ST feature proposed by the baseline and our proposed feature. We use Fisher Vector encoding as described in [14] to represent videos. We train multi-class SVMs [22] to perform multi-class action classification. For comparison, we made experiments with our proposed features coded by the conventional bag-of-visual-words (BoV) as well as the proposed features coded by Fisher Vector (FV).

---

[1] http://crcv.ucf.edu/data/UCF50.php
[2] http://crcv.ucf.edu/data/UCF101.php

UCF50. As evaluation criteria for UCF50, we follow method as suggested by the author [23], that is "Leave One Group Out Cross Validation" which will lead to 25 cross-validations. The videos belonging to the same group are kept being separated in training and testing, since the videos in a group are obtained from single long video, sharing videos from same group in training and testing sets would give high performance. We compared our proposed methods with other state-of-the-arts [24,25,?,23]. The results are shown in Fig.1.

Our proposed methods was ranked in the second in this ranking. The top result was [23]. Note that [23] fused several kinds of features, while we used only one kinds of the proposed ST feature and it it possible to fuse other kinds of features to boost performance.

Table 1. Experimental results on UCF-50

Method	Average Precision (AP)
HOG/HOF [24]	47.9 %
Our (BoV)	51.1%
Action Bank [25]	57.9 %
Motion Interchange [26]	68.51 %
Our (Fisher)	**71.6%**
Reddy and Shah [23]	76.9%

HMDB51. As for HMDB51, we use three distinct training and testing splits generated by its authors. For each action category, a set of 70 training and 30 testing clips was selected so that they fulfill the 70/30 balance for each meta tag, and clips from the same video were not used for both training and testing. HMDB is more diverse and difficult dataset than UCF-50 and UCF-101. As shown in Fig.2, the state-of-the-art are still less than 30%, while the top result exceeded 70% for UCF-50. As results, our proposed methods outperformed other state-of-the-arts.

Table 2. Experimental results on HMDB-51

Method	Average Precision (AP)
Our (BoV)	17.3%
HOG/HOF [24]	20.44 %
Action Bank [25]	26.9 %
Motion Interchange [26]	29.17%
Our (Fisher)	**29.5%**

UCF101. We follow evaluation set up as suggested in the action workshop page[3]. We adopt the provided three standard train/test splits to evaluate our results. In each split, clips from 7 of the 25 groups are used as test samples, and the rest for training.

---

[3] http://crcv.ucf.edu/ICCV13-Action-Workshop/

The result of each experiment reported here is calculated as the mean of average accuracies over the three provided test splits. We compare our results to the result reported in [27].

According to [27], their result is provided by baseline approach of action recognition [27] using standard BoV. The results are shown in Table 1. Note that [27] follows the old experimental set up, that is "Leave One Group Out Cross Validation".

Although UCF101 is the largest action dataset up to date, there are too few reported results for it. As demonstrated in [13], their method of extracting ST features is comparable to state-of-the-arts. By applying Fisher Vector encoding, their method obtains slightly better results. We propose to improve their method and significantly boost the overall precision as shown in Table 1. The results demonstrate that, our method could select more representative points as well as explore better the visual characteristics of them.

Table 3. Experimental results on UCF-101

Method	Average Precision (AP)
[13] (Fisher)	38.2%
Our (BoV)	40.1%
[27] (BoV)	44.5%
Our (Fisher)	**60.1%**

6 Conclusions

In this paper, we propose a novel method of extracting spatio-temporal features which is able to efficiently select interest points and descriptorize their features. The proposed methods consists of dense-sampling of SURF and Delaunay triangulation of selected interest points, the idea of which is based on our previous work [13].

The experimental results show significant improvement of our method over the baseline regarding the three up-to-date large-scale action recognition datasets: UCF-50, HMDB51, and UCF-101. In future work, we will introduce detailed analysis on human-object interaction for action recognition, and recent approach of compensating camera motion in order to handle more complicated cases such as zooming.

References

1. Chaudhry, R., Ravichandran, A., Hager, G., Vidal, R.: Histograms of oriented optical flow and binet-cauchy kernels on nonlinear dynamical systems for the recognition of human actions. In: Proc. of IEEE Computer Vision and Pattern Recognition, pp. 1932–1939 (2009)
2. Wang, H., Klaser, A., Schmid, C., Liu, C.-L.: Dense trajectories and motion boundary descriptors for action recognition. International Journal of Computer Vision 103(1), 60–79 (2013)
3. Dollar, P., Cottrell, G., Belongie, S.: Behavior recognition via sparse spatio-temporal features. In: Proc. of Surveillance and Performance Evaluation of Tracking and Surveillance, pp. 65–72 (2005)

4. Laptev, I., Marszalek, M., Schmid, C., Rozenfeld, B.: Learning realistic human actions from movies. In: Proc. of IEEE Computer Vision and Pattern Recognition (2008)
5. Scovanner, P., Ali, S., Shah, M.: A 3-dimensional sift descriptor and its application to action recognition. In: Proc. of ACM International Conference Multimedia, pp. 357–360 (2007)
6. Kläser, A., Marszałek, M., Schmid, C.: A spatio-temporal descriptor based on 3d-gradients. In: Proc. of British Machine Vision Conference, pp. 995–1004 (2008)
7. Yeffet, L., Wolf, L.: Local trinary patterns for human action recognition. In: Proc. of IEEE International Conference on Computer Vision, pp. 492–497 (2009)
8. Jain, M., Jegou, H., Bouthemy, P.: Better exploiting motion for better action recognition. In: Proc. of IEEE Computer Vision and Pattern Recognition (2013)
9. Brox, T., Bregler, C., Malik, J.: Large displacement optical flow. In: Proc. of IEEE Computer Vision and Pattern Recognition, pp. 41–48 (2009)
10. Jensen, F.V., Christensen, H.I., Nielsen, J.: Bayesian methods for interpretation and control in multi-agent vision systems. In: Proc. of SPIE 1708, Applications of Artificial Intelligence X: Machine Vision and Robotics, pp. 536–548 (1994)
11. Nowak, E., Jurie, F., Triggs, B.: Sampling strategies for bag-of-features image classification. In: Leonardis, A., Bischof, H., Pinz, A. (eds.) ECCV 2006. LNCS, vol. 3954, pp. 490–503. Springer, Heidelberg (2006)
12. Willems, G., Tuytelaars, T., Van Gool, L.: An efficient dense and scale-invariant spatio-temporal interest point detector. In: Forsyth, D., Torr, P., Zisserman, A. (eds.) ECCV 2008, Part II. LNCS, vol. 5303, pp. 650–663. Springer, Heidelberg (2008)
13. Noguchi, A., Yanai, K.: A surf-based spatio-temporal feature for feature-fusion-based action recognition. In: Kutulakos, K.N. (ed.) ECCV 2010 Workshops, Part I. LNCS, vol. 6553, pp. 153–167. Springer, Heidelberg (2012)
14. Perronnin, F., Dance, C.: Fisher kernels on visual vocabularies for image categorization. In: Proc. of IEEE Computer Vision and Pattern Recognition, pp. 1–8 (2007)
15. Atmosukarto, I., Ghanem, B., Ahuja, N.: Trajectory-based fisher kernel representation for action recognition in videos. In: Proc. of IAPR International Conference on Pattern Recognition, pp. 3333–3336 (2012)
16. Laptev, I., Lindeberg, T.: Local descriptors for spatio-temporal recognition. In: Proc. of IEEE International Conference on Computer Vision (2003)
17. Matikainen, P., Hebert, M., Sukthankar, R.: Trajectons: Action recognition through the motion analysis of tracked features. In: ICCV Workshop on Video-Oriented Object and Event Classification (2009)
18. Ikizler-Cinbis, N., Sclaroff, S.: Object, scene and actions: Combining multiple features for human action recognition. In: Daniilidis, K., Maragos, P., Paragios, N. (eds.) ECCV 2010, Part I. LNCS, vol. 6311, pp. 494–507. Springer, Heidelberg (2010)
19. Shandong, W., Omar, O., Mubarak, S.: Action recognition in videos acquired by a moving camera using motion decomposition of lagrangian particle trajectories. In: Proc. of IEEE International Conference on Computer Vision, pp. 1419–1426 (2011)
20. Uijlings, J.R.R., Smeulders, A.W.M., Scha, R.J.H.: Real-time visual concept classification. IEEE Transactions on Multimedia (2010)
21. Jhuang, H.A., Garrote, H.A., Poggio, E.A., Serre, T.A., Hmdb, T.: A large video database for human motion recognition. In: Proc. of IEEE International Conference on Computer Vision (2011)
22. Tsochantaridis, I., Joachims, T., Hofmann, T., Altun, Y.: Large margin methods for structured and interdependent output variables. The Journal of Machine Learning Research 6, 1453–1484 (2005)
23. Reddy, K.K., Shar, M.: Recognizing 50 human action categories of web videos. Machine Vision and Applications 24, 971–981

24. Laptev, I., Marszalek, A., Schmid, C., Rozenfeld, B.: Learning realistic human actions from movies. In: Proc. of IEEE Computer Vision and Pattern Recognition (2008)
25. Sadanand, S., Corso, J.J.: Action bank: A high-level representation of activity in video. In: Proc. of IEEE Computer Vision and Pattern Recognition (2012)
26. Kliper-Gross, O., Gurovich, Y., Hassner, T., Wolf, L.: Motion interchange patterns for action recognition in unconstrained videos. In: Fitzgibbon, A., Lazebnik, S., Perona, P., Sato, Y., Schmid, C. (eds.) ECCV 2012, Part VI. LNCS, vol. 7577, pp. 256–269. Springer, Heidelberg (2012)
27. Khurram, S., Amir, R.Z., Shar, M.: UCF101: A dataset of 101 human actions classes from videos in the wild. CoRR, abs/1212.0402 (2012)

Resource Constrained Multimedia Event Detection

Zhen-Zhong Lan, Yi Yang, Nicolas Ballas, Shoou-I Yu,
and Alexander Haputmann

School of Computer Science, Carnegie Mellon University, Pittsburgh, PA 15213, USA
{lanzhzh,yiyang,iyu,alex}@cs.cmu.edu, nicolas.ballas@cea.fr

Abstract. We present a study comparing the cost and efficiency trade-offs of multiple features for multimedia event detection. Low-level as well as semantic features are a critical part of contemporary multimedia and computer vision research. Arguably, combinations of multiple feature sets have been a major reason for recent progress in the field, not just as a low dimensional representations of multimedia data, but also as a means to semantically summarize images and videos. However, their efficacy for complex event recognition in unconstrained videos on standardized datasets has not been systematically studied. In this paper, we evaluate the accuracy and contribution of more than 10 multi-modality features, including semantic and low-level video representations, using two newly released NIST TRECVID Multimedia Event Detection (MED) open source datasets, i.e. MEDTEST and KINDREDTEST, which contain more than 1000 hours of videos. Contrasting multiple performance metrics, such as average precision, probability of missed detection and minimum normalized detection cost, we propose a framework to balance the trade-off between accuracy and computational cost. This study provides an empirical foundation for selecting feature sets that are capable of dealing with large-scale data with limited computational resources and are likely to produce superior multimedia event detection accuracy. This framework also applies to other resource limited multimedia analyses such as selecting/fusing multiple classifiers and different representations of each feature set.

Keywords: Multimedia Event Detection, Limited Resource, Feature Selection.

1 Introduction

Multimedia data have proliferated in the past few years, ranging from ever-growing personal video collections to films and professional documentary archives. Numerous tools and applications have been invented to describe, organize, and manage video data. Previous research mainly focuses on recognizing scene, object and action, which are building blocks of events and defined as atomic concepts in this paper. However, these atomic concepts are too primitive to be used by users to

C. Gurrin et al. (Eds.): MMM 2014, Part I, LNCS 8325, pp. 388–399, 2014.
© Springer International Publishing Switzerland 2014

search videos from data collections. When searching through online video communities such as YouTube, Hulu etc., people tend to use event descriptions such as "birthday party", "performing a board trick" or "mountain climbing" instead of simple scene, object or action words such as "indoor", "cake" or "walk". In this paper, we define an event as a combination of various actions, scenes and objects, which is more descriptive and meaningful.

The TRECVID MED evaluation [16], which is hosted by National Institute of Standards and Technology (NIST) and part of the TRECVID evaluation, is aimed at addressing the above problems by assembling state-of-the-art technologies into a system that can quickly and accurately search a multimedia collection for user-defined events. Since 2010, NIST has collected one of the largest and most challenging labelled video datasets, which contains a total of 144049 video clips. These videos contain more than 30 events such as 'making a sandwich', 'parkour' and 'parade', which are illustrated in Fig. 1. Participants from various organizations have made significant progress on MED in terms of accuracy. However, most of the progress researchers have made comes from adding more and more features into their MED systems. While promising results can be achieved on such systems, they are too expensive to be deployed in real-world applications with large-scale data.

Another problem of TRECVID MED that has been discussed among participants and organizers for a long time is that NIST did not provide a validation set with labels, so researchers from around the world publish MED related papers with their own splits of TRECVID MED into testing and training sets. Because of these independent splits, comparing different research groups' results becomes very hard. To deal with these difficulties, NIST recently released two standard validation datasets, namely MEDTEST and KINDREDTEST. It is important to have some baseline results on these two datasets that can be compared by researchers from all over the world.

Fig. 1. Example Key-frame for Event in MED

This paper attempts to address the above issues by thoroughly evaluating more than 10 multimedia features' performances and their contributions on the MEDTEST and KINDREDTEST datasets. Relying on this evaluation, we also propose a framework to select a subset of features to make a trade-off between accuracy and computational cost.

The remaining sections are organized as follows. We discuss related work in Section 2 and we elaborate our MED system including features, feature representations and evaluation metrics section 3. In Section 4, we discuss experimental results. Finally, we summarize our paper in Section 5.

2   Related Work

Compared with action recognition, recognizing a "complex event" is a new topic that has been introduced to take multimedia analysis to its next level of difficulty. Previous work [14] [17] [6] [13] on video event recognition can be divided in two main categories whether they rely on low-level features or high-level semantic concepts. Yang et al. [20] and Tamrakar et al. [17] proposed to evaluate the individual performance of different low-level visual features (SIFT, STIP, Trajectories...) as well as their combination. Meler et al. [14], Ebadollahi et al. [6] and Liu et al. [13] focus on testing high-level features performance on event recognition. In contrast to those works, this paper evaluates the performance of multi-modal features extracted from different streams associated with the multimedia data (image, audio, text), including both low-level and high-level features, leading to a more complete description. Moreover, most previous work only evaluates each feature's single performance. In contrast, we focus on each feature's contribution to the combined system. We show that the single feature performance, although important, does not necessarily reflect its contribution to the overall performance.

In terms of efficiency, most previous work focuses on improving one component of a classification or recognition system with faster algorithms. For example, Bay et al. [3] introduced SURF as a faster alternative of SIFT. Moosmann et al. [15] proposed to use random forest to replace Support Vector Machine (SVM). Jiang [7] conducted an interesting study to evaluate and combine a number of speed-up strategies to get a fast event recognition system. Different from previous work, we offer a resource constrained solution that can be customized by users who have different needs and resources.

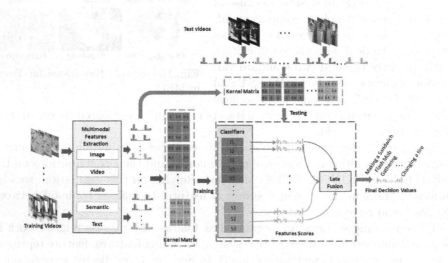

Fig. 2. MED system illustration

3 MED System

Fig. 2 shows a simplified version of our MED system which is used for this paper. Given a set of training and testing videos, we first extract features in different modalities from the videos and then train a χ^2 SVM classifier for each feature. A simple average late fusion is used to combine the prediction results each feature. More complex fusion methods may lead to better performance, but this is beyond the scope of this paper: providing a baseline for NIST's two newly released MED datasets and illustrating a framework for designing resource constrained MED system.

3.1 Features

To build a good MED system, it is important to have features that capture various aspects of an event. In our MED system, we explore five different feature modalities which are computed from different sources. Image features capturing appearance information are computed from key-frames. Video features are extracted from videos directly and collect motion information. Audio features characterizes acoustic information. Text features and semantic features can borrow domain knowledge from other datasets such as Flickr and give semantically meaningful representations for events.

Image Features: We use three image features that are computed from the keyframes extracted as described in [10]. The three images feature are SIFT, Color SIFT (CSIFT) and Transformed Color Histogram (TCH) [18].

After detecting key points using harris-laplace key point detectors from key frames, we use three different feature descriptors to generate SIFT, CSIFT and TCH features, which hopefully are complementary. From the key points descriptors, a k-means algorithm generates a codebook which has 4096 words for each feature. Next, a soft-mapping strategy, in which we choose the ten nearest clusters and assign a rank weight ($\frac{1}{rank}$) to them, maps key points into the codebook. Spatial pyramid matching as described in [9] compensates for spatial information lost in the bag-of-words representation. We then aggregate the image representation into video representations by averaging all the image representations in one video and normalize the video representation using an L2 normalization.

Video Features: We have three visual video features, namely Dense Trajectory (Traj) [19], MoSIFT [4] and STIP [11], which are computed directly from videos. Traj is a feature that uses dense optical flow to track feature points up to 15 frames to get trajectories, which are described by Histogram of Oriented Gradients (HOG) [5], Histogram of Optical Flow (HOF) and Motion Boundary Histogram (MBH) [19]. By computing MBH along the dense trajectories, Traj has an efficient solution to compensate for camera motion. MoSIFT, as a three dimensional extension of SIFT features, uses a Difference of Gaussian (DoG) based detector and is represented by a descriptor combining SIFT and HOF. STIP uses 3D Harris corner detectors and its interest points are represented as the combination of HOG and HOF. After getting the key point descriptors, the same bag-of words and spatial pyramid matching as with image features is adopted to cast the key point representation into a video-level representation.

Audio Features: Audio features are another important resource to detect events in videos. To represent general audio information, we use the Mel-frequency cepstral coefficients (MFCCs) feature, which is very popular in speech recognition systems. We compute 20 dimensional MFCCs for every 10ms over a 32ms sliding window. Given the raw features, we compute a 4096 word codebook and aggregate all MFCC features from one video into a 4096 dimensional bag-of-words representation. In addition to MFCC, we also use have Automatic Speech Recognition (ASR) features as described in [2] to capture semantic information in audio.

Semantic Features: In our MED system, three semantic features are used. The first one called SIN346 is defined by the TRECVID Semantic Indexing (SIN) track. This feature has 346 dimensions representing the 346 concepts in SIN [2]. The second one is the Object Bank feature (ObjBank) introduced by Li et.al. [12], in which we extended the original 176 objects to 1000 objects by using the Imagenet challenge 2012 dataset (ILSVRC2012) [8]. Another semantic feature that is also trained on the ILSVRC2012 dataset is the Deep Convolutional Neural Network feature (DCNN), in which we trained a Deep Convolutional Neural Network feature using the method introduced by Krizhevsky et al. [8] on a NVIDIA Tesla K20m GPU.

Text Features: Following Bao et al. [2], we also use Optical Character Recognition (OCR) features to represent the text feature. We use a commercial OCR system is used to recognize the text. As OCR rarely gets a complete word correct, we treat each trigram of characters as a token instead of each whole word as a token.

3.2   Evaluation Metrics

For performance comparison, three evaluation metrics are used: the first one is the Minimal Normalized Detection Cost(MinNDC) as indicated in Formula 1. It is an evaluation criteria for NIST to evaluate MED 2010 and MED 2011. Lower MinNDC indicates better performance.

$$NDC(S, E) = \frac{C_{MD} * P_{MD} * P_T + C_{FA} * P_{FA} * (1 - P_T)}{MINIMUM(C_{MD} * P_T, C_{MD} * (1 - P_T))} \tag{1}$$

where P_{MD} is the miss detection probability while P_{FA} is the false positive rate. $C_{MD} = 80$ is the cost for miss detection, $C_{FA} = 1$ is the cost for false alarm and $P_T = 0.001$ is a constant defining the priori rate of event instances.

Another metric that is related to MinNDC is $P_{MD}@TER = 12.5$, in which $TER = \frac{P_{MD}}{P_{FA}}$. Because MinNDC and $P_{MD}@TER = 12.5$ are only used for NIST evaluation and do not consider the ranking information or the detection result, we will also use mean average precision (MAP) as our evaluation criterion and use it to rank the performance of features because it is better at reflecting features' value due to its ranking sensitive characteristics.

4 Experiments

4.1 Data

We evaluate our system on two standard MED datasets, i.e., MEDTEST set and KINDREDTEST set, which both contain the same 20 events. The events and their ids are listed in Table 1. MEDTEST contains a total of 34051 video clips including 9094 training videos and 24957 testing videos. KINDREDTEST has the same training set but a testing set that only contains 12388 video clips.

Table 1. MED12 event ID and name

E06: Birthday Party	E21: Attempting a bike trick
E07: Changing a vehicle tire	E22: Cleaning an appliance
E08: Flash mob gathering	E23: Dog show
E09: Getting a vehicle unstuck	E24: Giving directions to a location
E10: Grooming an animal	E25: Marriage proposal
E11: Making a sandwich	E26: Renovating a home
E12: Parade	E27: Rock climbing
E13: Parkour	E28: Town hall meeting
E14: Repairing an appliance	E29: Winning a race without a vehicle
E15: Working on a sewing project	E30: Working on a metal crafts project

4.2 Computing Environment

For extracting features, we use the PSC blacklight [1] machine, which is an SGI UV 1000cc-NUMA shared-memory system comprising 256 blades. Each blade holds 2 Intel Xeon X7560 (Nehalem) eight-core 2.27 GHz processors. Compared to feature extraction, the classification and fusion time is minimal, so we will only take the feature extraction time into consideration in this paper.

4.3 Single Feature Performance and Contribution

Following the pipeline in Fig. 2, we study both single and combined features' performance using the three evaluation metrics described in Section 3.2.

Fig. 3 shows the single feature accuracies on our MEDTEST and KINDREDTEST sets. We order the features according to their MAP. From Fig. 3, we can see that although the rank varies for different metrics and datasets, in general, they are consistent with each other. Specifically, the top two features that significantly outperform other features are DCNN and Traj; the two features that are worse than other features are ASR and OCR; others have very similar performances and the rank order changes are due to minor performance difference. It is interesting to see that DCNN, a high-level feature, can significantly outperform low-level features. Also, our OCR has higher recognition accuracy than ObjBank and SIN, yet its overall performance is the worst among all eleven features. The reason is that these videos do not contain enough text to recognize.

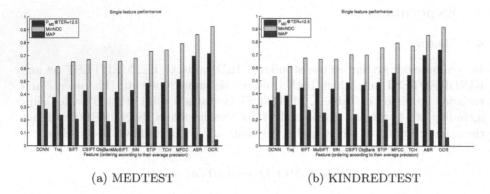

(a) MEDTEST   (b) KINDREDTEST

Fig. 3. Single feature accuracy for both datasets, ranked according to MAP. Lower score corresponds to better performance for MinNDC and $P_{MD}@TER = 12.5$, but higher is better for MAP.

The big difference between visual and audio features shows that in unconstrained videos visual information is more distinctive than audio information.

To determine the contribution of each feature, we first calculate the performance by combining all features, then we remove one feature from the set and recalculate the performance. Fig. 4 shows the performance drop (leave-one-feature-out accuracy) from removing each feature. This drop shows the importance of each feature to the overall combined system. The ranking by performance drop is quite different than the ranking of single feature performance. These two rankings are statistically uncorrelated. For example, MFCC has a very poor ranking as a single feature accuracy but is the highest ranking for leave-one-feature-out performance. This indicates that MFCC is orthogonal to the other features. While SIFT and CSIFT, align with most other features, they combine to reduce MAP because they reduce the overall weight per feature while not contributing additional information in the average late fusion method. More sophisticated fusion methods such as fusion by learning combination weights may be able to avoid this problem but will inevitably give smaller weights to those redundant features. Fig. 5 shows the Spearman rank coefficients for all of the features: it indicates MFCC and ASR are very different from the other features. Although the Spearman rank coefficients also show OCR is very different from the other features, its close to random individual performance indicates its negligible role in the system. Fig. 4 demonstrates that single feature accuracy alone does not indicate suitability for inclusion in the combined feature set. However, as long as the leave-one-feature-out accuracy is not negative, inclusion will increase the overall score. Unfortunately, including all features with a positive value in Fig. 4 will lead to a computationally expensive system that is generally unsuitable for a real world applications.

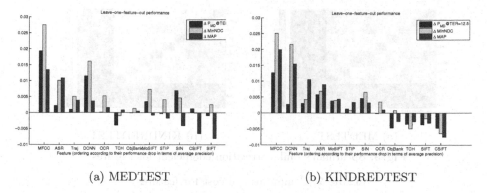

(a) MEDTEST   (b) KINDREDTEST

Fig. 4. Leave-one-out Accuracy for MED, Ranked According to ΔMAP. In all three metrics, higher values means higher *performance drop* when we leave the feature out, hence a higher contribution of the feature to the combined system.

4.4   Performance versus Cost Trade-off

In order to determine the performance versus cost trade-off, we first determine each feature's computational cost as shown in Table 2, which also shows the abbreviation of features for late usage in Table 3 through 6. For each feature, the time is the number of hours to process one hour of video. Let's assume our goal is to process one hour of video in one hour. We then determine, for a given number of CPUs, what the best possible performance is by a brute-force search across all features in Table 2. Fig. 6 shows the best possible performance for the given number of CPUs for all three metrics without using a GPU, thus excluding the DCNN feature. Tables 3 and 4 show the optimal feature sets for the given number of CPUs. Fig. 7 shows the best possible performance for the given number of CPUs for all three metrics using a GPU, including the DCNN feature. Tables 5 and 6 give the optimal feature sets for the given number of CPUs. As we can see from Tables 3 to 6, the MFCC feature appears in almost all configurations due to its low computational cost (Table 2) and relatively high contribution (Fig. 4). Although Traj has a high contribution, it does not show up in Tables 3 to 6 until we have a minimum of 16 CPUs due to its high computational cost. We can see from the tables that the optimal feature sets are very similar for the MEDTEST and the KINDREDTEST, which demonstrates that it is possible to select the optimal feature set from a smaller dataset like KINDREDEST and apply it to a larger dataset like MEDTEST. Likewise, these optimal feature sets are fairly similar across the three metrics. Further, we can also see from the figures that we get a diminishing return beyond 32 CPUs. In all cases, we can get more than 92 percent of the best performance by just using 32 cores.

Comparing Fig. 4 and Tables 5 and 6, we can see that in listing the importance of features, where importance in the table is measured by the ratio of the number of times the feature occurs to the number of possible occurrence given

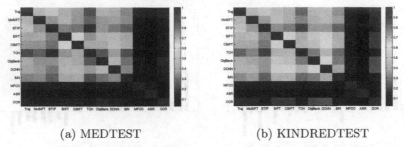

(a) MEDTEST   (b) KINDREDTEST

Fig. 5. Spearman's rank correlation coefficient for features

Table 2. Computational cost for features

Features (Abbrev.)	core hours	features (Abbrev.)	core hours
Traj(Tr)	12.38	Objbank(Ob)	28.43
MoSIFT(Mo)	11.23	DCNN(DC)	0.15 GPU
STIP(ST)	10.33	SIN(SIN)	78.92
SIFT(SI)	3.57	MFCC(MF)	1.36
CSIFT(CS)	5.05	ASR(AS)	4.99
TCH(TC)	2.12	OCR(OCR)	1.34

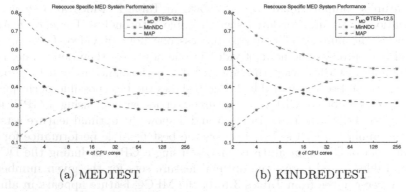

(a) MEDTEST   (b) KINDREDTEST

Fig. 6. Resource specific performance for MED (without DCNN)

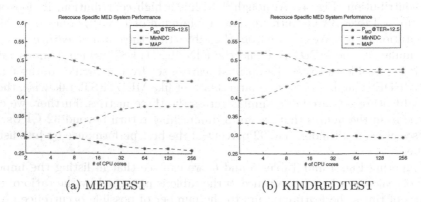

(a) MEDTEST   (b) KINDREDTEST

Fig. 7. Resource specific performance for MED (with DCNN on a GPU)

timing constraints, leave-one-feature-out performance is consistent with brute-force search results, hence very predictive in selecting the right feature set. For example, MFCC, DCNN, Traj and ASR, as the top 4 contributing features appear in almost all of the configurations as long as we have enough computational resources in terms of MAP. For other metrics, we have the same basic observation. The cost of computing the leave-one-feature-out accuracy is relatively inexpensive for late fusion, as all the components are already computed. In our system with 12 features, leave-one-feature-out accuracy computation is about 300 times faster than brute-force search.

Table 3. Resource specific feature sets for MEDTEST

CPUs	Optimal Sets in Real-time Performance		
	MinNDC	$P_{MD}@TER = 12.5$	MAP
2	MF	MF	MF
4	TC MF	SI	TC MF
8	TC SI MF	TC SI MF	TC SI MF
16	Tr TC MF	CS SI AS MF	Tr TC MF
32	Mo TC CS SI AS MF	Mo TC CS SI OCR AS MF	Tr TC SI OCR AS MF
64	Ob ST Mo TC CS AS MF	Ob Mo Tr SI OCR AS MF	Ob Mo Tr TC OCR AS MF
128	Ob ST Mo TC CS OCR AS MF	Ob Mo Tr CS SI OCR AS MF	Ob Mo Tr TC OCR AS MF
256	SIN Ob Mo Tr CS SI OCR AS MF	SIN Ob Mo Tr SI OCR AS MF	SIN Ob Mo Tr TC OCR AS MF

Table 4. Resource specific feature sets for KINDREDTEST

CPUs	Optimal Sets in Real-time Performance		
	MinNDC	$P_{MD}@TER = 12.5$	MAP
2	MF	MF	MF
4	SI	SI	SI
8	TC SI MF	TC SI MF	TC SI MF
16	Tr MF	Tr TC MF	Tr TC MF
32	ST Mo SI AS MF	Tr CS SI OCR AS MF	Tr TC SI AS MF
64	ST Mo Tr SI OCR AS MF	Ob Mo Tr SI OCR AS MF	ST Mo Tr TC SI OCR AS MF
128	SIN ST Mo Tr SI OCR AS MF	SIN ST Mo Tr SI OCR AS MF	SIN Mo Tr TC OCR AS MF
256	SIN Ob ST Mo Tr CS SI OCR AS MF	SIN Ob ST Mo Tr SI OCR AS MF	SIN Mo Tr TC OCR AS MF

Table 5. Resource specific feature sets for MEDTEST (with 1 additional GPU)

CPUs	Optimal Sets in Real-time Performance		
	MinNDC	$P_{MD}@TER = 12.5$	MAP
2	DC MF	DC MF	DC MF
4	DC MF	DC MF	DC TC MF
8	DC SI MF	DC SI MF	DC SI MF
16	DC Tr MF	DC TC SI OCR AS MF	DC Tr TC MF
32	DC Tr SI AS MF	DC Mo TC SI OCR AS MF	DC Tr CS OCR AS MF
64	DC Ob ST Mo TC SI AS MF	DC ST Mo Tr TC CS SI OCR AS MF	DC Ob Mo Tr TC OCR AS MF
128	SIN DC ST Mo TC SI OCR AS MF	SIN DC ST Mo SI OCR AS MF	DC Ob Mo Tr TC OCR AS MF
256	SIN DC Ob Mo Tr AS MF	SIN DC ST Mo SI OCR AS MF	DC Ob Mo Tr TC OCR AS MF

Table 6. Resource specific feature sets for KINDREDTEST(with 1 additional GPU)

CPUs	Optimal Sets in Real-time Performance		
	MinNDC	$P_{MD}@TER = 12.5$	MAP
2	DC MF	DC MF	DC MF
4	DC MF	DC MF	DC TC MF
8	DC AS MF	DC AS MF	DC SI MF
16	DC Tr MF	DC Tr MF	DC Tr MF
32	DC Tr AS MF	DC Mo Tr OCR AS MF	DC Tr SI AS MF
64	DC Tr AS MF	DC Mo Tr OCR AS MF	DC Mo Tr SI AS MF
128	SIN DC ST Mo Tr SI OCR AS MF	SIN DC Mo Tr OCR AS MF	SIN DC Mo Tr TC OCR AS MF
256	SIN DC ST Mo Tr SI OCR AS MF	SIN DC Mo Tr OCR AS MF	SIN DC Mo Tr TC OCR AS MF

5   Conclusion

In this paper, we systematically evaluated the performance and contributions of more than 10 multi-modality features for complex event detection on unconstrained videos over two newly released TRECVID MED datasets: MEDTEST and KINDREDTEST. These results can serve as a baseline for the community.

Based on the evaluation and computational cost of feature extraction, we propose a resource constrained video analysis framework that can meet different users' needs. More specifically, we select feature sets that have optimal real-time performance under various resource constraints by measuring leave-one-feature-out performance and brute-force search performance.

A particularly important insight from above experiments is that leave-one-feature-out feature performance is very predictive in selecting the right feature set.

We also found that in both datasets and across the three different metrics:

- DCNN and Trajectory features are very useful features in unconstrained video analysis. Especially DCNN, given its semantic and high accuracy characteristics, is a feature that is worth paying a lot of attention to.
- Even a less accurate feature such as MFCC, if it is cheap and complementary to other features, can be very useful.
- By selecting the right features, we can save a large amount of computational cost with a minimum accuracy drop. For example, in our experiments, by reducing the computational cost of 87 percent we still achieve 92 percent of optimal performance.

Acknowledgements. This work was partially supported by Intelligence Advanced Research Projects Activity (IARPA) via Department of Interior National Business Center contract number D11PC20068. The U.S. Government is authorized to reproduce and distribute reprints for Governmental purposes notwithstanding any copyright annotation thereon. Disclaimer: The views and conclusions contained herein are those of the authors and should not be interpreted as necessarily representing the official policies or endorsements, either expressed or implied, of IARPA, DoI/NBC, or the U.S. Government. This material is based in part upon work supported by the National Science Foundation under Grant No. IIS-1251187.

References

1. http://www.psc.edu/index.php/computing-resources/blacklight
2. Bao, L., Yu, S.-I., Lan, Z.Z., Overwijk, A., Jin, Q., Langner, B., Garbus, M., Burger, S., Metze, F., Hauptmann, A.: Informedia@ trecvid 2011. In: TRECVID 2011 (2011)
3. Bay, H., Tuytelaars, T., Van Gool, L.: Surf: Speeded up robust features. In: Leonardis, A., Bischof, H., Pinz, A. (eds.) ECCV 2006, Part I. LNCS, vol. 3951, pp. 404–417. Springer, Heidelberg (2006)
4. Chen, M.-Y., Hauptmann, A.: Mosift: Recognizing human actions in surveillance videos. CMU-CS-09-161 (2009)
5. Dalal, N., Triggs, B.: Histograms of oriented gradients for human detection. In: CVPR, vol. 1, pp. 886–893. IEEE (2005)
6. Ebadollahi, S., Chang, S.-F., Xie, L., Smith John, R.: Visual event detection using multi-dimensional concept semantics. In: ICME, pp. 881–884 (2006)
7. Jiang, Y.-G.: Super: Towards real-time event recognition in internet videos. In: ICMR, p. 7. ACM (2012)
8. Krizhevsky, A., Sutskever, I., Hinton, G.: Imagenet classification with deep convolutional neural networks. In: NIPS, pp. 1106–1114 (2012)
9. Lan, Z.-z., Bao, L., Yu, S.-I., Liu, W., Hauptmann, A.G.: Double fusion for multimedia event detection. In: Schoeffmann, K., Merialdo, B., Hauptmann, A.G., Ngo, C.-W., Andreopoulos, Y., Breiteneder, C. (eds.) MMM 2012. LNCS, vol. 7131, pp. 173–185. Springer, Heidelberg (2012)
10. Lan, Z.-Z., Bao, L., Yu, S.-I., Liu, W., Hauptmann, A.G.: Multimedia classification and event detection using double fusion. Multimedia Tools and Applications, 1–15 (2013)
11. Laptev, I.: On space-time interest points. IJCV 64(2-3), 107–123 (2005)
12. Li, L.-J., Su, H., Fei-Fei, L., Xing, E.P.: Object bank: A high-level image representation for scene classification & semantic feature sparsification. In: NIPS, pp. 1378–1386 (2010)
13. Liu, J., Yu, Q., Javed, O., Ali, S., Tamrakar, A., Divakaran, A., Cheng, H., Sawhney, H.S.: Video event recognition using concept attributes. In: WACV, pp. 339–346 (2013)
14. Merler, M., Member, S., Huang, B., Xie, L., Hua, G.: Semantic Model vectors for complex video event recognition. IEEE Trans. on Multimedia 14(1), 88–101 (2012)
15. Moosmann, F., Nowak, E., Jurie, F.: Randomized clustering forests for image classification. PAMI 30(9), 1632–1646 (2008)
16. Over, P., Awad, G., Michel, M., Fiscus, J., Sanders, G., Shaw, B., Kraaij, W., Smeaton, A.F., Quéenot, G.: Trecvid 2012 – an overview of the goals, tasks, data, evaluation mechanisms and metrics. In: TRECVID. NIST, USA (2012)
17. Tamrakar, A., Ali, S., Yu, Q., Liu, J., Javed, O., Divakaran, A., Cheng, H., Sawhney, H., International Sarnoff, S.R.I.: Evaluation of low-level features and their combinations for complex event detection in open source videos. In: CVPR, pp. 3681–3688 (2012)
18. Van De Sande, K.E.A., Gevers, T., Cees, G.M.S.: Evaluating color descriptors for object and scene recognition. PAMI 32(9), 1582–1596 (2010)
19. Wang, H., Klaser, A., Schmid, C., Liu, C.-L.: Action recognition by dense trajectories. In: CVPR, pp. 3169–3176. IEEE (2011)
20. Yang, J., Jiang, Y.-G., Hauptmann, A.G., Ngo, C.-W.: Evaluating bag-of-visual-words representations in scene classification. In: Workshop on ICMR, pp. 197–206. ACM (2007)

Random Matrix Ensembles of Time Correlation Matrices to Analyze Visual Lifelogs

Na Li, Martin Crane, Heather J. Ruskin, and Cathal Gurrin

School of Computing, Dublin City University, Ireland
na.li@dcu.ie, {mcrane,hruskin,cgurrin}@computing.dcu.ie

Abstract. Visual lifelogging is the process of automatically recording images and other sensor data for the purpose of aiding memory recall. Such lifelogs are usually created using wearable cameras. Given the vast amount of images that are maintained in a visual lifelog, it is a significant challenge for users to deconstruct a sizeable collection of images into meaningful events. In this paper, random matrix theory (RMT) is applied to a cross-correlation matrix C, constructed using SenseCam lifelog data streams to identify such events. The analysis reveals a number of eigenvalues that deviate from the spectrum suggested by RMT. The components of the deviating eigenvectors are found to correspond to "distinct significant events" in the visual lifelogs. Finally, the cross-correlation matrix C is cleaned by separating the noisy part from the non-noisy part. Overall, the RMT technique is shown to be useful to detect major events in SenseCam images.

Keywords: Random Matrix Theory, Cross-correlation Matrix, Eigenvalues and Eigenvectors, SenseCam.

1 Introduction

Lifelogging is the process of automatically recording aspects of one's life in digital form. This includes visual lifelogging using wearable cameras such as the SenseCam [1] and, more recently Google Glass [2]. The SenseCam, developed by Microsoft Research in Cambridge, UK, is a small wearable device that is worn via a lanyard suspended around the neck. The SenseCam incorporates a digital camera and multiple feeds, including sensors to detect changes in light levels, an accelerometer to detect motion, a thermometer to detect ambient temperature, and a passive infrared sensor to detect the presence of a person. The device takes pictures at VGA resolution, (480x640 pixels), and stores these as compressed JPEG files on internal flash memory. All sensor data and captured SenseCam images can be downloaded to a standard PC via a USB cable. SenseCam can collect a large amount of data even over a short period of time, with a picture typically taken every 30 seconds, an average of 2,000 images captured in any given day, together with associated sensor readings (logged every second).

Experience shows that the SenseCam can be an effective memory-aid device [3], as it helps users to improve recollection of an experience. However, given the large size of the dataset that is created by the SenseCam, refreshing one's memory just by browsing the vast corpus is a tedious, if not unacceptable task. Hence, techniques are required for

C. Gurrin et al. (Eds.): MMM 2014, Part I, LNCS 8325, pp. 400–411, 2014.

all users to manage, organise and analyse these large image collections, e.g., by automatically highlighting *key episodes* and, ideally, classifying them in order of importance to the life logger. Doherty et al. [4] address this challenge by identifying distinct events within a full day, (which typically consists of 2,000 lifelog images) *e.g. breakfast, working on PC, meeting, etc.*. However, their approach still contains a significant percentage of routine events. Li et al. [5] tackle the challenge by treating SenseCam images as time series. They show that these time series exhibit a strong long-range correlation, concluding that the time series is not a random walk, but is cyclical, with continuous low levels of background information picked up constantly by the device. Further, they adopt a cross-correlation matrix to highlight *key episodes*, thus identifying boundaries between different daily events.

However, due to the finite length of time series available to estimate cross correlations, the matrix contains much which corresponds to "random" contributions [6–9]. As a consequence, their technique results in the identification of a high percentage of noise or routine events. This phenomenon can also be observed in other domains such as the analysis of financial data, wireless communications and many other fields. A well-proven technique to handle this issue is the application of random matrix theory (RMT) [10].

In this paper, we investigate whether RMT can be used to distinguish routine events from important events. We argue that such routine events can then be removed from the cross-correlation matrix by applying RMT. Our goal is to segment the content of the cross-correlation matrix into two: (a) the part of the correlation matrix that conforms to the properties of random correlation matrices ("noise") and (b) the part of the correlation matrix that deviates from random (i.e. has "information" on important events).

In detail, we address this challenge as follows. First, we analyze the distribution of the correlation coefficients of the cross-correlation matrix C, which reveals an asymmetric, long positive tail with a high peak, implying that some background information is picked up constantly by the device. Next, we apply RMT methods to analyse the cross-correlation matrix C and show that a number ($\approx 20\%$) of the eigenvalues of C agree with RMT predictions, suggesting that C may have a "random band" matrix structure, i.e., "noise" element. Further, using the Inverse Participation Ratio concept, we analyze the eigenvectors of the cross-correlation matrix and find that both edges of the eigenvalue spectrum of C (smallest and largest eigenvalues) deviate from RMT prediction. We argue that these deviant components represent significant or unusual events in the data stream. Finally, the cross-correlation matrix is cleaned by separating the noisy part from the non-noisy part of C in order to remove the user's routine events.

This paper is organized as follows: In Section 2, we discuss related work in the fields of lifelogging and random matrix theory. Section 3 provides a brief overview of the data that are used within this work to study the research hypothesis. In Section 4, we discuss the use of random matrix theory in the extraction of information from a correlation matrix of SenseCam image time series. In Section 5, we discuss the statistics of cross-correlation coefficients and compare the eigenvalue distribution of the cross-correlation matrix with RMT results. In addition, we detail the analysis of the contents of eigenvectors that deviated from RMT. Finally, Section 6 provides conclusions on the work and anticipates future directions.

2 Related Work

This work builds on two research streams: (1) The creation and analysis of visual lifelogs and (2) the application of random matrix theory. In the remainder of this section, we briefly introduce both aspects.

2.1 Visual Lifelogging

As stated, visual lifelogging is the process of automatically capturing images and storing them in a personal repository. Although technologies for visual lifelogging has existed for several years, the use of such devices so far has been explored mainly by early adopters and researchers, in terms of studying its role as memory aid [3]. However, a recent study [11] suggests that lifelogging is an emerging trend with wider applications and an increasing member of devices will be available in the near future. A prominent example is Google Glass, which has received considerable media attention since it was first introduction to the public.

The majority of past research in the visual lifelogging domain has focused on issues of hardware miniaturisation [12] and also storage of images [13]. However, these challenges have been comparatively well addressed [1], resulting in improved wearability of devices and inexpensive storage [3]. The challenge is now that of retrieving *relevant* information from the vast quantities of captured data [3, 14–16].

A previous method, used to review images captured by the SenseCam is SenseCam Image View [1]. However, it can take two minutes or more to browse through a day's worth of SenseCam images, which means to 15 minutes to review all the images from one week. Therefore, we propose creation of one page 'visual summaries' of a day containing different images representing activities and events experienced by the user. Other research in this area has seen segmentation of the lifelog of SenseCam images into approximately 20 distinct events in a wearer's day, which translates to over 7,000 events per year [4]. Nevertheless, this large collection of personal information still contains a significant percentage of routine events. The objective still remains to determine which events are the most important or unusual for the lifelogger. In this paper, we address this issue by applying random matrix theory, described in the next subsection.

2.2 Random Matrix Theory

Random matrix theory (RMT) was first introduced by Wigner, Dyson, Mehta, and others [6–9] who aimed to study the energy levels of complex atomic nuclei [10]. Deviations from the *universal* predictions of RMT can be used to identify system specific, non-random properties of the system under consideration, providing clues about the underlying interactions [7–9].

Its successful application to atomic physics has stimulated its use in many other fields, including number theory and combinatorics [17], wireless communications [18], and in multivariate statistical analysis and principal components analysis [19, 20], as well as for financial and other large dimensional data analysis [10, 21] . Applications of RMT methods to analysis of the properties of the cross-correlation matrix C show that a large proportion of eigenvalues of C agree with RMT predictions, indicating a considerable degree of randomness in measured cross correlations. Deviations from RMT predictions are, however also observed.

In the context of visual lifelogging, it has been shown that SenseCam image time series reflect strong long-range correlation [5] which suggests continuous low levels of background information picked up constantly by the device. In this paper, we investigate the use of RMT applied to a cross-correlation matrix C to detect details on SenseCam lifelog data streams.

3 Data

For this study, we analyzed 2101 lifelog images, recorded using a SenseCam over the period of one day. The wearer of the camera, i.e., the lifelogger, experienced an average day: commuting to the office in the morning, sitting and working in the office at a desk, talking with colleagues and sharing lunch in the cafeteria, as well as commuting back home in the evening and so on. Fig. 1 shows some examples of SenseCam images that have been recorded on that day. Given the size of the test corpus and its content, we can argue that it is a typical visual lifelogging collection depicting a typical day of the subject's life.

As discussed above, a user will experience approximately 20 events per a day, but when exploring one's lifelog, reviewing routine or "boring" events has only limited interest, depending on the device purpose [22]. Efforts to determine automatically which events is most important or unusual (e.g., talking with a colleague as opposed to working in front of a computer), is an open research challenge. In order to distinguish routine or "boring" events from important events, we apply RMT methods to the cross-correlation matrix of the dataset, where such noise filtering has proved successful in many fields [10, 18–21]. In the next section, successful the method is outlined.

Fig. 1. Example of SenseCam Images

4 Methods

4.1 Random Matrix Theory

In order to optimize the calculation process and reduce the amount of memory required for our calculations, we first adopt an averaging method to decrease image size from 480×640 pixels to 60×80 pixels. Given pixels $G_i(t)$, $i = \{1, .., N\}$, of a collection of images, we normalized $G_i(t)$ in order to standardize the different pixels for the images as follows:

$$g_i(t) = \frac{G_i(t) - \overline{G_i(t)}}{\sigma_{(i)}} \tag{1}$$

Where $\sigma_{(i)}$ is the standard deviation of G_i for image numbers $i = \{1, .., N\}$ and $\overline{G_i}$ is the pixels' average of G_i over total pixel values T.

Then, the equal-time cross-correlation matrix [23] may be expressed in terms of $g_i(t)$

$$C_{ij} \equiv \langle g_i(t)g_j(t) \rangle \tag{2}$$

The elements of C_{ij} are limited to the domain $-1 \leq C_{ij} \leq 1$, where $C_{ij} = \pm 1$ defines perfect positive/negative correlation and $C_{ij} = 0$ corresponds to no correlation. In matrix notation, the correlation matrix can be expressed as

$$C = \frac{1}{T} GG^\tau \tag{3}$$

where τ is the transpose of a matrix, G is an $N \times T$ matrix with elements g_{it}, N is the number of images and T is the pixel size of an image.

The spectral properties of C may be compared to those of a "random" Wishart correlation matrix [10],

$$R = \frac{1}{T} AA^\tau \tag{4}$$

Where A is an $N \times T$ matrix with each element randomly distributed, with zero mean and unit variance. In particular, the limiting property for the sample size $N \to \infty$ and sample length $T \to \infty$, providing that $Q = T/N \geq 1$ is fixed, has been analysed to give the distribution of eigenvalues λ of the random correlation matrix R, given by:

$$P_{rm}(\lambda) = \frac{Q}{2\pi\sigma^2} \frac{\sqrt{(\lambda_+ - \lambda)(\lambda - \lambda_-)}}{\lambda} \tag{5}$$

for $\lambda_- \leq \lambda_i \leq \lambda_+$, where λ_- and λ_+ are the minimum and maximum eigenvalues of R, respectively, given by

$$\lambda_\pm = \sigma^2 \left(1 + \frac{1}{Q} \pm 2\sqrt{\frac{1}{Q}}\right) \tag{6}$$

Then, σ^2 is the variance of the elements of G and $\lambda_\pm$ are the bounds of the theoretical eigenvalue distribution. Eigenvalues that fall outside this region are said to deviate from the expected values of the Random Matrix. Hence, by comparing the empirical distribution of the eigenvalues of the correlation matrix to the distribution for a random matrix, as given in Eq. (5), we can identify those key eigenvalues which can be used to identify the specific information relating to the system. Eigenvector analysis enables identification of the specific information present, in terms of contributory components.

4.2 Eigenvector Analysis

Differences between the eigenvalues $P(\lambda)$ of C and RMT eigenvalues, $P_{rm}(\lambda)$ should also be displayed, therefore, in the statistics of the corresponding eigenvector components. In order to interpret this deviation of the eigenvectors, we note that the largest eigenvalue is an order of magnitude larger than the others, which constrains the remaining $N - 1$ eigenvalues, since the trace of C, $Tr[C]$ sums to N. Hence, in order to analyse the contents of the remaining eigenvectors, we need first remove the effect of the largest eigenvalue. To do this we can use the linear regression [10]

$$G_i(t) = \alpha_i + \beta_i G^{large}(t) + \epsilon_i(t) \tag{7}$$

Where $G^{large} = \sum_1^N u_i^{large} G_i(t)$ and N is the number of images in our sample. Here u_i^{large} corresponds to the components of the largest eigenvector. The cross-correlation matrix C, is then recalculated, using the residuals $\epsilon_i(t)$. If we quantify the 'remainder variance', (i.e., of the part not explained by the largest eigenvalue) as $\sigma^2 = 1 - \lambda_{large}/n$, this value can be used to recalculate our values of $\lambda_{\pm}$.

As suggested in [10], we also aim to assess whether random effects are less marked further away from RMT upper boundary λ_+. To do this we use the Inverse Participation Ratio (IPR). The IPR allows quantification of the number of components that participate significantly in each eigenvector and tells us more about the level and nature of the deviation from RMT. The IPR of the eigenvector u^k is given by $I^k = \sum_{l=1}^{N}[u_l^k]^4$ and allows us to compute the inverse of the number of eigenvector components that contribute significantly to each eigenvector.

5 Results

5.1 Statistics of Correlation Coefficients

In order to quantify correlations, we first analyse the distribution $P(C_{ij})$ of the elements $\{C_{ij} : i = j\}$ of the cross-correlation matrix C calculated by Eq. (3) and distribution $P(R_{ij})$ of the elements $\{R_{ij} : i = j\}$ of the random matrix calculated by Eq. (4). Fig. 2 shows that $P(R_{ij})$ is consistent with a Gaussian with zero mean, in contrast to $P(C_{ij})$. We note that $P(C_{ij})$ is asymmetric, with a long positive tail and has a high peak, implying that positively correlated behaviour is more prevalent than negatively correlated positively. This is consistent with our previous research [5]. We argue that the tail represents significant or unusual events in the data stream. In addition, we see that the $P(C_{ij})$ falls within the Gaussian curve for the control, suggesting the possibility that observed similarities with R in the cross-correlation matrix C may be an effect of randomness.

5.2   Eigenvalue Analysis

As stated above, our aim is to distinguish between information (major events) and noise in the cross-correlation matrix C, so we compare the eigenvalue distribution $P(\lambda)$ of C with the corresponding eigenvalue distribution predicted by RMT $P_{rm}(\lambda)$ for N = 2101 images, each containing T = 4800 pixels, Thus, $Q = T/N = 2.28$, and we obtain $\lambda_- = 0.11$ and $\lambda_+ = 2.76$ from Eq. (6). We compute the eigenvalues λ_i of C, where λ_i are rank ordered $(\lambda_{i+1} > \lambda_i)$. Fig. 3 compares the probability distribution $P(\lambda)$ with $P_{rm}(\lambda)$. We note the presence of a well-defined "bulk" of eigenvalues which fall within the bounds $[\lambda_-, \lambda_+]$ for $P_{rm}(\lambda)$. We also note deviations for ($\approx 80\%$) largest and smallest eigenvalues. Thus suggests that the cross-correlation matrix has captured most major events from the data streams, but still contains some percentage of noise ($\approx 20\%$).

5.3   Eigenvector Analysis

The deviations of $P(\lambda)$ from the RMT result $P_{rm}(\lambda)$ suggests that these deviations should also be observed in the statistics of the corresponding eigenvector components [24]. Accordingly, in this section, we analyse the distribution of eigenvector components. The distribution of the components $\{u_l^k; l = 1, ..., N\}$ of eigenvector U^k of a random correlation matrix R should conform to a Gaussian distribution with zero mean and unit variance. First, we compare the distribution of eigenvector components of C with a Gaussian distribution. We analyse $P(u)$ for C computed for the total of 2101 images. We choose one typical eigenvalue λ_k from the bulk $(\lambda_- \leq \lambda_k \leq \lambda_+)$ defined by $P_{rm}(\lambda)$ from Eq. (5). It shows that $P(u)$ for a typical U^k from the bulk shows reasonable total agreement with the RMT result $P_{rm}(u)$. Similar analysis on the other eigenvectors, belonging to eigenvalues within the bulk, yields consistent results, (in agreement with those of the previous sections) to random matrix predictions. We test the agreement of the distribution $P(U)$ with $P_{rm}(u)$ by calculating the kurtosis, which for a Gaussian has the value 3. We find that the largest eigenvector (≈ 4.07) significant deviates from the Gaussian value. The second and third Eigenvectors ($\approx 3.7, 3.8$) are also larger than the Gaussian value. The Eigenvector from the bulk is however consistent with the Gaussian value 3. These findings suggest that the largest eigenvalue (corresponding to the largest eigenvector) represent information from the image that reflects the largest change in the SenseCam recording.

In order to remove the effects of the largest eigenvalue we use the techniques described in Section 3.2. We remove the contribution of $G^{large}(t)$ to each time series $G_i(t)$, and construct C from the residuals $\epsilon_i(t)$ of Eq. (7). Fig. 5 shows that the distribution $P(C_{ij})$ thus obtained has a smaller average value $<C_{ij}>$, showing that a degree of cross correlations contained in C can be attributed to the influence of the largest eigenvalue and its corresponding eigenvector.

Having studied the largest eigenvalue and noting that if deviates significantly from RMT results, we conclude that it reflects the largest change in the SenseCam data streams. So next focus on the remaining eigenvalues to see whether these relate also to key sources or major events and what information these contribute additionally to the images. The Inverse Participation Ratio (IPR) quantities the reciprocal of the number of eigenvector components that contribute significantly. Fig. 6 (a) shows I^k for the

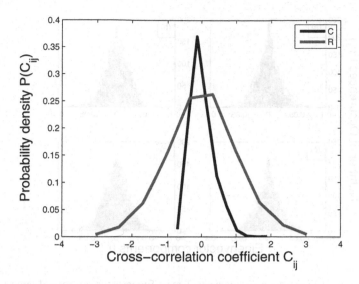

Fig. 2. Correlation Coefficients Distribution for Correlation Matrix C for SenseCam data (black) and Random Matrix R (red)

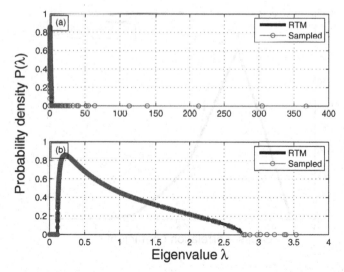

Fig. 3. Eigenvalue Distribution for the Correlation Matrix C for SenseCam data, Full spectral distribution (a) Partial spectral distribution (b)

case of the control of Eq. (4). The average value of I^k is $<I> \approx 0.0014 \approx 1/N$ with a very narrow spread, indicating that the vectors are extended [25] - i.e., almost all components elements (of the vector) contribute. Fluctuations around this average value are confined to a narrow range. Fig. 6 (b) shows I^k for the cross-correlation matrix constructed from the 2101 SenseCam images. The edges of the eigenvalue spectrum of

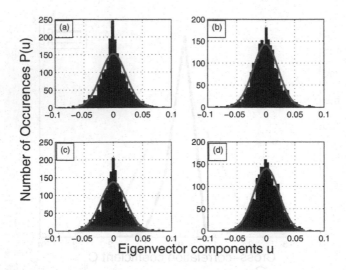

Fig. 4. Comparison of Eigenvector Components, largest Eigenvector (a), second largest Eigenvector (b), third largest Eigenvector (c) and Eigenvector from the bulk (d)

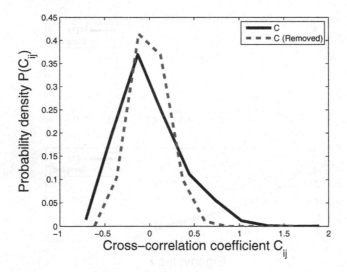

Fig. 5. Probability distribution P of the cross-correlation coefficients for data before (black) and after (red) removing the effect of the largest eigenvalue by linear regression method

C show significant deviations of I^k from $<I>$, indicating that there are major events contributing to these eigenvectors. In addition, we also find that there are a number of small eigenvalue deviations from the control case, which suggests that the vectors are localized [25] - i.e. only a few images contribute to them.

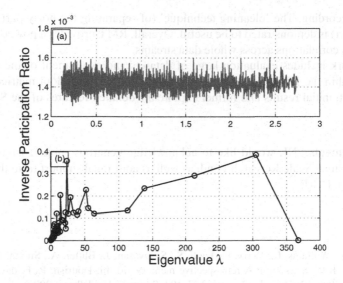

Fig. 6. Inverse Participation Ratio (IPR) as function of eigenvalue λ for the random cross-correlation matrix R (a) and cross-correlation matrix C (b)

Examination, of the eigenvalue and eigenvector content, indicates that the percentage of ($\approx 20\%$) noise for that period is described by the wearer working in front of the laptop for a long time without performing any other activities. The deviating eigenvalues from the RMT upper bound are described by the wearer morning from in front of the laptop and preparing to go home, with every image capturing different moments of this event. For example, considerable light is captured at the moment of standing up, different colours appear when she turns around, etc. Although all these images are visually very diverse, all have been captured at the same space, i.e., in the office. The deviating eigenvalues from the RMT lower bound involve several miner different activities, events such as commuting from home to the work place, the wearer talking with her colleague, the wearer sharing lunch with her colleague etc. Only a few images depict the same environment, but these appear on few images only. We argue that this confirms our observation that the IPR shows that the smallest eigenvalue deviations are localized, when only a few images contribute to them.

6   Conclusions

To summarise, we have illustrated that RMT, even with rather limited data, (2101 lifelog images depicting a typical day for a lifelogger) can be applied to extract the information (major events) and noise from a cross-correlation matrix. Significant deviations from RMT predictions are observed. In analysing these deviations find that (a) the largest eigenvalue and its corresponding eigenvector present information from the image that reflects the largest change in the SenseCam recording, and (b) the smallest eigenvalues, and their corresponding eigenvectors, represent short duration major events from the

SenseCam recording. The 'cleaning technique' (of separating the noisy part from the non-noisy part) is demonstrated to be useful. Overall, RMT provides a powerful tool to analyse cross correlations across whole data streams.

Future work includes evaluation large of datasets and assessment of the eigenvalues of C within the RMT bound for universal properties of random matrices, in order to confirm initial results and further explore the detailed features of the SenseCam images.

Acknowledgments. NL would like to acknowledge generous support from the Sci-Sym Centre Small Scale Research Fund, as well as additional support from the School of Computing, DCU.

References

1. Hodges, S., Williams, L., Berry, E., Izadi, S., Srinivasan, J., Butler, A., Smyth, G., Kapur, N., Wood, K.R.: Sensecam: A retrospective memory aid. In: Dourish, P., Friday, A. (eds.) UbiComp 2006. LNCS, vol. 4206, pp. 177–193. Springer, Heidelberg (2006)
2. Goldman, D.: Google unveils project glass virtual-reality glasses. Money (CNN) (2012)
3. Bell, G., Gemmell, J.: A digital life. Scientific American (2007)
4. Doherty, A.R., Smeaton, A.F.: Automatically segmenting lifelog data into events. In: WIAMIS, pp. 20–23 (2008)
5. Li, N., Crane, M., Ruskin, H.J., Gurrin, C.: Multiscaled cross-correlation dynamics on sensecam lifelogged images. In: Li, S., El Saddik, A., Wang, M., Mei, T., Sebe, N., Yan, S., Hong, R., Gurrin, C. (eds.) MMM 2013, Part I. LNCS, vol. 7732, pp. 490–501. Springer, Heidelberg (2013)
6. Wigner, E.P.: On the statistical distribution of the widths and spacings of nuclear resonance levels. Mathematical Proc. of the Cambridge Philosophical Society 47(10), 790–798 (1951)
7. Dyson, F.J.: Statistical theory of the energy levels of complex systems i. Journal of Mathematical Physics 3, 140–157 (1962)
8. Dyson, F.J., Mehta, M.L.: Statistical theory of the energy levels of complex systems iv. Journal of Mathematical Physics 4, 701–712 (1963)
9. Mehta, M.L., Dyson, F.J.: Statistical theory of the energy levels of complex systems v. Journal of Mathematical Physics 4, 713–719 (1963)
10. Plerou, V., Gopikrishnan, P., Rosenow, B., Nunes Amaral, L.A., Guhr, T., Stanley, H.E.: Random matrix approach to cross correlations in financial data. Physical Review E 65(6), 066126+ (2002)
11. Daskala, B., Askoxylakis, I., Brown, I., Dickman, P., Friedewald, M., Irion, K., Kosta, E., Langheinrich, M., McCarthy, P., Osimo, D., Papiotis, S., Pasic, A., Petkovic, M., Price, B., Spiekermann, S., Wright, D.: Risks and benefits of emerging life-logging applications. In: Final report, European Network and Information Security Agency (ENISA) (November 2011)
12. Mann, S.: Wearable computing: a first step toward personal imaging. Computer 30 (1997)
13. Gemmell, J., Bell, G., Lueder, R.: Mylifebits: a personal database for everything. Commun. ACM 49(1), 88–95 (2006)
14. Ashbrook, D., Lyons, K., Clawson, J.: Capturing experiences anytime, anywhere. IEEE Pervasive Computing 5(2), 8–11 (2006)
15. Lin, W.H., Hauptmann, A.: Structuring continuous video recordings of everyday life using time-constrained clustering. In: Multimedia Content Analysis, Management, and Retieval SPIE-IST Electronic Imaging, vol. 6073, pp. 111–119 (2006)

16. Lee, M.L., Dey, A.K.: Providing good memory cues for people with episodic memory impairment. In: ASSETS, pp. 131–138 (2007)
17. Conrey, J.B., Farmer, D.W., Keating, J.P., Rubinstein, M.O., Snaith, N.C.: Integral moments of l-functions. Proceedings of the London Mathematical Society 91, 33–104 (2005)
18. Tulino, A.M., Verdú, S.: Random matrix theory and wireless communications. Foundations and Trends in Communications and Information Theory 1(1) (2004)
19. Chirikjian, G.S.: Multivariate statistical analysis and random matrix theory. Applied and Numerical Harmonic Analysis 2, 229–270 (2012)
20. Ulfarsson, M.O., Solo, V.: Dimension estimation in noisy pca with sure and random matrix theory. IEEE Transactions on Signal Processing 56(12), 5804–5816 (2008)
21. Conlon, T., Ruskin, H.J., Crane, M.: Random matrix theory and fund of funds portfolio optimisation. Physica A: Statistical Mechanics and its Applications (2), 565–576 (2010)
22. Berry, E., Kapur, N., Williams, L., Hodges, S., Watson, P., Smyth, G., Srinivasan, J., Smith, R., Wilson, B., Wood, K.: The use of a wearable camera, scnsccam, as a pictorial diary to improve autobiographical memory in a patient with limbic encephalitis. Neuropsychological Rehabilitation 17(2), 580–601 (2007)
23. Conlon, T., Ruskin, H.J., Crane, M.: Multiscaled cross-correlation dynamics in financial time-series. Advances in Complex Systems (ACS) 12(04), 439–454 (2009)
24. Conlon, T., Ruskin, H.J., Crane, M.: Cross-correlation dynamics in financial time series. Papers 1002.0321, arXiv.org (February 2010)
25. Lee, P.A., Ramakrishnan, T.V.: Disordered electronic systems. Rev. Mod. Phys. 57, 287–337 (1985)

Exploring Distance-Aware Weighting Strategies for Accurate Reconstruction of Voxel-Based 3D Synthetic Models

Hani Javan Hemmat, Egor Bondarev, and Peter H.N. de With

Eindhoven University of Technology,
Den Dolech 2, 5612 AZ Eindhoven, Netherlands
{h.javan.hemmat,e.bondarev,p.h.n.de.with}@tue.nl
http://vca.ele.tue.nl/

Abstract. In this paper, we propose and evaluate various distance-aware weighting strategies to improve reconstruction accuracy of a voxel-based model according to the Truncated Signed Distance Function (TSDF), from the data obtained by low-cost depth sensors. We look at two strategy directions: (a) *weight definition* strategies prioritizing importance of the sensed data depending on the data accuracy, and (b) *model updating* strategies defining the level of influence of the new data on the existing 3D model. In particular, we introduce *Distance-Aware (DA)* and *Distance-Aware Slow-Saturation (DASS)* updating methods to intelligently integrate the depth data into the synthetic 3D model based on the distance-sensitivity metric of a low-cost depth sensor. By quantitative and qualitative comparison of the resulting synthetic 3D models to the corresponding ground-truth models, we identify the most promising strategies, which lead to an accuracy improvement involving a reduction of the model error by $10 - 35\%$.

Keywords: 3D Reconstruction, Voxel-Models, Weighting Strategy, Truncated Signed Distance Function (TSDF), Low-cost Depth Sensor.

1 Introduction

Accurate 3D reconstruction and mapping is a vital topic for currently prominent research domains such as 3D shape acquisition and modelling, surface generation and texturing, as well as localization and robot vision [2–4, 11, 12]. During recent years, the introduction of powerful general-purpose GPUs and low-cost, hand-held, and accurate 3D sensors [17–19], enable appearance of first solutions running a real-time 3D reconstruction process for relatively large and complex indoor and outdoor environments [7–12].

A 3D reconstruction process requires proper and accurate modelling structures, raw data acquisition methods, pose estimation, data fusion, surface reconstruction, and segmentation algorithms. Modelling volumetric structures for 3D data storage play a prominent role, influencing speed and quality of the process. Various model types have been introduced for 3D data representation.

C. Gurrin et al. (Eds.): MMM 2014, Part I, LNCS 8325, pp. 412–423, 2014.
© Springer International Publishing Switzerland 2014

These types include the *Signed Distance Function* (SDF) model aggregating the complete set of sensor data in a temporal manner [1], the *Octree-based* model providing an optimum structure in terms of memory usage efficiency [13, 14], the straightforward and expeditious *Points Cloud* structure [15], the *voxel-based* [3, 7, 8] and *surfel-based* [5, 6] models to represent volumetric data.

The introduction of low-cost depth sensing devices [17] including Kinect [18] and Xtion [19], have triggered the appearance of consumer 3D reconstruction applications, such as KinectFusion [7, 8], Kintinious [10], and the open source KinFu[15] and KinFu Large Scale[9]. These applications use a Truncated version of the SDF model (TSDF), in order to store 3D geometry of the environment. The TSDF model provides averaging of the complete set of sensor data over time. In the original model [1], a weighting strategy has been proposed to discriminate the newly upcoming sensed data by assigning a higher weight to less noisy data. This weight can be defined, based on both the *sensor-related* aspects (e.g. distance, lighting condition, surface material, etc.) and *method-related* (e.g. relative camera pose to the surface, pose estimation accuracy for the current frame, etc.) aspects. The KinectFusion and KinFu algorithms deploy *method-related* weight definition (e.g. weight associated with the surface normal), however, the 3D reconstruction results show minor or no improvement compared to the straightforward constant weight definition. Based on the intrinsic features of the Kinect as a depth sensor, we set the hypothesis that the 3D reconstruction result can be improved by an intelligent weighting strategy based on the *sensor-related* issues. While recent evaluation of the Kinect technology has revealed that it is relatively robust to ambient light, incidence angle, radiometric influences. The important limitation is that the sensor is less accurate for large distance measurements [23–25].

In the TSDF model, the captured environment is represented as a 3D array of voxels, forming a synthetic 3D model. For each depth frame, the newly extracted points are integrated into the synthetic 3D model. In the KinectFusion and KinFu, the weight value for each valid point is straightforwardly set to unity. Based on the intrinsic distance-sensitivity of the Kinect sensor, we expect a noticeable improvement in terms of model quality by introduction of a *sensor-related* weighting strategy. In the TSDF model, the newly sensed data is integrated into the model, regardless of its level of accuracy. As a consequence, the new depth data containing information about objects located at a distance exceeding 2 meters, deforms the corresponding objects in the synthetic 3D model. This occurs because the update mechanism corrupts the synthetic 3D model by fusing less accurate data with the more accurate data over time. To prevent this degradation, we introduce in this paper a weighting strategy for both *weight definition* and *updating method* of the TSDF model. Firstly, a *weight definition* is needed to assess the quality of each point of the new depth frame. Secondly, the *update method* makes a decision on how to update the synthetic 3D model by the new depth frame. The introduced weighting strategy has been implemented in the KinFu framework. A laser scanner has been used to generate the ground-truth point cloud. The final synthetic 3D model, in point cloud format, is compared to the ground-truth model. The results show a significant improvement in synthetic 3D model quality.

The paper is structured as follows. Section 2 proposes the weighting strategy. The experiments are elaborated in Section 3. Section 4 is focused on analysis and discussion of the results. Section 5 concludes the paper.

2 Weighting Strategy

2.1 Original KinFu and the TSDF Model

In TSDF model of the conventional applications [7, 8, 15], each voxel contains a pair of *distance value* (D_i) and *accumulated weight* (W_i), describing the truncated distance value to the closest surface and the weight for this value, respectively. This data structure enables averaging of the captured depth data, influencing the voxel model after i frames . For the *(i+1)*th depth frame, the model is updated by the corresponding pair of distance value (d_{i+1}) and weight (w_{i+1}) for voxel x by the following two equations:

$$D_{i+1}(x) = \frac{W_i(x)D_i(x) + w_{i+1}(x)d_{i+1}(x)}{W_i(x) + w_{i+1}(x)} , \qquad (1)$$

$$W_{i+1}(x) = W_i(x) + w_{i+1}(x) . \qquad (2)$$

The d_i is the calculated distance value for voxel x based on the corresponding newly sensed valid depth point. The w_i is the weight of the d_i. The d_i is integrated into the corresponding voxel x based on Equation (1). The weight for voxel x is accumulated in W_i according to Equation (2). Choosing $w_{i+1} = 1$ for each valid point found in the $(i + 1)$th frame, results in simple averaging over time. The constant value for the weight affects the synthetic model updating process in the following way. The objects in the model located close to the sensor (≤ 1 m) are created properly, while the objects located at a $\geq$ 2-m distance are being melted, significantly deformed or even completely destroyed. The Kinect provides more accurate data for closer ranges [23–25]. Due to this sensitivity to distance, the model is being deformed by overwriting the more accurate data (closer range) by the less accurate data (further range).

2.2 Weight Definitions

In this paper, we propose more adequate definitions of the weight factor and experiment with various approaches for updating the voxel weight. The prominent role of the *weight* in this paper, is to distinguish between the sensed data of closer and further distances. We have assessed different equations as weight definition (publicly available at http://vca.ele.tue.nl/demos/MMM14/mmm14.html). For all cases, assigning higher weights for closer distances is guaranteed. Based on the Kinect features, we consider a valid range for depth data, defined between a maximum and a minimum distance, d_{max} and d_{min}, respectively. In addition, the weight is bounded between 0 and a maximum weight W_{max}. In the following equation, we define a weight that has the best performance among various alternatives, which equals

$$weight_{depth\_point}(x) = [\frac{\frac{1}{d(x)^2} - \frac{1}{d_{max}^2}}{\frac{1}{d_{min}^2} - \frac{1}{d_{max}^2}}] * W_{max} \,. \tag{3}$$

For each depth value x with distance of $d(x)$ in the valid range between d_{min} and d_{max}, the corresponding weight is mapped to a value between 0 and W_{max}.

2.3 Updating Methods

In conventional TSDF model implementations, the model is straightforwardly updated with constant weight-value strategy (weight value is unity). The weight definition, introduced by Equation (3), enables us to distinguish between the closer and further distances. Therefore, we can exploit this weight definition to intelligently update the TSDF model. The intelligent update prevents more accurate values being overwritten by less accurate data. This updating strategy guarantees that the synthetic 3D model is updated by the most accurate data available during the update process. We propose two updating algorithms here.

Distance-Aware (DA) Updating Method. Each voxel value in the synthetic 3D model is updated based on a straightforward intelligent rule: *"if a voxel value has already been updated by a distance value with a higher weight, never update it again by a depth distance with a lower weight"*. The DA updating method is formulated below as:

$$Flag_{update}(v, x) = \begin{cases} Yes & \text{if } (weight_{new}(x) \geq r\% \times weight_{last\_max\_updated}(v)), \\ No & \text{otherwise}; \end{cases} \tag{4}$$

$$Update(v, x) = \begin{cases} \text{Integrate } x \text{ into the } v & \text{if } Flag_{update}(v, x) \text{ is Yes}, \\ \text{Discard } x, \text{ preserve } v & \text{otherwise}. \end{cases} \tag{5}$$

To make the updating method more robust to noise, we introduce a tolerance range, r with $0 \leq r \leq 100$. Since the distance values are compared to $r\%$ of the last maximum updated weight, the distance values less than the last maximum updated weight are therefore integrated into the synthetic 3D model. This method integrates the distance values close to the last maximum updated weight affected by noise.

   The $Update(v, x)$ function *conditionally* updates the model based on Equation (1). During the update process, the accumulated weight value for each voxel, $W_i(x)$, is calculated according to Equation (2). The weight of each new voxel value integrating into the model, w_i, is equal to the weight of the corresponding distance value (Equation (3)). Therefore, the w_i is always between 0 and W_{max}. When comparing the conventional TSDF implementation with $w_i = 1$, the proposed method grows faster with the *accumulated weight value* for each voxel. The accumulation of the distance values quickly grows out of the 1-Byte specification used for the conventional algorithm with unity weights. This can be circumvented by using a 2-Byte word for W_i, but this leads to a much larger memory requirement for the algorithm. For this reason, we present DASS below.

Distance-Aware Slow-Saturation (DASS) Updating Method. To stay in the framework of the conventional TSDF implementation and soften the impact of the fast accumulated weight saturation in the DA method, we propose the DASS method as an alternative. The DASS method performs similar to the DA method, except for the weight accumulation. The DASS uses the weight definition of Equation (3) for the $Update(v, x)$ function to conditionally update the synthetic 3D model, similar to the DA method. However, in contrast with the DA method, the DASS uses unity for the new weight w_{i+1}, to calculate the weight accumulation value W_{i+1}. This solution of $w_{i+1} = 1$ in the DASS method prevents the fast saturation of the accumulated weight value, and the $Update(v, x)$ function ensures the intelligent update.

It should be noted that both algorithms have the benefit of the distance-based updating feature compared to the conventional algorithms. The DA and DASS algorithms enable a choice in the memory complexity, while still employing this advantageous feature.

3   Experiments

3.1   Implementation

To implement the proposed updating methods, we have exploited the original framework of the open source KinFu implementation from the Points Cloud Library [15]. We have reused the original structure and only inserted the new definitions and updating algorithms as discussed above. We have used a previously allocated but unused byte in the original structure to store the W_i.

3.2   Experimental Dataset

For the dataset, we have prepared a room with various objects with distinctive 3D features. The ROS bag recorder [16] is used to capture the depth frames from Kinect. A laser scanner has been used to generate the ground-truth 3D model data of the room. The average distance between a point in the reconstructed 3D model and the corresponding point in the ground-truth 3D model is determined as *model error*. The drift error caused by the ICP camera pose estimation algorithm [22] overweighted the *model error* of mentioned algorithms. Therefore we have split the model into three different regions and compared them separately to the ground-truth models. Fig. 1 shows the room and its different regions.

Fig. 1. Dense point cloud of the dataset room captured by the laser scanner

3.3 Evaluation Approach

A. Quantitative comparison. We have used the CloudCompare [20] to compare the resulting point clouds to the ground truth. The CloudCompare calculates the model error for each reconstructed 3D model. Prior to calculation of the average distance, the clouds are aligned by using the implemented ICP algorithm. The comparison is performed in two levels. Firstly, the approximate distance between the clouds is calculated as a Chamfer distance. After this a local model is applied to the reference cloud to improve the precision of the cloud-to-cloud distance calculation. From the local models in CloudCompare, we have employed the Height function, as it is the slowest but the most precise method.

To evaluate the DA, DASS, and KinFu methods, we have compared their corresponding 3D models to the ground-truth model. In addition, we evaluate the performance of the algorithms with respect to three metrics, called *Truncation distance influence range*, *Tolerance range*, and *Always update* flag, which will be discussed in the next section.

B. Qualitative comparison. For a visual assessment of the reconstruction results, 30 reviewers have been asked to evaluate the untagged rendered snapshots of the 3D models obtained by the proposed methods.

4   Analysis and Discussion of Results

Comparing the Reconstructed 3D Models. We have found that the model error is reduced by exploiting the DA and DASS methods for all local models. Table 1 shows the comparison results of each reconstructed 3D model with the ground-truth model, in terms of mean distance and standard deviation. The average model error (height function) is reduced by 0.72 mm (10.3%) for the DA method, and 1.11 mm (15.9%) for the DASS method. The best improvement is

obtained in the Desk region, where the model error reduces by 1.66 *mm* (19.0%)
and 3.08 *mm* (35.2%) for the DA and DASS methods, respectively. The Desk
region was *heterogeneously* captured from both *close* and *far* distances. This
leads to the conclusion that the proposed methods perform best for the heterogeneously captured scenes and are marginally better for the homogeneously
captured scenes. Based on Table 1, the DASS outperforms the DA in all regions
and for all local models. We explain this by the reason that the maximum for
the accumulated weight value W_i, is set to 255 to limit the GPU memory usage.
The weight for each new voxel value w_i, is between 0 and W_{max} (=64 as optimal
setting). Therefore, the accumulated weight value quickly saturates in the DA
method (Equation (2)). In the DASS method, $w_i = 1$, so that there is sufficient
range prior to saturation.

As depicted in Fig. 2, the distribution of erroneous point-allocation deviations
for the DASS method shows that the majority of points have small deviations
and a narrow distribution, while for the KinFu method, the majority of points
are located on mid-deviation or large-deviation areas, giving a larger variance of
the error distribution.

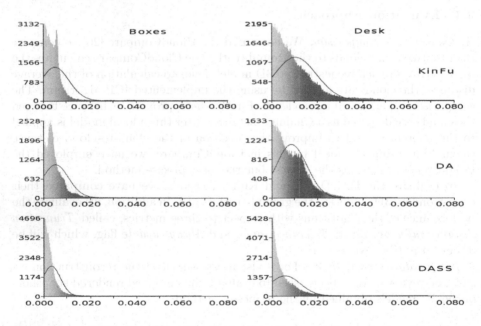

Fig. 2. Error distribution (Chamfer distance) on Desk and Boxes regions. The Y-axis
shows the number of points with the corresponding amount of errors of the X-axis (m).

Truncation Distance Influencing Range (r_{tdi}). This metric determines
the range, in which a valid depth point affects corresponding voxels in a 3D
model. The optimum value for r_{tdi} in the conventional TSDF implementation
equals 3 mm. We have found that the reduction of r_{tdi} reduces the model quality
for all methods. Based on Table 2, reducing the range by 1.0 mm increases the

Table 1. Comparison of obtained models by the DA, DASS and KinFu methods to ground-truth models. $r_{tdi} = 3$ mm and $r = 80\%$ (m.d. and s.d. are abbreviations of mean distance and standard deviation, respectively).

Method	Desk		Boxes		Bear		Average (mm)	
	m.d.	s.d.	m.d.	s.d.	m.d.	s.d.	m.d.	s.d.
KinFu	8.74	9.32	3.33	4.35	8.84	12.65	6.97	8.77
DA	7.08	6.79	3.36	4.00	8.32	11.99	6.25	7.59
DASS	5.66	6.64	3.29	3.95	8.65	11.63	5.86	7.41

model error in all regions (on the average 0.33 mm (4.7%), 0.19 mm (3.0%), and 0.50 mm (8.5%) for the KinFu, DA, and DASS methods, respectively). Whereas increasing the truncation distance by 1.0 mm can reduce the model error for the original KinFu by 0.70 mm (10.0%), the model error is increased by 0.20 mm (3.4%) for the DASS method. For the DA method, increasing the r_{tdi} has no impact on the model error (0.00 mm). Empirically, we have observed that the results are visually better for $r_{tdi} = 3$ mm in all cases (default settings for the original KinFu algorithm).

Table 2. Impact of the truncation distance influencing range on the 3D model quality obtained by the DA, DASS, and KinFu methods (height function, $r = 80\%$)

Method	Truncation distance influencing range	Desk		Boxes		Bear		Average (mm)	
		m.d.	s.d.	m.d.	s.d.	m.d.	s.d.	m.d.	s.d.
KinFu	02 mm	8.25	8.66	3.40	4.74	10.25	12.98	7.30	8.79
	03 mm	8.74	9.32	3.33	4.35	8.84	12.65	6.97	8.77
	04 mm	7.33	7.48	3.32	4.09	8.15	12.16	6.27	7.91
DA	02 mm	6.79	6.58	3.67	4.61	8.87	12.68	6.44	7.96
	03 mm	7.08	6.79	3.36	4.00	8.32	11.99	6.25	7.59
	04 mm	7.06	6.75	3.67	4.09	8.01	11.29	6.25	7.38
DASS	02 mm	6.85	7.24	3.36	4.28	8.87	11.69	6.36	7.74
	03 mm	5.66	6.64	3.29	3.95	8.65	11.63	5.86	7.41
	04 mm	6.61	6.24	2.77	3.39	8.82	12.64	6.06	7.43

Tolerance Range. This metric refers to parameter r in Equation (4). It is interesting to note that assigning $r = 80\%$ results in the most accurate 3D model for both the DA and DASS methods, in all regions. When inspecting the results of Table 3 more closely, it becomes clear that both large and low values of the tolerance range degrade the final 3D model. Increasing the range from 80% to 90%, increases the model error for both the DA and DASS methods by 0.57 mm (9.1%) and 0.70 mm (11.9%), respectively. Decreasing the range from 80% to 70% and 50%, increases the model error for the DA method by 0.25 mm (4.0%) and 0.70 mm (11.2%), and for the DASS method by 0.11 mm (1.8%) and 0.82 mm (13.9%), respectively.

Table 3. Impact of the tolerance range on quality of models obtained by the DA and DASS methods (height function, $r_{tdi} = 3$ mm)

Method	Tolerance range	Desk		Boxes		Bear		Average (mm)	
		m.d.	s.d.	m.d.	s.d.	m.d.	s.d.	m.d.	s.d.
DA	50%	8.42	8.27	3.44	4.70	9.00	12.83	**6.95**	**8.60**
	70%	7.36	8.39	3.65	4.46	8.49	12.41	**6.50**	**8.42**
	80%	7.08	6.79	3.36	4.00	8.32	11.99	**6.25**	**7.59**
	90%	7.79	8.40	3.97	4.51	8.70	12.37	**6.82**	**8.43**
DASS	50%	7.82	7.63	3.40	3.80	8.83	11.69	**6.68**	**7.71**
	70%	6.15	5.89	3.05	4.14	8.71	12.16	**5.97**	**7.40**
	80%	5.66	6.64	3.29	3.95	8.65	11.63	**5.86**	**7.41**
	90%	7.02	6.96	3.52	4.42	9.14	13.02	**6.56**	**8.13**

Another observation is that each region can be affected differently for different tolerance values. For example, for the DASS method, the model error for the Boxes region with $r = 70\%$ is 0.24 mm less than with $r = 80\%$, while the average model error for other regions with $r = 70\%$ is 0.28 mm more than with $r = 80\%$. We expect that dynamic identification and adaptation of the r value, depending on the distance to particular objects, can be useful to further improve the model accuracy.

Always Update. The original KinFu employs a threshold on camera pose for a model update. The new depth data is not integrated into the model, if the camera-movement metric does not exceed this threshold. The *Always update* flag disables this threshold. However, if the camera moves smoothly, most of the data is discarded. We have planned to keep all data and developed strategies to employ the data for updating. As reflected in Table 4, disabling the threshold increases the model error for all methods, by 0.39 mm (5.6%), 0.39 mm (6.2%), and 0.24 mm (4.0%), for the KinFu, DA, and DASS methods, respectively. We have found that while the DASS method suffers less than the original KinFu by disabling this updating threshold, it performs worse in comparison with the case that the threshold is enabled.

Table 4. Impact of the 'Always-update flag' on quality of models obtained by the KinFu, DA, and DASS methods (height function, $t_{tdi} = 3$ mm and $r = 80\%$)

Method	Desk		Boxes		Bear		Average (mm)	
(Always-update enabled)	m.d.	s.d.	m.d.	s.d.	m.d.	s.d.	m.d.	s.d.
KinFu	9.09	9.27	3.39	4.52	9.60	14.30	**7.36**	**9.36**
DA	7.46	7.13	3.75	4.20	8.72	12.92	**6.64**	**8.08**
DASS	5.77	6.75	3.29	3.95	9.23	12.51	**6.10**	**7.73**

Qualitative Evaluation of the Methods. The model obtained by the DA method is preferred because of preserving most of the details as indicated by 67% of the reviewers. The model resulting from the DASS method has been

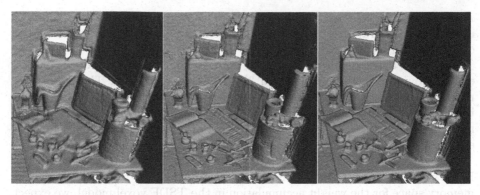

Fig. 3. From left to right, the snapshots of the Desk region of the final 3D meshes obtained by the original KinFu, DA, and DASS methods, respectively

Table 5. Visual evaluation of synthetic 3D models by 30 reviewers via a questionnaire

Question	KinFu	DA	DASS
Which one is more accurate and does contain more details?	0%	67%	33%
In which one, the boundaries between objects are more clear?	6%	37%	57%
In which one, the objects and their edges look sharper?	6%	40%	54%
Which one do you like the most?	10%	23%	67%
Which one do you like the least?	77%	20%	3%

ranked as the best in terms of object boundaries and sharp edges by 57% and 54% of the reviewers, respectively. We have found that 67% of the participants have perceived the model obtained by the DASS method the most attractive. Only 23% of the reviewers have found the model obtained by the DA method the most attractive, while 67% of them have perceived it as the most accurate and offering most visual detail. We explain this discrepancy by the fast saturation of the accumulated weight in the DA method, which prevents a proper averaging of all the sensed depth data.

5   Conclusion

In this paper, we have introduced novel intelligent distance-aware weighting strategies to improve the reconstruction quality of the Truncated Signed Distance Function (TSDF) voxel-based models, although the depth data is obtained by low-cost depth sensors. In conventional implementations of the TSDF model, every new depth value is straightforwardly integrated into the synthetic 3D model. As a consequence, considering the distance sensitivity of low-cost depth sensors, less accurate depth data can overwrite more accurate data. Our new proposed intelligent weighting strategies are developed into two directions: *weight definition* and *model updating* methods. Both strategies prevent the already-fused data of being corrupted by less accurate data. More specifically, we have introduced the

Distance-Aware (DA) and *Distance-Aware Slow-Saturation (DASS)* methods to intelligently integrate the depth data into the synthetic 3D model, based on the distance-sensitivity metric of the sensor.

According to the quantitative results, we have found that the proposed methods improve the quality of the final 3D model by $10 - 35\%$. The qualitative results show that 90% of the reviewers prefer the synthetic 3D model obtained by DA and DASS methods. Based on the current implementation, the DASS method delivers the best improvement of 16% on the average and 35% for heterogeneously captured scenes, in comparison with the original KinFu method.

It is clear that this research work can be further improved in several ways. First, we recommend to exploit memory usage in a better way. With sufficient memory space for the weight accumulation in the TSDF voxel model, we expect to obtain better result for the DA method, without suffering from fast saturation of the accumulated weight. Therefore, to dedicate more memory space for this, research should consider a memory-efficient model implementation such as Octree-based structures or the like. A second recommendation involves algorithm improvement aspects. One possibility is to use dynamic identification and adaptation of the influencing factors of the algorithms (such as truncation distance influencing range and tolerance range), depending on the distance to the particular objects. It is evident that this will further optimize the model accuracy. To extend the weighting strategies beyond the proposed distance-awareness, we may also explore new *angle-aware* and *texture-aware* strategies. We could utilize the surface normal and RGB information along with the depth data, for weight definition in angle-aware and texture-aware strategies, respectively. A third aspect for improvement relates to model quality. The model quality of the obtained models in this paper have been compared to the ground-truth model by individually aligning them, independent of each other. Instead, we suggest to focus on the impact of the proposed methods on pose estimation algorithms (pre-results are available at `http://vca.ele.tue.nl/demos/MMM14/mmm14.html`). Additionally, we recommend to use a *heat map* to emphasize the model accuracy by representing each point of the 3D model with its corresponding last maximum updated weight. We expect that this leads to the integration of the most accurate data for each scene by the user.

Acknowledgement. This research has been performed within the PANORAMA project, co-funded by grants from Belgium, Italy, France, the Netherlands, and the United Kingdom, and the ENIAC Joint Undertaking.

References

1. Curless, B., Levoy, M.: A volumetric method for building complex models from range images. In: ACM SIGGRAPH Conf. Proceedings, New Orleans, pp. 303–312 (1996)
2. Engelhard, N., et al.: Real-time 3D visual SLAM with a Hand-held Camera. In: RGB-D Workshop on 3D Perception in Robotics, European Robotics Forum (2011)
3. Newcombe, R.A., Lovegrove, S.J., Davison, A.J.: Dense Tracking and Mapping in Real-Time. In: IEEE Int. Conf. on Computer Vision (ICCV), Spain (2011)

4. Steinbruecker, F., et al.: Real-Time Visual Odometry from Dense RGB-D Images. In: Live Dense Reconstruction with Moving Cameras at the ICCV (2011)
5. Andersen, V., et al.: Surfel Based Geometry Resonstruction. In: Theory and Practice of Computer Graphics: UK Chapter of Eurographics Association, pp. 39–44 (2010)
6. Chang, J.Y., et al.: GPU-friendly multi-view stereo reconstruction using surfel representation and graph cuts. In: Comput. Vis. Image Understand. (2011)
7. Newcombe, R.A., et al.: KinectFusion: Real-time dense surface mapping and tracking. In: IEEE Int. Symp. Mixed and Augmented Reality, vol. 7, pp. 127–136 (2011)
8. Izadi, S., et al.: KinectFusion: Real-time 3D Reconstruction and Interaction Using a Moving Depth Camera. In: Proc. of 24th Annual ACM UIST 2011, pp. 559–568 (2011)
9. Bondarev, E., et al.: On Photo-realistic 3D Reconstruction of Large-scale and Arbitrary-shaped Environments. In: CCNC. IEEE Press, Las Vegas (2013)
10. Whelan, T., et al.: Kintinuous: Spatially Extended KinectFusion. In: RSS Workshop on RGB-D: Advanced Reasoning with Depth Cameras, Sydney, Australia (2012)
11. Whelan, T., et al.: Robust Tracking for Real-Time Dense RGB-D Mapping with Kintinuous. In: MIT technical report, MIT-CSAIL-TR-2012-031 (2012)
12. Whelan, T., et al.: Robust Real-Time Visual Odometry for Dense RGB-D Mapping. In: IEEE Int. Conf. on Robotics and Automation. ICRA, Germany (2013)
13. Hornung, A., Wurm, et al.: OctoMap: an Efficient Probabilistic 3D Mapping Framework Based on Octrees. J. Autonomous Robots 34(3), 189–206 (2013)
14. Zeng, M., Zhao, F., Zheng, J., Liu, X.: A Memory-Efficient KinectFusion Using Octree. In: Hu, S.-M., Martin, R.R. (eds.) CVM 2012. LNCS, vol. 7633, pp. 234–241. Springer, Heidelberg (2012)
15. A large scale, open project for point cloud processing, http://pointclouds.org/
16. Robotic Operationg System, http://www.ros.org/wiki/
17. PrimeSense^{TM}, http://www.primesense.com/
18. Microsoft Kinect, http://www.xbox.com/en-us/kinect/
19. Asus Xtion-PRO, http://www.asus.com/Multimedia/Xtion_PRO/
20. CloudCompare (version 2.4) (GPL software), http://www.danielgm.net/cc/
21. Meilland, M., Comport, A.I.: Simultaneous super-resolution, tracking and mapping. Sophia-Antipolis, France, CNRS-I3S/UNS (2012)
22. Besl, P.J., McKay, N.D.: A method for registration of 3-D shapes. IEEE Transactions on Pattern Analysis and Machine Intelligence 14(2), 239–256 (1992)
23. Chow, J., et al.: Performance Analysis of a Low-cost Triangulation-based 3D Camera: Microsoft Kinect System. The Int.l Archives of the Photogrammetry, Remote Sensing and Spatial Information Sciences 39(pt. B5), 175–180 (2012)
24. Khoshelham, K.: Accuracy Analysis of Kinect Depth Data. In: Lichti, D.D., Habib, A.F. (eds.) ISPRS Workshop Laser Scanning. IAPRS XXXVIII5/W12 (2011)
25. Khoshelham, K., Oude Elberink, S.J.: Accuracy and Resolution of Kinect Depth Data for Indoor Mapping Applications. Sensors 12(2), 1437–1454 (2012)

Exploitation of Gaze Data for Photo Region Labeling in an Immersive Environment

Tina Walber[1], Ansgar Scherp[2], and Steffen Staab[1]

[1] University of Koblenz-Landau, Germany
{walber,staab}@uni-koblenz.de
[2] University of Mannheim, Germany
ansgar@informatik.uni-mannheim.de

Abstract. Metadata describing the content of photos are of high importance for applications like image search or as part of training sets for object detection algorithms. In this work, we apply tags to image regions for a more detailed description of the photo semantics. This region labeling is performed without additional effort from the user, just from analyzing eye tracking data, recorded while users are playing a gaze-controlled game. In the game *EyeGrab*, users classify and rate photos falling down the screen. The photos are classified according to a given category under time pressure. The game has been evaluated in a study with 54 subjects. The results show that it is possible to assign the given categories to image regions with a precision of up to 61%. This shows that we can perform an almost equally good region labeling using an immersive environment like in *EyeGrab* compared to a previous classification experiment that was much more controlled.

1 Introduction

The management of digital images is a challenging task, and it is often performed based on metadata. For example, image search makes use of tags manually assigned to images or extracted from surrounding text information on web pages. A more detailed description of photo contents by region labeling can improve the search [4]. Different approaches were investigated for creating region labels. On the one hand, fully automatic approaches are far from delivering results that are on the level of human understanding of visual content [11]. On the other hand, manual labeling is a tedious task for users. The general idea behind our approach is to create image meta information without additional effort from the user. To reach this goal, we exploit the information gained from eye movements, while the user is viewing photos in the context of a specific task. In our first work [15], the data was collected in a controlled experiment. In this first experiment, a tag was first presented to the user, and afterwards, he/she had to decide whether an object described by this tag could be seen on the photo by pressing a key on the keyboard. We obtained a maximum precision of 65% at pixel level from comparing the calculated regions with manually created ground truth regions. In this work, we evaluate if region labeling is also possible in a very different

C. Gurrin et al. (Eds.): MMM 2014, Part I, LNCS 8325, pp. 424–435, 2014.
© Springer International Publishing Switzerland 2014

scenario, while the user is playing a game. As in the first experiment, the task is to decide whether an object, belonging to a given category, can be seen on a photo. While in the first study, the user had no time constraints and the photos were displayed full screen, the game *EyeGrab* [13] was developed to demand fast decision making from the participants and to break up the full concentration on photo viewing by bringing the user into the immersive situation of a game with distractions from the game setup, the gaze control, and the emotional pressure of success and failure. In *EyeGrab*, users classify and rate photos falling down the screen. Photos are selected by fixation them fixated with the eyes. Subsequently, the classification is performed by fixating specific objects on the screen. In addition, the classification comprises a personal rating of the photo. By analyzing the recorded gaze paths, we are able to automatically assign the given category, which describes a specific object like "car" or "tree" to an image region. To evaluate our approach, we have collected gaze data in a user study with 54 participants. All photos used in our evaluation had ground truth information concerning their classification and the depicted objects and image regions, respectively. We can state that the level of difficulty playing *EyeGrab* was not too high as only 7% of the shown images passed without classification and 90% of the classifications were correct with respect to the given category. In order to assign a given category to an image region, we apply two gaze measures and a baseline [15]. The measures predict which region of the photo is assumed to show an object, belonging to the category. This region is compared at pixel level with the ground truth image region from our data set. From the data collected in *EyeGrab*, we obtain a maximum precision of 61% of correctly labeled image region pixels. Thus, region labeling in the immersive environment of a game performs almost equally well as in previous work, where we considered non-moving full-screen images and could predict the regions with precision of about 65% [15]. We have also investigated different falling speed levels for analyzing the influence of speed in the region labeling results. We got a slightly higher number of photos passing without classification or those classified incorrectly for faster speed levels, but only small variations in the precision of the region labeling are observed. Overall, this study shows that one can obtain good region labels in an immersive game environment and that the results are comparable to those where the images were not moving and the experiment was much more restricted.

Please note that we provide the experiment images and gaze data on `http://west.uni-koblenz.de/Research/DataSets/gaze`.

2 Related Work

One approach for gaining information from gaze data is relevance feedback in image search. Kozma et al. [5] compared image selection by implicit gaze feedback with explicit user feedback by clicking on relevant images. Gaze information in combination with image segmentation also provides valuable information for photo cropping [9]. Klami et al. [3] identified heat-map-like image regions relevant in a specific task using gaze information. The given task is very general, and thus, the work does not aim at identifying single objects in the images from the generated heat map. Our previous work [15] showed that it is possible to assign given tags to image regions for describing depicted objects. However, the data was collected in a controlled experiment with static photos and without gaze control. Smith and Graham [10] described the advantages of gaze control in video games. They state that the use of gaze control can improve the game play experience. An example is *EyeAsteroids*[1], an eye-controlled arcade game presented by Tobii. The game is entertaining but does not have the goal to exploit the users' activities while playing. Games with a purpose (GWAPs) are computer games that have the goal to obtain information from humans in an entertaining way. The information is usually easy to create for humans, but challenging or impossible to be created by fully automatic approaches. An example of a GWAP is the game *Peekaboom* [12], presented by von Ahn et al. Two users playing together try to label the same image regions for a given tag. Ni et al. [7] introduced a game for explicitly labeling image regions. The users look for specific objects in photos taken from Flickr and mark them by drawing bounding boxes. The development of eye tracking hardware in the recent years supports the usage of gaze control in everyday device like laptops in the near future. Systems that can detect the eyes and can calculate the viewing direction from cameras integrated in common devices like tablet PCs are already on the market (e. g., Natural User Interface Technology, OKAO Vision[2]). Lin et al. [6] presented an eye tracking system using a web cam that is even working in real-time. Thus, the role of eye tracking as input device for controlling software and for collecting information from it's data is increasing.

3   Gaze-Based Measures for Labeling Image Regions

In this work, we apply two gaze-based measures for labeling image regions and one baseline measure, all introduced in previous work [15]. The two gaze-based measures are the segmentation measure (I) and the heat map measure (II). By means of these measures, we assign a given category to an image region for labeling this region. An overview of both measures is depicted in Figure 1. For all photos belonging to the given category, the input for the gaze analysis are (i) the given category and (ii) the gaze paths of all users who correctly classified the photo. The segmentation measure additionally takes (iii) automatically obtained (hierarchical) photo segments as input data. These photo segments are obtained from applying the bPb-owt-ucm algorithm [1]. The different hierarchy levels describe different levels of detail and are controlled by the parameter $k = 0, 0.1, \ldots 0.5$.

In the segmentation approach, the fixations on every region of the segmented photo are counted, which corresponds to the fixation measure *fixationCount*. The segment with the highest outcome is assumed to show the object for the

---

[1] http://www.tobii.com/en/gaze-interaction/
  global/demo-room/tobii-eyeasteroids/

[2] http://www.omron.com

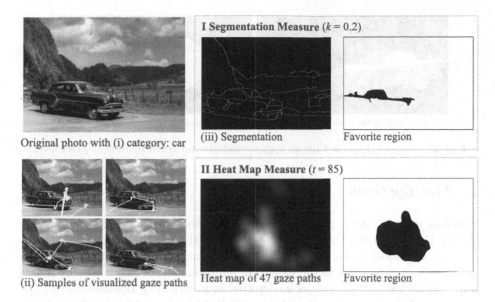

Fig. 1. Gaze-based region labeling with two measures I and II. Input data is (i) the given category, (ii) the users' gaze paths, and (iii) the segmented image (only for I).

given category. In order to take the inaccuracies in the eye tracking data into account, we apply region extension and weighting introduced earlier [15]. The region extension considers fixations in the surrounding of up to 13 pixels of an segment as being on the segment. Due to the weighting results for segments that are smaller than 5% of the photo are multiplied by a factor up to 4. Different segmentation levels $k = 0, 0.1, \ldots 0.5$ are considered in our analysis. The heat map approach identifies intensively viewed photo regions by summing up the fixations of all gaze paths at pixel level. A value of 100 is applied to the center of each fixation. In a radius of 50 pixels, linear decreasing values are applied to the surrounding pixels. From the created heat map, the object region is calculated by applying a threshold to the data, identifying the mostly viewed pixels. The parameter t indicates the percentage of viewing intensity (e.g. $t = 5$ indicates the 5% of all pixels with the highest values). After the thresholding, the biggest area of connected pixels is assumed to depict the object. The concrete parameter values for both approaches are determined based on the findings in our previous work [15]. The center baseline approach from earlier work [15] is also applied to the data. The element in the center of the segmented photo is considered as a depiction of the object.

By means of ground truth data for the image regions and labels (cf. Section 5), we are able to evaluate the computed object regions. For every pixel, we compare the ground truth with the label obtained from our measures by calculating precision, recall, and F-measure, with F-measure $= 2 \cdot \frac{precision \cdot recall}{precision + recall}$. An example photo with two object regions and their evaluation can be found in Figure 2.

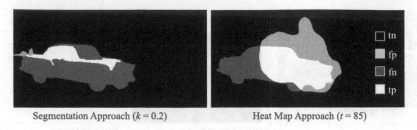

Segmentation Approach ($k = 0.2$)   Heat Map Approach ($t = 85$)

Fig. 2. Comparing labeled image regions and ground truth regions at pixel level

4   The *EyeGrab* Game

The task in *EyeGrab* to "clean up an aliens' universe" by categorizing and rating photos. Before starting the game, users have to calibrate the eye tracking device by fixating several points on the screen. Subsequently, a small introduction to the game's rules is given to the gamer. In addition, he/she has to choose a user name and to indicate his gender as depicted in Figure 3(a). Besides entering the gamer's nickname, the game is solely controlled by eye movements. Gaze-based interactions are triggered after a dwell time of 450 m. The ocular dwell time of fixations lies between 200 and 400 ms [2]. Hence, the selection dwell time lies above this value to avoid random selections. For example, the selection of the gender is done by focusing on a male or female character as shown in Figure 3(a). The gender information is used only for adapting the gaming environment, e.g., by changing some colors.

A game consists of several rounds. In each round, a set of photos has to be classified concerning a given category like "car", "person", and "sky". First, the category is presented to the user for 6 s. Subsequently, the photos fall down the screen as depicted in Figure 3(b) and are classified by the gamers. Each round has a different speed level at which the photos move. Several photos can be shown on the screen at the same time. The player selects an image by fixating it for longer than the dwell time of 450 ms. As soon as a photo is selected, it is highlighted by a thin frame, and the user can classify it into one of three categories. The classification takes place by fixating symbols on the screen as shown in Figure 3 (b). The categories are "not relevant" (symbolized by a trash can), "relevant & like" (symbolized by a hand pointing upward), and "relevant & dislike" (symbolized by a hand pointing downward). Playing *EyeGrab*, the gamer scores for each correctly categorized image, receives negative points for each wrong one, and no points for images that fell off the screen without classification. No scores are obtained for the ratings of "like" and "dislike". An acoustic feedback is given for each classification. An applause is played for correct classifications, while a booing sound signals incorrect classifications and missed photos. A high score list is presented to the user at the end of the game.

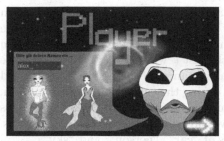 

(a) Entering of personal information   (b) Playing screen

Fig. 3. Screen shots from *EyeGrab*

5   Experiment Description

EyeGrab has been evaluated with 54 subjects (with 19 female). The subjects' ages were between 17 and 56 years (avg = 30 years, SD = 7.7). The majority of the participants were students or research fellows in computer science (70%), but students from other fields of study or members of other professional groups like restorers or psychotherapists participated in the experiment as well. Most subjects enjoyed playing the game. In a questionnaire subsequent to the experiment, 49 of the 54 subjects rated the statement "The game is fun." with a 4 or a 5 on a standard 5-point Lickert scale (avg = 4.22, SD = 0.72). The level of difficulty playing *EyeGrab* seems to have been adequate, as most of the participants did not agree with the statement "The game overexerts me." (M = 2.54, SD = 1). Most of the participants did not feel uncomfortable using the eye tracking device as shown by the low average agreement of 2.24 (SD = 1.15) to the statement "The eye tracker has a negative impact on my well-being."

Procedure. Every participant played four rounds of *EyeGrab*. The first round was a short test round consisting of only 12 photos. This test round with the category "tree" served as an introduction to the game. The data collected during this round was not used in the later analysis. The other three rounds with the categories "car", "person", and "sky" consisted of 24 photos each. The photos of each round were displayed in a randomized order. Different falling speeds were applied to each round. In the slowest pace (speed 1) the photos were falling with 3.6 pixels/ms, and they were visible on the screen for 5200 ms. In the medium pace (speed 2), the photos were visible for 4500 ms (pace = 4.3 pixels/ms). In the most challenging speed (speed 3) the photos were falling down within only 3800 ms (5 pixels/ms). A complete round took between 64,4 s (speed 1) and 50 s (speed 3). A Latin Square design was used in order to randomize the order of the three categories with the three different speed levels. The participants were asked to express their agreement to several statements on a 5-point Likert scale between 1 (strongly disagree) and 5 (strongly agree) in a questionnaire at the end of the experiment. The experiment was performed on a screen with a resolution of 1680 × 1050 pixels. The subjects' eye movements were recorded with a Tobii X60 eye tracker at a data rate of 60 Hz.

Data Set: Categories and Photos. The categories used in *EyeGrab* were taken from the top six of the list with the mostly used tags in LabelMe [8]. The LabelMe data set consists of photos, uploaded by the community, and has manually drawn region labels. The first two categories of this list ("window" and "building") are not taken into account because often not all instances of these objects are labeled on the photos. This could cause problems during the evaluation of our approach, as we need ground truth data with a complete labeling of all occurring objects belonging to the given category. Thus, we have taken the next top categories, which are the above-mentioned categories of "car", "person", and "sky".

In total, 84 photos (24 for each round and 12 for the test round) were selected from the image hosting page Flickr[3] and from LabelMe [8]. To create a challenge for the gamers, only 50% of the selected photos actually belonged to the given category. Thus, half of the photos were randomly chosen from the photos tagged with the given category, the other half from all other photos. An additional criterion for the selected photos was a minimum size of 450 pixels for one of the photo dimensions. All photos were scaled such that the longer edge has a length of 450 pixels. The 46 photos from Flickr belonged to the ones labeled as the most "interesting". For all photos in our experiment, we need ground truth information regarding the region labels. For the LabelMe images, manually drawn polygons describing the shapes of the depicted objects are part of the data set. Some photos had to be replaced after a manual check because not all occurrences of an object were labeled or an object described by the given category was depicted, although the photos were not labeled with it. For the Flickr images, the ground truth region labels were manually created by a volunteer not involved in the research.

6   Photo Classification Results

Excluding the test round, 72 photos in the three rounds were viewed by each subject. This makes a total of 3,888 photo views. In 260 cases (7%), the photo passed without classification, resulting in a total of 3,628 classified photos. 3,279 images (90%) were correctly classified. Overall, we had 1,624 correct classifications for photos belonging to the given category (true-positive), 1,655 correct classifications for photos not belonging to the given category (true-negative). Meanwhile, 241 classifications were false-negative (photo belonged to the category but was classified as not), and 108 classifications were false-positive, which leads to a precision of 94% and a recall of 87% over all users. The number of incorrect assignments per image lies between 2 and 40 with an average of 4. The three photos with the lowest error rate and the three photos with the highest error rate are depicted in Figure 4.

When comparing the error rate for different speed levels, we see that the number of unassigned or incorrectly assigned photos is increasing with the falling

---

[3] http://www.flickr.com/

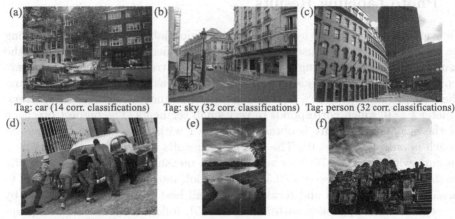

Tag: car (14 corr. classifications)   Tag: sky (32 corr. classifications)   Tag: person (32 corr. classifications)

Tag: car (52 corr. classifications)   Tag: sky (51 corr. classifications)   Tag: person (51 corr. classifications)

Fig. 4. Upper row: the three photos with the lowest number of correct classifications. Lower row: the photos with the highest number of correct classifications. All photos show an object described by the given category.

speed of the photos. The number of not-assigned photos is increasing from 7% to 12%. The number of incorrectly assigned photos is increasing from 4% to 11%. The number of unassigned photos is increasing more strongly than the incorrectly assigned photos. Thus, the subjects are still capable of deciding if an image belongs to a category or not, even with a higher speed level. However, they run out of time to focus each image for classification.

We compare the classification results of *EyeGrab* with results from the photo classifications in our previous experiment [15]. In the previous experiment, a specific tag was first presented to the subjects. Subsequently, a photo was presented to the user who had to decide whether an object described by the given tag is depicted. The decision was made by pressing a key on the keyboard. Of all classification, 5.4% were incorrect. In this work, 10% of all classifications are incorrect over all speeds. The slowest speed level with an error rate of 7% is close to the results observed in previous work [15].

The subjects were asked in the questionnaire how much effort they put into the subjective classification of the photos into "like" and "dislike". They answered this question with a mean value of 3.43 (SD = 1.35), which shows that their effort was not very high. Of the photos, 62% were rated as "like", the rest as "dislike". Of the Flickr images, 70%, were liked in comparison with 56% of the LabelMe images. As the Flickr photos were selected from the most interesting, we can assume that they are more attractive to most viewers than the LabelMe photos. This assumption is only reflected slightly in the rating results. In summary, the user gave a rating, but it does not seem to be of high quality. Thus, the rating information is not further considered in the remainder of the paper.

7   Photo Labeling Results

We evaluate our approach by analyzing the region labeling for all photos using the aggregated data of all users who correctly classified a photo. In Figure 5, the results for the region labeling using the different eye tracking-based measures are depicted by comparing precision and recall, as well as precision and F-measure. The best precision with 61% is obtained for the segmentation measure with parameter $k = 0$, which corresponds to very small segments. The highest precision for the heat map measure is obtained for $t = 1$ with 59%; for the baseline approach is only 19% ($k = 0$). The best recall results are 96% for the heat map measure with $t = 100$, 70% for segmentation measure with $k = 0.5$, and 53% for the baseline with also $k = 0.5$. We also look into the F-measure results to consider both, precision and recall. The overall best F-measure is obtained by the segmentation approach with 32% ($k = 85$), followed by the heat map approach with 31% ($k = 0.5$). The baseline approach clearly performs weaker, with a maximum result of 21% ($k = 0.5$). We applied a Friedman test to compare the results for the best performing parameters. We found that the differences are significant ($\alpha < .05$) for precision ($\chi^2(2) = 15.436, p = .000$) and F-measure ($\chi^2(2) = 18.048, p = .000$). A post-hoc analysis with pairwise Wilcoxon tests with a Bonferroni correction ($\alpha < .017$) showed two significant results for precision between heat map and baseline ($Z = -3.527, p = .000$) and segmentation and baseline ($Z = -3.704, p = .000$). No significance was measured in the post-hoc test for F-measures.

The results vary for the three categories "car", "person", and "sky". For example, the precision values for $k = 0$ are $p_{car} = 0.79$, $p_{person} = 0.28$, and $p_{sky} = 0.76$. This range of results seems to be caused by the sizes of the objects. The average size of the ground truth objects of the different categories are (compared with the whole image size) as follows: $size_{car} = 11.5\%$ (SD = 8.3%) , $size_{person} = 11.7\%$ (SD = 19.9%), and $size_{sky} = 42.8\%$ (SD = 23.1%). Although the $size_{car}$ and $size_{person}$ are similar, the high standard derivation for "person" shows that the object sizes vary strongly. Very small objects are known to complicate the region labeling [14].

In addition, we analyzed the region labeling results for the different falling speeds. A faster falling speed increases the pressure on the user to perform the classification. A summary of the results for the different speed levels can be found in Figure 6. It shows that the falling speed does not have a high impact on the precision and F-measure. For both eye tracking measures, the medium speed level delivers the best results. However, only minor differences can be noticed. Please note that the results for all speeds are not the average of all speed levels as the region labeling for the different speed levels is done with only one-third of the data. This is caused by the fact that every user played the game in three different speed levels (cf. Section 5). We conclude that the influence of the speed on the region labeling results is, at the least, not strong.

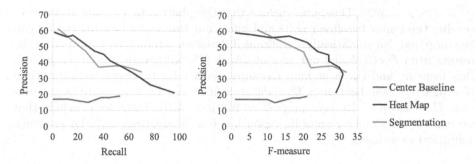

Fig. 5. Precision, recall, and F-measure results for the three labeling approaches. The curves are limited by the investigated parameters (e.g., the Center Baseline by the number of segmentation levels).

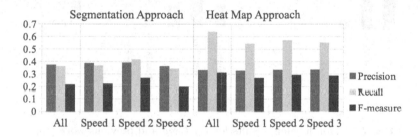

Fig. 6. Region labeling results for different falling speeds

8   Comparison with Previous Results

We have compared the results in terms of precision and F-measure from our *EyeGrab* experiment with the results obtained from our previous work [15]. The best performing parameters were determined in the previous work by means of a training set and applied to the test set of [15] and to the *EyeGrab* data. The parameters are $k = 0.1$ for the segmentation measure, $t = 95$ for the heat map measures, and $k = 0.4$ for the baseline. The results are depicted in Figure 7.

The segmentation measure performs best, while the baseline approach delivers clearly weaker results than both eye tracking methods. The F-measure results are more diverse. The differences between the two gaze-based measures and the baseline are less distinct for the *EyeGrab* data than for the data from the previous experiment [15] (i.e., the results between the measures and baseline in our earlier experiment differ more). Using the parameters from earlier work [15] for the *EyeGrab* analysis delivers only slightly better results for the segmentation approach than the baseline, whereas the heat map approach performs clearly better. We compare the center baseline results for photos of the first experiment [15] and *EyeGrab* data in a Mann-Whitney U test and do not obtain a significant difference, neither for precision ($U = 467, p = .291$) nor for F-measure

$(U = 446, p = .302)$. Thus, we conclude that the photo sets are comparable concerning the center baseline results and infer that the region labeling results can be compared. No statistically significant differences can be found comparing the results from *EyeGrab* and our previous work [15] with regard to the segmentation measure and the heat map measure, neither for precision (segmentation: $U = 528, p = .909$; heat map: $U = 480, p = .467$), nor for F-measure (segmentation: $U = 436, p = .19$; heat map: $U = 468, p = .376$). Thus, we conclude that we can obtain similar results in region labeling in *EyeGrab* and the previous, simplified experiment [15].

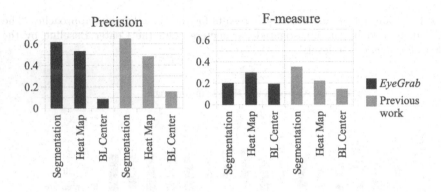

Fig. 7. Region labeling results for *EyeGrab* and previous work [15]

9 Conclusion

We have shown that the labeling of image regions is possible from data collected from subjects playing the immersive game-with-a-purpose *EyeGrab*. For one of two gaze-based measures, the results are comparable to those from a previous experiment [15]. This is quite interesting as the conditions for obtaining the gaze data are more difficult due to factors like time pressure and distraction by the gaming environment in *EyeGrab*. The region labeling results are only slightly influenced by different speed levels, which are forcing the subjects to make decisions on the photo classifications faster. As a broader spread of eye tracking hardware is assumed for the near future, it will become possible to use eye tracking technology in everyday tasks like image search on the web or for playing games. Thus, the results of our research may be applied for labeling image regions based on the gaze data obtained from users viewing the results of image search engines.

Acknowledgments. The research leading to this paper was partially supported by the EU project SocialSensor (FP7-287975). We thank our students D. Arndt, L. Buxel, K. Kramer, C. Neuhaus, C. Saal, H. Swerdlow, and M.-S. Usta.

References

1. Arbeláez, P., Maire, M., Fowlkes, C., Malik, J.: Contour detection and hierarchical image segmentation. IEEE TPAMI 33(5), 898–916 (2011)
2. Goldberg, J.H., Stimson, M.J., Lewenstein, M., Scott, N., Wichansky, A.M.: Eye tracking in web search tasks: design implications. In: Symposium on Eye Tracking Research & Applications, pp. 51–58. ACM (2002)
3. Klami, A.: Inferring task-relevant image regions from gaze data. In: Workshop on Machine Learning for Signal Processing. IEEE (2010)
4. Kompatsiaris, I., Triantafyllou, E., Strintzis, M.G.: A World Wide Web region-based image search engine. In: Conference on Image Analysis and Processing (2001)
5. Kozma, L., Klami, A., Kaski, S.: Gazir: Gaze-based zooming interface for image retrieval. In: Multimodal Interfaces, pp. 305–312. ACM (2009)
6. Lin, Y.-T., Lin, R.-Y., Lin, Y.-C., Lee, G.C.: Real-time eye-gaze estimation using a low-resolution webcam. In: Multimedia Tools a. Applications (2013)
7. Ni, Y., Dong, J., Feng, J., Yan, S.: Purposive Hidden-Object-Game: Embedding Human Computation in Popular Game. IEEE Transactions on Multimedia 14(5), 1496–1507 (2012)
8. Russell, B.C., Torralba, A., Murphy, K.P., Freeman, W.T.: Labelme: a database and web-based tool for image annotation. International Journal of Computer Vision 77(1-3), 157–173 (2008)
9. Santella, A., Agrawala, M., DeCarlo, D., Salesin, D., Cohen, M.: Gaze-based interaction for semi-automatic photo cropping. In: CHI, p. 780. ACM (2006)
10. Smith, J.D., Graham, T.C.: Use of eye movements for video game control. In: Advances in Computer Entertainment Technology. ACM (2006)
11. Viola, P., Jones, M.: Rapid object detection using a boosted cascade of simple features. In: Computer Vision and Pattern Recognition. IEEE (2001)
12. Von Ahn, L., Liu, R., Blum, M.: Peekaboom: a game for locating objects in images. In: Human Factors in Computing Systems, pp. 55–64. ACM (2006)
13. Walber, T., Neuhaus, C., Scherp, A.: Eyegrab: A gaze-based game with a purpose to enrich image context information. In: EuroHCIR (2012)
14. Walber, T., Scherp, A., Staab, S.: Identifying objects in images from analyzing the users's gaze movements for provided tags, pp. 138–148. Springer (2012)
15. Walber, T., Scherp, A., Staab, S.: Can you see it? two novel eye-tracking-based measures for assigning tags to image regions. In: Li, S., El Saddik, A., Wang, M., Mei, T., Sebe, N., Yan, S., Hong, R., Gurrin, C. (eds.) MMM 2013, Part I. LNCS, vol. 7732, pp. 36–46. Springer, Heidelberg (2013)

MR Simulation for Re-wallpapering a Room
in a Free-Hand Movie

Masashi Ueda, Itaru Kitahara, and Yuichi Ohta

University of Tsukuba, Department of Intelligent Interaction Technologies, Japan
{ueda,kitahara,ohta}@image.iit.tsukuba.ac.jp

Abstract. This paper proposes a wallpaper replacement method in a free-hand movie. Our method superimposes computer graphics (CG) wallpaper onto a real wall region in the free-hand movie looking around a room. To extract wallpaper planes, ordinarily a special and expensive 3D survey instrument is required. However, there is usually no such instrument in homes or offices where many users want to experience wallpaper replacement simulation. To solve this problem, we extract the wallpaper region by an image segmentation technique with user interaction. By applying our method, wallpaper replacement can be easily achieved using only a handy camera and a PC.

Keywords: Mixed-Reality, image segmentation, free-hand movie, wallpaper replacement, image stitching.

1 Introduction

When we purchase wallpaper, we imagine how a room will look with the new wallpaper. However, it often happens that when the selected wallpaper is installed, it is not suited for the room. This problem is caused by discrepancy between our imaginations and the actuality. To solve this problem, we propose a method that superimposes computer graphics (CG) wallpaper onto real wall regions in a free-hand movie while looking around a room. By using our method, a user can preview the room with the new wallpaper, and increase their chances of choosing the right one.

Figure 1 shows a result of our proposed method. To achieve such virtual redecoration by using Mixed Reality technique, ordinary, a 3D model of the target space is extracted in advance, and then a texture-mapping process is executed on the model. One of the most reliable ways to measure 3D shape is using a 3D survey instrument [1]. However, such a special and expensive instrument is typically not available in homes or offices where demand for wallpaper replacement simulation is greatest. Another way to conduct 3D reconstruction in a room is by applying the structure-from-motion technique to a captured video sequence of the target space [2]. However, in a small-scale space, accurate reconstruction is difficult, and the reconstruction error caused by clutters of the room seriously damages the quality of generated video. These days, RGB-D cameras that extract 3D information of an environment have become available to ordinary users [3], and it is possible to reconstruct the 3D

C. Gurrin et al. (Eds.): MMM 2014, Part I, LNCS 8325, pp. 436–448, 2014.
© Springer International Publishing Switzerland 2014

information by using a monocular camera [4]. However, when we consider the task in re-wallpapering, it is necessary to segment out thin objects attached to the wall, such as posters, as non-wall regions. Thus, we have to develop a segmentation method that is little dependent on the depth information of the environment. To solve these problems, we developed a method to apply the CG wallpaper without reconstructing the 3D information, but using 2D image segmentation techniques [5]-[8]. Our proposed method aims to accurately extract and replace the wallpaper region in a free-hand movie by an image segmentation technique with user interaction.

Fig. 1. MR Wallpaper replacement simulation: (left) an input free-hand movie, (right) the virtual wallpaper is applied

2   Wallpaper Replacement Method

An intuitive way to realize MR wallpaper replacement is to overlay virtual wallpapers generated by CG onto the real wall in front of users. Thanks to the excellent development of the camera tracking methods, real-time pose estimation of a camera capturing the user's view is available [18][19]. However, due to the high computational cost, it is still difficult to realize real-time generation of fine virtual wallpapers that cannot be distinguished from a real one. The indirect AR approach solves such problem by precisely overlaying fine quality virtual models in advance [20][21]. However, it is difficult to generate 3D virtual models that utilize the advantage of the approach. Thus, we develop a wallpaper replacement method without generating virtual models, but just extracting and replacing the wallpaper region in a video sequence.

Although our method extracts a wallpaper region by using an image segmentation technique utilizing simple user interaction, it is hard for users to perform segmentation at every frame of a free-hand movie. Therefore, we generated a panoramic image from the free-hand movie, and the user inputs the segmentation seeds for all frames at once in the panoramic image [9]. By projecting the seeds to each frame, segmentation seeds for each frame are estimated. Finally, the image segmentation process is executed at each frame.

Figure 2 shows the processing flow of the proposed system. In Fig. 2 (a), a user captures a free-hand movie of the target room with a hand-held camera. In Fig. 2 (b), a panoramic image is generated by using corresponding points in the video sequence. The user inputs seeds in the panoramic image to segment wallpaper and other regions. We use an interface similar to Interactive Graph Cuts. In Fig. 2 (c), the seeds from the panoramic image are projected to each frame of a captured video, and then the wallpaper region is estimated using GrabCut. If the user is not satisfied with the segmentation result, the segmentation regions are back-projected to the panoramic image, and additional seeds are added. Finally, in Fig. 2 (d), the CG wallpaper is superimposed on the estimated wallpaper regions in each frame.

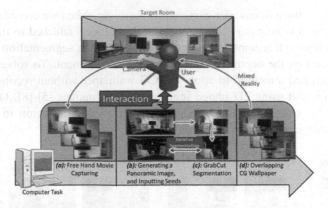

Fig. 2. Overview of proposed method

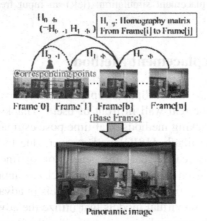

Fig. 3. Generating a panoramic image using homography matrix

3   Generating a Panoramic Image from a Free-Hand Movie

The seeds for image segmentation used in the GrabCut process are inputted in a pano-ramic image generated from a free-hand movie. Figure 3 shows generation of a pano-ramic image using homography matrices. A homography matrix is computed by using corresponding points between a target frame and the next frame. We use features of SURF [10], which is a robust and fast local feature detector, to acquire the corres-ponding points. RANSAC [11] is applied to estimate the homography matrix in order to reduce the influence of outliers. Then, each frame is projected to a base frame that is the middle of the captured sequence by using an integrated homography. If the camera motion is limited to the rotation around its optimal center or the shape of the target space is a single distant plane, the homography-based panoramic image stitch-ing works well [12]-[14]. However, it is difficult to achieve such conditions while capturing a hand-held movie in an indoor space like a home or office. To deal with these problems, we developed an adjusting method, in which we must always consider the quality of a panoramic image (especially, consistency of appearance), since it is important for users to input accurate information for image segmentation in the latter part.

3.1   Generating a Panoramic Image with Key Frames

To construct a panoramic image of a non-planar environment using the homography-based stitching approach, the camera motion should be limited to rotation around its optimal center. However, we use a hand-held camera for capturing, whose motion is not only rotation but also parallel translation. Therefore, the estimated homography matrix does not accurately transform all points in the image. In other words, transformation by the matrix has some error. Additionally, when each sequential frame is stitched to a panoramic image, the homography matrices are simultaneously integrated. In this process, the errors are also integrated. As a result, it causes serious damage to the generated panoramic image. To solve this problem, we extract key frames from a video sequence by calculating the overlapping regions of each frame. Our key frame approach is the same as Steedly's [13]. By using the key frame, it is possible to reduce the number of integration of homography matrices. As a result, the accumulation error is also reduced. Our algorithm for extracting the key frame is outlined in Figure 4. Here, the threshold ρ is set to 0.8.

The generated panoramic images are shown Figure 5. By comparing the panoramic image stitching with or without key frames, you can see that the appearance of the panoramic image with the key frame is improved.

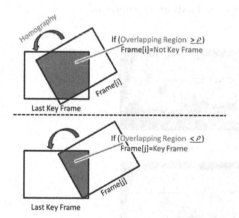

Fig. 4. Algorithm of our key frame extraction

Fig. 5. Generated panoramic image. (Top): Panoramic image without key frames. The red squared regions have distortion caused by stitching. (Bottom): Panoramic image with key frames.

3.2   Improving Panoramic Image Quality Using Graphcut Textures

The key frame-based image stitching effectively reduces the discontinuity of the generated panoramic image. However, when the parallel translation of the mobile camera becomes so large, the non-wall objects are cut into pieces in the panoramic image as shown in the upper stitching image of Figure 6. The discontinuity makes it difficult for users to input accurate information for image segmentation. Our method employs Graphcut Textures [15] proposed by Kwatra to obscure the discontinuity by using continuity of natural objects.

Graphcut Textures stitches multiple images by using the Graphcut algorithm. As illustrated in Figure 7, it formulates a graph that has nodes as pixels in the overlapping regions and edges connecting 4-connected neighbors' pixels. When we stitch image-$\mathbf{A}$ and image-$\mathbf{B}$, an edge connecting pixels s and pixels t has an energy $M(s,t,\mathbf{A},\mathbf{B})$ given by the following equation (1). Here, $\mathbf{A}(s)$, $\mathbf{B}(s)$, $\mathbf{A}(t)$ and $\mathbf{B}(t)$ are the pixel values at the points s and t of the images $\mathbf{A}$ and $\mathbf{B}$.

$$M(s,t,\mathbf{A},\mathbf{B}) = \|\mathbf{A}(s) - \mathbf{B}(s)\| + \|\mathbf{A}(t) - \mathbf{B}(t)\| \tag{1}$$

The boundary between the wall and non-wall region is estimated as a pass that cuts the graph with minimum energy. This energy minimization problem is solved by the Graphcut algorithm. When the Graphcut Texture stitches another image to the generated panoramic image, it is possible to obscure the discontinuity of the stitched image by remembering the energy values along with the boundaries estimated in the previous segmentation process.

Comparing the upper and bottom stitching images of Figure 6, you can see that the appearance of the panoramic image by Graphcut Textures is improved.

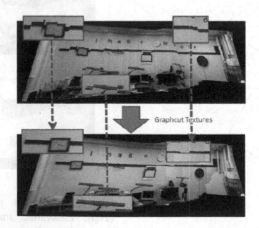

Fig. 6. Constructed panoramic images. (top): Panoramic image without Graphcut Textures. Red squared regions have an object cut off by the image border. (bottom): Panoramic image with Graphcut Textures.

3.3   Improving Panoramic Image Quality Using Poisson Blending

As an additional process to obscure the discontinuity in the stitched panoramic images, we apply Poisson Blending [16] that gradates the pixel values near the stitching boundary. As illustrated in Figure 8, the region-S is a region of the target image and the region is of the source image and $\partial\Omega$ is the boundary of region. Ω

The function f that gives a blended pixel value at the pixel p in region Ω is estimated by solving the Poisson equation formularized as equation (2).

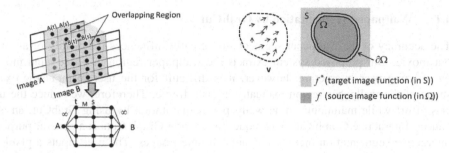

Fig. 7. Graph formulation for Graphcut Textures

Fig. 8. Vector field v in Ω and source / target regions

The function f that gives a blended pixel value at the pixel p in region Ω is estimated by solving the Poisson equation formularized as equation (2).

$$|N_p|f_p - \sum_{q \in N_p \cap \Omega} f_q = \sum_{q \in N_p \cap \partial\Omega} f_q^* + \sum_{q \in N_p} v_{pq} \qquad (2)$$

for all $p \in \Omega$,

Here, Np is the number of 4-connected neighbors of pixel p. pq denotes a pixel pair such that $q \in Np$. f_p^* is the value at p in the source image, and f_p is the value of f at p. vpq is the edge strength between p and q. By solving the Poisson equation, the target image and the source image are naturally stitched along with the boundary $\partial\Omega$ while preserving the edges (texture information) inside of region Ω .

The generated panoramic images are shown in Figure 9. By comparing with Figure 7, you can see that the discontinuity along with the boundary of stitching is well blurred without any harmful influence upon the other regions.

Fig. 9. Panoramic image with Poisson Blending

4   Wallpaper Region Extraction Using Image Segmentation

We use GrabCut to extract the wallpaper region in a free-hand video [17]. This section describes how to use GrabCut to estimate the wallpaper region in each frame, and how to project the segmentation seeds in a panoramic image to each frame of the video sequence. The segmentation results are back-projected to the panoramic image to add supplemental seeds, if the user is not satisfied with the segmentation result.

4.1   Wallpaper Segmentation by GrabCut

The accuracy of the wallpaper segmentation greatly influences the quality of an output movie. Our proposed system extracts the wallpaper region in a free-hand movie by utilizing user interaction; however, it is difficult for the user to input the exact shape of the wallpaper region manually in each frame. Therefore, we reduce the user's effort while maintaining fine wallpaper segmentation by using GrabCut, an advanced Interactive Graph Cuts technique. Interactive Graph Cuts is a general-purpose interactive segmentation method for monochrome images. The user inputs a pixel as "object" or "background". Afterwards, the histogram of gray value is learned and Graph Cuts finds the globally optimal segmentation. GrabCut extends Graph Cuts to color images, and the learning phase of pixel color is improved with the Gaussian Mixture Model. Additionally, GrabCut alternates between estimation and parameter learning to solve the min-cut problem. Therefore, the user interaction can be more relaxed and simple. Figure 10 shows wallpaper region extraction by GrabCut.

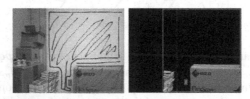

Fig. 10. GrabCut Segmentation. (left): The green rectangle is a segmentation area. The red and blue seeds are user input of foreground (object) and background (wallpaper). (right): Segmentation result by GrabCut. The black region in the green rectangle is a segmented wallpaper region.

4.2   Wallpaper Segmentation Seeds in Captured Video Sequence

The segmentation seeds inputted in the panoramic image are projected to each frame of a captured video sequence by using the inverse matrix of the homography that is given when the panoramic image is stitched. GrabCut segmentation extracts the wallpaper region in each frame by referring to these seeds.

When a user inputs the seeds only in regions of key frames that are visible in the panoramic image, only parts of the labels are projected to each frame since most frames are invisible in the panoramic image (i.e. occluded by key frames). On the other hand, when all seeds are projected to every frame just referring to the inputted position in the panoramic image, accuracy of the segmentation in each frame is damaged by the projection errors because there is parallax between the key frames and the invisible frames. To achieve accurate projection of these seeds, we adjust the position by detecting the correspondence between the panoramic image and every frame.

Figures 11 and 12 illustrate the idea of the adjustment process. A visible frame (key frame) on the panoramic image is Frame[f], and an invisible frame is Frame[b]. Homography matrices projecting a point in Frame[f], Frame[b] to the panoramic image are *Hf, Hb* respectively. The i-th feature points detected by SURF in Frame[f] and Frame[b] are *Ptf,i, Ptb,i* (i.e. *Ptf,i* is visible on the panoramic image, and   *Ptb,i* is not).

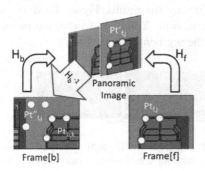

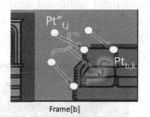

Fig. 11. Feature point projections in hidden frame

Fig. 12. Seed adjustment using corresponding feature points. (Blue dotted line): Projected user input. (Red line): Adjusted user input. (Purple arrows): Parallel translation vectors.

Here, *Ptf,j* and *Ptb,k* are corresponding feature points between Frame[f] and Frame[b]. When *Ptf,j* are projected to a panoramic image by using *Hf*, the projected points *Pt'f,j* are also visible in the panoramic image.

$$Pt'_{f,j} = H_f Pt_{f,j} \tag{3}$$

However, when *Pt'f,j* are projected to Frame[b] by using H_b^{-1}, the projected points *Pt''f,j* do not correspond with *Ptb,k* but *Ptf,j*. This is the parallax between the key frame (Frame[f]) and an invisible frame (Frame[b]).

$$Pt''_{f,j} = H_b^{-1} Pt'_{f,j} \tag{4}$$

This parallax is caused by an approximation error to fit Frame[b] to Frame[f] (key frame) (as described in Section 3.1). Then, when *Pt'f,j* are the inputted segmentation seeds by a user in the panoramic image, the projected point *Pt''f,j* should be adjusted to *Ptb,k*. In our method, as shown in Figure 11, the parallel translation vector *Pt''f,jPtb,k* is estimated by the corresponding relationship between *Ptf,j* and *Ptb,j* and the translation relationship between *Ptf,j* and *Pt''f,j* (via *Pt'f,j*). By calculating the weighted average of the translation vectors of the three nearest neighbors to the target point, it is possible to translate all pixels in Frame[b].

Figure 13 shows the result of our adjustment method. You can see that segmentation seeds are accurately projected to a frame.

4.3 Extracting Wallpaper Region by Iterative Interaction

GrabCut can improve the accuracy of image segmentation by inputting additional seeds. However, it is time-consuming to check the segmentation results in all captured frames. Thus, our system generated a panoramic image by stitching the segmented images, and the user can input additional seeds while confirming the results. Figure 14 shows the iterative interaction processes for image segmentation. You can see that the process, especially in the dotted-circle region, improves the segmentation accuracy.

Fig. 13. Comparison between non-adjusted and adjusted seeds. (top): Panoramic image and inputted seeds. (bottom left): Projected seeds without adjustment. (bottom right): With adjustment.

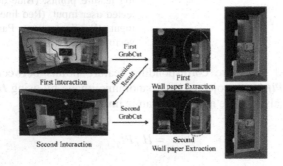

Fig. 14. Extracting wallpaper region by iterative interaction

5   Virtual Wallpaper Superimposing

5.1 Superimposing on a Panoramic Image

As shown in Figure 15, virtual wallpaper is manually superimposed on a real wall observed in a panoramic image while matching the four corners. If parts of the corners are not observed in the panoramic image, we estimate the position by using the edge information of the wall and other artificial objects such as a poster, shelf and desk.

Fig. 15. Superimposing virtual wallpaper onto a real wall in a panoramic image. The red points denote vertices of virtual wallpaper.

5.2   Projecting Virtual Wallpaper to Each Frame of a Free-Hand Movie

As shown in Figure 16, the virtual wallpaper (tentatively) superimposed on a panoramic image is projected to each frame of the captured free-hand movie by referring to the homography matrices estimated in Section 3. The projected virtual wallpaper is divided into visible and invisible regions, and then only the visible region is finally superimposed onto a frame.

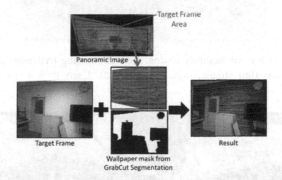

Fig. 16. Superimposing virtual wallpaper onto each frame of captured free-hand movie

6   MR Simulation for Re-wallpapering a Room

We developed a pilot system by implementing our proposed method and conducted a questionnaire survey (usability testing) with three subjects who participated in a trial of MR re-wallpapering. The experimental environment is shown in Figure 17. We conducted the experiments with two different layouts of the room. One was a simple layout in which all objects existed near the wall so that little occlusion occurred in the free-hand movie. The other was more complicated (i.e. some objects existed far from the wall).

At first, we observed the user who captured the free-hand movie. In the simple layout environment, all users captured the wall in a standing position. On the other hand, in the complicated environment, two of three users chose to capture the wall while walking around. In the questionnaire, they mentioned that they thought it was better to capture the wall region as much as possible without any occlusion. In such a situation, we confirm that our image stitching approach works better. As a result, panoramic images are generated as shown in Figure 18.

In the second survey, we observed subjects inputting seeds for image segmentation. As shown in Figure 18, although the discontinuity of the generated panoramic image is well disguised by applying Graphcut Textures and Poisson Blending, the image of the complicated layout is distorted by the motion parallax among the captured frames. We are concerned that the distortion makes it difficult to input image segmentation seeds. However, according to the questionnaire, the users did not think it had a harmful influence, and the image segmentation also supports their position. This result shows that if the discontinuity is well disguised, the users can input accurate seeds for image segmentation by using their own visual ability.

Finally, the re-wallpapering process is achieved as shown in Figure 19. By using our approach, even in a complicated environment, it is possible to accurately segment the wall region and superimpose the virtual wallpaper.

Fig. 17. Environment for our usability testing. Re-wallpapering trial was conducted with two different room layouts (left simple, right: complicated). Users took a free-hand movie from around the blue line.

Fig. 18. Generated panoramic images. Left: simple layout, Right: complicated layout.

Fig. 19. Examples of re-wallpapering in a room. Upper: simple layout, Bottom: complicated layout.

7   Conclusion

This paper proposed a re-wallpapering method in a free-hand movie. Our method superimposed CG wallpaper onto a real wall region in a movie while looking around a room. To extract wallpaper planes without a special and expensive 3D survey instrument, we extracted the wallpaper region by an image segmentation technique while utilizing user interaction and image stitching algorithms. By applying our method, wallpaper replacement visualization was easily achieved using only a handy camera and a PC.

References

1. Ohta, Y., Tamura, H.: Mixed Reality-Merging Real and Virtual Worlds, Ohmsha (1999)
2. Kanbara, M., Okuma, T., Takemura, H., Yokoya, N.: Real-time Composition of Stereo Images for Video See-through Augmented Reality. In: Proceedings of IEEE International Conference on Multimedia Computing and Systems, vol. 1, pp. 213–219 (1999)
3. Newcombe, R.A., Izadi, S., Hilliges, O., Molyneaux, D., Kim, D., Davison, A.J., Kohli, P., Shotton, J., Hodges, S., Fitzgibbon, A.: Kinect-fusion: Real-time dense surface mapping and tracking. In: Proceedings of IEEE International Symposium on Mixed and Augmented Reality, pp. 99–107 (2011)
4. Snavely, N., Seitz, S.M., Szeliski, R.: Modeling the world from Internet photo collection. International Journal of Computer Vision 80(2), 189–210 (2007)
5. Mortensen, E., Barrett, W.: Intelligent scissors for image composition. In: Proceedings of SIGGRAPH, pp. 191–198 (1995)
6. Kass, M., Witkin, A., Terzopoulos, D.: Snakes: Active contour models. In: Proceeding of IEEE International Conference on Computer Vision, pp. 259–268 (1987)
7. Chuang, Y., Curless, B., Salesin, D.H., Szeliski, R.: A Bayesian Approach to Digital Matting. In: Proceedings of IEEE Computer Vision and Pattern Recognition, vol. 2, pp. 264–271 (2001)
8. Rother, C., Komogorov, V., Blake, A.: GrabCut: Interactive Foreground Extraction Using Iterated Graph Cuts. ACM Transactions on Graphics 23(3), 309–314 (2004)
9. Rav-Acha, A., Kohli, P., Rother, C., Fitzgibbon, A.: Unwarp Mosaics: A new representation for video editing. ACM Transactions on Graphics 27(3) (2008)
10. Bay, H., Tuytelaars, T., Van Gool, L.: SURF: Speed Up Robust Features. In: Leonardis, A., Bischof, H., Pinz, A. (eds.) ECCV 2006, Part I. LNCS, vol. 3951, pp. 404–417. Springer, Heidelberg (2006)
11. Fischler, M.A., Bolles, R.C.: Random Sample Consensus: A Paradigm for Model Fitting with Applications to Image Analysis and Automated Cartography Communications of the ACM 24(6), 381–395 (1981)
12. DiVerdi, S., Wither, J., Höllerer, T.: All around the map: Online spherical panorama construction. Computers and Graphics 33(1), 73–84 (2009)
13. Steedly, D., Pal, C., Szeliski, R.: Efficiently Registering Video into Panoramic Mosaics. In: Proceeding of IEEE International Conference on Computer Vision, pp. 1300–1307 (2005)
14. Brown, M., Lowe, D.: Automatic Panoramic Image Stitching using Invariant Features. International Journal of Computer Vision 74(1), 59–73 (2007)
15. Kwatra, V., Schodl, A., Essa, I., Turk, G., Bobick, A.: Graphcut Textures: Image and Video Synthesis Using Graph Cuts. Proceedings of ACM Transactions on Graphics 22(3), 277–286 (2003)
16. Perez, P., Gangnet, M., Blake, A.: Poisson Image Editing. ACM Transactions on Graphics 22(3), 313–318 (2003)
17. Boykov, Y., Jolly, M.-P.: Interactive Graph Cuts for Optimal Boundary & Region Segmentation of Objects in N-D Images. In: Proceedings of International Conference on Computer Vision, vol. 1, pp. 105–112 (2001)
18. Klein, G., Murray, D.: Parallel Tracking and Mapping for Small AR Workspaces. In: Proceedings of International Symposium on Mixed and Augmented Reality (ISMAR 2007), pp. 225–234 (2007)

19. Newcombe, R.A., Davison, A.D., Izadi, S., Kohli, P., Hilliges, O., Shotton, J., Molyneaux, D., Hodges, S., Kim, D., Fitzgibbon, A.: KinectFusion: Real-time dense surface mapping and tracking. In: Proceedings of International Symposium on Mixed and Augmented Reality (ISMAR 2011), pp. 127–136 (2011)
20. Jason, W., Tsai, Y., Azuma, R.: Indirect Augmented Reality. Computers & Graphics 35(4), 810–822 (2011)
21. Okura, F., Kanbara, M., Yokoya, N.: Interactive Exploration of Augmented Aerial Scenes with Free-Viewpoint Image Generation from Pre-Rendered Images. In: Proceedings of International Symposium on Mixed and Augmented Reality (ISMAR 2013), pp. 279–280 (2013)

Segment and Label Indoor Scene
Based on RGB-D for the Visually Impaired

Zhe Wang, Hong Liu, Xiangdong Wang, and Yueliang Qian

Key Laboratory of Intelligent Information Processing &&
Beijing Key Laboratory of Mobile Computing and Pervasive Device
Institute of Computing Technology, Chinese Academy of Sciences
Beijing 100190, China
{wangzhe01,hliu,xdwang,ylqian}@ict.ac.cn

Abstract. The growing study in RGB-D sensor and 3D point cloud have made new progress in obstacle avoidance for the visually impaired. However, it remains a challenging problem due to the difficulty in design a robust and real-time algorithm. In this paper, we focus on scene segmentation and labeling. As man-made indoor scene contains many planar area and structure, plane segmentation and classification is important for further scene analysis. This work propose a multiscale-voxel strategy to reduce the effects of noise and improve plane segmentation. Then the segmentation result is combined with depth data and color data to apply graph-based image segmentation algorithm. After that, a cascaded decision tree is trained to classify different segments into different semantical type. The method is tested on part of the NYU Depth Dataset. Experimental results show that the proposed method combines the advantages of depth data and the geometry characteristics of the scene, and improves scene segmentation and obstacle detection.

Keywords: RGB-D, Plane Segmentation, Scene Segmentation, Obstacle Detection, Blind Navigation.

1 Introduction

In recent years, obstacle avoidance system or Electronic Travel Aids(ETA) [1] have been developed to help the visually impaired to walk safely. ETA takes advantage of modern sensing devices to obtain information of the surrounding environment and the detected information is converted to non-visual signal that can be received and understood by the visually impaired. Among different sensors for ETA, image sensors can provide the most abundant surrounding visual information. Traditionally, stereo vision is used to obtain the depth data of the scene. But it requires large amount of calculation and sensitive to changes in illumination, occlusion, shadow, etc. Coming with the RGB-D sensors, like the Microsoft Kinect, it becomes very easy to obtain both RGB and depth data. Using the Kinect sensors in scene segmentation and obstacle detection becomes a key research topic.

C. Gurrin et al. (Eds.): MMM 2014, Part I, LNCS 8325, pp. 449–460, 2014.

Fig. 1. Output of our system

However, RGB-D image segmentation is not an easy task. The depth data captured by the Kinect sensors is very noisy and the noise is not uniformly distributed. In addition, many image segmentation and object classification algorithms for 2D image processing are not suitable for 3D point cloud. For pointwise classification, it is also very hard to design effective 3D features that is invariant to translation, rotation and scaling. Furthermore, the lack of labeled dataset that is suitable for scene analysis for the visually impaired is another problem. So the classification method should be effective using a small training dataset.

This paper proposes a framework to semantically segment and detect object for indoor scene. Fig. 1 shows the sample output of the system. The original result is generated on point cloud and we project it to 2D image for better presentation. There are two major contributions in our work. Firstly, we propose a multiscale-voxel plane segmentation algorithm which build two different voxel grids on the point cloud. A dense voxel grid is built on the near area of the scene and a sparse voxel grid is built on the far area of the scene. Using multiscale voxel could solve the problem of un-uniformly distributed noise of depth data. Secondly, we extract segment-wise geometry and spatial features to train a cascaded decision tree. The extraction of segment-wise feature is fast and robust in point cloud. And compared with pointwise features, the training on geometry feature and spatial feature requires less data. The experimental results show the proposed algorithms perform well and are applicable in obstacle detection tasks for the visually impaired.

2   Related Work

In recent years, computer vision based ETA have been developed very fast and achieved new progress. Lin et al. [2] develop a wearable stereo vision system composed of an eyeglasses and an embedded processing device to help avoid obstacles in real-time. Zoellner et al. [3] uses the Microsoft Kinect and optical marker tracking to help visually impaired people find their way inside buildings. It provides continuous vibration feedback on the user's waist to give an impression of the environment. And the optical markers can be used to tag points of interest to enable synthesized voice instructions. However, these systems failed to give any semantical description of the scene. So the user could only accept "instruction" and not aware of what the whole scene is like. Tian et al. [4] proposed a computer vision-based wayfinding aid for blind people. Firstly, doors, elevators, and cabinets are detected on their general geometric shape. Then the intra-class objects are recognized using optical character recognition (OCR) software on extracted text regions. Lee et al. [5] incorporate visual odometry and feature based metric-topological Simultaneous Localization And Mapping (SLAM) into the navigation system. Then a vicinity map based on dense 3D data obtained from RGB-D camera is built to do path planning. These methods only focus on some specific part of the scene. So the user can't receive the information about the whole scene.

Semantical scene analysis could help the visually impaired know better of the surrounding environment. And there have been many development in RGB-D scene analysis method recently. Silberman et al. [6] use depth for bottom-up segmentation and use context features to infer support relationships in the scene. Ren et al. [7] use kernel descriptors on superpixels and use a Markov Random Field (MRF) on superpixel with segmentation tree to model the context of the scene. Choi et al. [8] use 3D geometric phrase model to capture the semantic and geometric relationship between objects which frequently co-occur in the same 3D spatial configuration and then understand the indoor scenes. Gupta et al. [9] propose algorithms for object boundary detection and hierarchical segmentation. Their algorithms visit the segmentation problem afresh from ground-up and develop a gPb like machinery to combine depth information naturally. Wang et al. [10] propose a label propagation method to utilize the existing massive 2D semantic labeled datasets such as ImageNet. Koppula et al. [11] parse the indoor scene with RGB-D data in a mobile robots. A full 3D reconstruction is applied with multiple views of the scene acquired with a Kinect sensor. Then the 3D point cloud is over-segmented and used as underlying structure for a MRF model. These methods focus on the algorithm for general scene segmentation and labeling, while lacking specific analysis for the visually impaired. Wang et al. [12] use hough transform to extract the concurrent parallel lines on the RGB channels and then use depth information to distinguish stairs from pedestrian crosswalks. Then stairs are be recognized as upstairs and downstairs. These methods are mainly focus on the accuracy of scene segmentation while neglect the efficiency of the algorithm which, however, is a key factor in our work. Liu et al. [13] use a graph-based segmentation algorithm which combines the result of plane segmentation and RGB-D data. The method is more focused on the efficiency of the algorithm. However, in order to help the visually impaired to know better of the scene, more semantical analysis, like the type of different structures, should be conducted.

In man-made indoor environment, there exist many planes which contain much structural information. Extracting these planes could be very helpful in scene segmentation. There are many plane segmentation algorithms in literature. One way to extract planes is applying 2D segmentation methods on 3D data. However, this approach performs badly if two planes are very close to each other. In order to take advantage and make full use of 3D data, many new methods have been proposed. Holz et al. [14] compute local surface normal of point clouds using integral images. And then the points are clustered, segmented, and classified in both normal space and spherical coordinates. This method achieves a frame rate at $30Hz$ at the resolution of 160×120 pixels. Dube et al. [15] use Randomized Hough Transformation to extract planes from depth images. This algorithm could run in real-time on a mobile platform. However it can only detect planes, and cannot segments out the planes. Wang et al. [16] propose a two-step fast plane segmentation algorithm which combines the speed of voxel-wise cluster and the accuracy of pixel-wise process. The algorithm is fast and robust if the 3D data is accurate and don't have much noise. However, due to the limitation of Kinect sensor, the 3D data that is more than 3 meters away from the sensor is inaccurate and very noisy. So, this algorithm could not properly extract planes 3 to 5 meters away from the sensor.

3   Overview of the System

The framework of our whole system is shown in Fig. 2. At first, the input RGB-D data is preprocessed. The position of the RGB camera and the depth camera are not the same, so they should be aligned to the same camera coordinate. The second step is scene segmentation, including plane extraction and graph-based image segmentation. Firstly, the proposed multiscale-voxel algorithm is applied to do plane extraction. Then the result is introduced into the graph-based image segmentation. The third step is segment classification. At first the segment is classified as planes and obstacles. Then planes are classified into different types. Segment classification involves training and testing. In training, the dispersion for each segments is calculated and the multilevel thresholds is determined. And then a cascaded decision tree is trained on geometry and spatial features. In testing, each segment is classified by dispersion into planes or obstacles. Then planes are further classified into different types using the trained cascaded decision tree. Thus, the whole scene is semantically labeled.

4   Scene Segmentation

Scene segmentation is the first step in our system. We propose a fast plane segmentation algorithm based on multiscale voxels. The algorithm is fast and robust to the noise in depth data. After plane segmentation, the results are used as part of the weights in graph-based image segmentation which also combines RGB data and depth data.

4.1   Fast Plane Segmentation with Multiscale Voxels

This paper proposes a novel plane segmentation algorithm to improve the two-step fast plane segmentation algorithm.

The two-step plane segmentation algorithm has two stages [16]. At the first stage, a voxel grid is created on the 3D point cloud data to achieve sparse sampling and noise suppression. For each voxel, the least square method is used to fit a plane and get the plane normal vector. Then region growing algorithm

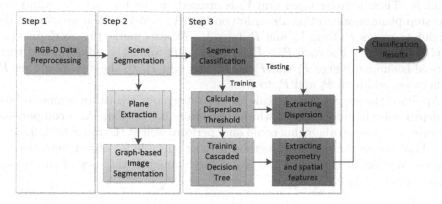

Fig. 2. System architecture

is applied to cluster the voxels according to the normals. At the second stage, accurate judgment is done for each point in un-clustered voxels. After plane segmentation, if the angle between the normal of two different planes is close to zero, these two planes is merge as one.

The size of the voxel grid directly affects the result of plane segmentation. Using an inappropriate side length of the voxel would probably obtain seldom planes. However, as mentioned before, the quality of the 3D point cloud data varies according to the distance. The depth data near the sensor would have less noise and higher resolution than the depth data far away from the sensor. As a result, using one voxel grid with fixed size cannot achieve the best segmentation result. For example, using a set of fine parameters, planes near the sensor can be extracted out well, but further planes may not be well extracted out because they would be considered as non-planar area under this condition.

This paper proposes a multiscale-voxel strategy to solve this problem. Firstly, a dense voxel grid, which has short side, is created on the whole 3D point cloud data. Using this voxel grid, we apply the two-step plane segmentation algorithm and get several planes. Secondly, a sparse voxel grid, which has comparatively long side, is created on the further part of the 3D point cloud. In each voxel grid, the two-step segmentation algorithm would extract different planes. Finally, overlapping areas are merged together.

Formally, a dense voxel grid V_d is created on the whole point cloud C. Then a threshold T is set to separate the point cloud into two parts: the near part C_n and the far part C_f. That is,

$$\forall p_i(x, y, z) \in C, p_i \in \begin{cases} C_n, & p_{iz} < T \\ C_f, & p_{iz} \geq T \end{cases} \quad (1)$$

where $p_i(x, y, z)$ represents the 3D coordinate of a point, p_{i_z} is the depth of point p_i. Then a sparse voxel grid V_s is created on the far part C_f. Apply the two-step plane segmentation algorithm on both V_d and V_s. Then we can get the candidate planes P_d from V_d and P_s from V_s. At last, merge the overlapping or duplicated planes. For each $P_i \in P_d, P_j \in P_s$, if P_i and P_j is adjacent and their normal is similar, merge P_i and P_j as P_n and add P_n to the final plane set P. Otherwise, add both P_i and P_j to P.

Applying the proposed multiscale voxel strategy, our algorithm is more robust to depth noise in indoor scenes which range from $0.5m \sim 5m$. As a comparison, the original two-step algorithm could only perform well at the range from $0.5m \sim 3m$. That means the proposed method is more suitable for object detection for the visually impaired because $3m$ is not safe enough if the user of the system moves relatively fast.

4.2 Graph-Based Image Segmentation

After plane segmentation, we apply graph-base image segmentation [17] on the image. The graph-based image segmentation algorithm is a fast unsupervised segmentation method, which is suitable in our task. The approach in Liu [13] is adopt to combine the result of plane segmentation in graph-based image segmentation. The weight for RGB and depth data is defined as

$$\omega(v_i, v_j) = \alpha \cdot \omega_{RGB}(v_i, v_j) + \delta \qquad (2)$$

where $\omega_{RGB}(v_i, v_j)$ is the difference of the RGB values between two pixels. And if v_i and v_j belongs to the same plane, α is set a small value in order to reduce the RGB difference in weight. Otherwise, α is set a value that is close to 1 in order to let the RGB difference become a main factor in weight. δ is used to balance the value under different condition so that the weight would not be too small or too big. In this way, the plane segmentation result is semantically combined with RGB and depth data, which can improve the result of image segmentation.

5 Segments Classification

After image segmentation, the scene has been divided into several segments. The next step is to classify these segments into different semantic blocks. In indoor scene, there are many planes which usually come from ground, walls and the surface of tables. These information is very important in understanding the scene. Other non-planar parts can treat as obstacles. So the first step is to divide the segments into planes and obstacles and then planes are classified into different types. In addition, small segments is ignored because they usually are fragments of a large objects and don't have significant affects in scene understanding. So only large segments would be classified and small segments would be merged into adjacent large segments.

5.1 Plane-Obstacle Classification

Although in plane extracting we have already known the extracted segments are planes, after image segmentation these segments would changed and many new segments are generated. So each segments should be classified again. The dispersion of all the points in a segment indicates whether the segment is a plane or not. Planar segments have low degree of dispersion and Non-planar segments have high degree of dispersion. In order to quantify the degree of dispersion, the least square estimation is used to fit a plane for all the points of each segment. It is obvious that if a segment is a plane, the average distance between the points to the plane is small and if a segment is not a plane, the average distance would be large. So, the average distance between the points to the fitting plane is used to represent a segment's dispersion. Formally, we assume that the plane equation of a segment is $Ax + By + Cz + 1 = 0$. Then the least square estimation for A, B, C is

$$\begin{bmatrix} A \\ B \\ C \end{bmatrix} = \begin{bmatrix} \sum x_i^2 & \sum x_i y_i & \sum x_i z_i \\ \sum x_i y_i & \sum y_i^2 & \sum y_i z_i \\ \sum x_i z_i & \sum y_i z_i & \sum z_i^2 \end{bmatrix}^{-1} \begin{bmatrix} \sum x_i \\ \sum y_i \\ \sum z_i \end{bmatrix} \qquad (3)$$

where $(x_i, y_i, z_i), i = 1, 2, 3, ..., N$ represents the coordinates of points in the segment. Secondly, the dispersion of the segment is calculated as

$$D = \frac{\sum_{i=1}^{N} F(p_i, P)}{N} \qquad (4)$$

where $F(p_i, P)$ represents the distance of point p_i to plane P.

The dispersion of every segment is calculated. If the dispersion is large, the segment would be an obstacle. Otherwise, the segment is more likely to be a plane. Then a threshold of dispersion is set to divide the segments into plane and obstacle. However, a fixed threshold is not suitable because of noise and imprecise image segmentation. We know that the depth data from Kinect sensors are very noisy and the image segmentation cannot be so precise. Moreover, larger segments tend to have more noise and the larger the segment the more imprecise points it would contain. Considering this, a multilevel threshold strategy is applied. In practice, we conduct statistic analysis on the labeled dataset and determine different thresholds for segments according to their size.

5.2 Plane Classification

In indoor environment, there exists many planes. The ground plane, wall plane, and table are the largest ones and they are the major factors that define the structure of the scene. Therefore, as we have got planes from the former step, they should be classified into different types so that useful information could be provided to the visually impaired. In this paper, we define three types of plane: ground, wall, and table. In addition, the side face of large cabinets are considered as wall because they impede walking just like walls. And the top face of beds, sofa or small cabinets are considered as table because people would put things on them just like tables.

As the obstacle detection system should run at a high frame rate, the plane classification algorithm should be efficient. In classification, feature extracting usually is the most time consuming step. Therefore, extracting features on pixel scale is not appropriate. On the other hand, ground, wall and table have significant structural differences which reflected in their geometry features and spatial features. It is obvious that the ground plane is horizontal and the wall plane is vertical and the table plane is much higher than the ground. So we extract the angle between the normal and the horizontal plane to represent the geometry feature of that plane.

Before that, we should determine the horizontal plane. We know that in the camera coordinate, the vertical axis may not be vertical in real world because the camera is always moving and rotating. In order to know the real horizontal plane, a two-step horizontal-plane-finding strategy is proposed. Firstly, the angle a_i between the normal n_i of each segment (vector (A, B, C) in Equation 3) and the vertical vector N_v of the camera coordinate (usually one of the axis in the coordinate system) is calculated. And all the normals is divided into approximate vertical and approximate horizontal according to the angle. More specifically,

$$n_i \in \begin{cases} N_v, & a_i < \frac{\pi}{2} \\ N_h, & a_i \geq \frac{\pi}{2} \end{cases} \tag{5}$$

Secondly, the average vertical normal n_v is calculated of the approximate vertical normals set N_v. Then n_v is the normal of real horizontal plane. Next, the angle between the segment normal and the vertical normal n_v is calculated as one of the features for classification. Another feature is the vertical distance between the segment and the lowest point of the scene. It can be easily calculated since we already have the plane equation of each segment and the vertical normal of the scene.

After extracting these features, we train decision trees on the training data. As the geometry feature and spatial feature have semantical meanings, we could use them as priors when training the decision trees. The geometry feature could indicate whether a plane is vertical or horizontal and all vertical planes in the indoor scene are considered as walls as discussed before. So at the first step, a decision tree is trained just on geometry feature to classify vertical plane and horizontal plane. Next step, for all horizontal planes, the spatial feature could indicate whether a plane is ground or table. Therefore, another decision tree is trained on spatial feature to classify ground and table. At last, the two decision trees are cascaded as one classifier. Taking advantage of the semantical meaning of the features, the manually intervened decision tree is more rapid and accurate. And the structure of the decision tree is simple and explainable, which makes the analysis of the training and testing results easily and clearly.

6   Experimental Results

6.1   Dataset

The data used in our experiments come from the NYU Depth Dataset [6]. The
NYU Depth Dataset is comprised of video sequences from a variety of indoor
scenes recorded by both the RGB and Depth cameras from the Microsoft Kinect.
The resolution of RGB images and depth data are both 640×480 pixels. In this
paper, 89 pairs of RGB and Depth images are selected. The dataset is manually
labeled for ground, wall, table and obstacle.

6.2   Plane Segmentation

We test the proposed multiscale-voxel segmentation algorithm on the whole
dataset. In our experiments, a dense voxel grid with side of $20cm$ is created
on the area range from $0.5m$ to $3m$ and a sparse voxel grid with side of $30cm$ is
created on the area range from $3m$ to $5m$. The segmentation result is compared
with the original fixed-size-voxel segmentation algorithm [16], which is also applied on our dataset. Some of the result are shown in Fig. 3. The first row is the
original RGB images, the second row is the segmentation result using fixed-size-
voxel algorithm and the third row is the segmentation result using the proposed
multiscale-voxle algorithm. The difference between the two segmentation algorithm is mark with yellow circles. We can see that in the near area both of the
two algorithms perform well in extracting planes. However, in the far area, the
fixed-sized-voxel fails to extract the planes. In comparison, the multiscale-voxel
algorithm could also extract the planes in far area as well as in the near area.
The experimental results indicate that the multiscale strategy is more robust to
noise. It is because when the depth of field get larger the depth data get more
noisy. This proves the proposed multiscale strategy is suitable for the depth data
generated by the Kinect sensors and have better plane extracting results.

6.3   Segments Classification

In order to determine the appropriate thresholds for plane-obstacle classification,
we calculate the dispersion for every plane on the training dataset after above
segmentation. Then the ground truth type of each segment can be determined
by the annotation. For each segment, if it is annotated as a plane, the dispersion
is calculated using Equation 4. Then the segments is divided into three different
set according to their size (the number of points in a segment). Empirically, we
define them as: small segments (size < 5000), medium segments ($5000 \leq$ size
< 10000) and big segment (size$\geq$ 10000). For each set, the average $\bar{d}$ of top
10% largest dispersion is calculated. However, due to the imprecision of image
segmentation, the average number would be too strict for classification. So we use
a compensation coefficient α to adjust it. Then the threshold for a set is $(1+\alpha)\cdot\bar{d}$.
In our experiment, α is 0.1. After the thresholds have been determined, we run
plane-obstacle classification on the testing dataset. We have 89 labeled images

Fig. 3. Plane segmentation results of fixed-size voxels and multiscale voxels. Different planes are randomly colored.

for our test, which consist 262 segments of wall, 112 segments of ground, and 91 segments of table. Half of the segments is used for training, and the others for testing.

After plane-obstacle classification, we test the proposed plane type classification. We conduct three set of experiments. The first is training decision tree only on geometry feature. The second is training decision tree only on spatial feature. And the third is the cascaded decision tree that combines the decision tree trained on geometry feature and the decision tree trained on spatial feature. Fig. 4 shows the confusion matrix for each set of the experiments. From Fig. 4(a) we can see that the recognition rate for wall and ground is very high while all the table have been classified to ground. It's because the normal of wall is horizontal and the normal of ground and table are both vertical. So the geometry feature could not classify ground and table since they have the same geometry feature. Fig. 4(b) shows that most wall and ground are correctly classified but most table have been wrongly classified as wall. It is obvious that tables are much higher above the ground just like the upper part of walls. So table would have the same spatial feature as part of the wall. Therefore it is reasonable that tables are classified to walls. In Fig. 4(c), all the plane types achieve very high recognition rate. This indicates that the cascaded decision tree combines both the advantage of the geometry feature and spatial feature. Theoretically, using plane normal could recognize walls from the other two kind of planes as only walls have horizontal normals. Then tables and ground could be separate by their height above the lowest point of the scene. The experiments proves the combination of the proposed geometry feature and spatial feature are mutually complementary. Then the results of plane type classification and plane-obstacle classification are integrated to analyze the overall classification result. We apply Liu's approach [13] on our dataset and compare the result with our method as Fig. 5 shows. We can see that our method significantly improves the classification result in wall, ground and table. This prove the spatial feature and geometry

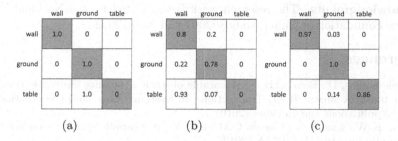

(a)   (b)   (c)

Fig. 4. Confusion Matrix. (a) is the confusion matrix of decision tree trained on geometry feature. (b) is the confusion matrix of decision tree trained on spatial feature (c) is the confusion matrix of the cascaded decision tree.

feature could well describe the character of large planar segments. However, when classifying obstacles, the proposed features performs not well. That's why our method haven't achieve a higher classification rate in obstacles.

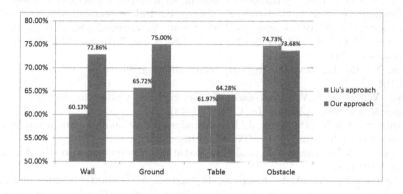

Fig. 5. Comparison between Liu's approach and our approach

7   Conclusions

For indoor scene analysis based on RGB-D for the visually impaired, we propose a multiscale-voxel strategy which improves the accuracy of scene segmentation. Compared with fixed-size-voxel method, our method is more robust to noise and have better performance for the far area of the scene. We also propose a cascaded decision tree based plane classification algorithm to structurally label the indoor scene. Our algorithm extract geometry and spatial feature from the segments after image segmentation. Experimental results show that the proposed method significantly improves the scene structural labeling. The whole system is fast and robust, which meet the requirement of obstacle detection for the visually impaired. In the future, we will apply our method on image sequence or videos and detect more structural area that the visually impaired may care.

Acknowledgments. The research work is supported by the National Nature Science Foundation of China No.60802067 and No.61202209.

References

1. Dakopoulos, D., Bourbakis, N.: Wearable obstacle avoidance electronic travel aids for blind: A survey. IEEE Transactions on Systems, Man, and Cybernetics, Part C: Applications and Reviews (2010)
2. Lin, K.W., Lau, T.K., Cheuk, C.M., Liu, Y.: A wearable stereo vision system for visually impaired. In: ICMA (2012)
3. Zöllner, M., Huber, S., Jetter, H.-C., Reiterer, H.: NAVI – A proof-of-concept of a mobile navigational aid for visually impaired based on the microsoft kinect. In: Campos, P., Graham, N., Jorge, J., Nunes, N., Palanque, P., Winckler, M. (eds.) INTERACT 2011, Part IV. LNCS, vol. 6949, pp. 584–587. Springer, Heidelberg (2011)
4. Tian, Y., Yang, X., Yi, C., Arditi, A.: Toward a computer vision-based wayfinding aid for blind persons to access unfamiliar indoor environments. Machine Vision and Applications (2012)
5. Lee, Y.H., Medioni, G.: A rgb-d camera based navigation for the visually impaired. In: RSS 2011 RGB-D: Advanced Reasoning with Depth Camera Workshop (2011)
6. Silberman, N., Hoiem, D., Kohli, P., Fergus, R.: Indoor segmentation and support inference from RGBD images. In: Fitzgibbon, A., Lazebnik, S., Perona, P., Sato, Y., Schmid, C. (eds.) ECCV 2012, Part V. LNCS, vol. 7576, pp. 746–760. Springer, Heidelberg (2012)
7. Ren, X., Bo, L., Fox, D.: Rgb-(d) scene labeling: Features and algorithms. In: CVPR (2012)
8. Choi, W., Chao, Y.W., Pantofaru, C., Savarese, S.: Understanding indoor scenes using 3d geometric phrases. In: CVPR (2013)
9. Gupta, S., Arbelez, P., Malik, J.: Perceptual organization and recognition of indoor scenes from rgb-d images. In: CVPR (2013)
10. Wang, Y., Ji, R., Chang, S.F.: Label propagation from imagenet to 3d point clouds. In: CVPR (2013)
11. Koppula, H.S., Anand, A., Joachims, T., Saxena, A.: Semantic labeling of 3d point clouds for indoor scenes. In: Advances in Neural Information Processing Systems, vol. 24 (2011)
12. Wang, S., Tian, Y.: Detecting stairs and pedestrian crosswalks for the blind by rgbd camera. In: BIBM Workshops (2012)
13. Liu, H., Wang, Z., Wang, X., Zhao, G., Qian, Y.: Adaptive scene segmentation and obstacle detection for the blind. Journal of Computer-Aided Design & Computer Graphics (2013)
14. Holz, D., Holzer, S., Rusu, R.B., Behnke, S.: Real-time plane segmentation using RGB-D cameras. In: Röfer, T., Mayer, N.M., Savage, J., Saranlı, U. (eds.) RoboCup 2011. LNCS, vol. 7416, pp. 306–317. Springer, Heidelberg (2012)
15. Dube, D., Zell, A.: Real-time plane extraction from depth images with the randomized hough transform. In: ICCV Workshops (2011)
16. Wang, Z., Liu, H., Qian, Y., Xu, T.: Real-time plane segmentation and obstacle detection of 3D point clouds for indoor scenes. In: Fusiello, A., Murino, V., Cucchiara, R. (eds.) ECCV 2012 Ws/Demos, Part II. LNCS, vol. 7584, pp. 22–31. Springer, Heidelberg (2012)
17. Felzenszwalb, P.F., Huttenlocher, D.P.: Efficient graph-based image segmentation. IJCV (2004)

A Low-Cost Head and Eye Tracking System for Realistic Eye Movements in Virtual Avatars

Yingbo Li, Haolin Wei, David S. Monaghan, and Noel E. O'Connor

CLARITY: Center for Sensor Web Technology, Dublin City University, Ireland

Abstract. A virtual avatar or autonomous agent is a digital representation of a human being that can be controlled by either a human or an artificially intelligent computer system. Increasingly avatars are becoming realistic virtual human characters that exhibit human behavioral traits, body language and eye and head movements. As the interpretation of eye and head movements represents an important part of nonverbal human communication it is extremely important to accurately reproduce these movements in virtual avatars to avoid falling into the well-known "uncanny valley". In this paper we present a cheap hybrid real-time head and eye tracking system based on existing open source software and commonly available hardware. Our evaluation indicates that the system of head and eye tracking is stable and accurate and can allow a human user to robustly puppet a virtual avatar, potentially allowing us to train an A.I. system to learn realistic human head and eye movements.

Keywords: Avatar, Eye tracking, Head movement, Uncanny valley.

1 Introduction

In this paper we present a hybrid head and eye tracking technique for use in both understanding and reconstructing realistic human head and eye movement in virtual avatars. An avatar, put simply, is any graphical representation of a person or user. This representation can take many forms: from simple icons, personalized cartoon characters, pictorial mock ups to full 3D humanoid representations. Often in computer gaming, virtual avatars can be fictional or fantasy based characters where a user will personalize the avatar based on how that user wishes to be represented. However, recently there has been a shift towards realistic personal avatars that accurately represent the user. This shift has been brought about, in part, by readily available and cheap data capture platforms such as the Microsoft Kinect depth sensor.

In the fields of animation and robotics there is a commonly used term known as the "uncanny valley" [1], which describes a sharp dip in familiarity or comfort levels that humans have towards virtual humans the closer they come to mimicking or looking like real human beings. This refers to the phenomenon whereby we typically feel very comfortable with a virtual avatar that we know for certain is a cartoon or animation, however we become very uncomfortable with a virtual avatar that is almost, but not quite, human. Thus it is important

C. Gurrin et al. (Eds.): MMM 2014, Part I, LNCS 8325, pp. 461–472, 2014.

to ensure a virtual avatar exhibits realistic and believable human features and behaviors. Among the most important of these features and behaviors are realistic facial, head and eye movements, as these contribute significantly to nonverbal communications. Indeed, tracking only one of these, e.g. tracking the head but ignoring eye movement within the eye socket, will lead to visually disturbing virtual representations. To address this, in this paper we propose a simple system for human eye tracking using a common HD webcam and a Microsoft Kinect. The cheap hardware and open source software toolkits used are currently widely available and we fuse them in a simple and efficient way to achieve real-time performance. To the best of our knowledge, the application of open-source and non-commercial systems for resolving the problem of eye tracking and thereby helping to avoid the "uncanny valley" has not been widely investigated in the state of the art.

The proposed system is low-cost, real-time and accurate. The cost of the hardware is around 200euro of a Kinect for Windows and a Logitech Webcam, which requires much less budget than the normal commercial system of eye tracking. Furthermore, this system could achieve real-time performance and satisfy the real-time requirement. And in the given environment, the detected eye center is only 1-2 pixel away from the real eye center in the captured video.

2   An Overview of the Proposed Approach

We differentiate three aspects of our approach: the macro module, the micro module and the fusion module. The first deals with robust head tracking by utilizing the Microsoft Kinect depth sensor. In this way, the head position, orientation and gaze direction can all be approximated with a high level of accuracy. The second module then leverages this to perform eye tracking utilizing a modified version of the open source ITU Gaze Tracker. For this we use a cheap high definition (HD) USB web camera (Logitech 1080p, whose price is around 70euro). The eye or gaze direction can be obtained by approximating the position of the pupil and the two corners of the eye. It should be noted that the RGB camera integrated into the Kinect hardware was found, via experimental observation, to be of too low a quality and resolution to be used robustly for the purpose of eye tracking within our scenario. And this is also the reason to exploit an additional HD Webcam. The outputs of these two modules are integrated and the pupil position and head movement and orientation are subsequently fused to puppet a human controlled avatar. The proposed approach is illustrated in Fig. 1. We describe each module in detail in the following sections.

3   Eye and Gaze Tracking

The believability and realism of virtual autonomous agents/avatars will, in part, depend on human-like gaze (eye) movement and attention within a virtual

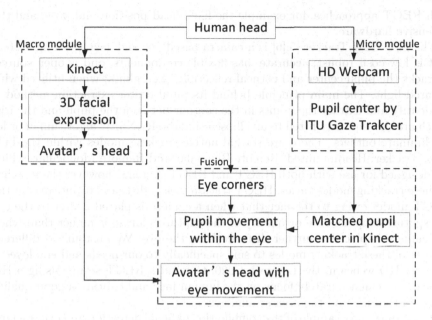

Fig. 1. Overview of the proposed approach

environment. Furthermore, when a human-avatar interaction is taking place the gaze of an avatar should be able to mimic the human operator. Visual attention has been widely researched for over 100 years. The study of this originated from the field of psychology has now gained popularity within the field of computer science and in particular, Human-Computer Interaction (HCI). The fundamental goal of eye and gaze tracking is to estimate the direction of gaze of a person [3], and its output should be an estimation of the projected point of view with regards to the person [2]. An ideal eye gaze tracker should be accurate, reliable, robust, non-intrusive, operate as close to real-time as possible, be robust to head motion vibrations, and not require complex calibration procedures [3].

Depending on the application, Eye and Gaze Trackers (EGTs) are separated into two groups, diagnostic applications and interactive applications [4]. In diagnostic applications EGTs are used to quantify a user's attention, while in interactive applications EGTs are used to interact with the user for use in a device or computer [3]. According to the distance between the EGT capture device and the human subject's eyes, EGT systems are categorized into Remote Eye and Gaze Tracking systems (REGT) and Head Mounted Eye and Gaze tracking systems (HEGT). For the purposes of our work we are concerned primary with REGT systems that are non-invasive and that do not require expensive and fragile head mounted hardware. It should, however, be noted that the current state-of-the-art REGT systems demonstrate the significant challenges to be addressed, specifically with respect to head motion and they typically exhibit lower accuracy than HEGT. By using our novel hybrid system that combines robust head tracking and eye tracking we overcome some of these existing problems with REGT approaches, for example the fixed head position and pose, and the expensive hardware.

The ITU Gaze Tracker [5][6] is a camera based eye and gaze tracking system that is low-cost, robust, accurate, has flexible components, and is open source. It tracks the pupil center and corneal reflections (i.e., glints) when utilized with infrared light. The main principle behind its pupil center extraction method is to identify and extract the points in the contour between the pupil and the iris, which are subsequently fitted to an ellipse using the RANSAC algorithm in order to eliminate outliers. There are several native tracking modes within the ITU Gaze Tracker: Headmounted, Remote Binocular and Remote Monocular, which are designed for use with both HEGT and REGT systems, however the stability of these tracking modes varies. HEGT tracking has a distinct advantage over the REGT modes owing to the fact that when a camera is placed nearer to the eye the spatial sampling rate or quality of the captured image is higher than when the camera is placed a further distance from the eye. We customized different aspects of these tracking modes to suit specifically to our needs and employed a low-cost HD webcam, the Logitech C920, to build a REGT setup. Using a HD webcam we can accurately focus on the human face and capture a good quality image of the face and the area around eyes.

Fig. 2 shows an example of the combination of facial detection and eye tracking from the eye tracker when the subject is 40-50cm away from the webcam. It can be seen that we can obtain good image quality and pixel resolution, and that the eyes are well focused by the HD webcam. Fig. 3 shows a close up of the subject's eyes.

Fig. 2. Eye location identification utilizing a modified version of the ITU Gaze Tracker for REGT

The ITU Gaze Tracker mode used in our approach does not consider if the detected pupil is from the left eye or right eye. To facilitate this we created and integrated a function that detects the presence of an eye pair by means of a Haar

Fig. 3. Close up segmentation of the tracked eye pair

Cascade [7], shown in Fig. 3 by the white rectangle. Then the position of the detected pupil in the eye pair, nearer to left border or right border, is used to calculate whether it is a left or right eye.

Alongside the modified and new functions added to the eye tracker, additional data outputs are leveraged in our hybrid approach. This computed data includes, left/right pupil tracking, pupil pair identification, pupil position, region of interest around pupil as well as the native ITU Gaze Tracker outputs of time stamp, and gaze position. This additional data is fed into the hybrid tracking system and used to robustly combine the eye tracking and head tracking systems outputs.

4   Head Position and Orientation Estimation

Head pose estimation [8] and facial expression tracking [9][10] have recently received much attention in the literature, in particular in fields of research pertaining to human-computer interaction, facial animation, virtual avatar creation etc. Facial expression analysis provides the important features and data on the human face which enable the further study of face animation and realistic virtual avatar creation.

Within the context of this work we define head pose estimation to be the estimation of the orientation of a person's head in relation to the position of the camera being used to capture that person. Head pose estimation is also used as an initial input into gaze tracking algorithms because it can offer a good estimation of the gaze direction.

To the best of our knowledge, most research on head pose and facial expressions are based on two dimensional image processing, which contains few distinct facial features and can be easily influenced by the capturing environment. However, the emergence of the low cost Kinect depth sensor platform provides us with a new and cheap platform to easily ascertain head pose and facial expression information [11] [12]. The Kinect consists of an RGB camera, an infrared (IR) emitter and an IR camera, and by using a reconstruction process known as the deformation of structured light, it can capture the depth of a scene with an effective range of 0.45 - 7.0 meters. The Microsoft Kinect toolkit provides a 3D face model and its corresponding data, shown here in Fig. 4(a) and 4(b).

In this work we use the Kinect for Windows system in the near mode, which enables us to capture the face with the minimum distance of 40cm between

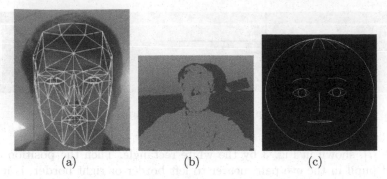

(a)　　　　　　　　　(b)　　　　　　　　　(c)

Fig. 4. (a) 3D facial tracking model from Kinect; (b) Depth map image; (c) A sample avatar's head from the face tracking toolkit

the person and the Kinect, while Xbox Kinect needs a much further minimum distance of approximately 1.2 meters. The output data from the Kinect face tracking can be used in facial animation, as the position of eye corners are very useful in eye and pupil tracking, as will be explained in Section 5. The Kinect face tracking visualization, shown in Fig. 4(c), can demonstrate the animation of an avatar's head corresponding to the motion of the person in real time and this forms a good basis for a hybrid tracking system. Furthermore, in Fig. 4(c) we can see that the pupil size in the orbit is too large to move in the orbit, so we shrink the pupil size in the avatar's head. We also enable this module to receive the additional data through UDP from the eye tracking module.

5 Hybrid Head and Eye Tracking System

In Section 3 we presented eye and pupil tracking and in Section 4 we presented head and facial tracking. Both of these systems have been uniquely modified, improved and adapted in order to fuse the data and animate the eyes in an avatar's head using a hybrid tracking system. In this section, we describe our novel approach to achieve real-time head and eye tracking.

The modified avatar head from Section 4 responds well to head movements and facial expression but the pupils cannot move. If we want to animate the pupils, we need to know the movement of the pupils from the subject at any given point in time. From Section 3, we only know the position of the pupil and the movement of the pupil in the context of the head is not known. To address this we firstly analyze the movement of the pupil. As shown in Fig. 5(a), we can see that the pupil is normally in the center of the eye socket. When the pupil moves, as shown in Fig. 5(b), there is a distance between the pupil and the center of the socket, which can be defined by eye corners (canthus).

From Section 3, we know the real-time position of the pupil, so the positions of eye corners are required. Some successful algorithms for the detection of eye corners [13] [14] already exist, however, many of them are implemented offline and the others cannot promise the robustness that is required in this system.

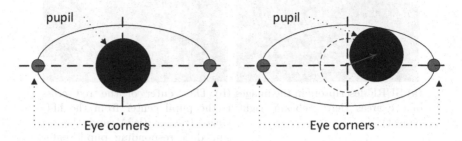

Fig. 5. The model of pupil motion: (a) The pupil is in the eye center; (b) The pupil moves away from the eye center

In our case, the robust and real-time coordinates of eye corners are required. We have found that the Kinect can provide this kind of data and satisfies our criteria for robust and real-time processing.

As previously mentioned, it was found that the RGB camera from Kinect is of too low quality to be robustly used for eye tracking, so we use the eye information from a separate HD web camera, as described in Section 3. To handle rectification between the two camera systems it is typical to find the corresponding pixels by homography and camera calibration. However, we have found that the accurate homography computations would not allow for the system to operate in real-time. On the other hand, if isolated segmentation of an eye pair from each camera system is used, real time performance is achieved but insufficient matching points (via SURF descriptors [16].) for the computation of a homography [15] are obtained. It is a challenge for homography computation to find a good compromise between computation time and sufficient SURF descriptors. Camera calibration before eye tracking is of course possible but this makes the hardware setup more complex and less accessible in general.

The SURF algorithm can detect distinctive locations in the image, such as corners, blobs, and so on. Since the pupil is a very distinctive blob in the eye images, the pupil center is one of the detected key points, as shown in Fig. 6(a). For the image from Kinect we need to sharpen the image as a preprocessing step to detect SURF descriptors, because the image from Kinect is not focused and is often blurred. The image size of eye region from webcam is 41 − 51 pixels for both width and height and the width and height of eye region image from Kinect camera is set as twice the distance of two eye corners.

From Fig. 6(a) we can see that pupil centers are detected as a good match of SURF points. The eye tracker from Section 3 provides us the pupil center in the webcam image, then the SURF point nearest to it is the detected blob center (pupil center) in the webcam image and the matched point in Kinect image is the corresponding pupil, illustrated in Fig. 6(b). The distance between the detected blob center and the pupil center by eye tracking is 0-2 pixels in Fig. 6(b) left, then in Fig. 6(b) right the distance between blob center and the assumed pupil center should be less than 1 pixel, because of a smaller eye captured by the

(a) SURF descriptors in the images of eye areas (left: webcam, right: Kinect);

(b) The center of the red cross is the pupil center from the ITU Gaze Tracker (left), and the assumed corresponding pupil center in the Kinect image (right);

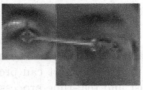

(c) Another example of the face with the glasses;

(d) The matched pupil center of another example in (c)

Fig. 6. SURF matched points

wide-angle RGB lens of Kinect. Therefore, we can get an accurate pupil center in the Kinect image.

Consequently, we can determine the pupil movement within the eye as illustrated in Fig. 5 and then we can animate the pupil on the avatar's head in real time, 4 snap shots of which are shown in Fig. 7 to show the effects in different head poses and eye positions. A demonstration video can also be viewed here: http://goo.gl/b0bhD. With the animated eye on the face, it is clear to see that the virtual avatar is more like a human face than the face without the eye motion in the face tracking demo of Kinect SDK. In the sense of eye motion, we partially improve the solution for "uncanny valley".

6 Evaluation

Fig. 7 and its corresponding video provide a subjective evaluation of the proposed approach. Objective evaluation, on the other hand, is not straightforward. Benchmarking against existing head mounted or remote systems is difficult as typically the operation of such systems will interfere with our set-up meaning that it is difficult to perform tracking simultaneously using two different systems. Thus, for objective evaluation we compare the tracked pupil centers in the Kinect video frames by our system with the pupil centers manually annotated by human beings.

We asked 4 people to select the pupil centers from 700 Kinect frames in total, captured from 2 different people, with and without glasses. Glasses typically pose

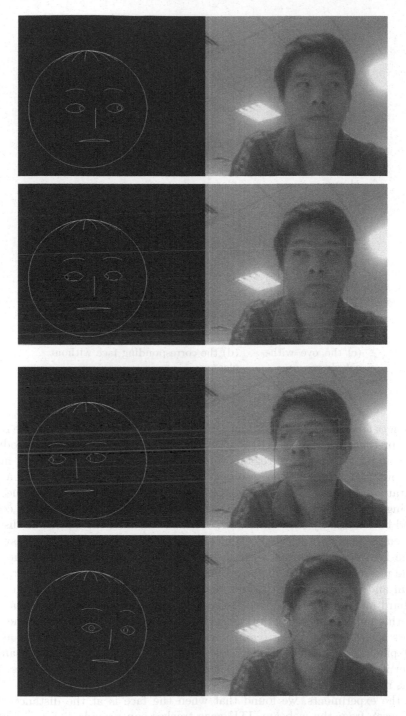

Fig. 7. Examples of avatar's head with animated eyes for various head and eye positions

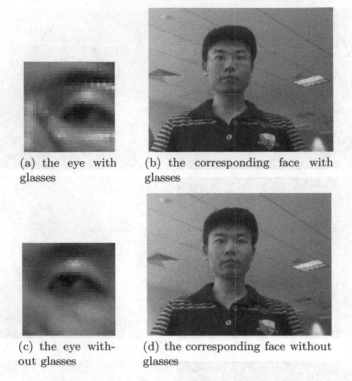

(a) the eye with glasses

(b) the corresponding face with glasses

(c) the eye without glasses

(d) the corresponding face without glasses

Fig. 8. The eye regions with and without glasses

challenges for most existing tracking systems. We require our system to be robust to the presence/absence of glasses, or even to glasses being removed/introduced. The example frames of eye regions with and without glasses are shown in Fig. 8. As our goal was to obtain a sample of manual annotations that was as representative as possible, each annotator could choose some subset of frames, e.g. ignoring very similar frames, with the constraint that at least 50 pupil centers should be selected by each annotator. We then calculate the Euclidean distance between the manually selected centers and the centers as recorded by our system.

The Euclidean distances, with and without the glasses, are shown in Table 1. The mean values of the distances between the pupil centers by the proposed system and the ground truth are 2.46 pixels and 2.87 pixels. Considering that it is hard for people to exactly pick the pupil centers to the accuracy of a single pixel this indicates good performance. Furthermore, the distances of the pupil centers at around 2.5 pixels, compared to the size of the eye region at around 41-51 pixels, are accurate. Finally, the proposed system performs the same no matter if the person wears the glasses or not, so the proposed system is stable in this regard.

In the experiments, we found that when the face is at the distance of 40-60cm away from the webcam, ITU gaze tracker can provide the real-time eye image to the proposed system. When the face is further, the higher resolution of

Table 1. The Euclidean distance between the human annotated pupil centers and the centers tracked by our system in Kinect frames

(pixel)	Person 1	Person 2	Person 3	Person 4	Mean value
with glasses	2.8438	2.7699	2.1916	2.0407	2.46
without glasses	2.8937	3.0857	2.8496	2.6514	2.87

the webcam frame for ITU gaze tracker is necessary, but PC in the experiment cannot perform the real-time processing for the computation of ITU gaze tracker. Therefore, a further distance is possible for the real-time and robust performance of the proposed system, but a higher standard PC is necessary.

7   Conclusion

In this paper we have proposed a hybrid system for capturing eye and head motions by fusing together modified Kinect and HD webcam systems. The system runs in real time and is demonstrated by animating the head and eye in a customized Kinect avatar. The proposed system is low cost and easy to setup. A demonstration video shows our proposed system working in real time and a subjective evaluation shows that the system is robust and accurate for situations when both eye and head are in different positions or poses. Furthermore, we have measured the accuracy of the proposed hybrid system against a human annotated ground truth. Thus, we have gone some way towards resolving the specific problems associated with pupetting eye movements in virtual avatars in a realistic manner, which is a novel and important application of our proposed system. The head and eye movements could in the future be recorded and fed into a machine learning classifier to train an artificial intelligent computer system to control a virtual avatar, using features from both head and eye movements. This is a target for our future work.

Acknowledgement. The research that lead to this paper was supported in part by the European Commission under the Contract FP7-ICT-287723 REVERIE.

References

1. Mori, M.: The uncanny valley. Energy 7(4), 33–35 (1970)
2. Duchowski, A.T.: Eye tracking methodology: Theory and practice, vol. 373. Springer (2007)
3. Morimoto, C.H., Mimica, M.R.M.: Eye gaze tracking techniques for interactive applications. Computer Vision and Image Understanding 98(1), 4–24 (2005)
4. Brolly, X.L.C., Mulligan, J.B.: Implicit calibration of a remote gaze tracker. In: Proceedings of the 2004 Conference on Computer Vision and Pattern Recognition Workshop, vol. 8, p. 134 (2004)

5. San Agustin, J., Skovsgaard, H., Mollenbach, E., Barret, M., Tall, M., Hansen, D.W., Hansen, J.P.: Evaluation of a low-cost open-source gaze tracker. In: Proceedings of the 2010 Symposium on Eye-Tracking Research & Applications (2010)
6. ITU Gaze Tracker, http://www.gazegroup.org/downloads/23-gazetracker
7. Castrilln-Santana, M., Dniz-Surez, O., Antn-Canals, L., Lorenzo-Navarro, J.: Face and facial feature detection evaluation. In: Third International Conference on Computer Vision Theory and Applications, VISAPP 2008 (2008)
8. Murphy-Chutorian, E., Trivedi, M.M.: Head pose estimation in computer vision: A survey. IEEE Transactions on Pattern Analysis and Machine Intelligence 31(4), 607–626 (2009)
9. Shan, C., Gong, S., Mcowan, P.W.: Facial expression recognition based on Local Binary Patterns: A comprehensive study. Image and Vision Computing 27(6), 803–816 (2009)
10. Terzopoulos, D., Waters, K.: Physically-based facial modelling, analysis, and animation. The Journal of Visualization and Computer Animation 1(2), 73–80 (1990)
11. Cruz, L., Lucio, D., Velho, L.: Kinect and rgbd images: Challenges and applications. In: SIBGRAPI Tutorial (2012)
12. Zhang, Z.: Microsoft Kinect Sensor and Its Effect. IEEE Multimedia 19(2), 4–10 (2009)
13. Santos, G., Proenca, H.: A robust eye-corner detection method for real-world data. In: 2011 International Joint Conference on Biometrics, IJCB (2011)
14. Xu, C., Zheng, Y., Wang, Z.: Semantic feature extraction for accurate eye corner detection. In: ICPR (2008)
15. Sukthankar, R., Stockton, R.G., Mullin, M.D.: Smarter presentations: Exploiting homography in camera-projector systems. In: ICCV (2001)
16. Bay, H., Ess, A., Tuytelaars, T., Van Gool, L.: SURF: Speeded Up Robust Features. Computer Vision and Image Understanding (CVIU) 110(3), 346–359 (2006)

Real-Time Skeleton-Tracking-Based Human Action Recognition Using Kinect Data

Georgios Th. Papadopoulos, Apostolos Axenopoulos, and Petros Daras

Information Technologies Institute
Centre for Research & Technology - Hellas
Thessaloniki, Greece

Abstract. In this paper, a real-time tracking-based approach to human action recognition is proposed. The method receives as input depth map data streams from a single kinect sensor. Initially, a skeleton-tracking algorithm is applied. Then, a new action representation is introduced, which is based on the calculation of spherical angles between selected joints and the respective angular velocities. For invariance incorporation, a pose estimation step is applied and all features are extracted according to a continuously updated torso-centered coordinate system; this is different from the usual practice of using common normalization operators. Additionally, the approach includes a motion energy-based methodology for applying horizontal symmetry. Finally, action recognition is realized using Hidden Markov Models (HMMs). Experimental results using the Huawei/3DLife 3D human reconstruction and action recognition Grand Challenge dataset demonstrate the efficiency of the proposed approach.

Keywords: Action recognition, skeleton-tracking, action representation, depth map, kinect.

1 Introduction

Action recognition constitutes a widely studied field and a very active topic in the computer vision research community [7]. This is due to the wide set of potential fields where the research outcomes can be commercially applied, such as surveillance, security, human computer interaction, smart houses, helping the elderly/disabled, to name a few. In order to develop a robust action recognition system, the respective algorithm needs to efficiently handle the differences in the appearance of the subjects, the human silhouette features, the execution of the same actions, etc. Additionally, the system should incorporate the typical rotation, translation and scale invariances. Despite the fact that multiple research groups focus on this topic and numerous approaches have already been presented, significant challenges towards fully addressing the problem in the general case are still present.

Action recognition approaches can be roughly divided into the following three categories [10], irrespectively of the data that they receive as input (i.e. single-camera videos, multi-view video sequences, depth maps, 3D reconstruction

C. Gurrin et al. (Eds.): MMM 2014, Part I, LNCS 8325, pp. 473–483, 2014.

data, etc.): Space-Time Interest Point (STIP)- [6][5][2], spatio-temporal shape-[13][14][3] and tracking-based [4][12][8][9]. STIP-based methods perform analysis at the local-level; however, they typically exhibit increased computational complexity for reaching satisfactory recognition performance. Spatio-temporal shape approaches rely on the estimation of global-level representations for performing recognition, using e.g. the outer boundary of an action; however, they are prone to the detrimental effects caused by self-occlusions of the performing subjects. On the other hand, the performance of tracking-based approaches, which rely on the tracking of particular features or specific human body parts in subsequent frames (including optical-flow-based methods) depends heavily on the efficiency of the employed tracker. Nevertheless, the advantage of the latter category of methods is that they can allow the real-time recognition of human actions.

In this paper, a real-time tracking-based approach to human action recognition is proposed. The method receives as input a sequence of depth maps captured from a single kinect sensor, in order to efficiently capture the human body movements in the 3D space. Subsequently, a skeleton-tracking algorithm is applied, which iteratively detects the position of 15 joints of the human body in every captured depth map. Then, a new action representation is introduced, which is based on the calculation of spherical angles between selected joints and the respective angular velocities, for satisfactorily handling the differences in the appearance and/or execution of the same actions among the individuals. For incorporating further invariance to appearance, scale, rotation and translation, a pose estimation step is applied prior to the feature extraction procedure and all features are calculated according to a continuously updated torso-centered coordinate system; this is different from the usual practice of using normalization operators during the analysis process [4]. Additionally, the approach incorporates a motion energy-based methodology for applying horizontal symmetry and hence efficiently handling left- and right-body part executions of the exact same action. Finally, action recognition is realized using Hidden Markov Models (HMMs). Experimental results using the Huawei/3DLife 3D human reconstruction and action recognition Grand Challenge dataset[1] demonstrate the efficiency of the proposed approach.

The paper is organized as follows: Section 2 outlines the employed skeleton-tracking algorithm. The proposed action recognition approach is described in Section 3. Experimental results are presented in Section 4 and conclusions are drawn in Section 5.

2   Skeleton-Tracking

Prior to the application of the proposed action recognition approach, the depth maps captured by the kinect sensor are processed by a skeleton-tracking algorithm. The depth maps of the utilized dataset were acquired using the OpenNI API[2]. To this end, the OpenNI high-level skeleton-tracking module is also used

---

[1] http://mmv.eecs.qmul.ac.uk/mmgc2013/
[2] http://www.openni.org/

for detecting the performing subject and tracking a set of joints of his/her body. More specifically, the OpenNI tracker detects the position of the following set of joints in the $3D$ space $G = \{g_i, \ i \in [1, I]\} \equiv \{Torso, \ Neck, \ Head, \ Left \ shoulder, \ Left \ elbow, \ Left \ wrist, \ Right \ shoulder, \ Right \ elbow, \ Right \ wrist, \ Left \ hip, \ Left \ knee, \ Left \ foot, \ Right \ hip, \ Right \ knee, \ Right \ foot\}$. The position of joint g_i is implied by vector $\mathbf{p}_i(t) = [x \ y \ z]^T$, where t denotes the frame for which the joint position is located and the origin of the orthogonal XYZ co-ordinate system is placed at the center of the kinect sensor. An indicative example of a captured depth map and the tracked joints is given in Fig. 1.

The OpenNI skeleton-tracking module requires user calibration in order to estimate several body characteristics of the subject. In recent versions of OpenNI, the 'auto-calibration' mode enables user calibration without requiring the subject to undergo any particular calibration pose. Since no calibration pose was captured for the employed dataset, the OpenNI's (v. 1.5.2.23) 'auto-calibration' mode is used in this work. The experimental evaluation showed that the employed skeleton-tracking algorithm is relatively robust for the utilized dataset. In particular, the position of the joints is usually detected accurately, although there are some cases where the tracking is not correct. Characteristic examples of the latter are the inaccurate detection of the joint positions when very sudden and intense movements occur (e.g. arm movements when performing actions like 'punching') or when self-occlusions are present (e.g. occlusion of the knees when extensive body movements are observed during actions like 'golf drive').

3 Action Recognition

In this section, the proposed skeleton-tracking-based action recognition approach is detailed. The developed method satisfies the following two fundamental principles: a) the computational complexity needs to be relatively low, so that the real-time processing nature of the algorithm to be maintained, and b) the dimensionality of the estimated action representation needs also to be low, which is a requirement for efficient HMM-based analysis [11].

3.1 Pose Estimation

The first step in the proposed analysis process constitutes a pose estimation procedure. This is performed for rendering the proposed approach invariant to differentiations in appearance, body silhouette and action execution among different subjects, apart from the typically required invariances to scale, translation and rotation. The proposed methodology is different from the commonly adopted normalization procedures (e.g. [4]), which in their effort to incorporate invariance characteristics they are inevitably led to some kind of information loss. In particular, the aim of this step is to estimate a continuously updated orthogonal basis of vectors for every frame t that represents the subject's pose. The calculation of the latter is based on the fundamental consideration that the orientation of the subject's torso is the most characteristic quantity of the subject during the execution of any action and for that reason it could be used as

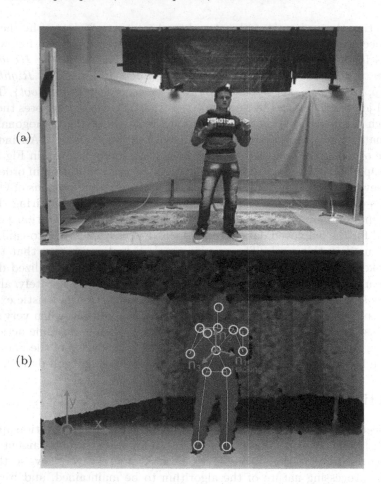

(a)

(b)

Fig. 1. Indicative pose estimation example: (a) examined frame and (b) captured depth map along with the tracked joints and the estimated orthogonal pose vectors

reference. For pose estimation, the position of the following three joints is taken into account: *Left shoulder*, *Right shoulder* and *Right hip*. These comprise joints around the torso area, whose relative position remains almost unchanged during the execution of any action. The motivation behind the consideration of the three aforementioned joints, instead of directly estimating the position of the torso joint and the respective normal vector, is to reach a more accurate estimation of the subject's pose. It must be noted that the *Right hip* joint was preferred instead of the obvious *Torso* joint selection. This was performed so that the orthogonal basis of vectors to be estimated from joints with bigger in between distances that will be more likely to lead to more accurate pose estimation. However, no significant deviation in action recognition performance was observed when the *Torso* joint was used instead. In this work, the subject's pose comprises the following three orthogonal vectors $\{\mathbf{n}_1, \mathbf{n}_2, \mathbf{n}_3\}$ that are calculated as follows:

$$\mathbf{n}_1 = \frac{\mathbf{p}_7 - \mathbf{p}_4}{\|\mathbf{p}_7 - \mathbf{p}_4\|}, \quad \mathbf{u} = \frac{\mathbf{p}_7 - \mathbf{p}_{13}}{\|\mathbf{p}_7 - \mathbf{p}_{13}\|}$$

$$\mathbf{n}_3 = \frac{\mathbf{n}_1 \times \mathbf{u}}{\|\mathbf{n}_1 \times \mathbf{u}\|}, \quad \mathbf{n}_2 = \mathbf{n}_3 \times \mathbf{n}_1 \tag{1}$$

where subscript t is omitted in the above expressions for clarity, $\|.\|$ denotes the norm of a vector and $\times$ denotes the cross product of two vectors. An indicative example of the proposed pose estimation procedure is illustrated in Fig. 1.

3.2 Action Representation

For realizing efficient action recognition, an appropriate representation is required that will satisfactorily handle the differences in appearance, human body type and execution of actions among the individuals. For that purpose, the angles of the joints' relative position are used in this work, which showed to be more discriminative than using e.g. directly the joints' normalized coordinates. In order to compute a compact description, the aforementioned angles are estimated in the spherical coordinate system and the radial distance is omitted, since it contains information that is not necessary for the recognition process. Additionally, building on the fundamental idea of the previous section, all angles are computed using the *Torso* joint as reference, i.e. the origin of the spherical coordinate system is placed at the *Torso* joint position. For computing the proposed action representation, only a subset of the supported joints is used. This is due to the fact that the trajectory of some joints mainly contains redundant or noisy information. To this end, only the joints that correspond to the upper and lower body limbs were considered after experimental evaluation, namely the joints *Left shoulder*, *Left elbow*, *Left wrist*, *Right shoulder*, *Right elbow*, *Right wrist*, *Left knee*, *Left foot*, *Right knee* and *Right foot*. Incorporating information from the remaining joints led to inferior performance, partly also due to the higher dimensionality of the calculated feature vector that hindered efficient HMM-based analysis. For every selected joint g_i the following spherical angles are estimated:

$$\varphi_i = \arccos(\frac{\langle(\mathbf{p}_i - \mathbf{p}_0), \mathbf{n}_2\rangle}{\|\mathbf{p}_i - \mathbf{p}_0\| \cdot \|\mathbf{n}_2\|}) \in [0, \pi]$$

$$\theta_i = \arctan(\frac{\langle(\mathbf{p}_i - \mathbf{p}_0), \mathbf{n}_1\rangle}{\langle(\mathbf{p}_i - \mathbf{p}_0), \mathbf{n}_3\rangle}) \in [-\pi, \pi] \tag{2}$$

where subscript t is omitted for clarity, φ_i is the computed polar angle, θ_i is the calculated azimuth angle and $\langle\cdot, \cdot\rangle$ denotes the dot product of two vectors.

Complementarily to the spherical angles, it was experimentally shown that the respective angular velocities provide additional discriminative information. To this end, the polar and azimuth velocities are estimated for each of the selected joints g_i, using the same expressions described in (2). The difference is that instead of the position vector $\mathbf{p}_i$ the corresponding velocity vector $\mathbf{v}_i$ is used. The latter is approximated by the displacement vector between two successive frames, i.e. $\mathbf{v}_i(t) = \mathbf{p}_i(t) - \mathbf{p}_i(t-1)$.

The estimated spherical angles and angular velocities for frame t constitute the frame's observation vector. Collecting the computed observation vectors for all frames of a given action segment forms the respective action observation sequence h that will be used for performing HMM-based recognition, as will be described in the sequel.

3.3   Horizontal Symmetry

A problem inherently present in action recognition tasks concerns the execution of the same action while undergoing either a right- or left-body part motion. In order to address this issue, a motion energy-based approach is followed in this work, which builds upon the idea of the already introduced subject's pose (Section 3.1). The proposed method goes beyond typical solutions (e.g. measuring the length or the variance of the trajectories of the left/right body joints) and is capable of identifying and applying symmetries not only concerning common upper limb movements (e.g. left/right-hand waving actions), but also more extensive whole-body actions (e.g. right/left-handed golf-drive). In particular, all joints defined in G are considered, except from the *Torso*, *Head* and *Neck* ones. For each of these joints, its motion energy, which is approximated by $\|\mathbf{v}_i(t)\| = \|\mathbf{p}_i(t) - \mathbf{p}_i(t-1)\|$, is estimated for every frame t. Then, the calculated motion energy value $\|\mathbf{v}_i(t)\|$ is assigned to the Left/Right Body Part (LBP/RBP) according to the following criterion:

$$if \ \langle(\mathbf{p}_i(t) - \mathbf{p}_0(t)), \mathbf{n}_1(t)\rangle = \begin{cases} > 0, \|\mathbf{v}_i(t)\| \to RBP \\ < 0, \|\mathbf{v}_i(t)\| \to LBP \end{cases} \tag{3}$$

where $\to$ denotes the assignment of energy value $\|\mathbf{v}_i(t)\|$ to LBP/RBP. By considering the overall motion energy that is assigned to LBP/RBP for the whole duration of the examined action, the most 'active' body part is estimated. Then, a simple reallocation of the extracted feature values and application of horizontal symmetry attributes is performed. This is realized as follows: If the most 'active' part is LBP, the representation described in Section 3.2 remains unchanged. In case that RBP is the most 'active' one, the feature values of horizontally symmetric joints (e.g. *Left shoulder-Right shoulder*) are exchanged in all observation vectors of the respective action observation sequence h, while the horizontal symmetry attribute is applied by inverting the sign of all estimated azimuth angles and azimuth angular velocities. In this way, efficient horizontal symmetry that covers both limb motions as well as more extensive body movements is imposed.

3.4   HMM-Based Recognition

HMMs are employed in this work for performing action recognition, due to their suitability for modeling pattern recognition problems that exhibit an inherent temporality [11]. In particular, a set of J HMMs is employed, where an individual HMM is introduced for every supported action a_j. Each HMM receives as input the action observation sequence h (described in Section 3.2) and at the evaluation stage returns a posterior probability $P(a_j|h)$, which represents the observation sequence's fitness to the particular model.

Regarding the HMM implementation details, fully connected first order HMMs, i.e. HMMs allowing all possible hidden state transitions, were utilized for performing the mapping of the low-level features to the high-level actions. For every hidden state the observations were modeled as a mixture of Gaussians (a single Gaussian was used for every state). The employed Gaussian mixture models (GMMs) were set to have full covariance matrices for exploiting all possible correlations between the elements of each observation. Additionally, the Baum–Welch (or Forward–Backward) algorithm was used for training, while the Viterbi algorithm was utilized during the evaluation. Furthermore, the number of hidden states of the HMMs was considered a free variable. The developed HMMs were implemented using the software libraries of [1].

4   Experimental Results

In this section, experimental results from the application of the proposed approach to the Huawei/3DLife 3D human reconstruction and action recognition Grand Challenge dataset are presented. In particular, the second session of the first dataset is used, which provides RGB-plus-depth video streams from two kinect sensors. In this work, the data stream from the frontal kinect was used. The dataset includes captures of 14 human subjects, where each action is performed at least 5 times by every individual. Out of the available 22 supported actions, the following set of 17 dynamic ones were considered for the experimental evaluation of the proposed approach: $A = \{a_j, \ j \in [1, J]\} \equiv$ {$Hand\ waving$, $Knocking\ the\ door$, $Clapping$, $Throwing$, $Punching$, $Push\ away$, $Jumping\ jacks$, $Lunges$, $Squats$, $Punching\ and\ kicking$, $Weight\ lift-ing$, $Golf\ drive$, $Golf\ chip$, $Golf\ putt$, $Tennis\ forehand$, $Tennis\ backhand$, $Walking\ on\ the\ treadmill$}. The 5 discarded actions (namely $Arms\ folded$, $T-Pose$, $Hands\ on\ the\ hips$, $T-Pose\ with\ bent\ arms$ and $Forward\ arms\ raise$) correspond to static ones that can be easily detected using a simple action representation; hence, they were not included in the conducted experiments that aim at evaluating the performance of the proposed approach for detecting complex and time-varying human actions. Performance evaluation was realized following the 'leave-one-out' methodology, where in every iteration one subject was used for performance measurement and the remaining ones were used for training; eventually, an average performance measure was computed taking into account all intermediate recognition results.

In Fig. 2, quantitative action recognition results are presented in the form of the estimated confusion matrix (Fig. 2 (a)) and the calculated recognition rates (Fig. 2 (b)), i.e. the percentage of the action instances that were correctly identified. Additionally, the value of the overall classification accuracy, i.e. the

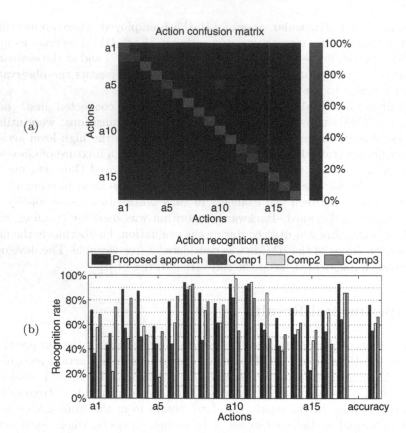

Fig. 2. Obtained action recognition results: (a) Estimated action confusion matrix and (b) calculated action recognition rates. Supported actions: a_1: Hand waving, a_2: Knocking the door, a_3: Clapping, a_4: Throwing, a_5: Punching, a_6: Push away, a_7: Jumping jacks, a_8: Lunges, a_9: Squats, a_{10}: Punching and kicking, a_{11}: Weight lifting, a_{12}: Golf drive, a_{13}: Golf chip, a_{14}: Golf putt, a_{15}: Tennis forehand, a_{16}: Tennis backhand, a_{17}: Walking on the treadmill.

percentage of all action instances that were correctly classified, is also given. For performance evaluation, it has been considered that $\arg\max_j(P(a_j|h))$ indicates the action a_j that is assigned to observation sequence h. From the presented results, it can be seen that the proposed approach achieves satisfactory action recognition performance (overall accuracy equal to 76.03%), which demonstrates the capability of the developed method to combine real-time processing with increased recognition rates. Examining the results in details, it can be seen that there are actions that exhibit high recognition rates (e.g. *Jumping jacks*, *Punching and kicking* and *Walking on the treadmill*), since they present characteristic motion patterns among all subjects. However, there are also actions for which the recognition performance is not that increased (e.g. *Punching*, *Knocking the door* and *Golf drive*). This is mainly due to these actions presenting very similar motion patterns over a period of time during their execution with

Table 1. Time efficiency evaluation (per frame average processing times in msec)

Kinect sensor	Steps of the proposed approach		
Capturing period	Skeleton tracking	Feature extraction	HMM-based recognition
46.821	45.350	0.010	0.106

other ones (i.e. *Punching and kicking*, *Hand waving* and *Golf chip*, respectively). Moreover, the careful examination of the obtained results revealed that further performance improvement was mainly hindered due to the following two factors: a) the employed tracker sometimes provided inaccurate joint localizations (especially in cases of rapid movements or self-occlusions) and b) significant ambiguities in the execution of particular action pairs (e.g. some *Golf drive* instances presented significant similarities with corresponding *Golf chip* ones that even a human observer would be difficult to discriminate between them).

The proposed approach is also quantitatively compared with the following variants: a) use of normalized Euclidean distance of every selected joint from the *Torso* one (*Comp*1), b) use of normalized spherical angles of each joint from the *Torso* one (*Comp*2), and c) variant of the proposed approach, where the estimated spherical angles and angular velocities are linearly normalized with respect to the maximum/minimum values that they exhibit during the execution of a particular action (*Comp*3). Variants (a) and (b) follow the normalization approach described in [4], i.e. the human joint vector is position and orientation normalized without considering the relative position of the joints. From the presented results, it can be seen that the proposed method significantly outperforms both (a) and (b) variants. The latter demonstrates the usefulness of estimating the pose of the subject during the computation of the action representation, compared to performing a normalization step that does not take explicitly into account the relative position of the detected joints. The proposed method also outperforms variant (c), which suggests that performing a normalization of the angles/velocities within the duration of a particular action leads to decrease in performance.

The time efficiency of the proposed approach is evaluated in Table 1. In particular, the per frame average processing time, i.e. the time required for processing the data that correspond to a single frame, are given for every algorithmic step of the proposed method. More specifically, the following steps were considered: a) skeleton-tracking, described in Section 2, b) feature extraction, which includes the pose estimation, action representation computation and horizontal symmetry application procedures that are detailed in Sections 3.1, 3.2 and 3.3, respectively, and c) HMM-based recognition, outlined in Section 3.4. The duration of the aforementioned procedures is compared with the capturing period of the employed kinect sensor, i.e. the time interval between two subsequently captured depth maps. The average processing times given in Table 1 are obtained using a PC with Intel i7 processor at 2.67 GHz and a total of 6 GB RAM, while for their computation all video sequences of the employed dataset were taken into account. From the presented results, it can be seen that the employed skeleton-tracking algorithm constitutes the most time-consuming part of the proposed approach, corresponding to approximately 99.74% of the overall processing. The latter highlights the increased time efficiency (0.26% of the overall processing) of the proposed action recognition methodology (Section 3), characteristic that received particular attention during the design of the proposed method. Additionally, it can be seen that the overall duration of all algorithmic steps is shorter that the respective kinect's capturing period; in other words, all proposed algorithmic steps are completed for the currently examined depth map before the next one is captured by the kinect sensor. This observation verifies the real-time nature of the proposed approach. It must be highlighted that the aforementioned time performances were measured without applying any particular code or algorithmic optimizations to the proposed method.

5   Conclusions

In this paper, a real-time tracking-based approach to human action recognition was presented and evaluated using the Huawei/3DLife 3D human reconstruction and action recognition Grand Challenge dataset. Future work includes the investigation of STIP-based approaches for overcoming the inherent limitations of skeleton-tracking algorithms and the incorporation of multimodal information from additional sensors.

Acknowledgment. The work presented in this paper was supported by the European Commission under contract FP7-287723 REVERIE.

References

1. Hidden Markov Model Toolkit (HTK), http://htk.eng.cam.ac.uk
2. Ballan, L., Bertini, M., Del Bimbo, A., Seidenari, L., Serra, G.: Effective codebooks for human action representation and classification in unconstrained videos. IEEE Transactions on Multimedia 14(4), 1234–1245 (2012)
3. Gorelick, L., Blank, M., Shechtman, E., Irani, M., Basri, R.: Actions as space-time shapes. IEEE Transactions on Pattern Analysis and Machine Intelligence 29(12), 2247–2253 (2007)
4. Gu, J., Ding, X., Wang, S., Wu, Y.: Action and gait recognition from recovered 3-d human joints. IEEE Trans. on Systems, Man, and Cybernetics, Part B: Cybernetics 40(4), 1021–1033 (2010)
5. Haq, A., Gondal, I., Murshed, M.: On temporal order invariance for view-invariant action recognition. IEEE Transactions on Circuits and Systems for Video Technology 23(2), 203–211 (2013)
6. Holte, M.B., Chakraborty, B., Gonzalez, J., Moeslund, T.B.: A local 3-d motion descriptor for multi-view human action recognition from 4-d spatio-temporal interest points. IEEE Journal of Selected Topics in Signal Processing 6(5), 553–565 (2012)

7. Ji, X., Liu, H.: Advances in view-invariant human motion analysis: a review. IEEE Trans. on Systems, Man, and Cybernetics, Part C: Applications and Reviews 40(1), 13–24 (2010)
8. Junejo, I.N., Dexter, E., Laptev, I., Pérez, P.: View-independent action recognition from temporal self-similarities. IEEE Trans. on Pattern Analysis and Machine Intelligence 33(1), 172–185 (2011)
9. Papadopoulos, G.T., Briassouli, A., Mezaris, V., Kompatsiaris, I., Strintzis, M.G.: Statistical motion information extraction and representation for semantic video analysis. IEEE Transactions on Circuits and Systems for Video Technology 19(10), 1513–1528 (2009)
10. Poppe, R.: A survey on vision-based human action recognition. Image and Vision Computing 28(6), 976–990 (2010)
11. Rabiner, L.R.: A tutorial on hidden markov models and selected applications in speech recognition. Proc. of the IEEE 77(2), 257–286 (1989)
12. Song, B., Kamal, A.T., Soto, C., Ding, C., Farrell, J.A., Roy-Chowdhury, A.K.: Tracking and activity recognition through consensus in distributed camera networks. IEEE Transactions on Image Processing 19(10), 2564–2579 (2010)
13. Turaga, P., Veeraraghavan, A., Chellappa, R.: Statistical analysis on stiefel and grassmann manifolds with applications in computer vision. In: IEEE Conf. on Computer Vision and Pattern Recognition, pp. 1–8. IEEE (2008)
14. Weinland, D., Ronfard, R., Boyer, E.: Free viewpoint action recognition using motion history volumes. Computer Vision and Image Understanding 104(2), 249–257 (2006)

Kinect vs. Low-cost Inertial Sensing for Gesture Recognition

Marc Gowing[1], Amin Ahmadi[1], François Destelle[1], David S. Monaghan[1],
Noel E. O'Connor[1], and Kieran Moran[2]

[1] INSIGHT: Centre for Data Analytics, Dublin City University, Ireland
[2] Applied Sports Performance Research, School of Health and Human Performance,
Dublin City University, Ireland
{marc.gowing,anim.ahmadi,francois.destelle,david.monaghan,
noel.oconnor,kieran.moran}@dcu.ie

Abstract. In this paper, we investigate efficient recognition of human gestures / movements from multimedia and multimodal data, including the Microsoft Kinect and translational and rotational acceleration and velocity from wearable inertial sensors. We firstly present a system that automatically classifies a large range of activities (17 different gestures) using a random forest decision tree. Our system can achieve near real time recognition by appropriately selecting the sensors that led to the greatest contributing factor for a particular task. Features extracted from multimodal sensor data were used to train and evaluate a customized classifier. This novel technique is capable of successfully classifying various gestures with up to 91 % overall accuracy on a publicly available data set. Secondly we investigate a wide range of different motion capture modalities and compare their results in terms of gesture recognition accuracy using our proposed approach. We conclude that gesture recognition can be effectively performed by considering an approach that overcomes many of the limitations associated with the Kinect and potentially paves the way for low-cost gesture recognition in unconstrained environments.

Keywords: Gesture recognition, Decision tree, Random forest, Inertial sensors, Kinect.

1 Introduction

There is a growing trend towards moving away from the traditional keyboard and mouse as the primary computer interaction tools. In the past decade, a wealth of research in academia and industry [16,7] has focused on finding new and more intuitive methods by which humans can interact with computers and computer-based content. Many such initiatives have been aimed at devising new algorithms and technologies for recognizing moving objects as well as human gestures and actions. Nonverbal behaviors such as hand movements, head gestures, body language, facial expression and eye contact play an import role within human communications. The recording and reconstruction of these human activities and gestures is one of the fundamental core building blocks in realizing

C. Gurrin et al. (Eds.): MMM 2014, Part I, LNCS 8325, pp. 484–495, 2014.

any advanced human/computer interaction system that is free from a keyboard and mouse.

Increasingly, this effort is driven by new application opportunities in 3D multimedia computing and modeling. The natural output of 3D multimedia capturing, processing, and scene reconstruction are novel virtual immersive environments that require more sophisticated control/interaction mechanisms than a simple point and click. In such scenarios, interaction or control based on human gestures seems a more intuitive and comfortable approach for end users [11,8]. In certain applications, we will wish to visualize the real-time motion and actions of other users in the same immersive environment in order to experience a truly shared collaborative experience [4,5]. In either case, approaches to real-time human gesture recognition in the real world are required so that the resulting gestures can be used to produce the required effect in the virtual world.

In recent years, the Microsoft Kinect has been a key driver for a new form of hands free interaction. As a low-cost and widely available approach to human motion sensing, the Kinect and the associated open source libraries have enabled researchers to target a range of next generation novel multimedia applications [3,17]. However, the Kinect is not without its own limitations. Whilst it constitutes a practical approach to breaking the tether to mouse/keyboard, it is still rather limited in terms of its practical application, restricting movement sensing to indoor and to a limited spatial volume (typically 3m x 3m) [18]. In this paper, we consider other sensors that in theory allow movement sensing outside of these constraints, potentially opening up the possibility of human-computer interaction *"in the wild"* i.e. in unconstrained environments, that could then subsequently be mapped to novel multimedia experiences in immersive environments.

In general, the recording and reconstruction of human motion is referred to as Motion Capture (or MoCap for short). MoCap is a well studied and broad research area that has been explored in multiple different research fields including computer vision, computer graphics and even body sensor networks [15,12,19,14]. The various different approaches are numerous and include approaches based on mechanical, inertial, magnetic, acoustic and visual sensing etc. After a suitable MoCap system has been identified for a particular requirement and human motion has been captured, the next step is to perform gesture recognition. That is, it is required to infer some semantic meaning to the movements being performed. This can often be accomplished by manually annotating the captured motion followed by machine learning, i.e. a human observer will watch the action being performed and decide what movement or gesture has been recorded and then a suitable machine learning technique can be used to enable an automatic system to recognize similar actions for previously unseen data. It should be noted that the complexity and speed of such systems often increase exponentially when additional gestures are added to the system. The key tenets of the work that we present here are the need for real time applications that remain low-cost.

Recent advancements in microelectronics and other technologies mean that inertial sensors are gaining popularity to monitor human movements in a num-

ber of sporting [2], rehabilitation [6] and everyday activities [9]. MEMS inertial
sensing technology is already integrated by default into many consumer devices
- virtually every smart phone and many computer games controllers (e.g. the
Nintendo Wii). MEMS inertial sensors are being widely used in MoCap research
due to the following reasons:

- They are miniaturized and lightweight so they can be placed on any part or
  segment of a human body without hindering performance.
- The cost of such sensors is falling dramatically as they start to persuade
  mass market consumer devices.
- They can be utilized to capture human movement/actions in real unconstrained environments (e.g. outdoor environments with variable lighting conditions) to obtain accurate results.
- They can be used to provide real time or near real time feedback.

On the other hand, Microsoft Kinect comes with an RGB camera and a depth
sensor, which in combination provide full-body 3D motion capture capabilities
and gesture recognition. This inexpensive technology is also widely being used
for gesture recognition mainly due to the following reasons:

- Using Kinect allows users to avoid wearing body sensors when performing
  the movements.
- We can extract skeleton data using off-the-shelf software such as Kinect SDK
  and OpenNI.
- Kinect sensors can be used to obtain real time feedback.

In order to investigate the relative benefits of both approaches in terms of
gesture recognition, in this paper we have investigated the use of wearable inertial sensors and a Microsoft Kinect depth sensor to classify a wide range of
activities performed by five different subjects. We compare the gesture recognition results obtained from inertial sensors with that from the Kinect sensor
using a customized Random Forest decision trees. In addition, we simulated an
ultra low-cost system where only a small number of low rate inertial sensors are
available by using down sampled data (from 256 Hz to 32Hz) from only three
sensors (three out of eight worn sensors).

The paper is organized as follows. In Section 2 we explain the dataset and also
describe the sensor modalities used in this paper. In Section 3 we fully explain
the methodology. We then provide our results and discussion section and finally
conclude and highlight our contributions.

2   Dataset

We use a gesture recognition dataset that includes recordings of human subjects performing various gestures and activities (17 in total) such as simple actions, training exercises and diverse sports activities. The dataset encompasses

recordings of five subjects whose actions were captured using eight wearable inertial sensors and one Microsoft Kinect. The inertial system was mounted on eight different places on the body and then s/he was asked to carry out random movements to ensure that s/he felt comfortable with the system and that the system was not limiting their movements. Next, the subject was asked to perform a series of actions. The performed actions can be divided into the following categories:

1. Simple actions (hand waving, knocking on the door, clapping, throwing, punching, push away by both hands).
2. Training exercises (jumping jacks, lunges, squats, punching and kicking).
3. Sports activities (golf drive, golf putt, golf chip, tennis forehand, tennis backhand, weight lifting, walking).

Once the sensors were switched on and worn by a subject, they were tapped separately to use the acceleration spike to synchronize all the inertial sensors. After the synchronization process, all participants performed each action/movement five times before starting the next action.

One Kinect camera was also setup about two meters away in front of the subject to capture the front body part movement. The inertial sensors were synchronized with the recorded images by clapping three times before each recording so that a specific event could be identified. The data set along with all annotations is available for download from: http://mmv.eecs.qmul.ac.uk/mmgc2013.

2.1 Sensors

We chose wearable inertial sensors and the Microsoft Kinect since they are low-cost and are each gaining in popularity in the area of human movement monitoring and gesture recognition due to their accuracy and potential for real time applications.

Kinect. Since very recently, computer game users can enjoy a novel gaming experience with the Xbox, thanks to the introduction of the Microsoft Kinect sensor, where *your body is the controller* [1]. Like the Nintendo Wii sensor bar, the Kinect device is placed either above or below the video screen. However, the Kinect adds the capabilities of a depth sensor to those of a RGB camera, recording the distance from all objects that lie in front of it. The depth information is then processed by a software engine that extracts, in real time, the human body features of players, thus enabling the interaction between the physical world and the virtual one. The Kinect dataset recordings of subjects' activities were captured using the OpenNI drivers/SDK. We employed the widely known NiTE framework to track the 3D skeleton for each subject from the Kinect sensor which in turn can be used to extract subjects' joint positions and angular velocities. Estimation of 3D joints position and orientation is illustrated in Fig 1.

---

[1] http://www.xbox.com/en-US/KINECT

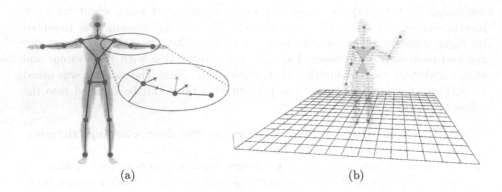

(a)   (b)

Fig. 1. Estimation of 3D joint positions from the Kinect sensor. (a) The Kinect skeleton and the local coordinate system at each joint. (b) A real scene point cloud and the visualization of its skeleton computation in real time.

Table 1. Technical specifications of the inertial sensor units

Features	Values
Resolution (Acc, Gyr, Mag)	12 bit, 16 bits, 12 bits respectively
Sampling rate	Scalable up to 512 Hz
Sensor range	Acc: scalable up to 8G Gyro: scalable up to 2000/s Mag: scalable up to 8.1G
Connectivity	Bluetooth-Class 1(100m range), Micro SD card
Dimension	57 × 38 × 21mm
Weight	49g including housing and battery

WIMU. In general, a Wireless/Wearable Inertial Measurement Unit, or WIMU, is an electronic device consisting of a microprocessor board, on-board accelerometers, gyroscopes and magnetometers and a wireless connection to transfer the captured data to a receiving client. WIMUs are capable of measuring linear acceleration, angular velocity, and gravitational forces and are often used in MoCap systems. Technical specifications of the WIMUs we have utilized are summarized in Table 1. In the dataset eight WIMUs were attached to different parts of the subjects to capture their activities. In particular, sensors were attached on the left/right wrist, left/right ankle, left/right foot, waist and chest of all participants. Placement of inertial sensors on a subject's body is depicted in Fig 2.

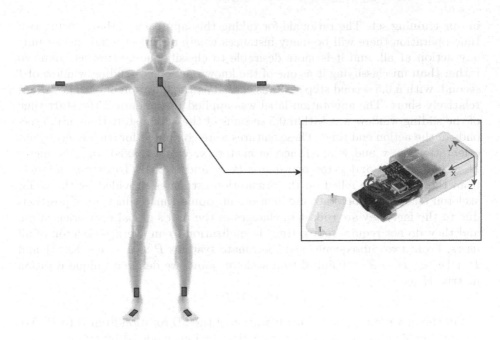

Fig. 2. Placement of inertial sensors on subjects body is shown. The lighter colored sensor indicates that the sensor is attached to the subject's lower back.

3   Methodology

To facilitate real time gesture recognition, we chose to extract features based on a sliding window approach. A fixed window size of one second was chosen based on an experimentally derived average duration of all gestures, with a 50% overlap. At each step, we compute temporal features from each modality.

For the final recognition system, different features were calculated from the one-second windows to compress the information of interest in the data. We initially calculate a large set of features for each sensor signal, and apply a standard sequential feature selection technique [1] to identify the most discriminative features prior to training our classification model. The following sections outline the procedure in more detail.

3.1   Feature Selection

In the dataset, each subject performs the full list of actions as one long continuous sequence of motion. Therefore, each sequence contains many instances of unlabeled data, where the subject is deemed to not be performing one of the predefined actions. Rather than segmenting the motion sequence into examples of predefined actions only, we instead chose to include the full sequence in our training phase. We use any instances of unlabeled data as negative examples in our training set. The rationale for taking this approach is that, during real time operation there will be many instances where a subject is not performing any action at all, and it is more desirable to classify the gesture as unknown rather than misclassifying it as one of the known actions. A sliding window of 1 second, with a 0.5 second step size was chosen as the duration of each gesture is relatively short. The annotation label was applied to the data if the start time of the sliding window was within 0.5 seconds of the action start time or 0.5 seconds of the action end time. These features were computed for each Kinect joint orientation (u, v and w axis), accelerometer (x, y and z axis), magnetometer (x, y and z axis) and gyroscope sensor (x, y and z axis). To extract features from the Kinect, we relied on the orientation estimates provided by the NiTE skeleton tracker. We opted to use joint orientation estimates instead of positions due to the fact they are robust to changes in the user's global root orientation, and they do not require retargeting/normalization to an average skeleton of all users. From two subsequent local coordinate systems $P : \{u, v, w \in \mathbb{R}^{3 \times 3}\}$ and $P' : \{u', v', w' \in \mathbb{R}^{3 \times 3}\}$ linked to a skeleton joint, we define a unique rotation matrix M as:

$$P' = M.P. \tag{1}$$

Let the quaternion q be the modelization of the 3D rotation from P to P'. We divide this 3D rotation in two separate 2D rotations modeled by quaternions:

$$\begin{cases} q_u : \{A_u^x, A_u^y, A_u^z, \varphi_u\} \\ q_v : \{A_v^x, A_v^y, A_v^z, \varphi_v\} \end{cases} \tag{2}$$

where $A_u, A_v \in \mathbb{R}^3$ are the axis of rotation from u to u' and from v to v', while $\varphi_v, \varphi_v \in \mathbb{R}$ are the deviation angles:

$$\begin{cases} A_u \equiv u \times u' \\ \varphi_u = u.u' \end{cases} \begin{cases} A_v \equiv v \times v' \\ \varphi_v = v.v' \end{cases} \tag{3}$$

We now compose these quaternion, $q = q_u.q_v$ and create the unique rotation matrix M from P to P'. From there, we directly obtain the Euler angles prior to feature extraction, to provide the quantity of rotation about the local x, y and z-axis.

The feature selection process involved extracting a large number of features for every signal in the motion sequence, and then reducing this using the well-known standard sequential forward selection technique to reduce the number of features and thereby to improve computational cost and to obtain near real time performance. Heuristic features including SMA [10] and Inter axis correlation [20,13] were derived from a fundamental understanding of how a specific movement would produce a distinguishable sensor signal. For instance, there are obvious correlations, using Pearson correlation test, between left and right wrist movements during all golf swings, walking, pushing with two hands, weight lifting and clapping. Correlation in x and y signals between the left wrist and the right wrist during clapping action (5 times) is shown in Figure 3.

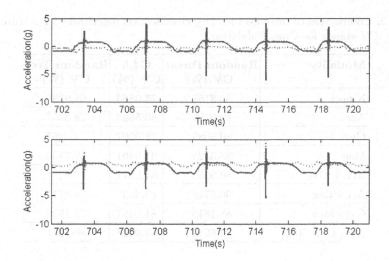

Fig. 3. Correlation in x (solid line) and y (dashed-line) acceleration signals between the right wrist (top) and left wrist (bottom) during clapping action is shown

3.2 Classification

We investigated a number of different fast decision trees in order to choose one technique to classify all the activities in the dataset using both modalities. We examined Random Forest, Random Tree and C4.5 algorithms and compared the results. Results are shown in Table 2. We have chosen the random forest method, as it provided the best accuracy among the algorithms investigated. In general, random forest does not over fit and it is fast, which makes it suitable for our near real time application. In addition, in this work, we extracted features for all subjects from triple axis inertial sensors and a Microsoft Kinect which results in a large amount of data. Therefore, we needed to apply a method such as random forest, which runs efficiently on large datasets. This method is also capable of providing feedback on what inputs/features are more important so we could enhance our model by removing low-priority features to speed up and implement towards real time application. We found that a random forest consisting of 30 trees provided the best results on our dataset. The results to classify different activities using wearable inertial sensors and Kinect are shown in Table 3.

3.3 Reducing the Number of Sensors

By observing all the gestures and activities in the dataset, it can be seen that all the activities have upper body movement component so they can be distinguished by upper body movements. Therefore, we only extracted features from three upper body (right wrist, left wrist and chest) sensors to classify the gestures. The advantages of employing a smaller number of sensors are as follows:

Table 2. Comparison of three decision tree classification algorithms for gesture classification. CV stands for Cross Validation.

Modality	Random Forest CV (%)	C4.5 CV (%)	Random Tree CV (%)
Kinect	80.3157	73.0883	68.9196
Acc	89.0725	80.5624	78.8357
Gyr	86.4085	78.9097	74.0257
Mag	88.3325	81.5491	77.7750
Acc + Gyr	89.9359	81.6971	79.6251
Acc+Mag	90.6759	83.4238	79.8717
Gyr+Mag	88.1845	81.5491	77.1830
Acc+Gyr+Mag	90.6512	83.9418	80.7351

Table 3. Results of activity recognition using a range of multimodal sensors

Modality	Sub. 1 CV (%)	Sub. 2 CV (%)	Sub. 3 CV (%)	Sub. 4 CV (%)	Sub. 5 CV (%)	All Subjs. CV (%)
Kinect	85.1544	82.797	86.5340	87.6494	87.7039	80.3157
Acc	90.0238	94.5545	90.5152	90.4382	87.9548	89.0725
Gyr	87.6485	92.9455	88.2904	86.8526	83.1870	86.4085
Mag	89.3112	93.1931	92.1546	88.1806	87.9548	88.3325
Gyr+ Mag	89.4299	93.0693	92.3888	88.9774	86.7001	88.1845
Gyr + Acc	91.4489	93.8119	91.3349	89.7742	89.5859	89.9359
Acc+ Mag	90.3800	94.5545	92.0375	91.8991	90.2133	90.6759
Acc+Gyr+Mag	90.9739	94.8020	92.2717	91.8991	90.3388	90.6512

- The entire gesture recognition system would be cheaper and less prone to set up/synchronization error.
- Reducing the number of sensor nodes allows us to reduce the amount of features to be extracted during the training phase. This can lead to less computational cost which in turn can result in providing near real time feedback.
- This can address a typical scenario for many applications - where end-users may only have a small number of sensors at their disposal.

Results obtained from utilizing three upper body sensors are illustrated in Table 4.

Table 4. Activity classification using reduced number of inertial sensors

Modality	Sub. 1 CV (%)	Sub. 2 CV (%)	Sub. 3 CV (%)	Sub. 4 CV (%)	Sub. 5 CV (%)	All Subs. CV (%)
Selected ACC+Gyr+Mag	90.7363	93.2693	89.8126	90.4382	88.0803	88.7025

Table 5. Activity classification using down-sampled data

Modality	Sampling Frequency	Sub. 1 CV (%)	Sub. 2 CV (%)	Sub. 3 CV (%)	Sub. 4 CV (%)	Sub. 5 CV (%)	All Subs. CV (%)
Selected ACC+Gyr+Mag	32 Hz	89.5487	93.1931	88.8759	89.1102	87.8294	87.8999

3.4 Using Down Sampled Data

In a further attempt to reduce the cost of the system amd the amount of computation at each window, we down sampled the inertial sensors data from 256Hz to as low as 32Hz, effectively reducing the computation time by a factor of 8. The results are shown in Table 5. Down sampling can simulate the output of inertial sensors systems manufactured to lower specification.

4 Results

We have compared three different fast decision tree techniques including random forest, random tree and C4.5 to classify activities in our dataset. We trained all the classifiers on each modality and performed 10 fold cross validation to test the performance. The comparisons of these techniques are summarized in Table 2. It is clearly shown that the highest accuracy can be obtained by using random forest followed by the C4.5 technique.

As can be seen in Table 3, using accelerometers, gyroscopes or magnetometers can provide more overall accuracy to classify activities than using the Kinect sensor. Table 3 also shows that by combining the data from accelerometers, gyroscopes and magnetometer, maximum accuracy can be achieved. However, even using any one of these modalities on its own its still outperforms the Kinect. The output from the random forest classifier for the selected sensors (left/right wrist sensor and chest sensor) is illustrated in Table 4. As shown, reducing the number of sensor nodes does not degrade the overall accuracy since all the activities studied in this work contain upper body movement components. In addition, we have investigated the effect of down-sampled data to simulate utilizing low-cost inertial sensors for gesture recognition. Not only does it simulate the low-cost scenario, it also decreases the computational cost to achieve near real time application. We have down sampled the data from 256 Hz to 32 Hz for this experiment. Table 5 summarizes the results obtained from down-sampled inertial sensors. As can be seen, this does not significantly affect the accuracy as normal human movements are not as fast as 32 repetitions per second (32Hz) and thus cheap inertial sensors with 32Hz sampling rate can be considered to be used to capture human activities.

Finally, decision trees are relatively fast to classify activities and therefore are suitable techniques for near real time or real time applications. Once the features were selected, it took between 0.5 seconds to 0.76 seconds to produce a random forest of 30 trees model for each subject's dataset on a MacBook Pro 2.33GHz framework. In the future, further optimization techniques are required to enhance the performance of the devised system to achieve real time feedback.

5   Conclusions

In this paper we described a novel inertial-based system that automatically classifies a large range of activities (17 different gestures) using a customized random forest decision tree. Our system achieved near real time gesture recognition by appropriately selecting the sensors that led to the greatest contributing factor for a particular task. Our technique is capable of classifying various gestures successfully using inertial sensors with up to 91% overall accuracy, making it extremely competitive with the MS Kinect. We have fully analyzed our system for a wide range of MoCap solutions thus providing a look up table to enable potential researchers to choose an appropriate MoCap solution based on their specific accuracy requirements. We managed to achieve a high level of accuracy for a low-cost system which is capable of providing feedback in near real time.

Our results point to the fact that the Kinect is clearly not the only option to be considered for applications requiring MoCap. Its attraction is rooted in its low-cost and lack of instrumentation but it is inherently limited in terms of the scenarios in which it can be implemented. Low-cost inertial sensors, on the other hand, do not suffer from many of the limitations associated with the Kinect and can operate in outdoor unconstrained environments. We have shown in this paper that very similar or even higher accuracy to Microsoft Kinect can be achieved with a very small amount of human instrumentation. This potentially paves the way for novel future multimedia applications whereby human motion and interaction can be captured in a range of challenging environments, not just indoors in front of a computer.

Acknowledgments. The research leading to these results has received funding from the European Community's Seventh Framework Programmes (FP7/2007-2013) under grant agreement no. ICT-2011-7-287723 (REVERIE project) and (FP7/2013-2016) under grant agreement no. ICT-2011-8.2.601170 (REPLAY project).

References

1. Aha, D.W., Bankert, R.L.: A comparative evaluation of sequential feature selection algorithms. In: Learning from Data, pp. 199–206. Springer (1996)
2. Ahmadi, A., Rowlands, D.D., James, D.A.: Development of inertial and novel marker-based techniques and analysis for upper arm rotational velocity measurements in tennis. Sports Engineering 12(4), 179–188 (2010)
3. Alexiadis, D.S., Kelly, P., Daras, P., O'Connor, N.E., Boubekeur, T., Moussa, M.B.: Evaluating a dancer's performance using kinect-based skeleton tracking. In: Proceedings of the 19th ACM International Conference on Multimedia, pp. 659–662. ACM (2011)
4. Bowman, D.A., Hodges, L.F.: User interface constraints for immersive virtual environment applications. Tech. rep. Atlanta: Graphics, Visualization and Usability (1995)
5. Bowman, D.A., Hodges, L.F.: Formalizing the design, evaluation, and application of interaction techniques for immersive virtual environments. Journal of Visual Languages & Computing 10(1), 37–53 (1999)
6. Culhane, K., O'Connor, M., Lyons, D., Lyons, G.: Accelerometers in rehabilitation medicine for older adults. Age and Ageing 34(6), 556–560 (2005)
7. Dix, A.: Human computer interaction. Pearson Education (2004)
8. Jaimes, A., Sebe, N.: Multimodal humancomputer interaction: A survey. Computer Vision and Image Understanding 108(1), 116–134 (2007)
9. Junker, H., Amft, O., Lukowicz, P., Tröster, G.: Gesture spotting with body-worn inertial sensors to detect user activities. Pattern Recognition 41(6), 2010–2024 (2008)
10. Karantonis, D.M., Narayanan, M.R., Mathie, M., Lovell, N.H., Celler, B.G.: Implementation of a real-time human movement classifier using a triaxial accelerometer for ambulatory monitoring. IEEE Transactions on Information Technology in Biomedicine 10(1), 156–167 (2006)
11. Lyons, D.M.: System and method for permitting three-dimensional navigation through a virtual reality environment using camera-based gesture inputs, uS Patent 6,195,104 (February 27, 2001)
12. Mannini, A., Sabatini, A.M.: Machine learning methods for classifying human physical activity from on-body accelerometers. Sensors 10(2), 1154–1175 (2010)
13. Mathie, M.: Monitoring and interpreting human movement patterns using a triaxial accelerometer. Ph.D. thesis, The University of New South Wales (2003)
14. Menache, A.: Understanding motion capture for computer animation and video games. Morgan Kaufmann (2000)
15. Moeslund, T.B., Granum, E.: A survey of computer vision-based human motion capture. Computer Vision and Image Understanding 81(3), 231–268 (2001)
16. Myers, B.A.: A brief history of human-computer interaction technology. Interactions 5(2), 44–54 (1998)
17. Ren, Z., Meng, J., Yuan, J., Zhang, Z.: Robust hand gesture recognition with kinect sensor. In: Proceedings of the 19th ACM International Conference on Multimedia, pp. 759–760. ACM (2011)
18. Roth, H., Vona, M.: Moving volume kinectfusion. In: BMVC, pp. 1–11 (2012)
19. Sturman, D.J.: A brief history of motion capture for computer character animation. In: SIGGRAPH 1994, Character Motion Systems, Course notes 1 (1994)
20. Yang, J.Y., Wang, J.S., Chen, Y.P.: Using acceleration measurements for activity recognition: An e ective learning algorithm for constructing neural classifiers. Pattern Recognition Letters 29(16), 2213–2220 (2008)

Yoga Posture Recognition for Self-training

Hua-Tsung Chen[1], Yu-Zhen He[2], Chun-Chieh Hsu[2], Chien-Li Chou[2],
Suh-Yin Lee[2], and Bao-Shuh P. Lin[1]

[1] Information & Communications Technology Lab,
National Chiao Tung University, Hsinchu, Taiwan
huatsung@cs.nctu.edu.tw, bplin@mail.nctu.edu.tw
[2] Department of Computer Science, National Chiao Tung University, Hsinchu, Taiwan
{hoyc,hsucch,fallwind,sylee}@cs.nctu.edu.tw

Abstract. Self-training plays an important role in sports exercise, but improper training postures can cause serious harm to muscles and ligaments of the body. Hence, more and more researchers are devoted into the development of computer-assisted self-training systems for sports exercise. In this paper, we propose a Yoga posture recognition system, which is capable of recognizing what Yoga posture the practitioner is performing, and then retrieving Yoga training information from Internet to remind his/her attention to the posture. First, a Kinect is used for capturing the user body map and extracting the body contour. Then, *star skeleton*, which is a fast skeletonization technique by connecting from centroid of target object to contour extremes, is used as a representative descriptor of human posture for Yoga posture recognition. Finally, some Yoga training information for the recognized posture can be retrieved from Internet to remind the practitioner what to pay attention to when practicing the posture.

Keywords: sports training, posture analysis, Yoga, star skeleton, feature extraction, Kinect.

1 Introduction

Sports and exercises have always been attractive to the general public, inspiring numerous researchers to be devoted to this area. With the rapid advance in computer vision and video processing technologies, sports professionals thirst for automatic/semi-automatic systems to assist players to improve their performance. One of the compelling research topics is automatic tactic analysis. Chen et al. [1-3] track the ball motion in different kinds of sports videos and provide manifold trajectory-based applications, such as pitching evaluation in baseball, shooting location estimation in basketball, and set type recognition in volleyball. To recognize tactical patterns in soccer video, Zhu et al. [4] analyze the temporal-spatial interaction among the ball and the players to construct a tactic representation, *aggregate trajectory*, based on multiple trajectories. There are also research works focusing on tactic analysis in basketball video [5, 6], which perform camera calibration to obtain 3D-to-2D

C. Gurrin et al. (Eds.): MMM 2014, Part I, LNCS 8325, pp. 496–505, 2014.

transformation, map player trajectories to the real-world court model, and detect the *wide-open* event or recognize the offensive tactic pattern–*screen*.

Another emerging trend is computer-assisted self-training in sports exercise. In general, most sports players are required to take additional time to exercise on their own. However, players may get injured during self-training due to improper postures. In this paper, we take Yoga as our research subject and develop a Yoga posture recognition system for assisting the practitioner in exercising Yoga by himself/herself.

As an ancient Indian art, Yoga exercise not only promotes physical health but also helps to purge the body, mind, and soul. Therefore, Yoga is becoming more and more popular. Patil et al. [7] describe the project "Yoga tutor" which uses Speeded Up Robust Features (SURF) to detect and visualize the postural difference between a practitioner and an expert. However, only the contour information seems unable to describe and compare the posture appropriately. Luo et al. [8] propose a Yoga training system based on motion replication technique (MoRep). Though precise body motions can be captured by the InterfaceSuit, which comprises Inertial Measurement Units (IMUs) and tactors, it does influence the practitioner's exercising. Wu et al. [9] develop a Yoga expert system, which instructs training technique based on images and text. However, the practitioner's posture is not analyzed.

To overcome the limitations of existing works, a Yoga posture recognition system is proposed in this paper. The schematic diagram is illustrated in Fig. 1. First, a Kinect located at a distance of more than 200 cm away from the practitioner is used to capture the body map, as shown in Fig. 1(a). Then, we extract the body contour and compute the star skeleton to describe the posture currently performed by the practitioner, as shown in Fig. 1(b). On the other hand, we pre-build the template star skeletons of standard Yoga postures from the body maps of the static postures performed by Yoga experts. A distance function is designed to evaluate the dissimilarity between two skeletons, and the practitioner's posture is recognized by choosing the template with the most similar skeleton, as shown in Fig. 1(c). The proposed system is capable of recognizing up to twelve Yoga postures, including: (1) *Tree*, (2) *Warrior III*, (3) *Downward-Facing Dog*, (4) *Extended Hand-to-Big-Toe*, (5) *Chair*, (6) *Full Boat*, (7) *Warrior II*, (8) *Warrior I*, (9) *Cobra*, (10) *Plank*, (11) *Side Plank*, and (12) *Lord of the Dance*. For more details about each Yoga posture, please refer to [10]. Finally, some Yoga training information for the recognized posture can be retrieved from Internet to remind the practitioner what to pay attention to when practicing the posture, as shown in Fig. 1(d).

The paper is organized as follows. In the next section, the main processing modules of the proposed system, including user body map capturing, contour extraction, star skeleton computation, and Yoga posture recognition, are elaborated. Section 3 presents the experimental results with discussion. Finally, we conclude this paper in section 4.

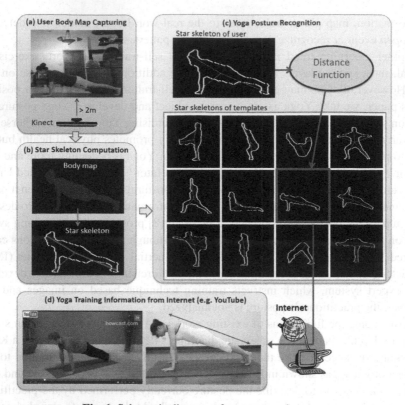

Fig. 1. Schematic diagram of our proposed system

2   Yoga Posture Recognition Using Star Skeleton

Yoga comprises various *asanas*, each of which denotes a static physical posture. Body contour is a good way to represent a posture. However, using the whole body contour to describe a pitch posture is inefficient because each contour point is very similar to its neighbors. On the other hand, simple information like human width and height may be rough to represent a posture. Consequently, representative features, such as human skeleton, must be extracted to describe a posture. There exist many standard skeletonization techniques, e.g. thinning and distance transformation. However, these techniques are computationally expensive and even susceptible to noise in the target boundary. Since the computational efficiency and good representability of *star skeleton* have been validated in human action analysis [11, 12], we use *star skeleton* as the posture descriptor in our Yoga posture recognition system. In the following sub-sections, the processing steps of Yoga posture recognition will be explained in detail.

2.1   User Body Map Capturing and Contour Extraction

For feature extraction, the binary map of user body is first captured by a Kinect using OpenNI library [13]. The user body map is smoothed by the morphological "open" operation, and then the contour can be extracted (c.f. OpenCV [14]). The processing steps are illustrated in Fig. 2, wherein Figs. 2(a) and (b) show the original RGB frame and the obtained body map. Fig. 2(c) shows the smoothed body map, and the extracted contour is presented in Fig. 2(d). With the contour extracted, we are now ready for the subsequent star skeleton computation.

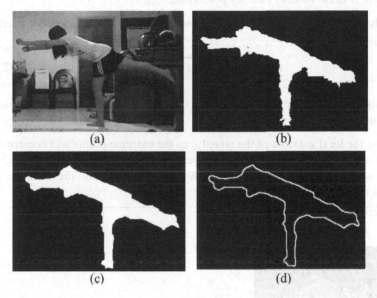

Fig. 2. Illustration of user body map capturing and contour extraction. (a) Original RGB frame. (b) User body map. (c) Smoothed body map.

2.2   Star Skeleton Computation

The concept of star skeleton is to connect from centroid to gross extremities of a body contour. To derive the gross extremities of body contour, the distance from the centroid to each contour point p_i is computed in a clockwise order. Then, we smooth the distance signal d_i by Gaussian smoothing filter and locate extremities at the representative local maxima of the distance signal, as shown as Fig. 3. Note that we modify the original star skeleton algorithm [11] via calculating the centroid of the whole body, which is more robust to posture change, instead of the centroid of the body contour. The algorithm of star skeleton computation is presented as Algorithm 1.

Algorithm 1: (Star Skeleton Computation)
Input: Human body map
Output: A skeleton in star fashion

1. Define the centroid (x_c, y_c) of the human body map:

$$x_c = \frac{1}{N}\sum_{i=1}^{N} bx_i \ , \ \ y_c = \frac{1}{N}\sum_{i=1}^{N} by_i \qquad (1)$$

where N is the number of body pixels (bx_i, by_i), $i = 1 \ldots N$.

2. Calculate the distance d_i from the centroid (x_c, y_c) to each contour pixel (x_i, y_i).

$$d_i = \sqrt{(x_i - x_c)^2 + (y_i - y_c)^2} \qquad (2)$$

3. Smooth the distance signal d_i to s_i by Gaussian smoothing filter for noise reduction.

4. Take the local maxima of the signal s_i as the extreme points and construct the star skeleton by connecting extreme points to the centroid (x_c, y_c). Local maxima are detected by finding zero-crossing of the difference function.

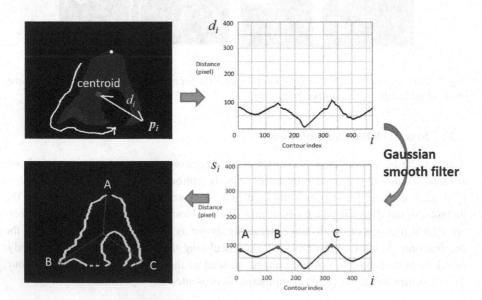

Fig. 3. Process flow of star skeleton computation

2.3   Yoga Posture Recognition

After extracting star skeleton, we define the feature as a set of vectors from the centroid to shape extremities, as shown in Fig. 4. Since the lengths of the feature vectors vary with the human size and shape, normalization is required to get relative distribution of the feature vectors. We perform normalization via dividing the vectors by the length of the longest vector in the star skeleton feature. For the example in Fig. 4, we obtain the feature $V = \{v_1, v_2, v_3, v_4, v_5\}$ to describe the posture. On the other hand, we pre-build the template star skeletons of standard Yoga postures from the body maps of the static postures performed by Yoga experts, as shown in Fig. 1(c), and obtain for each template posture the feature $T = \{t_1, t_2, t_3, \dots, t_m\}$, where m is the number of vectors in T.

To recognize what Yoga posture the practitioner is performing, we compare his/her star skeleton feature V with each template T, and evaluate the difference by the following distance function $D(V, T)$.

$$D(V,T) = \sum_{i=0}^{n} \min_{t_j \in T} d(v_i, t_j) \ , \quad d(v_i, t_j) = \sqrt{(v_i - t_j)^2} \tag{3}$$

However, here comes a problem. For example, as shown in Fig. 5, Warrior I Pose may be misrecognized as Warrior II Pose because only the sum of the distances between the *matched* skeleton vectors is considered but the *mismatched* vectors are neglected. To overcome this problem, some penalty mechanism should be incorporated for the mismatched vectors. Since the feature vector with larger length is more important and the extremities of the star skeleton are mainly from the head, two hands, and two legs, we take into account the longest five vectors of each template skeleton. Then, a penalty value is added to the distance function for each mismatched vector. The distance function is revised, termed $D_P(V, T)$, by adding the penalty term p, as follows.

$$D_P(V,T) = D(V,T) + p \ , \quad p = \sum_{k} \|t_k\|_2 \tag{4}$$

In (4), t_k represents a mismatched vector of T. Consequently, the practitioner's posture can be recognized by choosing the template T with the minimal $D_P(V, T)$.

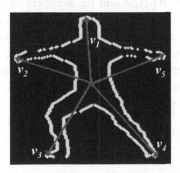

Fig. 4. The feature vectors of star skeleton

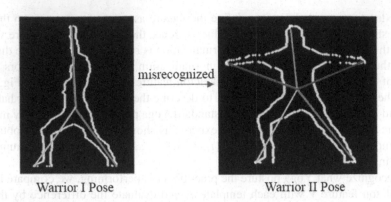

Warrior I Pose   Warrior II Pose

Fig. 5. Example of misrecognition: Warrior I Pose will be misrecognized as Warrior II Pose, if the mismatched vectors (in red) are neglected

3   Experimental Results

The proposed YogaST system is implemented in C++ with OpenNI 1.5.4.0 [13] and OpenCV 2.3.1[14] libraries, and run on an Acer notebook (Intel Core i5 CPU M430 @2.27GHz, 4GB RAM, Windows 7 64-bit OS). Twelve typical Yoga postures are selected in our system. For brevity, these postures are abbreviated to its code name P_i ($i = 1~12$), as illustrated in Fig. 6.

Our experiments are conducted in such a way that five practitioners perform each of the twelve Yoga postures five times. That is, we have 25 video clips for each posture, and totally 300 clips to be recognized. For each clip, an observation window of forty frames is used and the practitioner's posture in each frame within the window is recognized individually. Then, majority voting is applied to decide what Yoga posture the practitioner is performing in the video clip. The confusion matrix of Yoga posture recognition is presented in Table 1, wherein the left column indicates the ground truth postures, and the top row indicates the recognition results. The elements on the diagonal of the confusion matrix represent the numbers of correct recognitions. Since these Yoga postures are quite different from each other in appearance, almost all clips can be correctly recognized, except for two error cases, as given in Fig. 7. The two error cases are due to the reason that the posture performed by the practitioner is too far from the standard template that the star skeleton feature is more similar to the other posture. Overall, the proposed system can achieve a quite high accuracy of 99.33% (298/300) in Yoga posture recognition.

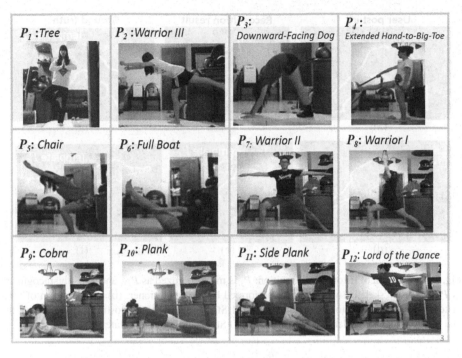

Fig. 6. Illustration of the twelve Yoga postures used in our system

Table 1. Confusion matrix of Yoga posture recognition

Ground truth	Recognition results											
	P_1	P_2	P_3	P_4	P_5	P_6	P_7	P_8	P_9	P_{10}	P_{11}	P_{12}
P_1	25	0	0	0	0	0	0	0	0	0	0	0
P_2	0	24	0	0	0	0	0	0	0	0	0	1
P_3	0	0	24	0	0	0	0	0	0	0	1	0
P_4	0	0	0	25	0	0	0	0	0	0	0	0
P_5	0	0	0	0	25	0	0	0	0	0	0	0
P_6	0	0	0	0	0	25	0	0	0	0	0	0
P_7	0	0	0	0	0	0	25	0	0	0	0	0
P_8	0	0	0	0	0	0	0	25	0	0	0	0
P_9	0	0	0	0	0	0	0	0	25	0	0	0
P_{10}	0	0	0	0	0	0	0	0	0	25	0	0
P_{11}	0	0	0	0	0	0	0	0	0	0	25	0
P_{12}	0	0	0	0	0	0	0	0	0	0	0	25

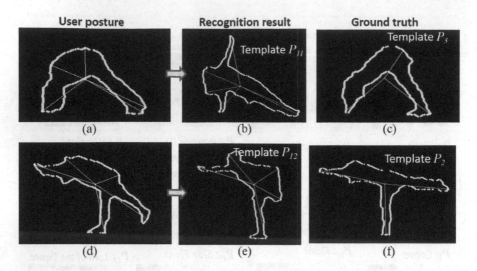

Fig. 7. Illustration of incorrect recognition: P_3 is misrecognized as P_{11}, and P_2 is misrecognized as P_{12}. (a) Star skeleton of P_3 posture performed by the practitioner. (b) Star skeleton of template P_{11}. (c) Star skeleton of template P_3. (d) Star skeleton of P_2 posture performed by the practitioner. (e) Star skeleton of template P_{12}. (f) Star skeleton of template P_2.

4　Conclusions

Computer-assisted self-training in sports exercise is an ever-growing trend. In this paper, we develop a preliminary system capable of recognizing the practitioner's posture when he/she is practicing Yoga, and then retrieving Yoga training information from Internet to remind his/her attention to the posture. We use a Kinect to capture the user body map, and choose star skeleton as posture descriptor due to its computational efficiency and good representability. A distance function is designed to evaluate the dissimilarity between practitioner's posture and the pre-built templates of standard Yoga postures, and posture recognition is achieved by choosing the template with the most similar skeleton.

Currently, we are working on enhancing the proposed system by providing visualized instruction for posture rectification, as presented in the work [15], which is capable of analysing only three Yoga postures so far and more postures need to be considered. Also, the proposed scheme will be adapted to more sports exercises. It can be expected that the effectiveness of sports learning will thus be significantly improved.

Acknowledgements. This research is supported in part by projects ICTL-102-Q528, ATU-102-W958, NSC-101-2221-E-009-087-MY3, and NSC-102-2221-E-009-031.

References

1. Chen, H.T., Chen, H.S., Hsiao, M.H., Tsai, W.J., Lee, S.Y.: A Trajectory-Based Ball Tracking Framework with Enrichment for Broadcast Baseball Videos. Journal of Information and Science Engineering 24(1), 143–157 (2008)
2. Chen, H.T., Tien, M.C., Chen, Y.W., Tsai, W.J., Lee, S.Y.: Physics-Based Ball Tracking and 3D Trajectory Reconstruction with Applications to Shooting Location Estimation in Basketball Video. Journal of Visual Communication and Image Representation 20(3), 204–216 (2009)
3. Chen, H.T., Tsai, W.J., Lee, S.Y., Yu, J.Y.: Ball Tracking and 3D Trajectory Approximation with Applications to Tactics Analysis from Single-Camera Volleyball Sequences. Multimedia Tools and Applications 6(3), 641–667 (2012)
4. Zhu, G., Xu, C., Huang, Q., Rui, Y., Jiang, S., Gao, W., Yao, H.: Event Tactic Analysis Based on Broadcast Sports Video. IEEE Transaction on Multimedia 11(1), 49–67 (2009)
5. Hu, M.C., Chang, M.H., Wu, J.L., Chi, L.: Robust Camera Calibration and Player Tracking in Broadcast Basketball Video. IEEE Transaction on Multimedia 13(2), 266–279 (2011)
6. Chen, H.T., Chou, C.L., Fu, T.S., Lee, S.Y., Lin, B.S.P.: Recognizing Tactic Patterns in Broadcast Basketball Video Using Player Trajectory. Journal of Visual Communication and Image Representation 23(6), 932–947 (2012)
7. Patil, S., Pawar, A., Peshave, A., Ansari, A.N., Navada, A.: Yoga Tutor Visualization and Analysis using SURF Algorithm. In: Proc. IEEE Control and System Graduate Research Colloquium (ICSGRC), pp. 43–46 (2011)
8. Luo, Z., Yang, W., Ding, Z.Q., Liu, L., Chen, I.M., Yeo, S.H., Ling, K.V., Duh, H.B.L.: Left Arm Up! Interactive Yoga Training in Virtual Environment. In: Proc. IEEE Virtual Reality Conference (VR), pp. 261–262 (2011)
9. Wu, W., Yin, W., Guo, F.: Learning and Self-Instruction Expert System for Yoga. In: Proc. 2nd International Workshop on Intelligent Systems and Applications (ISA), pp. 1–4 (2010)
10. Yoga Journal, http://www.yogajournal.com/
11. Fujiyoshi, H., Lipton, A.J.: Real-Time Human Motion Analysis by Image Skeletonization. In: Proc. Fourth IEEE Workshop on Applications of Computer Vision (WACV), pp. 15–21 (1998)
12. Chen, H.S., Chen, H.T., Chen, Y.W., Lee, S.Y.: Human Action Recognition Using Star Skeleton. In: Proc. 4th ACM International Workshop on Video Surveillance and Sensor Networks (VSSN), pp. 171–178 (2006)
13. OpenNI (2011), http://openni.org/
14. OpenCV (2013), http://opencv.org/
15. Chen, H.T., He, Y.Z., Chou, C.L., Lee, S.Y., Lin, B.S.P., Yu, J.Y.: Computer-Assisted Self-Training System for Sports Exercise using Kinects. In: Proc. IEEE International Conference on Multimedia and Expo (ICME)

Real-Time Gaze Estimation Using a Kinect and a HD Webcam

Yingbo Li, David S. Monaghan, and Noel E. O'Connor

CLARITY: Center for Sensor Web Technology, Dublin City University, Ireland

Abstract. In human-computer interaction, gaze orientation is an important and promising source of information to demonstrate the attention and focus of users. Gaze detection can also be an extremely useful metric for analysing human mood and affect. Furthermore, gaze can be used as an input method for human-computer interaction. However, currently real-time and accurate gaze estimation is still an open problem. In this paper, we propose a simple and novel estimation model of the real-time gaze direction of a user on a computer screen. This method utilises cheap capturing devices, a HD webcam and a Microsoft Kinect. We consider that the gaze motion from a user facing forwards is composed of the local gaze motion shifted by eye motion and the global gaze motion driven by face motion. We validate our proposed model of gaze estimation and provide experimental evaluation of the reliability and the precision of the method.

Keywords: Gaze estimation, Gaze tracking, Eye tracking, Kinect.

1 Introduction

Novel forms of human-computer interaction are becoming more and more essential in our increasingly highly computerised modern world. This is also evident by an increasing volume of research being carried out in the field of eye and gaze tracking, for example the newly emerging Google Glasses. Human eye and gaze direction can reflect the natural and stable attention of end users and even give insights into mood and human affect. By performing real-time and robust gaze tracking we can track the objects of interest to a user and even design customised and interesting content for the user.

Currently the most popular gaze estimation approaches can be classified into two groups [1] [2] [3] [4] [5]: model-based gaze estimation and screen-based gaze estimation. The former dealing with the graphical object and model before rendering, while the latter manipulates the image at the pixel level. In model-based gaze estimation, researchers estimate the gaze by analysing the model of the eye and the model of the gaze. The authors in [6] estimate the gaze by only a single image of the eye through the iris circle. Kim et al. [7] estimates the gaze by exploiting the position of two eyes in relation to a simple 2D mark. In [8] the authors propose a new method for estimating eye/gaze direction based on appearance-manifolds by utilising the linear interpolation among a small subset

C. Gurrin et al. (Eds.): MMM 2014, Part I, LNCS 8325, pp. 506–517, 2014.

of samples to approximate the nearest manifold point. Head pose and facial expression has also be employed to enhance gaze tracking estimation [9] [10]. The low-cost 3D capturing device, the Microsoft Kinect, has also been exploited to estimate the gaze by modelling the head and eye [11].

In comparison to model-based methods, screen-based methods try to calibrate and build the relation between the eye in 3D space and the gaze on the screen. For example, a video-oculographic gaze tracking system is proposed in [17]. The Microsoft Kinect has also been used as a source for screen gaze tracking [13]. Recently it has been demonstrated how the gaze detection of a user can be used to animate the gaze shifts of virtual characters within virtual environments on the screen [14].

In addition to these ongoing research efforts, the open source software for eye and gaze tracking in both remote or eye-wearing modes has become more and more prevalent in recent years. The most popular of these open sources eye trackers include OpenEyes [18], AsTeRICS [19], Opengazer [20], TrackEye [21] and ITU Gaze Tracker [22]. Within the scope of the work presented here we have utilised a customised version of the ITU Gaze Tracker.

2   Motivation

The authors in [12] have proposed a real-time method to robustly detect the eye (pupil) motion by HD webcam and Microsoft Kinect with the assistance of ITU Gaze Tracker [22]. The authors calibrate and identify the matched pupil centres, by SIFT descriptor, between the images from the Kinect and the Webcam after obtaining the pupil centre from ITU Gaze tracker. By utilising real-time information about the pupil motion and the 3D movement of the head, obtained from the Kinect, the authors can accurately animate the eye motion in a virtual avatar. The resultant avatar thus appears to be more realistic with accurately animated eyes.

It reasonably follows, from the successful detection of the user's eye motion, to be able to estimate the gaze direction from the moving eyes, i.e. The exact place where is the person (or avatar) is looking. In [12] the authors accomplish the task of eye tracking for a user positioned in front of a computer screen and then animate an avatar on the screen. In this scenario the gaze is the focused point on the screen. The ITU Gaze Tracker, exploited in [12], was employed to do the gaze tracking on the screen when the users head/face is a fixed position. However, even a slight facial movement can corrupt the performance of the gaze tracking of ITU Gaze Tracker, based on experimental results.

In [13] the authors proposed gaze estimation by using the face information from the Kinect but this method only tracks the gaze caused by head motion and pose, and does not take into account the motion of the pupils. It is often the case when a user is positioned in front of a computer screen that the gaze is often shifted by moving the pupils, as well as the head position.

In this paper we propose a novel approach to gaze estimation for the case when a user is positioned in front of a computer screen by analysing both the movement of the pupils and the translational movement of the head using a low cost webcam and a Kinect depth sensor. We have not seen the similar contribution for gaze tracking in front of a computer monitor and capturing devices, which achieves the same accurate gaze tracking by the low-cost devices for around 200euro. The proposed system is very practical because it is often for a person to work and play in front of a computer with capturing device in the current world, while it is meaningful to track his gaze on the computer monitor to learn the real-time attention.

3   An Overview of the Proposed Approach

A HD webcam was used in this approach as it was found, through experimental observation, that the RGB camera in the Kinect was of too low quality for our purposes. The webcam and Kinect are arranged in front and underneath the computer screen, as shown in Figure 1. The eye tracking approach of obtaining the pupil position from Kinect frames follows on from previous work [12]. Firstly the factors influencing the gaze position, including the pitch, yaw and raw of the face are analysed.

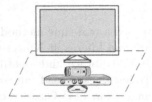

Fig. 1. The designed hardware system

Since the eye and the pupil is on the face, the factors involved in shifting the gaze position can be classified into two groups. The motion of the face, we have termed the global gaze motion and the motion of the pupil, that we have termed the local gaze motion. The global motion captured by the Kinect in 3D includes the translation, in metres, and three angles of 3D head pose, in degrees. The translation is measured by a 3D coordinate with the optical centre of Kinect sensor as the origin, illustrated in Figure 2, while the 3D head pose, measured in degrees, is illustrated in Figure 3, which consists of the pitch, yaw and roll. Whereas the head pose for the face is an important factor in gaze orientation, in this paper we limit the research to the movement of a face orientated forwards. In this paper the global motion consists only of the translational movement and an amalgamation of both translation and rotation under the heading of global motion is currently being researched for future work.

The local motion refers to the pupil motion in planar orbit about two dimensions, X and Y, as compared to the centre of the orbit, shown in Figure 5(a) [12]. In this work it is however assumed that the motions of the left eye and right eye are synchronous.

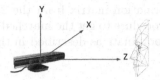

Fig. 2. The 3d space of the Kinect

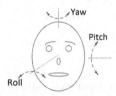

Fig. 3. 3D head pose of pitch, yaw and roll

We summarize the approach via a flowchart that explains the global and local motions shifting the gaze position on the screen (or other objects) in Figure 4.

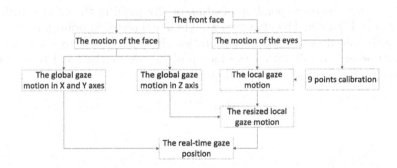

Fig. 4. The flowchart of the proposed model of the gaze estimation

4 Local Gaze Motion

Assuming that the face is fixed and only the local motion of the pupil shifts the gaze position, then the pupil motion can be modelled as the motion in a rectangle, shown in Figure 5(a). While, if the eyes are watching the screen, the screen can be modelled as the other rectangle as in Figure 5(b). So the relation between the gaze position on the screen and the pupil position in the orbit, in two 2D coordinates, can be represented by an affine transformation. In order to get the corresponding position of the pupil in the coordinate system of the screen we need to calibrate two coordinates of the orbit and the screen to obtain its affine transformation matrix. Thus, we require at least 3 matched point pairs to compute the affine transformation matrix from the 2D coordinates. Therefore, we propose a calibration procedure to get the matched points between the screen coordinate and the orbit coordinate as described in the next section.

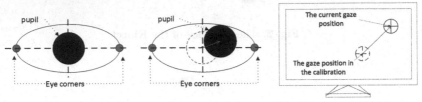

(a) The local motion of the pupil in the orbit

(b) The gaze shifted by the local motion

Fig. 5. The local gaze estimation model

4.1 Calibration between the Screen and Orbit Coordinate Systems

In the calibration of two coordinate systems, the method of 9 points is the most popular [3], which is also used in the ITU Gaze Tracker. Simply speaking, the 9-point eye calibration process is to sequentially and randomly display the points one by one on the screen and record the positions of the points in the screen coordinate and the corresponding position of the pupil in the orbit coordinate, illustrated in Figure 6. Therefore, we would have 9 matched point pairs between two coordinates after the 9-point calibration. It is important to note that during the 9-point eye calibration the face position should be fixed to avoid any corrupting influence from the global gaze motion.

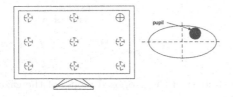

Fig. 6. The 9-point calibration

4.2 Affine Transformation between the Orbit and Screen Coordinates

To compute the affine transformation matrix, 3 matched points pairs are adequate, while from the 9-point calibration we can obtain 9 matched point pairs. Some of the 9 point pairs may not be accurate and would lead to an inaccurate affine transformation. It is necessary to select or obtain 3 accurate point pairs, which are vertices of a triangle, from the 9 points pairs. We propose to exploit two ways to select or construct the 3 points pairs from 9 point pairs in order to ascertain the affine transformation.

1. These 3 point pairs for the affine transformation matrix should be the most common ones, so we can consider that these 3 point pairs are inliers. To select inliner point pairs, RANdom SAmple Consensus (RANSAC) [16] is a popular method. Since RANSAC is a well-known classical algorithm, we will not describe it in detail here and ask the interested reader to consult the reference provided.
2. Since some of the gaze points are considered to be inaccurate, we propose to use the average value to compensate for the inaccurate points. When we display the points on the screen, the positions of the points are exact, but the corresponding points in the orbit coordinates are considered as inaccurate. So we acquire 3 virtual pupil points in the orbit coordinate by averaging 9 pupil points in the orbit coordinate. This procedure is illustrated in Figure 7 and called VIRSELECT by us, and the position of the virtual point $V(m, n)$ is formulated from 9 real pupil points $R(m, n)$:

$$\begin{cases} V(m,n)_x = \frac{\sum_n R(m,n)_x}{3}; \\ V(m,n)_y = \frac{\sum_m R(m,n)_y}{3} \end{cases} \tag{1}$$

where $R(m, n)$ are 9 positions of the real pupil points. m and n are separately 3 rows in y axis and 3 columns in x axis, such that $m = 1, 2, 3$ and $n = 1, 2, 3$. $V(m, n)$ are the constructed virtual points.

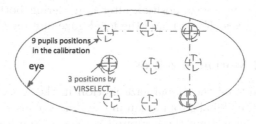

Fig. 7. The principle of VIRSELECT

We compare these two approaches in the experiment described in Section 6.

4.3   Local Gaze Motion

The 2×3 homogeneous matrix M of the affine transformation can be computed by 3 point pairs in the calibration from the previous section, Section 4.2. We then subsequently get the real-time gaze position on the screen $L(x, y)$, caused by the local gaze motion as compared to the coordinate centre in the calibration, obtained from the corresponding pupil position in the orbit $P(x, y)$ by the affine homogeneous matrix.

$$L(x, y) = M \cdot P(x, y, 1); \tag{2}$$

5   Global Gaze Motion

In the previous section, we have ascertained the gaze position that is associated with the local gaze motion, which is caused by the movement of the pupil in the orbit. In this section we discuss the global gaze motion shifted by the facial movement. Since we are only concerned with the face translation here we consider the motion of up, down, left, right, forwards and backwards as compared to the position of the Kinect. When a user is looking at the screen, the face is mostly and naturally forwards facing watching the screen in the wide-angle capturing area of the Kinect and the HD Webcam. Therefore, we assume that the face is almost parallel with the screen, the Kinect, and HD Webcam, which means that the pitch, yaw, and roll of the front face in Figure 3 is not obvious.

5.1   The Global Motion in X and Y Axes

When the face together with the eyes moves in X and Y axes as the directions in Figure 2 (up, down, left, and right), the gaze on the screen moves with the same value in the X and Y axes, due to the face being parallel to the screen. The position of the gaze after considering the global motion in X and Y axes following the local gaze motion can be formulated as follows:

$$G_{xy}(x, y) = L(x, y) + K(x, y) \tag{3}$$

where G_{xy} is the current gaze position after considering both the global gaze motion in X and Y axes, $K(x, y)$, and the local gaze motion $L(x, y)$.

5.2   The Global Motion in Z Axis

We have only considered the global gaze motion in the X and Y axes. In this section we discuss the global gaze motion in the Z axis, caused by the face/head moving towards or away from the computer screen. In the frames captured by the HD Webcam and the Kinect, the sizes of the eyes, the pupils and the orbits on the face are resized by the distance between the face and the Kinect or HD Webcam. When the face is nearer to the capturing devices and the screen, the size of the orbits in the frame is larger, and vice versa.

Since the sizes of the pupils and the orbits change together with the distance of the face in the Z axis, the local gaze motion would be different when the face is at a distance that is different then it was at the time of calibration. However, if the relative pupil motion in the orbit is the same as the calibration shown in Figure 8,

$$\frac{p_c}{S_c} = \frac{p_r}{S_r} \tag{4}$$

where p_c and p_r are the pupil motions in a calibration and a real-time test, and S_c and S_r are the sizes of the orbit in the calibration and real-time test, we argue that the local gaze motion in the test is the same as the calibration. Therefore, we can resize the real-time pupil motion in the orbit to its corresponding pupil motion in the calibration. Thus, we can use the resized pupil motion to get the local gaze motion by the affine transformation matrix obtained at calibration time. Thus Eq. 2 becomes:

$$L'(x,y) = M \cdot P(\frac{x * S_r}{S_c}, \frac{y * S_r}{S_c}, 1) \qquad (5)$$

And consequently Eq. 3 can be reformulated as

$$G_{xyz}(x,y) = L'(x,y) + K(x,y) \qquad (6)$$

where $G_{xyz}(x,y)$ is the gaze position, according to our proposed model, on the screen considering both the local gaze motion and global gaze motion in the X, Y and Z axes.

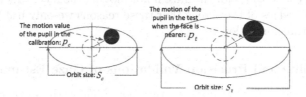

Fig. 8. The pupil motions in the orbit

6   Experimental Results

In this section, we validate the proposed approach of gaze estimation. The experimental arrangement is as shown in Figure 1. The test subjects used for the calibration and the validation sit facing forward to the screen and two capturing devices. During calibration the test subjects are required to maintain a static face position in order to avoid global gaze motion, while during the validation process they are required to randomly move their eyes and their facial position.

Inspired by [15], we propose two simple but effective approaches to evaluate the reliability and the precision of the proposed model of estimating the gaze. Before that, we first compare two methods of RANSAC and VIRSELECT to select 3 points pairs from 9 points for the affine transformation in the local gaze motion.

6.1   RANSAC and VIRSELECT

In Section 4.2 we have proposed two methods, RANSAC and VIRSELECT, to select or construct 3 point pairs from 9 point pairs between the orbit and the screen coordinate systems during calibration. Here we propose a method to validate and compare these two approaches.

The reliability validation is the following: After the calibration, 9 points will be uniformly and sequentially displayed on the screen, each point for 3 seconds, as in Figure 6. The test subject is required to look at the displayed point at that time. If the estimated gaze on the screen is within a circle of diameter 10cm with the displayed point as the centre for more than half of the display time, we consider the gaze reliable to focus on this point, as we need to consider the transition and focusing time for the gaze to move on the screen.

For VIRSELECT, we found that all the 9 points are well focused by the test subjects. However, only the lower 6 points can be focused using RANSAC and the upper 3 points are totally lost. To explain this we argue that the upper area of the screen is more uncomfortable to be focused on than then the lower areas, so the quality of the gaze at the upper area in the calibration is worse than the lower area. Consequently, RANSAC always selects 3 points from the lower area to get the affine transformation. The mean values of the gaze in VIRSELECT compensate the fade zones for the gaze if not moving the face, which are near the corners and the edges of the screen. For these reasons we only use VIRSELECT for the affine transformation.

6.2   Reliability and Precision Validations of Gaze Estimation

The reliability evaluation of the gaze estimation model is the same as described in Section 6.1. However, we present more test data of the gaze falling into differently sized circles with 5 displayed points as the centers on the screen. These 5 points are $(\frac{1}{4}T_w, \frac{1}{4}T_h)$, $(\frac{1}{4}T_w, \frac{3}{4}T_h)$, $(\frac{3}{4}T_w, \frac{1}{4}T_h)$, $(\frac{3}{4}T_w, \frac{3}{4}T_h)$, and $(\frac{1}{2}T_w, \frac{1}{2}T_h)$ where T_w is the width of the screen, and T_h the height of the screen. The reason of using these 5 points is to be far from the fade zones on the screen and to reduce time taken during the experiment – the longer the calibration takes the higher the probability of calibration error due to human error. The reliability validation is shown in Table 1. We measure the reliability by the percentage of the time the gaze falls into the sized circle around the displayed point in the total display duration.

Table 1. The data of reliability validation

Diameter		5cm	10cm	20cm
G_{xy}	center point	6%	14%	68%
	the other 4 points	20%	55%	87.5%
G_{xyz}	center point	19.8%	78.1%	100%
	the other 4 points	33.3%	66.7%	83.3%

In Table 1 we separately show the reliability data of the center point and the other 4 points, where G_{xy} does not consider the distance between the face and the screen and where G_{xyz} does consider this. The reliability values of G_{xyz} are mostly higher than G_{xy}, but the values for G_{xy} become better when the circle diameter is larger because the gaze is still nearby without considering the face motion in the Z axis. We can see that the reliability data on the center point is mostly higher than the other 4 points, because the center point is further from the fade zones around the edges and the corners of the screen. When the diameter is 5cm, the reliability of the center point is less than the other points, which is caused by averaging and compensating the low reliability values by lots of data from 4 points.

For the precision we consequently show 6 points in a triangle, shown in Figure 9. The reason of displaying fewer points in the lower area is to avoid the pitch action of the face. Each point is shown for a short time, approximately 0.5 seconds. If the gaze can hit a point within the short time to within a tolerance of 5cm or 10cm diameter, the precision value is increased by 1/6. In Table 2 we show the precision of G_{xyz} and G_{xy}. It can be seen that the precision of the gaze estimation model is very high.

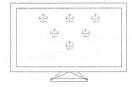

Fig. 9. The precision validation

Table 2. The data of precision validation

Diameter	5cm	10cm
G_{xy}	66.7%	100%
G_{xyz}	83.3%	100%

Comparing our results with those from the literature [15] our proposed method performs exceptionally well for the reliability and precision. Compared to [23], we can see that we achieve accurate gaze tracking too considering the distance to the capturing devices. Since the hardware and facing problems are so distinguished, it is impossible to make the quantitative comparison. It should be noted that in our method the motion of the face in the X and Y axes should be less than 7cm because due to the capturing ability of the Kinect and HD webcam. Additionally the face motion in the Z axis should be limited in a range of 20cm as has been detailed in [12].

7   Conclusions and Future Work

In this paper we have proposed a real-time gaze estimation model based on simple capturing devices for a desktop computer user. We have separated the gaze motion on the screen into local motion and global motion. The local motion refers to the gaze motion driven by pupil movement and the global motion to motion driven by head movement. By analysing the face movement in the Z axis, moving towards and away from the screen, our proposed method resizes the orbit together with the motion of the pupil, and correspondingly rescales the local gaze motion. Our proposed gaze estimation model uses only simple capturing devices, a HD webcam and a Kinect, but can achieve real-time and accurate gaze estimation, as demonstrated by the presented experimental results.

In our future research we are working to expand our method to incorporate the 3D head pose rotations of a user, i.e. the yaw, pitch and roll. By incorporating the user head rotations into our method we hope to make our gaze estimation model more robust to account for naturalistic human head motion and also to allow for larger computer displays where head rotation is needed to view around the screen. The current proposed system is only for one person, but Kinect and ITU gaze tracker is probable to process multiple faces. Therefore, we are considering the probability of the proposed system for multiple heads.

Acknowledgement. The research that lead to this paper was supported in part by the European Commission under the Contract FP7-ICT-287723 REVERIE.

References

1. Duchowski, A.T.: A breadth-first survey of eye-tracking applications. Behavior Research Methods, Instruments & Computers 34(4), 455–470 (2002)
2. Hansen, D.W., Ji, Q.: In the eye of the beholder: A survey of models for eyes and gaze. IEEE Transactions on Pattern Analysis and Machine Intelligence 32(3), 478–500 (2010)
3. Duchowski, A.T.: Eye tracking methodology: Theory and practice, vol. 373. Springer (2007)
4. Morimoto, C.H., et al.: Eye gaze tracking techniques for interactive applications. Computer Vision and Image Understanding 98(1), 4–24 (2005)
5. Bohme, M., Meyer, A., Martinetz, T., et al.: Remote eye tracking: State of the art and directions for future development. In: Proc. of the 2006 Conference on Communication by Gaze Interaction (COGAIN), pp. 12–17 (2006)
6. Wang, J.G., et al.: Eye gaze estimation from a single image of one eye. In: Proceedings of the Ninth IEEE International Conference on Computer Vision, pp. 136–143 (2003)
7. Kim, K.N., Ramakrishna, R.S.: Vision-based eye-gaze tracking for human computer interface. In: IEEE SMC 1999 Conference Proceedings of the IEEE International Conference on Systems, Man, and Cybernetics, vol. 2, pp. 324–329. IEEE, MLA (1999)
8. Tan, K.H., et al.: Appearance-based eye gaze estimation. In: Proceedings of the Sixth IEEE Workshop on Applications of Computer Vision, pp. 191–195. IEEE (2002)
9. Reale, M., et al.: Using eye gaze, head pose, and facial expression for personalized non-player character interaction. In: IEEE Computer Society Conference on Computer Vision and Pattern Recognition Workshops (CVPRW), pp. 13–18 (2011)

10. Langton, S.R.H., Honeyman, H., Tessler, E.: The influence of head contour and nose angle on the perception of eye-gaze direction. Perception & Psychophysics 66(5), 752–771 (2004)
11. Funes Mora, K.A., Odobez, J.-M.: Gaze estimation from multimodal Kinect data. In: 2012 IEEE Computer Society Conference on Computer Vision and Pattern Recognition Workshops (CVPRW), pp. 25–30 (2012)
12. Li, Y., Wei, H., Monaghan, D.S., OConnor, N.E.: A Hybrid Head and Eye Tracking System for Realistic Eye Movements in Virtual Avatars. In: The International Conference on Multimedia Modeling (2014)
13. Jafari, R., Ziou, D.: Gaze estimation using Kinect/PTZ camera. In: IEEE International Symposium on Robotic and Sensors Environments (ROSE), pp. 13–18 (2012)
14. Andrist, S., Pejsa, T., Mutlu, B., Gleicher, M.: A head-eye coordination model for animating gaze shifts of virtual characters. In: Proceedings of the 4th Workshop on Eye Gaze in Intelligent Human Machine Interaction (2012)
15. Ciger, J., et al.: Evaluation of gaze tracking technology for social interaction in virtual environments. In: Proc. of the 2nd Workshop on Modeling and Motion Capture Techniques for Virtual Environments (CAPTECH 2004) (2004)
16. Fischler, M.A., Bolles, R.C.: Random sample consensus: a paradigm for model fitting with applications to image analysis and automated cartography. Communications of the ACM 24(6), 381–395 (1981)
17. Villanueva, A., Cabeza, R.: Models for gaze tracking systems. Journal on Image and Video Processing 2007(3), 4 (2007)
18. Li, D., et al.: openEyes: A low-cost headmounted eye-tracking solution. In: Proceedings of the ACM Eye Tracking Research and Applications Symposium (2006)
19. Nussbaum, G., Veigl, C., Acedo, J., et al.: AsTeRICS-Towards a Rapid Integration Construction Set for Assistive Technologies. In: AAATE Conference (2011)
20. Zielinski, P.: Opengazer: open-source gaze tracker for ordinary webcams (software), Samsung and The Gatsby Charitable Foundation,
    http://www.inference.phy.cam.ac.uk/opengazer/
21. Savas, Z.: TrackEye: Real time tracking of human eyes using a webcam,
    http://www.codeproject.com/KB/cpp/TrackEye.aspx
22. San Agustin, J., Skovsgaard, H., Hansen, J.P., et al.: Low-cost gaze interaction: ready to deliver the promises. In: CHI 2009 Extended Abstracts on Human Factors in Computing Systems, pp. 4453–4458. ACM (2009)
23. San Agustin, J., et al.: Evaluation of a low-cost open-source gaze tracker. In: Proceedings of the 2010 Symposium on Eye-Tracking Research & Applications. ACM (2010)

A Framework of Video Coding for Compressing Near-Duplicate Videos

Hanli Wang[1,2], Ming Ma[1,2], Yu-Gang Jiang[3], and Zhihua Wei[1,2]

[1] Department of Computer Science and Technology,
Tongji University, Shanghai, China
[2] Key Laboratory of Embedded System and Service Computing,
Ministry of Education, Tongji University, Shanghai, China
[3] School of Computer Science, Fudan University, Shanghai, China
{hanliwang,2012mingma,zhihua_wei}@tongji.edu.cn, ygj@fudan.edu.cn

Abstract. With the development of multimedia technique and social network, the amount of videos has grown rapidly, which brings about an increasingly substantial percentage of Near-Duplicate Videos (NDVs). It has been a hot research topic to retrieve NDVs for a number of applications such as copyright detection, Internet video ranking, etc. However, there exist a lot of redundancies in NDVs, and to the best of our knowledge it is an untouched research area on how to efficiently compress NDVs in a joint manner. In this work, a novel video coding framework is proposed to effectively compress NDVs by making full use of the relevance among them. Experimental results demonstrate that a significant storage saving can be achieved by the proposed NDV coding framework.

1 Introduction

The surging of the multimedia technology is exercising a great influence on our daily life. The convenience and popularization of the video capture devices give rise to a lot of creation and uploading of original videos. Naturally, the providers of network video service have to face the challenge of the video explosion. Thus, how to store and transmit videos effectively is an urgent problem that needs to solve. Especially in large-scale network video databases such as Youtube, Google Video, Yahoo Video, Youku, etc., there exist a huge amount of Near-Duplicate Videos (NDVs). NDVs can be considered as approximately identical videos that might differ in encoding parameters, photometric variations (color, lighting changes), editing operations (captions, logo insertion), or audio overlays [1]. It is a hot research topic to retrieve NDVs which is quite crucial to a number of applications such as copyright detection, video monitoring, Internet video ranking, video recommendation, etc.

At present, these NDVs are individually compressed with some video coding standard and then stored in video servers, which requires a tremendously large storage to keep them. The relevance among these NDVs has not been fully explored and utilized for a more efficient compression. If there is a method capable

C. Gurrin et al. (Eds.): MMM 2014, Part I, LNCS 8325, pp. 518–528, 2014.

of investigating the redundancy in NDVs and jointly compress the video data, much space can be saved in video servers. This is the aim of the proposed work.

Relevant to the proposed work, there are several researches dedicated to compressing similar images together. In [2], the term of set redundancy was proposed to describe the redundancy in images and the Set Redundancy Compression (SRC) algorithm was developed to jointly compress similar images. Then, a number of SRC variants were developed such as the min-max differential method [3] and the centroid-based method [4]. A comparison of SRC algorithms is made in [5]. On the other hand, Chen *et al* [6] raised a new prediction structure to divide the whole image set into different groups for compression. For grouping the similar images more efficiently, the Minimum Spanning Tree (MST) method [7] has been designed. In [8], Zou *et al* introduced the intra coding of the newly developed video coding standard High Efficiency Video Coding into image set compression and achieved an improved performance. Based on discrete cosine transform pyramid multi-level low frequency template, Li *et al* [9] proposed a method to apply subband approximation as the prediction template for similar images compression. In [10], Yue *et al* utilized the scale invariant feature transform descriptor to group and store similar images for compression.

The methods mentioned-above are designed to compress similar images, however, how to jointly compress NDVs is not exploited in the literature to the best knowledge of us. The main challenge on NDV joint compression is how to design an analysis and coding framework which can fully explore the redundancies in NDVs and then reduce these redundancies by an efficient video coding structure, which is the focus of this work. The rest of this paper is organized as follows. Related works are described in Section 2, including a brief overview of NDV retrieval and multiview video coding, which are useful to elicit the proposed NDV joint compression framework being proposed in Section 3. Experimental results are given in Section 4. Finally, Section 5 concludes the paper and presents future research prospects.

2 Related Work

A concise overview of NDV retrieval and Multiview Video Coding (MVC) is presented below. The proposed NDV joint compression framework applies NDV retrieval methodology to locate NDVs and analyze the similar images among NDVs. The MVC is extended as the prototype of the proposed NDV joint compression framework.

2.1 Near-Duplicate Video Retrieval

NDV retrieval has become a hot research topic in recent years, and the research efforts may be divided into two categories. One is for the retrieval speed and the other for the retrieval precision. As for faster retrieval, global features are widely adopted, such as the color [11], the edge [12], and the ordinal [13]. This category of algorithms gets a good score in retrieving videos which only have little variations, e.g., to add subtitle or logo, or to shift the pixel lightness, etc. However, if videos are not nearly identical, the global features will be unstable and the retrieval precision is unsatisfied. Therefore, local feature-based methods are proposed for retrieval precision improvement, such as [14]-[15] to name a few. In these methods, the Bag-of-Features technique [16] is widely used. On the other hand, Sarkar *et al* [17] introduced a vector quantization-based descriptor method for NDV retrieval. A three-dimensional structure tensor model was designed for online retrieval system in [18]. In [19], Chiu *et al* proposed to speed up the NDV retrieval process by efficiently skipping unnecessary subsequence matches. Huang *et al* [20] employed a sequence of compact signatures called linear smoothing functions as the video's feature to speed up retrieval. In [21], the content fingerprint is used for NDV retrieval by making use of the spatial-temporal relevance among NDVs. Cheng *et al* [22] developed a stratification-based key frame method to enhance the precision and efficiency for copy detection.

2.2   Multiview Video Coding

Multiview videos have attracted much attention in a wide range of multimedia applications, such as three-dimensional television, free-viewpoint television, etc. A multiview video consists of video sequences of the same scenario captured by multiple cameras, but from different angles and locations, resulting in the need to store and/or transmit tremendous amounts of data. In order to compress the multiview videos effectively, Multiview Video Coding (MVC) [23] was designed for exploring not only temporal but also inter-view redundancies and thus providing higher coding performance than the independent mono-view coding. The joint compression of NDVs or in another word, Near Duplicate Video Coding (NDVC), is quite analogous to MVC if each video of NDVs is considered as one of the video views in a multiview video system. The analogy between MVC and NDVC is illustrated in Fig. 1.

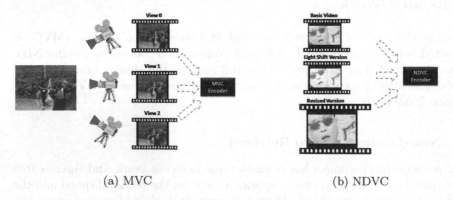

(a) MVC   (b) NDVC

Fig. 1. Analogy between MVC and NDVC

In order to explore both the temporal and inter-view redundancies among multiview videos, sophisticated prediction structures are employed in MVC with a typical example shown in Fig. 2. Among all views, the first view (i.e., Camera 0 in Fig. 2, also known as the base view) is coded independently; while for the other views, all frames are predicted with temporal motion estimation and/or inter-view disparity estimation.

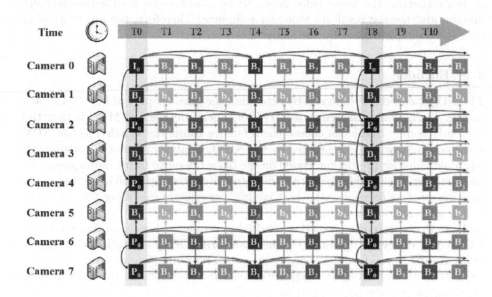

Fig. 2. A typical 8 view prediction structure in MVC [24]

3   Proposed Near-Duplicate Video Coding Framework

The proposed NDVC framework is described below, including four major components: analyzer, encoder, assembler and decoder.

3.1   Analyzer

The functionality of the analyzer is to identify NDVs and make prediction reference among them. For the convenience of discussion, the case of two NDVs is presented in the rest of this paper, which can be extended to multiple NDVs' cases. The analyzer employs some NDV detection approach to judging whether two videos are near-duplicate. If the NDVC condition holds true, the analyzer will further investigate the prediction relation between these two NDVs by generating reference indices. In other words, one video will be assigned as the basic video and the other as the dependent video which is encoded/decoded with the reference of the basic video. For each frame of the dependent video, the analyzer will detect its most similar frame in the basic video. If the similarity score between these two frames is higher than a predefined threshold, it indicates that these two frames are similar enough to become a reference pair and the dependent frame may be coded by referencing the corresponding basic frame. However, generally speaking, the basic video and the dependent video usually exhibit an indirect reference relation. For example, the basic video and the dependent video record the same scenery by different cameras from different view points. Under such a situation, the basic video needs to be transformed with a homography matrix, and use the resultant video for reference. Therefore, the pre-processing task is also implemented by the analyzer.

3.2   Encoder

The NDVC encoder is designed by extending the encoder part of MVC as shown in Fig. 3. The main difference of NDVC encoder from that of MVC is that the currently coding frame can choose the reference frame (as signaled as the NDV reference index which is generated by the NDVC analyzer) from the corresponding basic video for prediction. Based on the basic reference frame index passed by the analyzer, the encoder performs the rate-distortion optimization-based mode decision process to choose the best coding parameters including the reference index, encoding mode, motion vectors, etc. Then, motion compensation is carried out to generate the prediction residual to remove video signal redundancies, followed by the typical Discrete Cosine Transform (DCT), Quantization (Q), entropy coding to produce the coded bitstream. The coding components of Inverse DCT (IDCT) and Inverse Q (IQ) are employed to reconstruct the video frames for subsequent reference prediction.

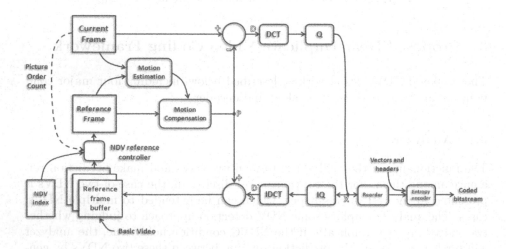

Fig. 3. Flowchart of NDVC encoder

3.3 Assembler

After encoding, each video is compressed into one bitstream. In view of the convenience of storing and decoding, the proposed NDVC framework utilizes an assembler to combine these individual bitstreams into a single bitstream as MVC does. However, the details of the NDVC assembler is different from that of MVC as shown in Fig. 4. In MVC, the inter-view reference exists at the same temporal position as show in Fig. 2. Therefore, the MVC assembling flow as illustrated in Fig. 4(a) is reasonable, since the reference frame will have been decoded ahead of the current frame to be coded/decoded. Whereas in NDVC, it is possible that every frame in the basic video can be referred by the current frame, so the assembler has to ensure that the whole basic video has been decoded before decoding the dependent video as shown in Fig. 4(b).

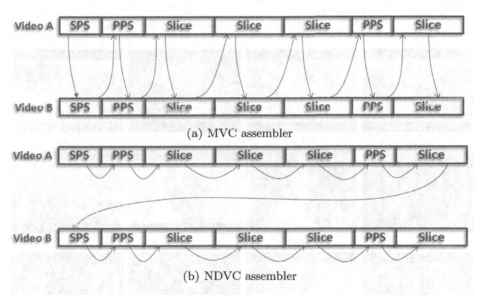

(a) MVC assembler

(b) NDVC assembler

Fig. 4. Assemblers of MVC and NDVC

3.4 Decoder

The NDVC decoder is shown in Fig. 5 which can be well understood by considering it as the reconstruction part of NDVC encoder as illustrated in Fig. 3.

4   Experimental Results

In order to evaluate the proposed NDVC framework, it is designed based on the MVC reference software JMVC with version of 8.5 [23]. Sixteen sets of NDVs

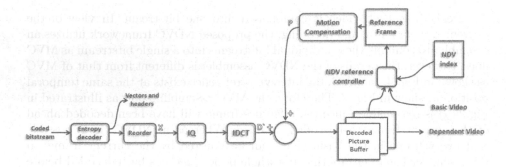

Fig. 5. Flowchart of NDVC decoder

Fig. 6. Illustration of test video sequences

are downloaded from the Youku video website [25] according to their popularity ranking by Google Zeitgeist [26] and Youku, including the Top 10 popularly queried videos at Google Zeitgeist as illustrated in Figs. 6(a)-(j) and six popular videos at Youku as shown in Figs. 6(k)-(p).

Six scenarios are applied to test the proposed NDVC framework, which are achieved by generating the dependant video with modifications on the original (i.e., basic) video, including (1) insertion of subtitle and logo, (2) addition of Gaussian white noise, (3) slowing down video play speed to half in the first 2/3 video segment and bursting up video play speed to two times in the last 1/3 video segment, (4) adjustment of 25% lightness up, (5) adjustment of 25% lightness down, and (6) resizing video resolution to half horizontally and vertically. All the test video sequences and their variations are in the format of 4:2:0 YUV. The resolution of the original videos is 320×240, and the number of frames to be coded for each video is 1575. The frame rate is 25 frames per second.

The NDV retrieval method [15] is utilized to determine whether two videos are similar enough for NDVC. Two evaluation criteria are used including Peak-Signal-to-Noise Ratio (PSNR) degradation ΔP and compression ratio $C_\%$, which are defined as:

$$\Delta P = P_{NDVC} - P_{org}$$

$$C_\% = \frac{S_{NDVC}}{S_{org}} \times 100\% \tag{1}$$

where P_{NDVC} and P_{org} stand for the PSNR values of the dependant video encoded with the proposed NDVC framework and the original mono-view MVC encoder, respectively; S_{NDVC} and S_{org} are the resultant bitrates or bitstream sizes of the dependent video with the proposed NDVC framework and the original mono-view MVC encoder, respectively.

Table 1. Average results of ΔP and $C_\%$ under the six test scenarios

Scenario	S_{org} (kbps)	$C_\%$	P_{org} (dB)	ΔP (dB)
Subtitle and logo	123.36	58.82	36.89	-0.32
Gaussian noise	118.88	55.19	36.82	-0.04
Playing speed change	139.36	59.90	37.74	-0.11
Lightness up	110.16	92.51	36.38	-0.20
Lightness down	95.44	96.80	37.94	-0.02
Resize	158.64	43.24	36.84	-0.05

The summary of the experimental results are given in Table 1. Due to the space limit, only the average results on the sixteen test video sequences are presented. From the results, it can be observed that for most cases, the proposed NDVC framework will save nearly 45% of the bitrates at the cost of insignificant PSNR loss. In fact, under the current test configuration, the same Q_p value is applied to encode dependent videos by both NDVC and mono-view MVC. It has been demonstrated that by adjusting Q_p both the bitrate and PSNR performances resulted from the proposed NDVC can be better than that of mono-view MVC, which is to be presented below. On the other hand, when dealing with the test scenario of lightness shifting, the results are not so good since pre-processing lightness shifting is currently not handled by the NDVC analyzer, which is open as one of our future research directions.

In order to further illustrate the advantage of the proposed NDVC framework and considering the space limit, the comparison of Rate-Distortion (R-D) performances of six video sequences including Videos (k)-(p) under the 'Subtitle and logo' scenario is shown in Fig. 7, where the R-D curves of NDVC and the independent mono-view MVC are both plotted for comparison. From the results,

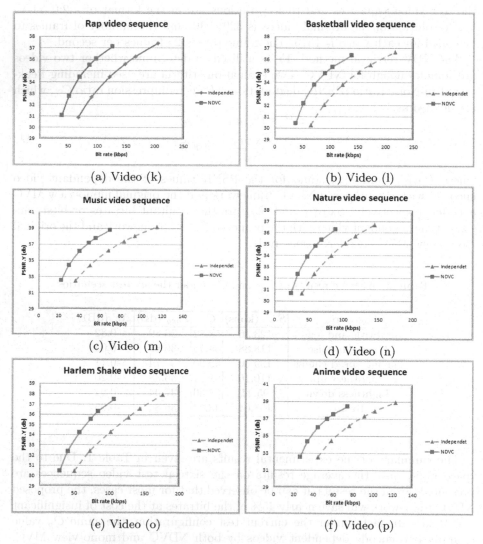

(a) Video (k)   (b) Video (l)

(c) Video (m)   (d) Video (n)

(e) Video (o)   (f) Video (p)

Fig. 7. Comparison of R-D performances between mono-view MVC and NDVC

it be easily observed that the proposed NDVC framework achieves significantly better R-D performances than independent video coding. Similar observations can be made for other test video sequences and scenarios.

5 Conclusion and Future Work

In this paper, a novel video coding framework NDVC is proposed to jointly compress NDVs aiming to effectively trimming down the storage of video servers, including four major parts: analyzer, encoder, assembler and decoder. Sixteen popular video sets have been applied to evaluate the proposed NDVC framework. The experimental results have demonstrated that the proposed NDVC framework is very efficient in reducing the redundancy in NDVs and thus saving the storage or network bandwidth resources.

In the future, the NDVC analyzer will be further enhanced by incorporating more pre-processing functionalities, such as adjustment of lightness and calculation of homography transform matrix, for generating better references. It is also desired to design more efficient approaches to locating similar frames between basic and dependent videos. Within the proposed NDVC framework, fast NDVC coding technologies and rate control algorithms are also preferred to improving the NDVC performances.

Acknowledgments. This work was supported in part by the National Natural Science Foundation of China under Grant 61102059, the "Shu Guang" project of Shanghai Municipal Education Commission and Shanghai Education Development Foundation under Grant 12SG23, the Program for Professor of Special Appointment (Eastern Scholar) at Shanghai Institutions of Higher Learning, the Program for New Century Excellent Talents in University of China under Grant NCET-10-0634, the Fundamental Research Funds for the Central Universities under Grants 0800219158 and 1700219104, and the National Basic Research Program (973 Program) of China under Grant 2010CB328101.

References

1. Cherubini, M., Oliveira, R., de Oliveira, N.: Understanding Near-Duplicate Videos: A User-Centric Approach. In: ACM MM 2009, pp. 35–44 (2009)
2. Karadimitriou, K.: Set Redundancy, the Enhanced Compression Model, and Methods for Compressing Sets of Similar Images. PhD Thesis. Louisiana State University (1996)
3. Karadimitriou, K., Tyler, J.M.: Min-Max Compression Methods for Medical Image Databases. In: ACM SIGMOD 1997, pp. 47–52 (1997)
4. Karadimitriou, K., Tyler, J.M.: The Centroid Method for Compressing Sets of Similar Images. Pattern Recognition Lett. 19(7), 585–593 (1998)
5. Samy, A.-A., Abdelhalim, G.: A Comparison of Set Redundancy Compression Techniques. EURASIP Journal Applied Signal Process., 1–13 (2006)
6. Chen, C.-T., Chen, C.-S., Chung, K.-L., Lu, H., Tang, G.Y.: Image Set Compression through Minimal Cost Prediction Structure. In: IEEE ICIP 2004, pp. 1290–1292 (2004)

7. Nielsen, C., Li, X.: MST for Lossy Compression of Image Sets. In: DCC 2006, p. 463 (2006)

8. Zou, R., Au, O.C., Zhou, G., Li, S., Sun, L.: Image Set Modeling by Exploiting Temporal-Spatial Correlations and Photo Album Compression. In: APSIP ASC 2012, pp. 1–4 (2012)

9. Li, S., Au, O.C., Zou, R., Sun, L., Dai, W.: Similar Images Compression based on DCT Pyramid Multi-level Low Frequency Template. In: MMSP 2012, pp. 255–259 (2012)

10. Yue, H., Sun, X., Wu, F., Yang, J.: SIFT-based Image Compression. In: IEEE ICME 2012, pp. 473–478 (2012)

11. Kasutani, E., Yamda, A.: The MPEG-7 Color Layout Descriptor: A Compact Image Feature Description for High-Speed Image/Video Segment Retrieval. In: IEEE ICIP 2001, pp. 674–677 (2001)

12. Bertini, M., Del Bimbo, A., Nunziati, W.: Video Clip Matching Using MPEG-7 Descriptors and Edit Distance. In: Sundaram, H., Naphade, M., Smith, J.R., Rui, Y. (eds.) CIVR 2006. LNCS, vol. 4071, pp. 133–142. Springer, Heidelberg (2006)

13. Kim, C., Vasudev, B.: Spatiotemporal Sequence Matching for Efficient Video Copy. IEEE Trans. Circuits Syst. Video Technol. 15(1), 127–132 (2005)

14. To, J.L., Chen, L., Joly, A., Laptev, I., Buisson, O., Brunet, V.G., Boujemaa, N., Stentiford, F.: Video Copy Detection: A Comparative Study. In: ACM CIVR 2007, pp. 371–378 (2007)

15. Zhao, W., Wu, X., Ngo, C.-W.: On the Annotation of Web Videos by Efficient Near-Duplicate Search. IEEE Trans. Multimedia 12(5), 448–461 (2010)

16. Csurka, G., Dance, C., Fan, L., Willamowski, J., Bray, C.: Visual Categorization with Bags of Keypoints. In: ECCV SLCV 2004, pp. 1–22 (2004)

17. Sarkar, A., Singh, V., Ghosh, P., Manjunath, B.S., Singh, A.: Efficient and Robust Detection of Duplicate Videos in A Large Database. IEEE Trans. Circuits Syst. Video Technol. 20(6), 870–885 (2010)

18. Zhou, X., Chen, L.: ASVTDECTOR: A Practical Near Duplicate Video Retrieval System. In: ICDE 2013, pp. 1348–1351 (2013)

19. Chiu, C.-Y., Li, S.-Y., Hsieh, C.-Y.: Video Query Reformulation for Near-Duplicate Detection. IEEE Trans. Inf. Forensics Security 14(4), 1220–1233 (2012)

20. Huang, Z., Shen, H.T., Shao, J., Cui, B., Zhou, X.: Practical Online Near-Duplicate Subsequence Detection for Continuous Video Streams. IEEE Trans. Multimedia 12(5), 386–398 (2010)

21. Esmaeili, M.M., Fatourechi, M., Ward, R.K.: A Robust and Fast Video Copy Detection System Using Content-Based Fingerprinting. IEEE Trans. Inf. Forensics Security 6(1), 213–226 (2011)

22. Cheng, X., Chia, L.-T.: Stratification-Based Keyframe Cliques for Effective and Efficient Video Representation. IEEE Trans. Multimedia 13(6), 1333–1342 (2011)

23. Vetro, A., Pandit, P., Kimata, H., Smolic, A., Wang, Y.-K.: Joint Draft 8.0 on Multiview Video Coding. JVT-AB204 (2008)

24. MerKle, P., Müller, K., Smolic, A., Wiegand, T.: Efficient Compression of Multi-View Video Exploiting Inter-View Dependencies based on H.264/MPEG4-AVC. In: ICME 2006, pp. 1717–1720 (2006)

25. Youku Webiste, http://www.youku.com/

26. Google Zeitgeist, http://www.google.com/zeitgeist/

An Improved Similarity-Based Fast Coding Unit Depth Decision Algorithm for Inter-frame Coding in HEVC

Rui Fan[1], Yongfei Zhang[1,2], Zhe Li[1], and Ning Wang[1]

[1] Beijing Key Laboratory of Digital Media, School of Computer Science
and Engineering, Beihang University, Beijing 100191, China
[2] State Key Laboratory of Virtual Reality Technology and Systems,
Beihang University, Beijing 100191, China
zhangyf.ac@gmail.com, frwjxwx@163.com

Abstract. The emerging High Efficiency Video Coding (HEVC) aims to achieve significantly improved compression performance with respect to the state-of-the-art H.264/AVC high profile. This better compression performance is achieved at a much higher computational complexity, which makes it difficult for real-time video systems. To reduce the encoding complexity, this paper presents an improved similarity-based fast coding unit depth decision algorithm for inter-frame coding. Firstly, a fast and precise depth information acquisition method is proposed to improve the accuracy of depth prediction. Secondly, CTUs of the medium similarity degree is further divided into two similarity degree categories, and the complexity of coding the coding tree units of these two categories is reduced by different coding unit depth decision strategies. Experimental results show that proposed algorithm can save on average 35.72% of the encoding time with negligible loss on the rate-distortion performance compared with HM11.0 and consistently outperforms the state-of-the-art schemes.

Keywords: HEVC, inter prediction, coding tree unit, coding unit.

1   Introduction

The state-of-the-art video coding standard H.264/AVC has been widely used in video applications [1]. However, with the increasing requirement of the high-definition video applications, a new video coding standard called High efficiency video coding (HEVC) with better performance has been drafted by the Joint Collaborative Team on Video Coding (JCT-VC) in the early 2013 [2].

In HEVC, the input video is first divided into a sequence of coding tree units (CTUs). CTU is analogous to macroblocks in previous standards. HEVC supports a partitioning of the CTU using a quadtree structure, which allows recursively splitting into four equally sized nodes, starting from the CTU [3]. The leaf node of the quadtree structure is a coding unit (CU). CU is a basic unit which is a region sharing the same prediction mode (intra, inter or skip). This complex quadtree structure brings HEVC better coding performance, but it also makes HEVC have several times higher complexity than H.264/AVC [4]. In order to reduce the computational complexity in

C. Gurrin et al. (Eds.): MMM 2014, Part I, LNCS 8325, pp. 529–540, 2014.

HEVC, several CU depth decision algorithms have been proposed. These algorithms can be roughly classified into two categories as follows.

The first category is pruning-based schemes. Choi proposes a tree-pruning algorithm in [5] which exploits the observation where the sub-tree computations can be skipped if the coding performance of the current node is good enough (i.e., SKIP mode). This method can significantly reduce the inter-frame coding time and has been adopted by HM (HEVC Test Model). Another algorithm that collects relevant and computation-friendly feathers and utilizes Bayes discriminant analysis to decide whether to split the current CU or not by comparing the posterior probabilities is proposed in [6]. The CU sizes in motion regions are generally small and the time saving of pruning is not obvious. Besides, the efficiency of the algorithm based on Bayes decision in [6] is not stable. This is because that the threshold adopted is trained offline and the coding performance depends on the video characteristics in training set.

The other category of methods is eliminating impossible depth levels. A depth range selection mechanism algorithm is presented in [7] to select the coding depth levels by taking advantage of the skip mode. This method can effectively reduce the encoding time, while it produces an obvious loss of performance. Meanwhile, a fast encoding algorithm based on depth of coding unit is proposed in [8] which sorts the current CTU into three complexity groups, and then selecting one from them by using the depth information of the co-located CTU. Time saving is limited as there is at most one CU depth excluded in this method. Zhang proposes a fast CU depth decision algorithm for inter-frame based on similarity degree in [9]. This algorithm classifies the current CTU as low similarity degree, medium similarity degree or high similarity degree by utilizing the spatial-temporal correlation to reduce the depth search range more adaptively. Then different CU depth decision strategies are employed for CTUs of different similarity degrees. Shen provides an adaptive CU depth range determination method which predicts the optimal depth level through reference information given different weights in [10]. According to the predicted value of the optimal depth level, CTUs are divided into five types and each type has its own depth range. The methods presented in [9, 10] can effectively save encoding time on the premise of guaranteeing coding quality. However, the procedure of extracting depth information is based on CTU, and it is not precise for the edge of stationary areas and motion areas. Furthermore, there are still some redundant depth levels can be eliminated.

In this paper, we will address these problems found in [9, 10] and propose an improved similarity-based fast coding unit depth decision algorithm for inter-frame coding in HEVC. More specifically, a fast and precise CTU depth information extracting method is first presented, then the CTUs of medium similarity degree are classified into two subclasses and different CU depth decision strategies are proposed to further reduce the complexity of CU depth decision.

The rest of the paper is organized as follows. The original similarity-based fast coding unit depth decision algorithm in [9] is briefly introduced in Section 2. The proposed similarity-based fast coding unit depth decision algorithm is elaborated in Section 3. Performance evaluations and comparisons with state-of-art algorithm are presented in Section 4. Finally, the conclusion is drawn in Section 5.

2   Review of Original Algorithm Based on Similarity Degree[9]

A fast CU depth decision algorithm for inter-frame based on similarity degree proposed in [9] will be reviewed briefly below.

In [9], CTU is classified as one of three similarity degree categories, namely high similarity degree, medium similarity degree and low similarity degree. CTUs of each kind of Similarity degree exhibit respective CU depth characteristic and different CU depth decision methods are adopted for CTUs of different similarity degree. Similarity degree is determined according to the depth information among the referenced neighboring CTUs. The referenced neighboring CTUs are derived from two CTU sets shown in Fig. 1. The first CTU set is the basic predictive set labeled by α which contains CTU A, B, C and I. The other CTU set is the extended predictive set labeled by β in which CTU D, E, F, G and H are included. For the basic predictive set α, the extended predictive set β is an auxiliary reference set, because the correlation between the current CTU and the CTUs in set α is higher than that of set β. The basic predictive set α and the extended predictive set β are expressed in formula (1).

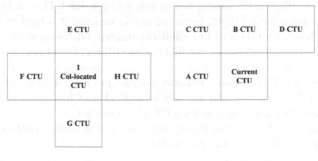

Reference Frame   Current Frame

Fig. 1. Neighboring CTUs of the current CTU

$$\alpha = \{A, B, C, I\} \quad \beta = \{D, E, F, G, H\} \tag{1}$$

Three classes of similarity degree are defined as follows.

High similarity degree means that all the CTUs in set α adopt the same one depth level. If all the CTUs in set β choose the same depth level, then that depth level will be determined as the final depth level to be searched. Otherwise, one more depth level will be searched.

Low similarity degree indicates that all depth levels are adopted by the overall CTUs in set α. The partitions are quite different among the neighboring CTUs of the current low similarity degree CTU. The co-located CTU I is adopted to predict the depth levels of the current CTU.

Medium similarity degree is the remaining similarity degree which has no enough evidence to determine the final depth levels like high similarity degree. Meanwhile, CTUs of medium similarity degree can relatively accurately predict the depth range compared with CTUs of low similarity degree. All the CTUs in set α are employed as the references to the determination of the current CTU.

CTUs of Low similarity degree and medium similarity degree utilize their own reference depth information to eliminate one most impossible depth level. The depth search range is 3 to keep prediction accurate when the similarity degree is low and medium. Meanwhile, the depth search range of high similarity degree is 1 or 2 to reduce the encoding time thanks to the high accuracy of prediction.

Though the above algorithm can effectively reduce the encoding time, there are still several drawbacks. Firstly the CTU depth information acquisition method presented in [9] detects the CTU partition depth recursively to get which depth levels are adopted. The time efficiency depends on CTU depth partition characteristics. It is appropriate for background regions or stationary regions which are usually chosen with larger CU sizes. However, it does not work well in motion complexity regions and much time is wasted in recursive procedures. Moreover, the method searches the whole CTU depth information of the neighboring CTUs. It is unnecessary and usually not accurate for regions which are located between stationary regions and motion complexity regions because incorrect depth information may be taken into consideration.

Besides, we analyze the feature of similarity degree and observe that it is difficult to further reduce the encoding complexity for CTUs of high similarity degree and low similarity degree. That is because it is efficient enough for CTUs of high similarity degree as there are at most 3 depth levels are eliminated and it is hard to further eliminate the depth levels for CTUs of low similarity degree, otherwise severe performance loss might be introduced. However, we find that the CTUs of medium similarity degree, which takes nearly half of all the CTUs, exhibits quite different characteristics. Some of them trend toward CTUs of high similarity degree and have a close correlation with neighboring CTUs, while the rest are similar to CTUs of low similarity degree and have a weak correlation with neighboring CTUs as contrast.

To solve these problems, this paper proposes an improved similarity-based fast coding unit depth decision algorithm in section 3.

3   Proposed Improved Fast Depth Decision Algorithm

In this section, we aim to improve the algorithm presented in [9] to further reduce the encoding time with negligible coding loss. Firstly, the method of extracting CTU depth information is redesigned to achieve higher accuracy of classification. Secondly, we divide the medium similarity degree into two similarity degree classes, medium_high similarity degree and medium_low similarity degree. Next, different CU depth decision strategies are applied to the two similarity degree classes by analyzing their respective characteristics. Finally, the overall proposed fast coding depth decision scheme is presented.

3.1   Improved Accurate CTU Depth Information Extraction

The CTU depth information acquisition method presented in [9] is not accurate enough and does not work well in regions located between stationary regions and motion complexity regions, since depth information extracting region size is at CTU level and it might bring in incorrect depth information. To improve the accuracy of depth information, this section proposes a precise CTU depth information acquisition method.

Above all, R is introduced to regulate the depth information extracting region size to find the most appropriate one. Fig. 2 shows the relationship with the depth information extracting region size and R for spatial-neighboring CTU A, B and C in the basic predictive set α and CTU D in the extended predictive set β.

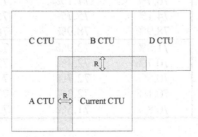

Fig. 2. The depth information extracting region decided by R

Zhang introduces formula (2) to assess the depth correlation degree between the current CTU and the neighboring CTUs in [9].

$$Cor = \frac{\sum_{i=0}^{i=N-1} (cur\_dep[i] == neigh\_dep[i])}{N} \tag{2}$$

where Cor ranges from 0 to 1, N represents the maximum depth of CTU partition, i is the depth level, $cur\_dep[i]$ and $neigh\_dep[i]$ indicate whether current CTU and neighboring CTU adopt the depth level i respectively. However, depth decision of the current CTU excluding spatial-neighboring regions is influenced mainly by the collocated CTU. With the decrease of R, Cor is not accurate any more. This is due to the fact that the whole current CTU might be much larger than the depth information extracting region. To guarantee the accuracy of spatial-neighboring depth correlation Cor, we redefine the Cor_st as

$$Cor\_st = \frac{\sum_{i=0}^{i=N-1} (cur\_dep[i] == (stneigh\_dep[i] | col\_dep[i]))}{N} \tag{3}$$

where Cor_st ranges from 0 to 1, which represents the depth correlation degree between the current CTU and the depth information extracting region of neighboring CTU. $stneigh\_dep[i]$ and $col\_dep[i]$ indicate whether depth information extracting region of neighboring CTU and collocate CTU adopt the depth level i respectively.

Extensive experiments are performed to compute the depth correlation Cor_st. The CU size in HEVC ranges from 64×64 to 8×8, therefore the tested values of R are set as 64, 32, 16 and 8. The experimental result is shown in Table 1. In order to ensure consistency, the depth correlations of temporal-neighboring CTUs are also calculated using formula 3.

Table 1. The depth correlations between the current CTU and the neighboring CTUs extracting regions with different values of R

Set	CTU	R=64	R=32	R=16	R=8
α	A	74.3%	76.3%	76.7%	76.7%
	B	76.4%	77.1%	77.2%	77.2%
	C	74.1%	75.7%	75.9%	75.9%
	I	78.9%	78.9%	78.9%	78.9%
β	D	71.8%	72.9%	73.5%	73.5%
	E	70.7%	72.4%	72.5%	72.5%
	F	70.8%	72.7%	72.9%	72.9%
	G	70.1%	72.3%	72.5%	72.5%
	H	70.8%	71.5%	72.2%	72.2%

From the results of Table 1, the fact that CTUs in set α have higher correlation than those in set β is validated. And we observe that horizontal CTU A and vertical CTU B have a higher depth correlation than diagonal CTU C and D, because CTU A and B are closer to the current CTU. Another reason is that the probability of horizontal and vertical movement is larger than that of diagonal movement for natural objects. Furthermore, we find that the depth correlation becomes larger with the decrease of R and this result validates the fact that it brings in incorrect depth information when the depth information extracting region size is large. Besides, whether 16×16 CU in extracting regions with R=16 is further split or not, the depth information is exactly the same when R=16 and R=8. That's why the depth correlation is consistent when R=16 and R=8 in above experiments. This paper employs the value of R which achieves the highest depth correlation and the least extracting time, namely 8. The depth information extracting region is employed implicitly in the following sections.

3.2   Similarity Degree-Based CTU Classification

Three similarity degree classes are defined through the depth information of CTU A, B, C and I in the basic predictive set α in [9]. However, the CTUs of medium similarity degree exhibit two quite different trends of characteristics. Some of them have an obviously higher correlation with neighboring CTUs than the others. Consequently, the original medium similarity degree can be further divided into two similarity degree classes which are designed in this section.

The method of determining the similarity degree-based CTU classification is redefined as follows.

In the first place, we employ a two-dimensional array $depth\_unused[i][j]$ ($i \in [0, size(\alpha)-1]$), $j \in [0, max\_depth]$, $i, j \in Z$) as a flag to indicate whether the depth level j is adopted or not for the ith element in the basic predictive set α. $depth\_unused[i][j]$ is defined as:

$$depth\_unused[i][j] = \begin{cases} 0 & the\ depth\ level\ j\ is\ adopted\ for\ a(i) \\ 1 & the\ depth\ level\ j\ is\ not\ adopted\ for\ a(i) \end{cases} \quad (4)$$

where $\alpha(i)$ represents the *i*th element in the basic predictive set α. As the set α is composed of CTU A, B, C and I, $size(\alpha)$ is equal to 4.

Then, *Simlevel* is expressed as

$$Simlevel = \sum_{j=0}^{max\_depth} (\prod_{i=0}^{size(\alpha)-1} (depth\_unused[i][j])) \tag{5}$$

Usually, the CU size changes from 64×64 to 8×8, so all the possible values of *Simlevel* are equal to 0, 1, 2 and 3. *Simlevel* represents the number of depth levels which are not adopted by any one of CTUs in set α and can be a measure of similarities. The larger the *Simlevel* value is, the more similar to the neighboring CTUs.

Eventually, we define *Similarity Degree* as

$$Similarity\ Degree = \begin{cases} low & Simlevel = 0 \\ medium\_low & Simlevel = 1 \\ medium\_high & Simlevel = 2 \\ high & Simlevel = 3 \end{cases} \tag{6}$$

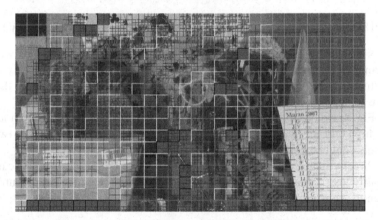

Fig. 3. Demonstration of similarity degree classification (Cactus, QP 32, 5th frame)

Fig.3 is an example of proposed similarity degree classification. The CTUs with green edges belong to high similarity degree, which is mostly the background. The CTUs painted with red represent the CTUs of low similarity degree, which are generally the edge of the background and moving objects. The CTUs with yellow edges are the CTUs of medium_high similarity degree which own higher correlation with neighboring CTUs than the CTUs of the medium_low similarity degree whose edges are painted purple. These two kinds of CTUs are usually located in moving regions.

For the CTUs of two new classifications and original medium similarity degree, we conduct experiments with several video sequences with various resolutions and QPs to computer the probability distribution of the numbers of depth levels adopted. The experimental results are shown in Fig. 4. From the results, we observe that the CTUs of medium_high similarity degree prefer to adopt one or two depth levels, while the

CTUs of medium_low similarity degree trends to adopt three depth levels. Thanks to their respective characteristics, the solutions to the CTUs of medium_high and medium_low similarity degree are introduced separately in Section 3.3 and 3.4.

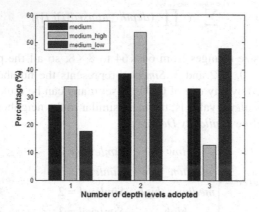

Fig. 4. The probability distribution of the numbers of depth levels adopted

3.3 CU Depth Decision Strategy for CTUs of Medium_high Similarity Degree

For CTUs of medium_high similarity degree, there are strong correlations with their neighboring CTUs. According to the definition of medium_high similarity degree, two depth levels are not adopted by any one of the candidate CTUs in the basic predictive set α.

From Fig. 4, we observe that the percentage of adopting one or two depth levels in medium_high CTUs is much higher than that of the other. Therefore, one or two more depth levels could be eliminated to further reduce the complexity.

If we directly ignore the two depth levels which are less used, it might cause obvious quality loss. To address this problem, the extended predictive set β is utilized to strengthen the constraint.

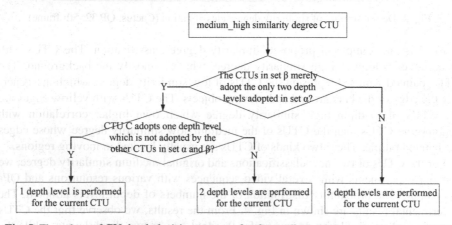

Fig. 5. Flowchart of CU depth decision strategy for the medium_high similarity degree CTU

If the CTUs in set β adopt the only two depth levels adopted in set α, we eliminate the other depth levels for the current CTU. Otherwise, besides the two depth levels adopted in set α, one more depth level is performed according to the frequency of its occurrence. In addition, sometimes CTUs of medium_high similarity degree adopt only one depth level. If one depth level only appears in CTU C which owns the lowest correlation in set α and all the CTUs in set β don't adopt that depth level, we eliminate that depth level in depth decision of the current CTU on the basis of adopting two depth levels. A flowchart of the above method is shown in Fig. 5.

3.4   CU Depth Decision Strategy for CTUs of Medium_Low Similarity Degree

As can be seen from Fig. 4, even for CTUs of medium_low similarity degree, only about half of them adopt three depth levels, while 35% of them adopts two depth levels and 15% adopt only one depth level. Therefore, one or two more depth levels could be eliminated for these CTUs to further reduce the complexity. To do so, this section refines the procedure of the original method of depth level elimination.

In order to express clearly, we first define *Adopted(T,d)* as

$$Adopted(T,d) = \begin{cases} 1 & depth\ level\ d\ is\ adopted\ by\ CTU\ T \\ 0 & Otherwise \end{cases} \tag{7}$$

where T is one of CTU A, B, C and I in set α, and d ranging from 0 to 3 represents depth level. Then *Probablity(d)* is expressed as

$$Probablity(d) = \frac{\sum_{i=0}^{size(\alpha)-1} Adopted(\alpha(i),d)}{size(\alpha)} \tag{8}$$

where *Probablity(d)* represents the probability that depth level d is adopted by CTU A, B, C and I and it ranges from 0 to 1. α(i) represents the ith CTU in set α and size(α) indicate how many CTUs in set α.

According to the definition of medium_low similarity degree, all the CTUs in set α adopt three depth levels in total. It means there must be only one depth level meeting the fact that *Probablity(d)* is equal to 0 and we eliminate this redundant depth level at first. Then if the remaining three depth levels all possess large values of *Probablity(d)*, it's difficult to eliminate one more depth level for the current CTU because the current CTU tends to adopt three depth levels. And it is also hard to eliminate redundant depth levels when values of *Probablity(d)* for the remaining three depth levels are all small, due to low correlation between the current CTU and its neighboring CTUs. Hence, we eliminate the most impossible depth level which is adopted merely by one CTU in set α, when there is depth level adopted by every CTU in set α, namely meeting *Probablity(d)=1*. If two most impossible depth levels exist, we eliminate the one which owns a larger depth distance from the depth level adopted by every CTU in set α.

3.5   Summary of the Proposed Fast Coding Depth Decision Algorithm

As a summary, the proposed algorithm takes the following steps:

1) Obtain the depth information from the extracting regions of CTUs in set α and β.
2) Determine the similarity degree for the current CTU by using depth information acquired in Step 1).
3) If the current CTU is medium_high similarity degree, go to Step 4); If the current CTU is medium_low similarity degree, go to Step 5); otherwise, go to Step 6).
4) **Medium_high similarity degree:** Based on the depth information of CTUs in set α and β, we perform 1, 2 or 3 depth levels for the current CTU. The procedure introducing which depth levels are eliminated is described in section 3.3 in detail. Then Go to Step 7).
5) **Medium_low similarity degree:** Eliminate the depth level only adopted by one CTU in set α when there is another depth level adopted by every CTU in set α. The procedure is described in section 3.4 in detail. Go to Step 7).
6) **High similarity degree or low similarity degree:** CU depth decision strategies in [9] are performed for these CTUs. Go to Step 7).
7) Return to Step 1) to repeat the procedure of the overall proposed algorithm for the next CTU.

4   Experimental Results

In order to evaluate the performance of the proposed algorithm, we implemented it into the latest HEVC test model 11.0 (HM11.0). And we also incorporated the algorithms in [9] and [10] into HM11.0 for comparison. The experimental condition is "low delay P, main" and test sequences include all the video sequences in class B (1920×1080), class C (832×480) and class D (416×240). Extensive experiments are performed on the recommended sequences under different QPs (22, 27, 32, 37). The experimental results are presented in Table 2, in which coding performance is measured by the BD-rate as defined in [11], and computational complexity is measured by the encoding time, both against HM11.0.

From the experimental results, it can be observed that our proposed algorithm can reduce on average 35.72% of the encoding time with only 0.65% bitrate increment. Compared with Zhang's algorithm [9], our proposed algorithm can further reduce on average 12% more encoding computational complexity with slight BD-Rate increment. This is because that the proposed algorithm can extract more precise depth information with less computational complexity than Zhang's [9] and further eliminates the redundant depth levels. Meanwhile, the proposed algorithm consistently outperforms Shen's algorithm [10], with much improved coding performance, averagely 0.65% against 1.89% as well as on average 5% more encoding time savings.

It is also interesting to notice that the proposed algorithm achieves more time savings for video sequences of higher resolutions. The reason is that there is a stronger correlation between neighboring CTUs in larger pictures and more accurate depth information can be extracted to speed up the encoding procedure. Nevertheless, it's

Table 2. Performance results of each algorithm compared with HM11.0

Class	Sequences	Zhang [9] BD-rate (%)	Zhang [9] Δ T (%)	Shen [10] BD-rate (%)	Shen [10] Δ T (%)	Proposed algorithm BD-rate (%)	Proposed algorithm Δ T (%)
B	Kimono	0.24	-25.53	0.69	-32.70	0.69	-41.24
B	ParkScene	0.54	-32.06	2.29	-36.67	0.90	-44.08
B	Cactus	0.39	-27.76	2.08	-32.90	1.07	-42.95
B	BasketballDrive	0.48	-29.64	1.89	-34.98	1.00	-42.77
B	BQTerrace	0.34	-27.06	1.30	-32.85	0.63	-41.70
B	**Average of B**	**0.40**	**-28.41**	**1.65**	**-34.02**	**0.86**	**-42.55**
C	BasketballDrill	0.64	-28.27	2.82	-33.18	1.37	-38.19
C	BQMall	0.27	-28.25	3.54	-35.95	1.00	-38.31
C	PartyScene	0.00	-18.11	2.03	-26.83	0.16	-32.27
C	RaceHorsesC	0.13	-17.50	2.40	-25.41	0.50	-30.88
C	**Average of C**	**0.26**	**-23.03**	**2.70**	**-30.34**	**0.76**	**-34.91**
D	BasketballPass	0.26	-26.86	2.24	-30.86	0.52	-34.74
D	BQSquare	-0.23	-17.28	0.27	-24.77	-0.10	-27.63
D	BlowingBubbles	0.15	-16.81	1.00	-23.69	0.36	-25.29
D	RaceHorses	0.20	-12.83	1.96	-19.76	0.41	-24.26
D	**Average of D**	**0.10**	**-18.45**	**1.36**	**-24.77**	**0.30**	**-27.98**
	Average of all	**0.26**	**-23.68**	**1.89**	**-30.04**	**0.65**	**-35.72**

not quite obvious for Shen's algorithm [10], because the depth information represented by the deepest depth level is not accurate enough and the eliminated depth levels are fixed, which leads to relatively fixed time saving with quality losses. Zhang's algorithm [9] has a better classification of CTUs than Shen's [10], but the CTU-based depth information extraction is still less accurate than the proposed one. Besides, there are still some redundant depth levels and more encoding time can be saved. Fig. 6 show the RD curves comparison between HM11.0 and proposed algorithm. It is seen that the RD difference is very small.

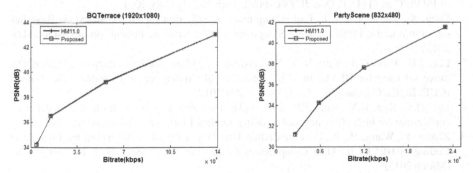

Fig. 6. RD performance curves of BQTerrace and PartyScene

5   Conclusion

This paper proposes an improved fast coding unit depth decision algorithm based on similarity degree to further reduce the inter-frame coding complexity of the HEVC encoder by improving the accuracy of extracting CTU depth information and further classifying medium similarity degree into two kinds of similarity degree. Extensive experimental results show that the proposed algorithm outperforms the HEVC test model (HM11.0) as well as state-of-the-art algorithms, and can significantly reduce the encoding time with negligible loss on the RD performance.

Acknowledgements. The authors would like to thank the anonymous reviewers for their potential valuable comments and suggestions that would help further improve the presentation of this paper .This work was partially supported by the 973 Program (2010CB327900), the National Natural Science Foundation of China (61272502) and the National Science Fund for Distinguished Young Scholars (61125206).

References

1. Wiegand, T., Sullivan, G.J., Bjontegaard, G., Luthra, A.: Overview of the H.264/AVC video coding standard. IEEE Transactions on Circuits and Systems for Video Technology 13(7), 560–576 (2003)
2. Bross, B., Han, W.J., Ohm, J.R., Sullivan, G.J., Wang, Y.K., Wiegand, T.: High Efficiency Video Coding (HEVC) Text Specification Draft 10. JCTVC of ISO/IEC and ITU-T, Doc. JCTVC-L1003, Geneva, Switzerland (January 2013)
3. Sullivan, G.J., Ohm, J., Woo-Jin, H., Wiegand, T.: Overview of the High Efficiency Video Coding (HEVC) Standard. IEEE Transactions on Circuits and Systems for Video Technology 22(12), 1649–1668 (2012)
4. Yoo, H., Suh, J.: Fast coding unit decision algorithm based on inter and intra prediction unit termination for HEVC. In: IEEE International Conference on Consumer Electronics, ICCE, Las Vegas, NV, pp. 300–301 (January 2013)
5. Choi, K., Park, S.H., Jang, E.S.: Coding tree pruning based CU early termination. JCTVC of ISO/IEC and ITU-T, Doc. JCTVC-F092, Torino, Italy (July 2011)
6. Shen, X., Yu, L., Chen, J.: Fast coding unit size selection for HEVC based on Bayesian decision rule. In: Picture Coding Symposium, PCS, Kraków, Poland, pp. 453–456 (May 2012)
7. Lee, J.H., Park, C.S., Kim, B.G.: Fast coding algorithm based on adaptive coding depth range selection for HEVC. In: IEEE International Conference on Consumer Electronics, ICCE, Berlin, Germany, pp. 31–33 (September 2012)
8. Lee, H.S., Kim, K.Y., Kim, T.R., Park, G.H.: Fast encoding algorithm based on depth of coding-unit for high efficiency video coding. Optical Engineering 51(6), 67402 (2012)
9. Zhang, Y., Wang, H., Li, Z.: Fast Coding Unit Depth Decision Algorithm for Interframe Coding in HEVC. In: Data Compression Conference, DCC, Snowbird, UT, pp. 53–62 (March 2013)
10. Shen, L., Liu, Z., Zhang, X., Zhao, W., Zhang, Z.: An Effective CU Size Decision Method for HEVC Encoders. IEEE Transactions on Multimedia 15(2), 465–470 (2013)
11. Bjontegard, G.: Calculation of average PSNR differences between RD-curves. VCEG of ITU-T, Doc. VCEG-M33, Austin, Texas, USA (April 2001)

Low-Complexity Rate-Distortion Optimization Algorithms for HEVC Intra Prediction

Zhe Sheng, Dajiang Zhou, Heming Sun, and Satoshi Goto

Graduate School of Information, Production and Systems,
Waseda University, Kitakyushu, Japan
shengzhe@fuji.waseda.jp

Abstract. HEVC achieves a better coding efficiency relative to prior standards, but also involves dramatically increased complexity. The complexity increase for intra prediction is especially intensive due to a highly flexible quad-tree coding structure and a large number of prediction modes.

The encoder employs rate-distortion optimization (RDO) to select the optimal coding mode. And RDO takes a great portion of intra encoding complexity. Moreover HEVC has stronger dependency on RDO than H.264/AVC. To reduce the computational complexity and to implement a real-time system, this paper presents two low-complexity RDO algorithms for HEVC intra prediction. The structure of RDO is simplified by the proposed rate and distortion estimators, and some hardware-unfriendly modules are facilitated. Compared with the original RDO procedure, the two proposed algorithms reduce RDO time by 46% and 64% respectively with acceptable coding efficiency loss.

Keywords: HEVC, intra prediction, RDO, video coding.

1 Introduction

High Efficiency Video Coding (HEVC) [1] is the latest video coding standard developed by JCT-VC. It adopts a flexible quad-tree coding structure using variable sizes of coding unit (CU), prediction unit (PU) and transform unit (TU). For intra prediction, there are several new features. The coding unit can vary from 64x64 down to 8x8 and up to 35 prediction modes are defined in HEVC. The RDO result, rate-distortion Lagrangian cost (R-D cost), is the criterion of choosing the best mode and optimal sizes of CU, PU, TU. Since HEVC has much more prediction modes than H.264/AVC, its dependency on RDO becomes stronger. And the traditional RDO-off method is not integrated in HEVC test model (HM) because of poor performance. The hardware implementation of HEVC encoder becomes critical as a result of the sequential processing of high-complexity RDO and the data dependency of neighboring blocks. Therefore, an efficient algorithm and simplified RDO structure become essential.

The related works about computational complexity reduction of RDO can usually be classified into two categories: (a) Some low-possibility candidate modes

C. Gurrin et al. (Eds.): MMM 2014, Part I, LNCS 8325, pp. 541–552, 2014.

or sizes are excluded based on the spatial or temporal features; (b) Full search of all candidate modes is maintained, while the calculation of each R-D cost value is simplified by estimating the rate and distortion in a low complexity way. Many works proposed methods of category (a), e.g. [2] developed a mode filter by the SAD of original pixels. This work is dedicated in (b), for which some works have been proposed for H.264/AVC, but hardly any for HEVC. In [3], a rate estimator used Generalized Gaussian Distribution to model the coded coefficients. An efficient R-D cost estimation method was proposed in [4], which used the l_1-norm of coefficients and the transform-domain distortion. In [5], the coordinate of non-zero coefficient helped to model rate estimation. Distortion is approximated when the DCT coefficients is modeled by Laplacian distribution in [6]. However, these proposals can not be directly applied to HEVC because they are mostly based on TU size of 4x4, while TU can be 4x4 to 32x32 in HEVC.

The target of this work is to develop a low-complexity RDO algorithm which is suitable for hardware implementation of HEVC encoder. Therefore, it is necessary to consider about replacing some parts of the sequential RDO with low-complexity algorithms. This paper presents two efficient RDO algorithms based on quantized coefficients and transformed coefficients. For rate calculation, the real number of bits coded by CABAC is estimated. For distortion calculation, the distortion caused by quantization is used to estimate the actual distortion. Thus, some of the high-complexity modules in the conventional RDO can be simplified, which benefits the hardware implementation.

The rest of this paper is organized as follows. Section 2 introduces the brief structures of original RDO and the proposed ones. Section 3 and 4 present the proposed RDO algorithms based on quantized coefficients and transformed coefficients, respectively. Section 5 shows the experimental results, followed by the conclusion in Section 6.

2   RDO in HEVC

Fig.1.(a) illustrates the flowchart of original RDO based mode decision. The area in dashed line means R-D cost calculation, and the wide grey line shows the latency for rate and distortion. The rate is counted after the quantized transform pixels are entropy coded. And the distortion is calculated after the block is reconstructed and compared with the original pixels.

In HEVC intra prediction, the PU can vary from 64x64 to 4x4. For each PU, several candidate prediction modes are selected from all the 35 directional prediction modes by calculating the SATD based cost, which is defined by

$$J_{pred,SATD} = SATD + \lambda_{pred} \cdot B_{pred} \quad [7], \qquad (1)$$

where $SATD$ is the sum of absolute difference of the Hadamard transformed coefficients, B_{pred} specifies bit cost of encoding the mode information.

Then the best prediction mode is selected from these candidates by calculating the R-D cost, which is defined by

$$J_{mode} = SSE + \lambda_{mode} \cdot B_{mode} \quad [7], \qquad (2)$$

where distortion SSE is the sum of square error between the original pixels and reconstructed pixels, B_{mode} specifies bit cost of encoding the whole block by CABAC. This cost is also the criterion of deciding the optimal CU size, PU size and TU size.

Even though only several number of candidates are selected to calculate the R-D cost, the procedure of rate-distortion optimization still occupies an essential portion in the computational complexity.

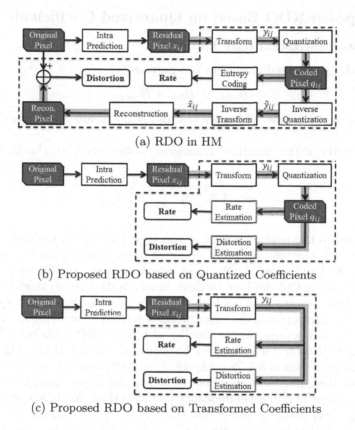

(a) RDO in HM

(b) Proposed RDO based on Quantized Coefficients

(c) Proposed RDO based on Transformed Coefficients

Fig. 1. The Flowchart of R-D cost Based Mode Decision

In Fig.1.(a), the successive modules including Transform (T), Quantization (Q), Inverse Quantization (IQ) and Inverse Transform (IT) would cause a strong data dependency, but relatively easy to be parallelized and pipelined in hardware. Whereas the module of Entropy Coding, in which coefficients are processed sequentially, is disadvantageous to hardware implementation.

The flowchart of the proposed RDO based on quantized coefficients is shown in Fig.1.(b). The module of entropy coding is replaced with a simple rate estimation module. The inverse quantization, inverse transform and pixel reconstruction are replaced with a low-complexity distortion estimation module. Therefore, the processing of RDO is simplified and the data dependency between blocks is alleviated, which brings a lot of convenience to hardware implementation.

The flowchart of the proposed RDO based on transformed coefficients is shown in Fig.1.(c). Compared with previous one, the module of quantization is also skipped. R-D cost calculation is further facilitated and the data dependency is further alleviated.

3 Proposed RDO Based on Quantized Coefficients

3.1 Post-quantization Rate Estimation

The rate B_{mode} can be divided into two parts as

$$B_{mode} = B_{hdr} + B_{coef},\qquad(3)$$

where B_{hdr} is the number of header bits and B_{coef} means the number of coefficient bits.

The l_p-norm of the quantized transform coefficients $q_{ij}(i, j = 0..N - 1)$ is defined by

$$E = \left(\sum_{i=0}^{N-1} \sum_{j=0}^{N-1} |q_{ij}|^p \right)^{1/p}.\qquad(4)$$

According to the residual coding syntax described in [8], the coefficient bits are coded from the information of non-zero map, greater-than-one map, sign, remaining level and so on. As the most significant part, the binarization of remaining level uses Golomb-Rice method, in which the rice parameter cRP may increase when scanning from the high frequency coefficients to low frequency coefficients. The prefix part of the binarization is derived by invoking the Truncated Rice binarization process, whose number of bins is $cRP + \lceil Level/2^{cRP} \rceil$. The suffix is derived using the k-th order Exponential Golomb (EGk) binarization, whose number of bins is $k + \lceil \log_2(Level/2^k + 1) \rceil$[9], where $k = cRP + 1$.

Based on these processes, we draw the conclusion that a small level would be coded into longer bins in lower frequency positions than that in higher frequency, and a large level would be coded into shorter bins in lower frequency positions than that in higher frequency. By adding the position information of non-zero coefficients into (4), we model the bit consumption E_{pos} as

$$E_{pos} = \left(\sum_{q_{ij} \neq 0} (|q_{ij}| + \theta \cdot |q_{ij}| (i + j))^p \right)^{1/p},\qquad(5)$$

where θ is a balance parameter of the position information. For computation reduction, we choose $p = 1$. Taking the fact that most non-zero coefficients are 1 into consideration, E_{pos} is simplified as

$$E_{pos} = \sum_{q_{ij} \neq 0} (|q_{ij}| + \theta \cdot (i + j)).\qquad(6)$$

Fig. 2 shows the relationship between the number of actual bits B_{coef} and E_{pos} when $\theta = 1$. There exists high correlation between the number of actual bits and E_{pos}, while the linearity is not perfect. And it is a problem that how to estimate the number of bits for different TU sizes. Based on the observation of Fig. 2, we propose a rate estimator designed for different TU sizes, which is formulated as

$$\hat{B}_{coef} = \alpha \cdot E_{pos}^{\beta} = \alpha \cdot \left(\sum_{q_{ij} \neq 0} (|q_{ij}| + \theta \cdot (i+j)) \right)^{\beta}, \qquad (7)$$

where $\hat{B}_{coef}$ represents the estimated number of coefficient bits, α and β are model parameters which are different for different TU size N and different quantization parameter (QP). With proper $\alpha(QP, N)$, $\beta(QP, N)$ and $\theta(QP, N)$, the proposed estimator model will achieve accurate prediction.

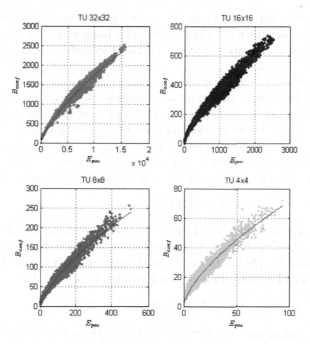

Fig. 2. The number of actual bits B_{coef} versus E_{pos}

Benefiting from the rate estimator, the module of Entropy Coding for R-D cost calculation is skipped. Since the estimator in (7) can be calculated by a look-up table, it is easy to be implemented in hardware. As $\hat{B}_{coef}$ is only calculated once for each block, a little additional complexity would be caused.

For intra prediction, the number of header bits B_{hdr} is usually much smaller than the coefficient bits. According to [8], the header bits of intra prediction mainly include the information about prediction mode, partition size, transform unit splitting and so on. And the prediction mode is coded based on the predicted modes inferred from the neighboring blocks. Since the calculation of header bits is not complex, we directly use the method in HEVC test model.

3.2   Post-quantization Distortion Estimation

As described in Section 2, the distortion is measured by the SSE between the original pixels and the reconstructed pixels. The data flows through modules of Intra Prediction, Transform, Quantization, Inverse Quantization, Inverse Transform and Reconstruction.

In forward transform, the transform coefficients y_{ij} are derived from the residual samples x_{ij}, which is formulated as

$$Y = T(X) = C_f X C_f^T \otimes E_f,   \tag{8}$$

where C_f is the transform matrix and E_f represents the scaling matrix. Although HEVC uses more accurate C_f than H.264, the integer matrix is still different from the theoretical transform matrix. So the forward transform would cause some distortion, and the module of inverse transform is similar. We define the distortion caused by T and IT as D_T.

In quantization module, the transform coefficients y_{ij} are scaled into smaller quantized coefficients q_{ij}. The division brings distortion because of the right shift operation. Let D_Q be the distortion resulted from quantization.

Each coefficient would be clipped when the value exceeds the data range. The distortion resulted from clipping is denoted by D_C. Then the total distortion D is a co-effect result of D_T, D_Q, and D_C, which can be formulated as

$$D = G(D_T, D_Q, D_C).   \tag{9}$$

However, G is not same as the sum function because of the counteraction. According to [4], the transform-domain distortion is close to the actual spatial-domain distortion D. That is to say, D_Q takes the majority part of D.

The conventional quantization can be formulated as

$$q_{ij} = (y_{ij} * f[QP\%6] + offset) \gg iQBits   \quad [7],   \tag{10}$$

where f is the scaling parameter related to QP, $iQBits$ is the number of shifted bits related to QP and TU size. And we define the scaled coefficients $s_{ij} = y_{ij} * f[QP\%6] + offset$.

In HEVC, the default quantization module is Rate-Distortion Optimized Quantization (RDOQ) [7], in which the quantized level is selected from three possible values by calculating each cost. The three possible quantized values are $0, l_{ij}^{floor}, l_{ij}^{ceil}$, defined by

$$l_{ij}^{ceil} = s_{ij} \gg iQBits, \quad l_{ij}^{floor} = l_{ij}^{ceil} - 1.   \tag{11}$$

We calculate the difference between the scaled coefficient and the quantized coefficient, noted by d_{qs}, as

$$d_{qs_{ij}} = |s_{ij} - q_{ij} \ll iQBits|.$$ (12)

And the total scaled quantization distortion D_{qs} can be derived as

$$D_{qs} = \sum_{i,j} d_{qs_{ij}}^2 = \sum_{i,j}(s_{ij} - q_{ij} \ll iQBits)^2.$$ (13)

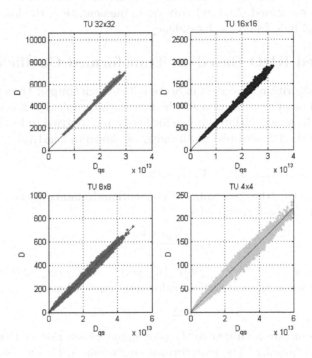

Fig. 3. The real distortion D versus D_{qs}

Fig.3 shows the relationship between the real distortion D and the scaled quantization distortion D_{qs}, which possesses strong linearity. Since inverse quantization is formulated as

$$\hat{y}_{ij} = (q_{ij} * scale + iAdd) \gg iShift,$$ (14)

where $scale$ is a scaling parameter correlates with the scaling parameter in quantization procedure, $iAdd$ is an offset and $iShift$ is the number of shifted bits related to TU size, we model the estimated transform-domain distortion $\hat{D}_Q$ as

$$\hat{D}_Q = \eta \cdot D_{qs} \gg 2(k - \log_2 N),$$ (15)

where η is the parameter caused by *scale*, k is the parameter caused by shifting and N is the TU width. Finally, the distortion estimator is formulated as

$$\hat{D} = \hat{D}_Q = \mu \cdot D_{qs} \cdot N^2, \tag{16}$$

where $\hat{D}$ represent the estimated distortion and μ ($\mu = \eta \gg 2k$) is the model parameter related to QP. With proper $\mu(QP)$, the total distortion would be estimated accurately.

With the help of distortion estimator, inverse quantization, inverse transform and pixel reconstruction can be skipped for R-D cost calculation. And reconstruction is only needed for the best mode. Therefore, the number of pixel processing steps is decreased. And in hardware implementation, the low-complexity estimator would hide the data dependency.

4   Proposed RDO Based on Transformed Coefficients

Although the algorithm described in Section 3 is much simpler than the original RDO, the module of quantization still brings much complexity. Since our target is to develop a RDO algorithm suitable for hardware, we tried to further simplify RDO based on the transformed coefficients, as shown in Fig.1.(c).

4.1   Pre-quantization Rate Estimation

From equation (10), the procedure of conventional quantization is actually a division, which can be formulated as

$$q_{ij} = \frac{|y_{ij}| + r}{\Delta}, \tag{17}$$

where the divisor $\Delta = 2^{iQBits}/f[QP\%6]$, and $r = offset/f[QP\%6]$ (in HM $r = \Delta/3$). And we define a threshold value function $T[k]$ as

$$T[k] = k \cdot \Delta - r \quad (k \geq 1). \tag{18}$$

The analysis of rate is similar to the post-quantization rate estimation model described in Section 3.1. The transformed coefficients with the absolute value smaller than $T[1]$ would become 0 after quantization. So these coefficients have no effect on the final rate and can be ignored in the rate estimation model. Take this fact into consideration, we model the bit consumption E_{pos} as

$$E_{pos} = \sum_{|y_{ij}| \geq T[1]} (|y_{ij}| + \theta \cdot (i + j)). \tag{19}$$

Similarly, the estimated number of coefficient bits $\hat{B}_{coef}$ is formulated as

$$\hat{B}_{coef} = \alpha \cdot E_{pos}^{\beta} = \alpha \cdot \left(\sum_{|y_{ij}| \geq T[1]} (|y_{ij}| + \theta \cdot (i + j)) \right)^{\beta}, \tag{20}$$

where $\alpha(QP, N)$, $\beta(QP, N)$ and $\theta(QP, N)$ are model parameters. The experimental results shows that the modified model can still estimate the rate accurately.

4.2   Pre-quantization Distortion Estimation

Since the distortion is mainly caused by quantization, the final distortion is not easy to be estimated without quantization. It is considered to calculate the real distortion for the majority of coefficients and discard some small portion of coefficients. It is observed that the quantized coefficients with smaller absolute values have much higher possibilities than those with higher absolute values. A comparator-based distortion estimator is proposed.

When the absolute value of transformed coefficient y_{ij} is smaller than $T[k]$, the actual distortion would be calculated after the comparators find out which quantized value q_{ij} it would be. Otherwise, an expectation of distortion would be added instead of the actual distortion. In this case, $\lceil \log_2(k+1) \rceil$ comparators are needed. Then the difference between the transformed coefficient y_{ij} and the inverse quantized coefficient $\hat{y}_{ij}$, noted by d_q, is formulated as

$$d_{q_{ij}} = \begin{cases} |y_{ij} - \hat{y}_{ij}|, & |y_{ij}| \leq T[k] \\ E[d_q], & otherwise \end{cases} \tag{21}$$

$$d_{q_{ij}}^2 = \begin{cases} |y_{ij} - \hat{y}_{ij}|^2 \\ E[d_q^2] \end{cases} = \begin{cases} (y_{ij} - \Delta q_{ij})^2, & |y_{ij}| \leq T[k] \\ \Delta^2/9, & otherwise \end{cases} \tag{22}$$

The expectation $E[d_q^2]$ would become $\Delta^2/9$ when using the plain distribution. Other distributions can also be considered, such as the Laplacian distribution as described in [6].

Table 1. The Arrangement of k

i', j'	0	1	2	3
0	15	7	7	3
1	7	7	3	3
2	7	3	3	1
3	3	3	1	1

To estimate the distortion accurately, the threshold value $T[k]$ is arranged with different k at different positions, as shown in Table 1. Position coordinate $i' = 4i/N$, $j' = 4j/N$, where N is the TU width. It is basically ensured that more than 90% of coefficients satisfy $|y_{ij}| \leq T[k]$ when $QP = 22$ and their actual distortions are calculated. The needed comparators would cost much less than the original multipliers. The total quantization distortion D_q can be derived by

$$D_q = \sum_{i,j} d_{q_{ij}}^2. \tag{23}$$

Similarly to equation (16), the distortion estimator is formulated as

$$\hat{D} = \hat{D}_Q = \mu \cdot D_q \cdot N^2, \tag{24}$$

where $\hat{D}$ represent the estimated distortion and μ is the model parameter related to QP. With proper $\mu(QP)$, the total distortion would be estimated accurately.

5 Experimental Results

5.1 Model Parameter Training

The model parameters are obtained under 4 QP values: 22, 27, 32 and 37 [10]. The rate estimation model parameters $\alpha(QP, N)$, $\beta(QP, N)$ and $\theta(QP, N)$ mentioned in Section 3.1 and 4.1 are trained by Levenberg-Marquardt algorithm, a nonlinear regression algorithm based on least squares as described in [11]. The distortion estimation model parameters $\mu(QP)$ mentioned in Section 3.2 and 4.2 are computed by the simple linear regression method.

Two sequences are used to train the model parameters. For the high resolution sequences in class A and B, one set of parameters is obtained by a randomly selected sequence *BQTerrace*. For the low resolution sequences in class C, D and E, another set of parameters is obtained by *BasketballPass*.

5.2 The Performance of Proposed Algorithm

The proposed R-D cost estimation algorithms are integrated with HM 8.0. In the experiment, all test sequences listed in [10] except the two sequences used to obtain the model parameters are encoded under configuration of all-intra, with QP of 22, 27, 32 and 37. The configuration of RDOQ is turned off. The coding efficiency and complexity is compared between our proposed RDO and that in HM 8.0. The coding efficiency is measured by bit-rate difference and PSNR difference using Bjontegard's method [12]. And the complexity reduction is measured by time reduction ratio of R-D cost calculation ΔT_{RDO}, which is defined by

$$\Delta T_{RDO} = \frac{T_{ProposedRDO} - T_{OriginalRDO}}{T_{OriginalRDO}} \times 100\%. \tag{25}$$

The experimental result of the proposed RDO based on quantized coefficients is shown in Table 2. The proposed algorithm causes an average quality loss of $0.109dB$ and increased BD-rate of 1.93%, which is acceptable and demonstrates that the R-D cost estimation is relatively accurate. Meanwhile it achieves an average of 46% RDO time saving. Almost the same complexity reduction can be achieved regardless of the resolution or the sequence characteristics, which indicates the potential to apply the proposed algorithm into real-time systems.

The experimental result of the proposed RDO based on transformed coefficients is shown in Table 3. The BD-PSNR decreases by $0.135dB$ and BD-rate increases by 2.41% in average. While about 64% of RDO time is reduced. It can achieve more RDO time saving than the first test because that the module of quantization is also skipped. Considering its contribution of complexity reduction, the coding efficiency loss is acceptable.

Note that the contribution of the proposed algorithms to hardware implementation should be larger than that reflected by ΔT_{RDO}. This is because the operations simplified by the proposed algorithm are hardware-unfriendly parts including entropy coding and reconstruction that involves a long pipeline latency. The remaining complexity of RDO is mainly from transform and estimation which are relatively easy for parallel processing and pipelining.

Table 2. Performance of Proposed RDO based on Quantized Coefficients

Class	Sequence	BD-rate (%)	BD-psnr (dB)	ΔT_{RDO} (%)
A	Traffic	1.75	-0.095	-45.9
(4K)	PeopleOnStreet	1.91	-0.108	-46.2
	Kimono	1.42	-0.049	-43.0
B	ParkScene	2.15	-0.093	-46.4
(1080p)	Cactus	1.92	-0.073	-46.7
	BasketballDrive	2.40	-0.105	-47.5
	RaceHorses	1.83	-0.120	-46.3
C	BQMall	2.28	-0.143	-47.2
(WVGA)	PartyScene	2.30	-0.184	-49.0
	BasketballDrill	1.32	-0.064	-45.0
	RaceHorses	1.62	-0.115	-47.7
D	BQSquare	2.31	-0.202	-48.7
(WQVGA)	BlowingBubbles	1.98	-0.123	-47.5
	Vidyo1	2.01	-0.096	-43.8
E	Vidyo3	1.74	-0.087	-42.8
(720p)	Vidyo4	1.91	-0.084	-43.2
Total Average		**1.93**	**-0.109**	**-46.1**

Table 3. Performance of Proposed RDO based on Transformed Coefficients

Class	Sequence	BD-rate (%)	BD-psnr (dB)	ΔT_{RDO} (%)
A	Traffic	2.16	-0.116	-58.6
(4K)	PeopleOnStreet	2.61	-0.147	-60.3
	Kimono	2.23	-0.077	-57.4
B	ParkScene	2.37	-0.102	-59.0
(1080p)	Cactus	2.52	-0.095	-59.4
	BasketballDrive	2.54	-0.157	-66.6
	RaceHorses	2.22	-0.145	-65.6
C	BQMall	2.71	-0.169	-69.7
(WVGA)	PartyScene	2.80	-0.204	-69.4
	BasketballDrill	1.75	-0.086	-67.3
	RaceHorses	1.98	-0.138	-68.8
D	BQSquare	2.93	-0.227	-70.2
(WQVGA)	BlowingBubbles	2.44	-0.154	-69.1
	Vidyo1	2.72	-0.129	-59.3
E	Vidyo3	2.28	-0.114	-59.9
(720p)	Vidyo4	2.27	-0.100	-62.6
Total Average		**2.41**	**-0.135**	**-63.9**

6   Conclusion

We proposed two low-complexity rate-distortion optimization algorithms for HEVC intra prediction with proper rate and distortion estimation models based on quantized coefficients and transformed coefficients, respectively. They facilitate some hardware-unfriendly modules in the original RDO, and the data dependency can be alleviated in hardware implementation. In software simulation, the computational complexity of RDO is reduced by 46% and 64% respectively with acceptable coding efficiency loss. Furthermore, these algorithms can also be used in inter prediction or combined with mode filtering algorithms for HEVC.

Acknowledgement. This work is partly supported by KAKENHI and STARC.

References

1. Sullivan, G.J., Ohm, J.R., Han, W.J., Wiegand, T., Wiegand, T.: Overview of the high efficiency video coding (hevc) standard. IEEE Transactions on Circuits and Systems for Video Technology 22(12), 1649–1668 (2012)
2. Sun, H., Zhou, D., Goto, S.: A low-complexity hevc intra prediction algorithm based on level and mode filtering. In: 2012 IEEE International Conference on Multimedia and Expo (ICME), pp. 1085–1090 (April 2012)
3. Zhao, X., Sun, J., Ma, S., Gao, W.: Novel statistical modeling, analysis and implementation of rate-distortion estimation for h.264/avc coders. IEEE Transactions on Circuits and Systems for Video Technology 20(5), 647–660 (2010)
4. Tu, Y.K., Yang, J.F., Sun, M.T.: Efficient rate-distortion estimation for h.264/avc coders. IEEE Transactions on Circuits and Systems for Video Technology 16(5), 600–611 (2006)
5. Wang, Q., Zhao, D., Gao, W., Ma, S.: Low complexity rdo mode decision based on a fast coding-bits estimation model for h.264/avc. In: IEEE International Symposium on Circuits and Systems, ISCAS 2005, vol. 4, pp. 3467–3470 (May 2005)
6. Moon, J.M., Moon, Y.H., Kim, J.H.: A computation reduction method for rdo mode decision based on an approximation of the distortion. In: 2006 IEEE International Conference on Image Processing, pp. 2481–2484 (2006)
7. JCT-VC: High efficiency video coding (hevc) test model 8 encoder description. JCTVC-J1002 (July 2012)
8. JCT-VC: High efficiency video coding (hevc) text specification draft 8. JCTVC-J1003_d7 (July 2012)
9. Bardone, D., Carotti, E., De Martin, J.: Adaptive golomb codes for level binarization in the h.264/avc frext lossless mode. In: IEEE International Symposium on Signal Processing and Information Technology, ISSPIT 2008, pp. 287–291 (December 2008)
10. JCT-VC: Common test conditions and software reference configurations. JCTVC-E700 (January 2011)
11. Seber, G., Wild, C.: Nonlinear Regression. Wiley Series in Probability and Statistics. Wiley (2003)
12. Bjontegard, G.: Calculation of average psnr differences between rd-curves. ITU-T VCEG-M33 (April 2001)

Factor Selection for Reinforcement Learning in HTTP Adaptive Streaming

Tingyao Wu and Werner Van Leekwijck*

Alcatel Lucent - Bell Labs, Copernicuslaan 50, B-2018 Antwerp, Belgium
{tingyao.wu,werner.van_leekwijck}@alcatel-lucent.com

Abstract. At present, HTTP Adaptive Streaming (HAS) is developing into a key technology for video delivery over the Internet. In this delivery strategy, the client proactively and adaptively requests a quality version of chunked video segments based on its playback buffer, the perceived network bandwidth and other relevant factors. In this paper, we discuss the use of reinforcement-learning (RL) to learn the optimal request strategy at the HAS client by progressively maximizing a pre-defined Quality of Experience (QoE)-related reward function. Under the framework of RL, we investigate the most influential factors for the request strategy, using a forward variable selection algorithm. The performance of the RL-based HAS client is evaluated by a Video-on-Demand (VOD) simulation system. Results show that given the QoE-related reward function, the RL-based HAS client is able to optimize the quantitative QoE. Comparing with a conventional HAS system, the RL-based HAS client is more robust and flexible under versatile network conditions.

Keywords: Reinforcement Learning, HTTP Adaptive Streaming, Machine Learning, Variable Selection.

1 Introduction

In recent years, video delivery over traditional best-effort Internet has attracted lots of attention. Among this, HTTP adaptive streaming (HAS) has become a key technology. Several instances, like IIS Smooth Streaming by Microsoft [1], HTTP Live Streaming (HLS) by Apple [2] and HTTP Dynamic Streaming by Adobe [3], of this technology are in the market today.

In HAS, video content is encoded in different qualities (bit-rates), and chunked into independent segments, typically 2-10 seconds. The encoded segments are hosted on an HTTP Web server, together with a playlist (or manifest) file describing the quality levels and available segments. A client first retrieves the playlist, and then requests the segments with different qualities from the Web server in a linear fashion and downloads them using plain HTTP progressive download. Because the segments are carefully encoded without any gaps or overlaps between them, the segments can be played back as a seamless video.

---

* This research was partially funded by the iMinds MISTRAL project (under grant agreement no. 10838).

C. Gurrin et al. (Eds.): MMM 2014, Part I, LNCS 8325, pp. 553–567, 2014.
© Springer International Publishing Switzerland 2014

The key feature of HTTP adaptive streaming is that the client is responsible for deciding for each consecutive segment, which quality to download. Typically a rate determination algorithm, also called heuristic, is responsible for the selection of the highest sustainable quality, and is also adaptive to the changing environment. The decision is made by taking into account one or many observed factors to adaptively select a bit-rate. These factors may include the perceived bandwidth (based on download temporal intervals of previous segments), the playback bufferat the client together its dynamics, user screen resolution, CPU load, etc. Usually when the playback buffer is low or the perceived bandwidth is limited, the heuristic tends to request low quality segments, in order to avoid freeze; when the playback buffer is high or the bandwidth is sufficient, the heuristic attempts to increase the quality level of; while when the bandwidth is stable, the heuristic maintains sustainable quality level. More details of adaptive streaming can be found in [1], [2] and [3].

It is desirable for a client-side bit determination algorithm that is able to balance multivariate variables, attempting to obtain optimal QoE given certain circumstances. However, recent research work has shown that it is not an easy task for the current video streaming heuristics to pick a suitable video stream rate. For instance, in [4], it is pointed out that the existing streaming heuristics can be both too conservative (not fully make use of available bandwidth) and too aggressive (not fully consider the fast decreasing buffer filling level); in [5], it is observed that the competition between two or more adaptive streaming players can lead to issues of stability, unfairness and potential bandwidth under-utilization. To optimize the bit-rate selection algorithm, there have been several studies. In [6], a rate adaptation algorithm for HAS was proposed to detect bandwidth changes using a smoothed HTTP throughput measure based on the segment fetch time. In [7], the authors presented a Quality Adaptation Controller(QAC), which uses a feedback control to drive stream-switching for adaptive live streaming applications. In [8], we have proposed a Q-learning based client quality selection algorithm for HTTP adaptive video streaming to dynamically learn the optimal behavior corresponding to the current network environment. Comparing to other adaptive bit-rate algorithms, reinforcement learning by nature is able to learn from its punishments or rewards by trying different actions in a certain environment state and thus moves towards the maximum of its defined reward function. Some promising results have been reported in [8]. Huang [9] et al., following his work in [4], argued that only observing and controlling the playback buffer, without having to estimate network capacity, is already able to avoid unnecessary rebuffering and achieve an average video rate equal to available capacity in steady state, with the assumption that the available bandwidth never goes below the lowest encoded bit-rate. However, based on our observation, most of freezes occur exactly when the bandwidth is extremely limited due to the burstiness property of TCP. To deal with such circumstance, an integration of different information resources seem indispensable to prevent the client from rebuffering.

In this paper, we extend our study on the use of reinforcement learning by carefully and incrementally integrating factors relevant to bit-rate selection step by step, using a forward variable selection strategy; we attempt to quantitatively identify which one or more factors are the most important and influential for the bit-rate selection algorithm, and how they are combined. To simplify our study, we do not consider external factors like screen resolution or CPU load, as these factors are not easy to control in a simulation setting; instead, we focus on 5 parameters that matter to the HAS player itself, namely, playback buffer, previous playback buffer, instant perceived bandwidth, average perceived bandwidth in a sliding window, and previous requested bit-rate. Our selection result shows that the playback buffer, the average bandwidth and the previous playback buffer are the three most valuable ones in descending order. We simulate the evolution of the learning process and compare with one commercial HAS client, e.g. Microsoft IIS Smooth Streaming, and demonstrate the robustness and flexibility of RL-based HAS client.

The rest of the paper is organized as follows. In section 2, we describe the adaptation and implementation of the RL algorithm for the HAS client. The forward variable selection strategy for identifying the most influential parameters is presented in section 3. Section 4 is about the experiments, including the experiment design, forward variable selection for the RL environment, and the comparison with a traditional HAS client. The conclusions and future work are given in section 5.

2   RL Based HAS Client

In the sense of machine learning, reinforcement learning is concerned with how an agent ought to take actions in an environment so as to maximize a given cumulative reward [10]. Reinforcement learning has been widely used in control theory, game theory, etc. The motivation of using RL in HAS client is that if we can define a QoE-oriented reward function for RL, then in a multi-variable controlled environment, a RL agent should be able to incrementally maximize the reward during its trial-and-error procedure, thus enhancing the quantitative QoE. A RL model consists of a set of environment states $\mathbf{S}$, a set of actions $\mathbf{A}$, a reward function $\mathbf{R}$ and a state-action paired Q-function $Q(\mathbf{S}, \mathbf{A})$.

2.1   Environment Variables

The environment that a HAS client encounters and interacts with may include factors like the playback buffer, the perceived bandwidth, the quality levels of previous segments, the speed of buffer filling/consuming, etc. But these state factors must be carefully chosen: integrating too many undiscriminative variables into the environment states may reduce the interpretability for the agent and may make the searching task tedious, while selecting too few variables may not be descriptive enough, preventing the agent from reaching higher rewards. Therefore, a variable selection strategy(see section 3) is adopted to deliberately select necessary factors for the HAS environment. An environment state is then represented by a combination of discretized states of chosen factors.

2.2   Action Set

The action set **A** in RL indicates what actions an agent could take at a certain state. In the RL-based HAS client, this corresponds to all possible bit-rates that a client may request for a segment. The probability of choosing action a $(a \in A)$ associated with state s $(s \in S)$ is described by the state-action paired function $Q(s, a)$.

2.3   Reward Function

A reward function **R** in the RL defines the reward that an agent would get when it takes a certain action and jumps to another state. To obtain a reliable delivery strategy, the reward function in the RL-based HAS client should be directly related to QoE. Certainly, the duration of freeze f_i between playing segment i and segment $i + 1$ and the quality q_i for segment i are two indispensable factors for the QoE when requesting segment i. Moreover, [11][12] report that alternatively degrading and upgrading the quality level in a short time can also result in a degraded QoE, implying that a frequency oscillation should be punished and a stable quality sequence is desired. Meanwhile [13] suggest that a linear combination of these QoE-relevant parameters is already good enough to model the mean opinion score (MOS) of viewers. As a result, the instant reward function is modeled as:

$$r_i = -A * f_i - B * q_i - C * o_i. \tag{1}$$

The freeze f_i can be detected by checking the playback buffer b_i and the time duration e_{i+1} for retrieving segment $i + 1$:

$$f_i = \begin{cases} 0 & b_i - e_{i+1} \geq 0 \\ e_{i+1} - b_i & b_i - e_{i+1} < 0 \end{cases} \tag{2}$$

The second item q_i of the reward function is quality related. Intuitively, when the highest encoded bit-rate a_{max} is chosen, no punishment is given; otherwise, lower qualities are penalized proportionally in terms of their encoded bit-rates:

$$q_i = \frac{a_{max} - a_i}{a_{max}}. \tag{3}$$

The last item o_i in the reward function is about quality oscillation, and is calculated based on the following rules. Suppose there is a quality switch at the i-th segment: $a_i \neq a_{i-1}$, implying that a potential quality oscillation occurs, we will look for the closest quality switch point within the last M segments. If no closest switch is found, or the closest switch is in the same direction as the switch at the i-th segment, then no oscillation is detected. Otherwise, within M consecutive segments, quality decreasing/increasing co-exist. Then the oscillation o_i is calculated as the average of these two quality changes. This calculation implies that a pair of quality decrease/increase occur for a period of more than $M + 1$ segments is not seen as an oscillation.

The positive weights A, B and C in Eq. 1 represent to what extent these factors influence the QoE. As there is no decisive conclusion about how the freeze, quality and oscillation impact the QoE of viewers, and how importance they are relatively, we arbitrarily set their weights to be $A = 100$, $B = 10$ and $C = 10$ in our preliminary experiments, assuming that much higher punishment should be given to the freeze [14].

2.4 State and Action-Paired Q-Value

Suppose at time t_i, the RL-based HAS client agent reaches environment state s_i, and in probability requests quality level a_{i+1} ($a_{i+1} \in \mathbf{A}$) for segment $i + 1$. After having received the segment at time t_{i+1}, the agent transits to environment state s_{i+1}. Between t_i and t_{i+1}, it is possible that due to buffer under-run, the client experiences a picture freeze with duration f_{i+1}. Together with the quality and the quality oscillation for requesting the a_{i+1}th version of segment $i + 1$, a reward r_{i+1} is calculated based on Eq. 1. Then the learned action-value function $Q(s_i, a_i)$ is updated using *one step Q-learning* method [10]:

$$Q(s_i, a_i) \leftarrow Q(s_i, a_i) + \alpha[r_{i+1} + \gamma \max_{\forall a} Q(s_{i+1}, u) - Q(s_i, u_i)], \quad (4)$$

where (s, a) is the state-action pair, and $\alpha \in [0; 1]$ and $\gamma \in [0; 1]$ are the learning rate and the discount parameter respectively. The learning rate α determines to what extent the newly acquired information overrides the old information, while the discount factor γ is a measure of the importance of future rewards. This online update runs for multiple episodes until converge, and the probability of choosing a_i at state s_i is calculated as: $Pr(a_i|s_i) = \frac{Q(s_i, a_i)}{\sum_j Q(s_i, a_j)}$.

3 Forward Selection for Environment Variables

The environment states, containing multi-variables, represent the variable space that an agent interacts with. Those variables must be carefully selected to sufficiently represent the learning environment, as not all candidate variables contribute to illustrate the environment states. For instance, [9] claims that only using the playback buffer is already able to avoid freeze, assuming that the perceived bandwidth is never less than the lowest encoded bit rate a_1. But we observed that in our simulation (and in many cases), this assumption is usually not held. In this sense, we conjecture that only relying on the playback buffer is seemly insufficient; some other factors may help the client to make a better selection decision. To this end, we use a forward selection procedure to select discriminative variables among some potential factors. The forward selection is a data-driven model which adds variables to the model one at a time. At each step, each variable that is not already in the model is tested for inclusion in the model. The most significant one of these variables is added to the model, so long as its p-value is below a pre-determined level.

The potential influential factors that we will test, composed of a variable set P, are listed in Table 1. As shown, b_i and pb_i are playback buffers when receiving segment i and segment $i-1$. w_i and $\bar{w}_i$ are the instant perceived bandwidth and the average of recently perceived bandwidths in a sliding window with the size M respectively. The last potential variable d_i indicates how many segments that the current quality change direction has been maintained, implying the quality switches.

Table 1. Potential environment variable set P

Variable	Representation	Unit
Play buffer	b_i	sec
Previous play buffer	$pb_i = b_{i-1}$	sec
Instant BW	w_i	kbps
ave. BW	$\bar{w}_i = \frac{\sum_{j=0}^{M-1} w_{i-j}}{M}$	kpbs
number of segments in the same trend	d_i	level

We begin with a model including the variable that is the most significant in the initial analysis, and continue adding variables until none of remaining variables is "significant". The significance is tested by the average reward obtained by different variable sets. Given a set of selected variables S ($S \subset P$) in the RL-based HAS client, the average reward for variable set S over all segments is $\bar{R}^S = \sum_{i=1}^{N} r_i/N$ (N denotes the number of segments that the client requests), then a variable v from the rest variable set S^C ($S \cup S^C = P$) will be incorporated into the existing variable set S, only if the following conditions are met:

$$\begin{cases} \bar{R}^{S+v} > \bar{R}^{S+j} & j \in S^C, j \neq v \\ \bar{R}^{S+v} > \bar{R}^S \\ H_0 : Pr(\bar{R}^{S+v} > \bar{R}^S) = 0.5 & rejected \end{cases} \qquad (5)$$

The above procedure is repeated until no variable is significant. For the hypothesis test in Eq. 5, a non-parametric *two-sided sign test* is performed as the distributions of the tested variables are unknown. If the null hypothesis test is rejected with the significance level 0.05, then two variable sets are considered to be statistically different.

4   Experiments

In this section, we first describe the network topology of our simulation and the design of the experiment. Then the forward variable selection method is performed to select influential factors for the RL environment. The performance of RL-based HAS client is compared with that of a standard HAS client [1] in the same condition of randomly generated bandwidth, from which we demonstrate how reinforcement learning gradually optimizes the request sequence along with the number of trials. Finally, we show that the RL based HAS client is superior in terms of obtained rewards in multiple weighting parameter setups.

4.1 Simulation Setup

The simulation design of a Video-on-Demand (VOD) system is demonstrated in Fig. 1. At the server side, a video clip, *Big_Buck_Bunny*, is hosted and is available for retrieval. This video trace, about 10 minutes long, consists of 299 segments ($N = 299$), each with a fixed length of 2 seconds. Each video segment is encoded in 7 quality levels: 300kbps, 427kbps, 608kbps, 866kbps, 1233kbps, 1636kbps and 2436kbps. This leads to the action set $A = \{300, 427, 608, 866, 1233, 1636, 2436\}$kbps, and $a_{max} = 2436kbps$. At the client side, a standard Microsoft IIS Smooth Streaming [1] or a RL-based adaptive streaming runs. The client either immediately starts the request for the next segment once the previous segment has been fully received in the buffering state, or starts the request for each 2 seconds in the steady state [1].

Besides a typical HAS server-client topology, a cross-traffic manager, which is implemented on Click Modular Router [15], is also present. The cross-traffic manager is used to limit bandwidth towards the server as a bottleneck and also generates a tunable random cross traffic by keeping sending packets to the server. The randomly generated cross-traffic is determined by two random values, namely, the sending rate of packets and the corresponding duration. As a result, the bandwidth between the client and the server is the difference between the bottleneck bandwidth and the cross traffic. Considering the encoded

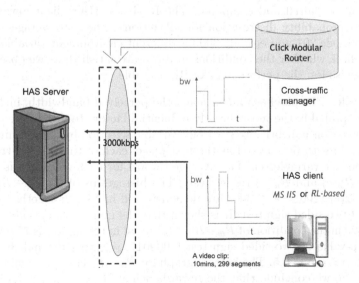

Fig. 1. Simulation design

bit-rates, the bottleneck bandwidth is set to be 3000kpbs (denoted as BW_{max}). To allow the perceived bandwidth at the client covering all bit-rates, the generated bandwidth by the cross traffic is in the range of 0kpbs and 2700kpbs. But note that because of the burstiness of TCP connections, the client could encounter very low bandwidth in a short period.

4.2 Forward Selection of Environment Variables

Parameter Discretization. Given that the maximum buffer length at the client is 32 seconds (which is the default maximum buffer length for the reference Microsoft IIS system), b_i and pb_i are discretized into 16 non-overlapping states, with one state spanning over each 2-second interval. Setting $M = 5$, $\bar{w}_i$ is calculated as the average of previous 5 perceived bandwidths. w_i and $\bar{w}_i$ are then discretized into 7 states, whose lower and upper boundaries for each state are relevant to the encoded bit-rates: $[0, a_2]$, $[a_2, a_3]$,...,$[a_7, BW_{max}]$. Note that the upper boundary for the first state is a_2, as the bandwidth less than a_2 (427kbps) corresponds to the lowest quality (300kbps). d_i is in the range between 0 and 5. When the direction of instant quality switch differs with the maintained direction, this variable is set to 0; when two consecutive segments are in the same quality or in the same change direction, this variable increases by 1, until 5.

Experiment Design for Variable Selection. A series of packet-sending rate and the corresponding durations are randomly generated and fed into the cross-traffic manager. The total duration of the cross traffic is 179,400 seconds (corresponding to 300 episodes). For each variable selection step, the RL-based client running with different environment variable sets will perceive almost identical bandwidth for the same episode; as a result, the average reward for the same episode can be paired and compared. The RL-based HAS client always starts with equal probability distribution for all actions. The 100 averaged rewards between episode 201 and episode 300 for different environment variable sets are taken to check whether the conditions in Eq. 5 are satisfied, as we observe that after 200 episodes, the performance of RL is stable.

1st **Variable.** The playback buffer b and the perceived bandwidth, including w and $\bar{w}$, are probably the most directly influential factors to decide the desirable quality level. Consequently, we start our variable selection by determining which one of the three to be selected at the first place: each of the three is treated as an independent variable set. The average reward for each episode e is denoted as $\bar{R}_b(e)$, $\bar{R}_{\bar{w}}(e)$ and $\bar{R}_w(e)$ respectively. The histograms of $\bar{R}_{\bar{w}}(e) - \bar{R}_b(e)$ and $\bar{R}_w(e) - \bar{R}_b(e)$ ($200 < e \leq 300$) are demonstrated in Fig. 2, together with the p-values of two-sided sign test. It is shown that the number of episodes between 201 and 300 in the condition of $\bar{R}_{\bar{w}} - \bar{R}_b < 0$ (in red in the figure) is 72 out of 100, with the p-value of two-sided sign test 0.000017, rejecting the null hypothesis $H_0 : Pr(\bar{R}_{\bar{w}} > \bar{R}_b) = 0.5$. The bottom graph for $\bar{R}_w - \bar{R}_b$ can be explained in the same way. So we conclude that the rewards achieved by w and $\bar{w}$ individually are significantly lower than that of b, and the most important variable to be selected is the playback buffer b: $S = \{b\}$.

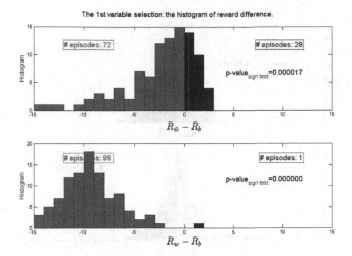

Fig. 2. The 1st variable selection

2nd **Variable.** The second variable will be chosen from the factors w and $\bar{w}$. To do so, these two variables are added into the existing set S separately and their performance is tested. The histograms of the difference between their average rewards and the benchmark, $\bar{R}_b$, are shown in Fig. 3. From the upper graph we can see that incorporating $\bar{w}$ into the environment variables ($\bar{R}_{b+\bar{w}}$) significantly increases the averaged reward (the number of episodes of positive difference is 69), while adding the instant bandwidth w does not provide additional benefits (shown in the bottom graph). As a consequence, in this step $\bar{w}$ is then added into the variable set: $S = \{b, \bar{w}\}$.

3rd **Variable.** The three remaining variables pb, w and d in S^C are tentatively incorporated into the existing environment variable set S, and their performance is compared with the set without any of them, as demonstrated in Fig. 4. Not surprisingly, adding pb provides helpful information for the agent to learn its tangible environment, while the other two variables do not improve the rewards. Actually, the combination of pb and b describes how quickly the play buffer fills/consumes. If the buffer increases/decreases too fast, the agent knows that the true bandwidth does not match the previous requested quality, thus it may adjust the requested quality for the next segment. The variable set S is then empirically set to be $S = \{b, \bar{w}, pb\}$.

No 4th Variable. Fig. 5 shows the comparison of the performance of the environment variable set with and without the remaining two variables w (upper) and d (bottom). It is clearly shown that neither of these two variables helps to

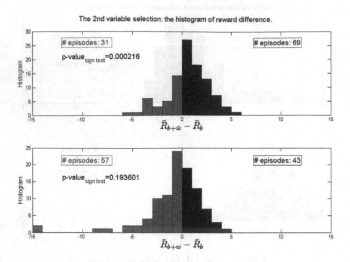

Fig. 3. The 2nd variable selection

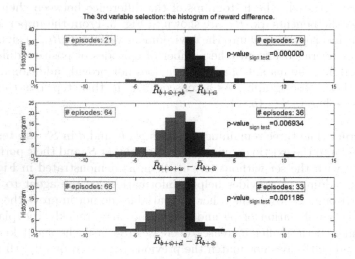

Fig. 4. The 3rd variable selection

increase the reward of RL. So finally, no further variable will be selected and the environment variable set is fixed to be $S = \{b, \bar{w}, pb\}$.

Discussion. The results of our selection of prominent variables can be explained in two-folds. First we confirm that the playback buffer is the most influential

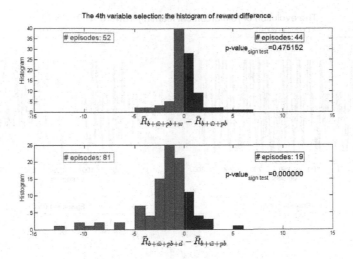

Fig. 5. The 4th variable selection

one, as shown in [9]. Second, we also see that the average bandwidth and the speed of filling/consuming of the buffer also play important roles, helping the RL-learner to obtain higher rewards by clearly stating the environment that the learner stays at. We also notice that $\bar{w}$ is more preferred than w in the selection. This probably could be explained that the HAS client should not be too sensitive to the instant perceived bandwidth, as the playback buffer acts as a container to keep the quality as stable as possible; the quality does not necessarily react instantly to a sudden bandwidth fluctuation. Actually $\bar{w}$, filtered by a low-pass filter, smooths the recent perceived bandwidths, giving the RL learner a clearer perception of the tendency of bandwidth change. Besides, it is somewhat surprising that the tendency monitor d does not benefit to the description of the RL environment. We conjecture that as the penalty of quality oscillation has been given in the reward function, knowing how long the current quality tendency has been maintained does not provide more information.

4.3 Comparison with Microsoft IIS Smooth Streaming

Requested Quality. In section 4.2, we have obtained the optimal environment representation $S = \{b, \bar{w}, pb\}$. To verify the performance of RL-based HAS client, the cross-traffic manager re-generates another series of cross traffic for 500 episodes such that the RL-based client and Microsoft IIS Smooth Streaming can be running on the same bandwidth condition for the same episode and can be compared.

Fig. 6 shows the requested qualities (blue lines)and the corresponding perceived instant bandwidths (the green dashed line, corresponding to the right

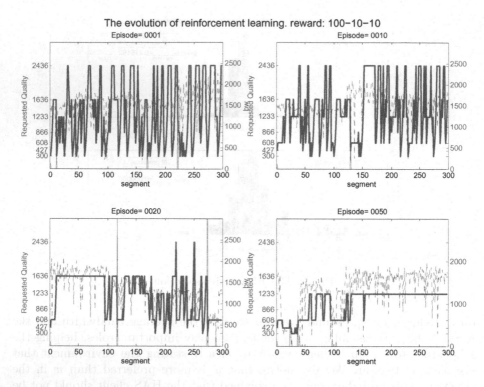

Fig. 6. The evolution of reinforcement learning. Selected episodes: 1, 10, 20 and 50.

vertical axis) for episodes 1, 10, 20 and 50 respectively. The horizontal axis is the index of segment, ranging from 1-299. In the graphs, we also show the normalized picture freeze (the red vertical line. The longer the line is, the longer the freeze is). As can be seen, the requested qualities are very random and irregular in the beginning (episode "0001"). But this randomness, together with oscillation, keeps diminishing as more episodes have been learned. At episode 50, the RL-based client has already learned to follow the fluctuation of bandwidth, which proves the validity of our approach.

Fig. 7 and 8 show respectively the requested quality levels by the RL-based client (the upper graph) and Microsoft IIS Smooth Streaming client (the bottom graph) in episode 490 and 491. Visually comparing with IIS Smooth streaming, both of figures demonstrate that the RL-based HAS client could adapt, without too sensitively, to the rapidly fluctuated bandwidth, achieving quite stable request list. Meanwhile, the RL-based client is also well protected from freeze, especially when the bandwidth is extremely low.

Various Weight Settings in Reward Function. Fig. 7 and 8 show the superior performance achieved by the RL-based HAS client, with the weights of picture freeze, quality and oscillation in the reward function being $A = 100$, $B = 10$ and $C = 10$ respectively. These QoE related parameters are intuitively

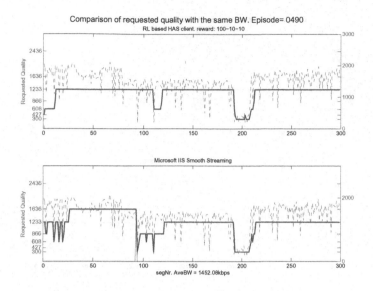

Fig. 7. Requested quality for RL based HAS client and Microsoft IIS Smooth Streaming. Episode: 490

Table 2. Average rewards with multiple parameter sets. Episode: 400-499. $A - 100$

Ave Reward		oscillation cost: C					
RL / IIS		2	6	10	14	18	22
quality cost: B	2	-4.9/ -6.3	-5.2/ -6.5	-5.3/ -6.8	-5.5/ -7.1	-5.6/ -7.4	-5.7/ -7.6
	6	-13.7/-16.3	-14.3/-16.5	-14.7/-16.8	-14.9/-17.1	-15.2/-17.4	-15.5/-17.6
	10	-22.5/-26.3	-23.2/-26.5	-23.5/-26.8	-23.9/-27.1	-24.5/-27.4	-24.5/-27.6
	14	-31.4/-36.3	-32.1/-36.5	-32.4/-36.8	-32.8/-37.0	-33.1/-37.2	-33.6/-37.6
	18	-40.2/-46.3	-41.0/-46.5	-41.3/-46.8	-41.7/-47.1	-42.2/-47.3	-42.6/-47.6
	22	-48.8/-56.1	-49.8/-56.5	-50.1/-56.8	-50.7/-57.1	-51.0/-57.3	-51.4/-57.6

chosen: the reason that the weight of freeze is 10 times bigger than the other two is that normally picture freeze leads to much worse QoE. However, it is still uncovered how these factors are combined to simulate the QoE perceived by a human. While looking for the mapping from QoE to the reward function is an interesting but out-of-scope topic, our target is to show that as long as the "optimal" QoE-oriented reward function is defined, the RL-based HAS client could automatically learn the "optimal" requesting strategy. In order to do this, we run the RL-based HAS client in multiple times with different weight sets of the reward function, still assuming that they are linearly combined. As a reference, the reward obtained by the RL learner with a set of parameters is then compared with the reward of the standard Microsoft IIS Smooth Steaming, pretending that the standard one were using the same parameter set for the reward. To this end, we fix the weight A for freeze to be 100, and alternatively

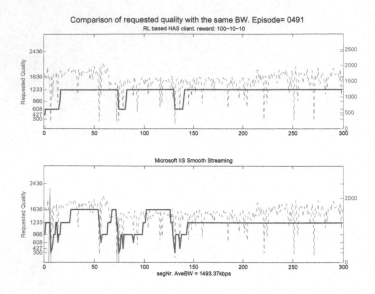

Fig. 8. Requested quality for RL based HAS client and Microsoft IIS Smooth Streaming. Episode: 491

change the weights of quality B and oscillation C, ranging from 2 to 22. The average rewards of episode 400-499 with different parameter sets for both types of clients are shown in Table 2. Note that for IIS, the requested quality sequence of an episode is independent of the concerned parameters A, B and C, thus the "achieved reward" is quite regular, like an arithmetic sequence. Nevertheless it can be seen that within this grid, the average rewards of the RL-based HAS client are uniformly higher than the ones obtained by Microsoft IIS Smooth Streaming. This implies that given a reward representation, the RL-based HAS client could learn to maximize the reward incrementally; if the perceived QoE can be represented as a reward function, then by maximizing the reward the reinforcement learning should be able to optimize the QoE progressively.

5   Conclusions and Future Work

In this paper, reinforcement learning was employed in the HAS client to demonstrate its robustness and adaptability for fast changing network conditions. Specifically, we identified the most influential factors to represent the RL environment, using the forward feature selection strategy. We then compared the requests given by the RL-based client and a standard HAS client and showed that the RL can be an alternative method for the bit-rate selection algorithm. We conclude that as long as the designed reward function, as an objective function, could map the true QoE, reinforcement learning should be able to learn the optimal request strategy in its trial-and-error procedure.

In the future, we will investigate the feasibility of other potential factors for the RL environment, such as recent requested quality bit-rates, the first/second order derivatives of play buffer, etc, without excessively expanding the state space. Some parameters, like the frequency of freeze, could also be incorporated into the reward function. One ongoing study is the co-operation of multiple RL agents for quality fairness in the framework of HTTP Adaptive Streaming.

References

1. Microsoft, Smooth streaming (2008),
   http://www.iis.net/downloads/microsoft/smooth-streaming (accessed July 2013)
2. Pantos, R., May, W.: HTTP live streaming overview (2012),
   http://tools.ietf.org/html/draft-pantos-http-live-streaming-10 (accessed July 2013)
3. Adobe, HTTP dynamic streaming: Flexible delivery of on-demand and live video streaming (2010),
   http://www.adobe.com/products/hds-dynamic-streaming.html (accessed July 2013)
4. Huang, T.-Y., Handigol, N., Heller, B., McKeown, N., Johari, R.: Confused, timid, and unstable: Picking a video streaming rate is hard. In: ACM Internet Measurement Conference, pp. 225–238 (November 2012)
5. Akhshabi, S., Anantakrishnan, L., Dovrolis, C., Begen, A.: What happens when HTTP adaptive streaming players compete for bandwidth? In: ACM NOSSDAV, pp. 9–14 (June 2012)
6. Liu, C., Bouazizi, I., Gabbouj, M.: Rate adaptation for adaptive http streaming. In: ACM MMSys, pp. 169–174 (2011)
7. De Cicco, L., Mascolo, S., Palmisano, V.: Feedback control for adaptive live video streaming. In: ACM MMSys, pp. 145–156 (February 2011)
8. Claeys, M., Latré, S., Famaey, J., Wu, T., Van Leekwijck, W., De Turck, F.: Design of a Q-learning-based client quality selection algorithm for HTTP adaptive video streaming. In: Proc. Conference on Autonomous Agents and Multiagent Systems (May 2013)
9. Huang, T.-Y., Johari, R., McKeown, N.: Downton abbey without the hiccups: Buffer-based rate adaptation for http video streaming. In: ACM FhMN (to appear, August 2013)
10. Sutton, R.S., Barto, A.G.: Reinforcement learning: an introduction. MIT Press, Cambridge (1998)
11. Mok, R., Chan, E., Chang, R.: Measuring the quality of experience of HTTP video streaming. In: Proc. IFIP/IEEE International Symposium on Integrated Network Management (IM), pp. 485–492 (May 2011)
12. Balachandran, A., Sekar, V., Akella, A., Seshan, S., Stoica, I., Zhang, H.: Developing a predictive model of quality of experience for internet video. In: ACM SIGCOMM (to appear, August 2013)
13. Vriendt, J.D., Vleeschauwer, D.D., Robinson, D.: Model for estimating qoe of video delivered using HTTP adaptive streaming. In: Proc. IFIP/IEEE Workshop on QoE CENTRIC Management (May 2013)
14. Hossfeld, T., Egger, S., Schatz, R., Fiedler, M., Masuch, K., Lorentzen, C.: Initial delay vs. interruptions: Between the devil and the deep blue sea. In: 2012 Fourth International Workshop on Quality of Multimedia Experience (QoMEX), pp. 1–6 (July 2012)
15. Click modular router (2010), http://read.cs.ucla.edu/click/click (accessed July 2013)

Stixel on the Bus: An Efficient Lossless Compression Scheme for Depth Information in Traffic Scenarios

Qing Rao[1], Christian Grünler[1], Markus Hammori[1], and Samarjit Chakraborty[2]

[1] Daimler AG, Research and Development
Benz-Str. Gate 16, 71063 Sindelfingen, Germany
[2] Technische Universität München
Arcisstr. 21, 80333 München, Germany
{qing.rao,christian.gruenler,markus.hammori}@daimler.com,
sarmajit@tum.de

Abstract. The modern automotive industry has to meet the requirement of providing a safer, more comfortable and interactive driving experience. Depth information retrieved from a stereo vision system is one significant resource enabling vehicles to understand their environment. Relying on the stixel, a compact representation of depth information using thin planar rectangles, the problem of processing huge amounts of depth data in real-time can be solved. In this paper, we present an efficient lossless compression scheme for stixels, which further reduces the data volume by a factor of 3.3863. The predictor of the proposed approach is adapted from the LOCO-I (LOw COmplexity LOssless COmpression for Images) algorithm in the JPEG-LS standard. The compressed stixel data could be sent to the in-vehicle communication bus system for future vehicle applications such as autonomous driving and mixed reality systems.

Keywords: stixel, lossless compression, LOCO-I.

1 Introduction

The growing demands for safety, comfort and interactivity in vehicles are motivating manufacturers and suppliers in the automotive industry to search for new sensor and vehicle architecture solutions. In recent years, the development of camera sensor techniques and stereo vision algorithms has enabled real-time retrieval of depth information. This has spawned new applications in the field of robotics, video games and driving assistance systems, and raises the issue of how to most effectively represent and utilize the huge amount of depth data generated as a result of these applications.

Stixel [1–4], a compact representation of a depth map, is an effective means of modeling depth information in traffic scenarios. The term "stixel" is the combination of "stick" and "pixel", indicating a thin rectangular area on the image, inside which the pixels have the same or similar depth properties, as shown in Fig. 1.

C. Gurrin et al. (Eds.): MMM 2014, Part I, LNCS 8325, pp. 568–579, 2014.

Fig. 1. Example of stixel representation in a typical traffic scene [3]. Distances to front objects are encoded by different color with red being close and green being far away. Ground area remains with its original gray scale intensity. The white rectangle frame without filling color indicates placeholder stixel for ground, which is introduced by us to fill the gap inside a stixel column in vertical direction, so that the adjacent constraint which will be explained later in subsection 2.1 can be satisfied.

Although a considerable amount of research has been conducted on both modeling and compression of depth information, a sophisticated combined solution has not yet been developed either in research or in the industry. In this paper, we present an efficient lossless compression scheme for stixel data. Our specific goal is to transmit compressed stixel data through in-vehicle communication bus systems such as CAN [5] or FlexRay [6], so that the depth information is available to multiple electronic control units (ECUs) in the vehicle's electronic system for use by different applications, such as those pertaining to driver assistance systems. Reduction of the data volume is realized through simplification of the source model and predictive coding. The prediction scheme we have designed regards the stixel column[1] as an atomic unit, resembling the pixel in the context of image compression. Simultaneous minimization of spatial and temporal redundancies in stixels is achieved through adaptation of the predictor of the LOCO-I algorithm [7, 8], a compression algorithm of the JPEG-LS standard for continuous-tone image.

The rest of this paper is structured as follows. In the next subsection, we briefly discuss related work in this area. Section 2 explains the proposed approach in detail, and Section 3 presents the experimental results. We conclude in Section 4 by outlining some directions for future work.

1.1   Related Work

Correlation-based stereo matching is a traditional method for retrieving depth information from stereo vision systems [9] with several efficient real-time implementations [10–12]. Generally, correlation-based methods suffer from blurred boundaries in the resulting depth map, due to the assumption of constant disparities within a local correlation window. The semi-global matching (SGM)

---

[1] The precise definition of stixel column is given in subsection 2.1.

algorithm [13] alternatively solves the stereo matching problem by optimizing a global cost function, taking both pixel-wise matching cost and local smooth constraint into consideration. The outcome of the SGM algorithm has proved to be robust and illumination insensitive. Relying on the low-cost implementation of the SGM algorithm on field programmable gate array (FPGA) [14], real-time computation of disparity images becomes possible.

The basic idea of stixel emerged in [1] with the aim of building a mid-level representation of the dense depth map computed through the SGM algorithm. Taking into account that in common traffic scenarios, objects such as cars, pedestrians and building facades can be represented through vertical planar surfaces, the authors introduced the novel term "stixel" as the atomic element for approximating object surfaces. A stixel represents a thin, standing rectangle area, which has a certain height, width and depth. In the authors early proposals [1, 2], the border between the road and the objects is firstly computed by means of occupancy grid [15]. The constraint that stixels have to "stand" on the ground had to be followed, as shown in Fig. 2. In their most recent proposal [3], this constraint is lifted and the computation of stixel is reformulated as a global segmentation problem. Every column in the disparity image is segmented into two classes: either object or ground. While object segments preserve constant depth value, ground segments tend to have a steady increment by depth, in accordance with the physical change of the road. The segmentation problem is treated as a max a posteriori probability (MAP) problem and is solved through dynamic programming (DP) [16] in real-time.

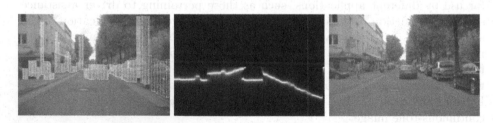

Fig. 2. Early approach of stixel computation [2]. The left image shows the stixel representation where only maximal one stixel appears per image column. The image in the middle shows the border between the ground and the objects standing on it. The green area in the right image represents the free space computed by means of occupancy gird.

In general, a lossless compression scheme consists of two major steps: analyzing the statistical model of the source data, and designing an optimal coding method that minimizes redundancy in the source model. Binary Huffman coding [17] is ideal if the probability of each input symbol is a power of half; Golomb coding [18] is optimal if the source satisfies one-sided geometric distribution. In the field of image compression, the modeling step is commonly realized through a prediction technique, by which image data is effectively decorrelated by using only a few prediction parameters. In the FELICS (Fast Efficient & Lossless

Image Compression System) algorithm [19] for example, the current pixel to be encoded is firstly modeled by the difference between the maximal and the minimal pixels in its closest causal neighborhood. Other lossless image compression algorithms based on predictive coding include the LOCO-I algorithm and the CALIC (Context-based Adaptive Lossless Image Codec) [20] algorithm. The LOCO-I algorithm predicts the current pixel through a simple fixed predictor taking four neighboring pixels as inputs, whereas the CALIC algorithm extends the neighborhood to eight pixels and uses a more complex prediction equation. With respect to both compression ratio and encoding time, the LOCO-I algorithm outperforms other lossless compression algorithms and has become the core of the JPEG-LS standard.

2   Proposed Method

In this paper, the latest multi-layer stixel representation is used for the experiments. The work in [3] results in a spatial redundancy in the stixel data, since neighboring columns in the disparity image are assumed to be independent in order to achieve real-time computation capability. Additionally, since stixels represent objects or environments in real world traffic scenarios, where motions are considered to be continuous, there is a certain temporal redundancy between the stixels in adjacent disparity image frames. In our approach, we reduce the spatial as well as the temporal redundancy.

In what follows, we explain in detail how our proposed method models and removes the redundancies in the stixel data. In subsection 2.1, we discuss the mathematical model of the source data. Subsequently, we explain the prediction workflow and the mechanism of entropy coding in subsection 2.2 and 2.3.

2.1   Source Data Model

A *Stixel* S is denoted as the tuple $S(w, u, d, v^t, v^b)$. The five parameters of S indicate the width, the horizontal coordinate of the center point, the disparity, the top and the bottom boundary of the "stick" rectangle, respectively. In the multi-layer stixel representation, width of a stixel is constrained to a fixed value. In this case, the stixel tuple can be degenerated to $S(\hat{u}, d, v^t, v^b)$, where $\hat{u} = \lfloor \frac{u}{w} \rfloor$ can be understood as the ordinal number of the stixel along the horizontal direction. To simplify notation, the hat symbol of $\hat{u}$ will be omitted in the following text.

A *Stixel Column* $C = S_0 S_1 \ldots S_{n-1}$ is a sequence of stixels ordered from top to bottom in the same column. The vertical boundary coordinates of stixels that belong to the same stixel column satisfy $v_0^t < v_0^b < v_1^t < v_1^b < \cdots < v_{n-1}^t < v_{n-1}^b$, where $n \in \mathbb{N}$ indicates the number of stixels inside the column. Additionally, we introduce the vertical adjacent constraint $v_{i-1}^b = v_i^t - 1, i = 1 \ldots n - 1$ to further simplify the stixel model. In case that a "ground area" appears vertically between two "objects areas", as shown in Fig. 1, a placeholder stixel S_g will be added in order to force the adjacent constraint to be valid. Relying on the vertical adjacent constraint, the stixel tuple can be further degenerated since only one of the parameters v^t and v^b is necessary. In the following text, v^t will be omitted and v^b will be denoted as v.

A *Stixel Frame* $\mathcal{F}$ is a sequence of stixel columns $\mathcal{F} = \mathcal{C}_0 \mathcal{C}_1 \ldots \mathcal{C}_{N-1}$ ordered from left to right, where $N \in \mathbb{N}$ denotes the number of stixel columns inside the stixel frame. Since the stixel width w is predefined, the number of stixel columns can be determined by $N = \lfloor \frac{W}{w} \rfloor$, where W indicates the width of the disparity image.

It is possible to differentiate two stixel columns if their word lengths are identical, i.e. if they have the same number of stixels. The difference between them is referred to as *Residual Column* and represented as $\mathcal{R} = \Delta d_0 \Delta v_0 \ldots \Delta d_{n-1} \Delta v_{n-1}$, where $\Delta d_i \in \mathbb{Z}, i = 0 \ldots n - 1$ indicates the difference of the corresponding stixel disparities and $\Delta v_i \in \mathbb{Z}, i = 0 \ldots n - 1$ the difference of stixel boundaries.

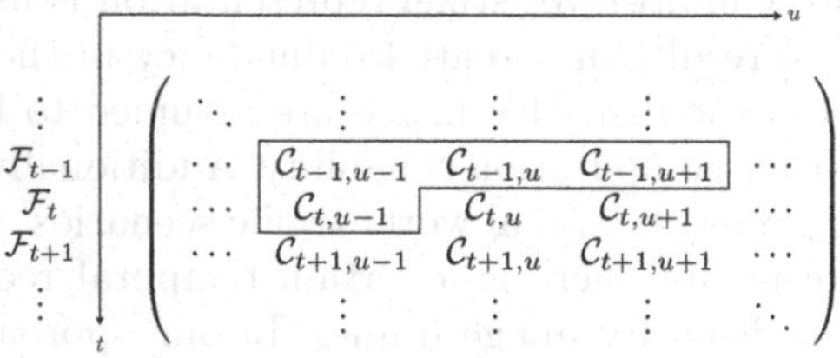

$$
\begin{array}{c}
\\
\mathcal{F}_{t-1} \\
\mathcal{F}_t \\
\mathcal{F}_{t+1} \\
\\
\end{array}
\left(
\begin{array}{ccccccc}
\ddots & & \vdots & \vdots & \vdots & & \\
\cdots & & \mathcal{C}_{t-1,u-1} & \mathcal{C}_{t-1,u} & \mathcal{C}_{t-1,u+1} & \cdots & \\
\cdots & & \mathcal{C}_{t,u-1} & \mathcal{C}_{t,u} & \mathcal{C}_{t,u+1} & \cdots & \\
\cdots & & \mathcal{C}_{t+1,u-1} & \mathcal{C}_{t+1,u} & \mathcal{C}_{t+1,u+1} & \cdots & \\
& & \vdots & \vdots & \vdots & & \ddots \\
\end{array}
\right)
$$

Fig. 3. Temporal-spatial neighborhood matrix of stixel columns. Each row of the matrix represents a stixel frame at certain time point. The four entries inside the black box, including three stixel columns $(\mathcal{C}_{t-1,u-1}, \mathcal{C}_{t-1,u}, \mathcal{C}_{t-1,u+1})$ in the previous frame and one stixel column $\mathcal{C}_{t,u-1}$ in the current frame, construct a causal reference area for predicting the current stixel column $\mathcal{C}_{t,u}$. Alternative reference area containing $\mathcal{C}_{t,u+1}$ instead of $\mathcal{C}_{t,u-1}$, and extended reference area containing columns in $\mathcal{F}_{t-2}$ can also be taken into consideration.

A stixel column correlates with its neighbor columns in both spatial and temporal respect. Let $\mathcal{C}_{t,u}$ be the current observed stixel column, its temporal-spatial neighborhood can be represented by the matrix shown in Fig. 3. The black box inside the matrix shows a causal reference area for predicting the current stixel column.

2.2   Prediction Scheme

We formulate the compression task as follows. The first step involves designing an optimal prediction scheme that minimizes the spatial and temporal redundancies in stixels. One of the challenges in decorrelating the stixel data is the selection of a proper neighborhood for prediction, since stixels on a disparity image do not lie in a rectangular grid as pixels on a normal image. To deal with this problem, we introduce the concept of a reference group, by which similar stixels in a reference area are first grouped. This however leads to the further problem that the number of stixels in a reference group can vary. A flexible predictor is thus necessary to deal with various input combinations. We decided to use the simple predictor of the LOCO-I algorithm for the following reasons. First of all, the prediction equation only takes three neighboring pixels as inputs (the fourth neighboring pixel is used for context modeling) and performs only addition/subtraction operations. This allows us to save a certain amount of computation and memory resource compared to the CALIC predictor. Furthermore, the LOCO-I predictor provides us with the possibility of detecting depth discontinuity between stixel columns, which appears to be a common case in traffic scenarios, so that the prediction quality can be improved.

To make the notation consistent with LOCO-I, the left reference column in the current frame will be denoted as $\mathcal{C}_A$, whereas the reference columns in the previous frame as $\mathcal{C}_C, \mathcal{C}_B$ and $\mathcal{C}_D$ from left to right. The current stixel column will be denoted as $\mathcal{C}_X$ and the predicted stixel column as $\mathcal{C}_P$.

Reference Group. We use the following expression to measure the dissimilarity of stixels:

$$\delta\left(\mathcal{S}_1, \mathcal{S}_2\right) = \lambda_1 |d_1 - d_2|^2 + \lambda_2 |l_1 - l_2|^2 + \lambda_3 \left(|v_1^b - v_2^b|^2 + |v_1^t - v_2^t|^2\right) \quad (1)$$

where $l_i = v_i^b - v_i^t, i = 1, 2$ indicates the length of stixel. The first term in Eq. 1 regards the disparity difference, while the second and the third term take the length difference and the coordinate difference of boundary into consideration. The three parameter λ_1, λ_2 and λ_3 are introduced to control the weight of the three terms and are set to satisfy $\lambda_1 > \lambda_2 > \lambda_3$. In other words, we consider disparity value to be the most significant indicator to measure the stixel dissimilarity and the boundary coordinates to be less important. If $\delta\left(\mathcal{S}_1, \mathcal{S}_2\right)$ is below a certain threshold τ_s, the two stixels $\mathcal{S}_1$ and $\mathcal{S}_2$ are considered to be similar and grouped together.

Simple Prediction. From each reference group, a stixel is predicted. We use the simple predictor in LOCO-I algorithm to predict disparity value d_P, as expressed in Eq. 2. The predictor tends to choose (i) d_A if a temporal discontinuity (a huge gap between d_A and d_C) is detected; (ii) d_B if a spatial discontinuity (a huge gap between d_B and d_C) is detected; (iii) the general prediction $d_A + d_B - d_C$ if no discontinuities are detected.

To predict boundary values v_P, the assumption that real world objects in common traffic scenarios have clear and sharp edges is made. Under this assumption, similar stixels in a neighborhood tend to form basic geometric shapes, such as large rectangles or trapezoids [4], and the missed boundary value can be efficiently predicted through linear interpolation. Although in practice this assumption is not always accurate due to noises in stixels, it is still a reasonable approach to predicting boundary values.

$$d_P = \begin{cases} \min(d_A, d_B), & \text{if } d_C \geq \max(d_A, d_B) \\ \max(d_A, d_B), & \text{if } d_C \leq \min(d_A, d_B) \\ d_A + d_B - d_C, & \text{otherwise} \end{cases} \qquad (2)$$

To deal with the problem of varying number of reference stixels, we classify the reference group according to the combination of reference stixels and assign a confidence score to each class. For example, the reference group containing three stixels from column $\mathcal{C}_A, \mathcal{C}_B$ and $\mathcal{C}_C$ is considered to be more reliable than the reference group merely containing one stixel from $\mathcal{C}_D$. The confidence score will be used for the final prediction step: the competitive fusion.

For a reference group containing less than three stixels, the predicted stixel will be straightforwardly given by the expectations of the disparities and the boundary coordinates, since the LOCO-I predictor requires at least three inputs. However, the confidence score of the predicted stixel will be typically low in this case.

Competitive Fusion. Two predicted stixels have to compete if they overlap each other vertically with more than τ_v pixels. The one with higher confidence score wins the competition. At last, all highly confident candidates are fused together back into a stixel column regarding the vertical adjacent constraint.

After experimenting with different combinations of the parameters, we choose $\lambda_1 = 4$, $\lambda_2 = 2$, $\lambda_3 = 1$, together with $\tau_s = 300$ and $\tau_v = 12$, which yields the best average compression factor. Fig. 4 illustrates the entire prediction workflow.

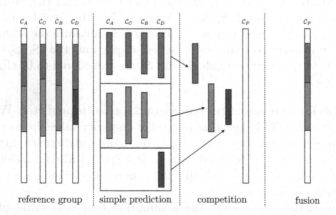

Fig. 4. Example of the prediction scheme. At first, stixels in the reference columns are grouped into three stixel groups. The simple prediction procedure will then be executed on every group, yielding three stixel candidates. The green stixel and the blue stixel have to compete since they are overlapping each other. The green one wins the competition for its reference group is more confident than the reference of the blue one. At last, the red one and the green one are fused together into a stixel column.

2.3   Entropy Coding

If the stixel numbers of the predicted column C_P and the current column C_X are identical, i.e. $n_P = n_X$, the encoder switches to *Residual Mode* and encodes the residual column with a modified version of Golomb code similarly as in the LOCO-I algorithm. Golomb coding uses a single parameter $m \in \mathbb{N}^+$ which divides the input symbol $x \in \mathbb{N}$ into two parts: the quotient q which is encoded using unary code, and the remainder r which is encoded using modified binary code. Theoretically, Golomb code is optimal if the source follows one-sided geometric distribution [21], i.e. $P(x) = (1 - \theta)\theta^x$ where $0 < \theta < 1$ denotes the parameter that affects the shape of the distribution curve. For a certain parameter θ, there exists an optimal Golomb divisor $m = \lceil -\log(1 + \theta) / \log \theta \rceil$ which minimizes the average code length. In practice, the divisor is usually chosen to be $m = 2^k, k \in \mathbb{N}$, so that the encoding process can be implemented more efficiently through binary arithmetic [8, 22].

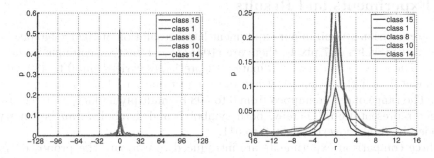

Fig. 5. Probability distributions of disparity residuals of different reference classes. The disparity residual follows approximately two-sided geometric distribution. The mapping $x \leftarrow 2x, x \geq 0$ and $x \leftarrow -2x - 1, x < 0$ is used to transform the source back to be one-sided.

To encode the residual stixel column, we use different Golomb divisors for different reference classes, since each reference class shows its particular statistical properties, as presented in Fig. 5 and Tab. 1. The Golomb divisor for a certain reference class is updated adaptively based on the "one liner" implementation in [7]. Furthermore, the Golomb divisors for disparity residual Δd and boundary residual Δv are calculated separately as well.

If the word lengths of C_P and C_X are not identical, the encoder switches to *Rescue Mode*. A stixel in rescue mode is modeled by its disparity d and its length l and will be simply binary encoded. A stixel column is still represented by the sequence of stixels ordered from top to bottom, only additionally with the number of stixels inside the column. Furthermore, the horizontal ordinal numbers u of all false predicted stixel columns are recorded and encoded in a header structure of the current stixel frame, in order to let the decoder be able to know on which column it should run in rescue mode.

Table 1. Statistical properties of the most frequent reference classes. Class index is related to the combination of reference stixels and assigned in a binary coding manner. Class frequency and residual entropy are calculated from 10, 000 stixel frames. Notice that nearly 15% of the stixel references belong to class 1, which contains only one stixel from column C_D and yields high residual entropy. This happens if objects on the disparity image move in the opposite direction against the predictor, i.e. from right to left. However, the predicted stixel from class 1 will in most cases be dropped during the competitive fusion step due to low confidence score.

Class index	C_A	C_B	C_C	C_D	Class frequency	Entropy of disparity residual
15	×	×	×	×	0.2691	2.1517
1				×	0.1490	6.5906
8	×				0.0968	4.7662
10	×		×		0.0760	4.7946
14	×	×	×		0.0744	2.7650

3   Experiments and Results

The stereo vision system in our experiment has a resolution of 1024×440 pixels. With a fixed width of 7 pixels, there are $\lfloor 1024/7 \rfloor = 146$ stixel columns inside a frame. The stixel boundary coordinate v theoretically varies from 0 to 439 with increment 1 and the disparity d has the range from 0 to 127 with increment 0.25. The range of d is remapped from 0 to 508 by multiplying factor 4 to ensure integer representation. Thereby, both symbol v and symbol d can be binary encoded by 9 bits covering from 0 to 511.

Three comparison experiments are introduced to evaluate the proposed approach. The *Raw* method refers to binary encoding without any compression mechanism. Method *Space+Time* indicates the approach described in this paper, where the reference stixels for prediction are taken from both spatial and temporal neighborhood. Method *Space* refers to the experiment where only spatial neighbor C_A is used for prediction. Additionally, we also compared our method to the general compression method using library *Zlib* [23], a combination of LZ77 [24] algorithm and Huffman coding. All experiments are implemented with C++ and executed on an Intel Core i7-2620M CPU, 2.7GHz with 4GB RAM. A dataset containing 25,000 stixel frames calculated from real traffic scenario is used for the evaluation and the results are presented in Tab. 2.

In addition to the evaluation from an average point of view, the compression performance on single stixel frame is analyzed as well, as illustrated in Fig. 6. The encoder performance is closely related to the traffic scenarios: The red curve remains low and flat when the test vehicle keeps driving straightforward, whereas sudden impulses appear when unexpected objects pass by or the test vehicle makes a rapid turn.

The bandwidth requirement on the vehicle bus is determined by the peaks in the curve instead of the average value. For example, the bandwidth of the uncompressed data could be greater than $2000B \times 25fps \approx 50KB/s$, which exceeds the maximal payload capability of a CAN bus (34KB/s with 11bit identifier, 30KB/s

Table 2. Experiment results. Column "false prediction rate" shows the ratio of false predicted columns which are encoded in rescue mode. Column "average processing time" shows the average encoding time per frame in *milliseconds*. Real-time capability of the encoder is guaranteed since the computation of dense disparity map only runs at 25 frames per second [3]. However, there are still over 20% false predicted columns with the reference area shown in Fig. 3. A more sophisticated mechanism for selecting reference area will be considered in future work.

Method	Average compression factor	False prediction rate	Average processing time
Raw	1	-	-
Space	3.1242	0.2124	0.272
Space+Time	3.3863	0.2023	2.920
Zlib	2.5063	-	0.267

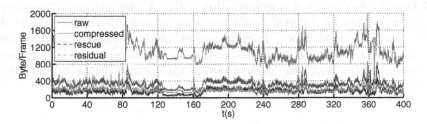

Fig. 6. Data volume along time axis. The magenta curve shows data volume with raw encoding, while the red curve shows compressed data volume. The green and the blue dashed curves illustrate data volume in residual mode and rescue mode respectively.

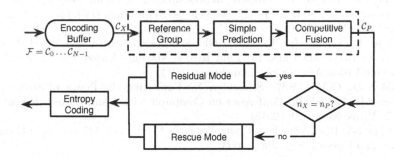

Fig. 7. Workflow of the encoder. At certain time point, a new stixel frame arrives at the encoding buffer. Each stixel column C_X of the frame will then go through the column predictor. If the stixel number of the predicted column C_P and C_X are not identical, the encoder switches into rescue mode. Otherwise, the encoder runs in residual mode.

with 29bit identifier) [25]. Therefore, the more expensive FlexRay with 500KB/s payload capability [25] should be used to transmit the uncompressed data. Although theoretically, the compressed stixel data with 800B × 25fps ≈ 20KB/s bandwidth requirement can already be transmitted through CAN bus, we are still considering introducing combination of lossless and lossy compression to specifically deal with those peaks in future work, since the vehicle bus is actually full of other signals and never complete idle for one particular signal.

4   Concluding Remarks

We present an efficient lossless compression scheme for stixel data. To compress stixel, we have designed the stixel column predictor which adapts the LOCO-I predictor in the JPEG-LS standard. The experiment results show the real-time capability of the proposed encoder with an average compression factor of 3.3863. The compressed stixel data can be theoretically transmitted through CAN bus, while the uncompressed data requires at least FlexRay for transmission.

In the future, we plan to improve the stixel column predictor by experimenting with different reference areas and prediction paths. A combination of lossy and lossless compression for different traffic scenarios could further reduce the peak data volume on the vehicle bus. Transmission of the compressed stixel data will be simulated to make depth information available to different vehicle electronic control units, for the next generation of vehicle applications such as autonomous driving and mixed reality systems.

References

1. Badino, H., Franke, U., Pfeiffer, D.: The Stixel World - A Compact Medium Level Representation of the 3D-World. In: Denzler, J., Notni, G., Süße, H. (eds.) Pattern Recognition. LNCS, vol. 5748, pp. 51–60. Springer, Heidelberg (2009)
2. Pfeiffer, D., Franke, U.: Efficient Representation of Traffic Scenes by Means of Dynamic Stixels. In: IEEE Intelligent Vehicles Symposium (IV), pp. 217–224. IEEE Press, New York (2010)
3. Pfeiffer, D., Franke, U.: Towards a Global Optimal Multi-Layer Stixel Representation of Dense 3D Data. In: 22nd British Machine Vision Conference. British Machine Vision Association, Manchester (2011)
4. Pfeiffer, D., Gehrig, S.K., Schneider, N.: Exploiting the Power of Stereo Confidences. To Appear. In: Conference on Computer Vision and Pattern Recognition. IEEE Press, New York (2013)
5. ISO 11898-1: Road Vehicles – Controller Area Network (CAN) – Part 1: Data Link Layer and Physical Signaling (2003)
6. FlexRay Consortium: FlexRay Communications System-Protocol Specification, Version 2.1, Revision A (2005)
7. Weinberger, M.J.: LOCO-I: A Low Complexity, Context-based, Lossless Image Compression Algorithm. In: Data Compression Conference, pp. 140–149. IEEE Press, New York (1996)
8. Weinberger, M.J., Seroussi, G., Sapiro, G.: The LOCO-I Lossless Image Compression Algorithm: Principles and Standardization into JPEG-LS. IEEE Transactions on Image Processing 9(8), 1309–1324 (2000)
9. Scharstein, D., Szeliski, R.: A Taxonomy and Evaluation of Dense Two-frame Stereo Correspondence Algorithms. International Journal of Computer Vision 47, 7–42 (2002)

10. Hirschmüller, H.: Improvements in Real-time Correlation-based Stereo Vision. In: IEEE Workshop on Stereo and Multi-Baseline Vision, pp. 141–148. IEEE Press, New York (2001)
11. Hirschmüller, H., Innocent, P.R., Garibaldi, J.M.: Real-time Correlation-based Stereo Vision with Reduced Border Errors. International Journal of Computer Vision 47, 229–246 (2002)
12. Ding, J., Du, X., Wang, X., Liu, J.: Improved Real-time Correlation-based FPGA Stereo Vision System. In: International Conference on Mechatronics and Automation, pp. 104–108. IEEE Press, New York (2010)
13. Hirschmüller, H.: Accurate and Efficient Stereo Processing by Semi-global Matching and Mutual Information. In: IEEE Computer Society Conference on Computer Vision and Pattern Recognition, vol. 2, pp. 807–814. IEEE Press, New York (2005)
14. Gehrig, S.K., Eberli, F., Meyer, T.: A Real-time Low-power Stereo Engine Using Semi-global Matching. In: Fritz, M., Schiele, B., Piater, J.H. (eds.) ICVS 2009. LNCS, vol. 5815, pp. 134–143. Springer, Heidelberg (2009)
15. Badino, H., Mester, R., Vaudrey, T., Franke, U.: Stereo-based Free Space Computation in Complex Traffic Scenarios. In: IEEE Southwest Symposium on Image Analysis and Interpretation, pp. 189–192. IEEE Press, New York (2008)
16. Bellman, R.: Dynamic Programming. Princeton University Press, Princeton (1957)
17. Huffman, D.A.: A Method for the Construction of Minimum-Redundancy Codes. Proceedings of the IRE 40(9), 1098–1101 (1952)
18. Golomb, S.W.: Run-length Encodings. IEEE Transactions on Information Theory 12(3), 399–401 (1966)
19. Howard, P.G., Vitter, J.S.: Fast and Efficient Lossless Image Compression. In: Data Compression Conference, pp. 351–360. IEEE Press, New York (1993)
20. Wu, X., Memon, N.: Context-based, Adaptive, Lossless Image Coding. IEEE Transactions on Communications 45(4), 437–444 (1997)
21. Gallager, R., Voorhis, D.V.: Optimal Source Codes for Geometrically Distributed Integer Alphabets. IEEE Transactions on Information Theory 21(2), 228–230 (1975)
22. Liu, Z., Qian, L., Bo, Y., Li, H.: An Improved Lossless Image Compression Algorithm LOCO-R. In: International Conference on Computer Design and Applications, vol. 1, pp. (V1)328–(V1)331. IEEE Press, New York (2010)
23. Zlib Home Page, http://www.zlib.net/
24. Ziv, J., Lempel, A.: A Universal Algorithm for Sequential Data Compression. IEEE Transactions on Information Theory 23(3), 337–343 (1977)
25. Zimmermann, W., Schmidgall, R.: Bussysteme in der Fahrzeugtechnik: Protokolle und Standards (2. Auflage) (German Edition). Vieweg Verlag, Wiesbaden (2007)

A New Saliency Model Using Intra Coded High Efficiency Video Coding (HEVC) Frames

Matthew Oakes and Charith Abhayaratne

Department of Electronic and Electrical Engineering, The University of Sheffield,
Sheffield, S1 3JD, United Kingdom

Abstract. The computation of visual attention is an exhaustive procedure to locate conspicuous regions within a frame, which contrast with the surrounding background. In this paper we propose a unique algorithm to estimate visual saliency in the compressed domain using intra-coded frames from High Efficiency Video Coding (HEVC) encoded video sequences. By exclusively combining data obtained from the coding unit structure, intra mode block predictions and the residual data, a visual saliency approximation is obtained. The proposed model can accurately detect salient regions without the need to fully decode the HEVC bitstream. Experimental results show the proposed algorithm compares positively against multiple methods in the literature, highlighting accurate saliency detection with minimal time additions to the video coding computation. The new methodology can provide aid to a wide variety of fields such as advertising, watermarking, video editing and spatial-temporal adaptation.

1 Introduction

Visual saliency is a subjective perceptual characteristic determining human focus toward stimulating regions within media. Most state-of-the-art saliency models rely upon the early feature integration theory [1]. The study theorises that object saliency is characterised from a parallel combination across multiple characteristics, such as, the contrast in orientation, intensity and colour. Itti proposed the first major accepted saliency model [2] based upon the early Treisman foundations, which adopts a center-surround multi resolution analysis approach. Riche [3] proposed combining orientation contrast with statistical probabilities to determine salient features. Due to the iterative nature of this method, realtime saliency estimation is unattainable. Ngau [4] predicts salient features by detecting colour deviation within the LL subband. Despite computational efficiency, this method provides a deficient saliency estimation relying solely upon variations within the luma and chroma channels.

The computation of visual attention is an exhaustive procedure. By incorporating a saliency model from directly within the video codec, significant computational savings are attainable. Kim proposes region of interest (ROI) based compression for sport videos [5], where the ROI is determined as a pre-process to video coding. Significant improvements could be made by determining the scene saliency parallel to the video coding process. Visually stimulating regions arise from the presence of conspicuous scene abnormalities, which can originate from multiple feature mechanisms. Neural stimuli are extremely sensitive to contrasts within scene brightness or orientation, which

C. Gurrin et al. (Eds.): MMM 2014, Part I, LNCS 8325, pp. 580–592, 2014.

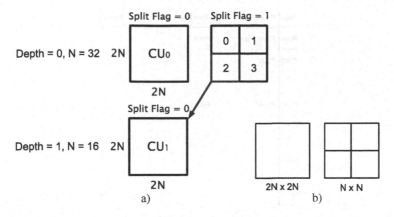

Fig. 1. a) CU partitioning b) Intra mode PU block partitioning

largely contribute to low level attentiveness. This is highly relatable within the modern video codecs, such as, Advanced Video Coding standard (H.264/AVC) [6] and High Efficiency Video Coding (H.265/HEVC) [7]. In HEVC codec, much of the coding gains have been obtained by using extensive analysis of frames for optimising the predictions. HEVC dissects each video frame to determine various coding decisions based upon the analysis of the frame characteristics. Some coding coding decisions involved in intra coding include a flexible quad-tree block size partitioning scheme, based on the regions homogeneousness, and a scene orientation approximation using arbitrary directional intra (ADI) prediction.

Our previous work detects salient regions from the uncompressed video domain, suitable for applications such as digital watermarking. [8] [9] The aim of this paper is to explore such intra coding decisions and other data available in HEVC bitstreams to estimate visual saliency in frames. In this paper, we explore the intra coding decisions and prediction errors (usually known as residuals) to propose an HEVC compressed saliency model. A major advantage of our proposed compressed domain saliency algorithm, is saliency maps can be generated without fully decoding the HEVC bitstream. For existing methodologies [2–5, 10, 11], HEVC bitstream data would first have to be fully decoded before a saliency model is applied. Therefore, our proposed model is highly suitable towards any application or device limited by low computational complexity, such as mobile phones.

The rest of the paper is arranged as follows. Section 2 shows an overview of the HEVC codec and describes some of the new features within the standard. Section 3 and Section 4 illustrate the proposed saliency estimation algorithm and consequent experimental results, respectively. Lastly, the concluding remarks are drawn within Section 5.

2   HEVC Coding Structure Overview

There is an increasing need to compress data efficiently as the demand for higher quality media distribution rises. The H.264/AVC standard was released in 2003, [6] nearly a

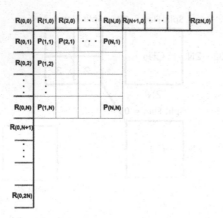

Fig. 2. Reference samples $R(x, y)$ used to predict prediction samples $P(x, y)$ for an $N * N$ block

decade ago, so further advancements within video codec structures are required to cope with the expansion of high definition media storage. The HEVC codec main intention is to increase the video coding gain by the implementation of content adaptive prediction schemes. [7] Compared with its H.264/AVC predecessor, HEVC can attain bitrate savings of up to 50% while maintaining the same video quality. [7]

The overall structure of HEVC bears resemblance to classical hybrid video codec architecture. However, there are numerous improvements compared with the H.264/AVC predecessor. New key features include an adaptive motion parameter prediction, an enhanced binary coding arithmetic scheme (CABAC) and an improved intra coding scheme, which is described in more detail in Section 2.2.

2.1 Block Partitioning

HEVC employs an improved, more flexible quad-tree block partitioning structure within the prediction and transform stages of the codec. This ensures a more efficient block partitioning scheme as a greater number of block sizes are available. The HEVC block structure is determined by 3 main components; coding unit (CU), prediction unit (PU) and transform unit (TU).

The CU contains a hierarchial structure, defining a maximum block depth of 4 sizes. The quadtree syntax allows for splitting the CU into an appropriate size dependant upon the region characteristics by signalling a split flag. The CU is further decomposed into a $PART\_2Nx2N$ or $PART\_NxN$ PU which can comprise of 4, 8, 16 and 32 block sizes only. Fig. 1a) and Fig. 1b) describe the CU and PU partitioning, respectively.

The block partitioning structure is highly related to the quantisation parameter (QP), used to encode each frame. To obtain an overall low output bit-rate, a high QP value greatly compresses the residual data. Less block partitioning occurs to ensure minimal data is required for encoding.

2.2   Intra Prediction Mode

HEVC incorporates a novel arbitrary directional intra (ADI) prediction scheme to reduce residual errors, which arise from inaccurate block estimations in previous coding
standards. Angular intra prediction is performed by interpolating the reference points
surrounding each intra block. The best approximation from 33 differing angular predictions plus 2 additional modes from a DC average or Planar prediction is chosen for the
overall block prediction. [12]

Each prediction mode utilizes the same set of reference points, $R_{x,y}$, which are depicted in Fig. 2. DC mode provides a suitable block estimation when predicting flat
homogeneous regions. The prediction samples, $P_{x,y}$, for an N by N block are calculated from a zero order equation, described by:

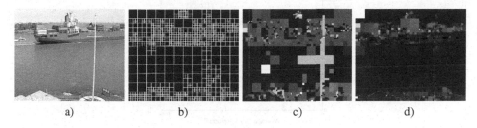

a)   b)   c)   d)

Fig. 3. a) Original frame b) PU block structure c) ADI prediction mode number (0-34) d) Average
block residual

$$P_{x,y} = \frac{\sum_{x=1}^{N} R_{x,0} + \sum_{y=1}^{N} R_{0,y}}{2N}. \tag{1}$$

Planar prediction provides a suitable block approximation for smooth sample surfaces
within homogeneous regions, without containing any prominent edge boundaries. [13]
The prediction comprises from a combination of 2 first order linear equations, $P1_{x,y}$
and $P2_{x,y}$, using 4 reference sample points:

$$P_{x,y} = P1_{x,y} + P2_{x,y} >> 1, \tag{2}$$
$$P1_{x,y} = L_{x,y}(R_{N,0} + R_{0,y}),$$
$$P2_{x,y} = L_{x,y}(R_{0,N} + R_{x,0}),$$

where $L_{x,y}$ is the linear interpolation between the respective reference samples.

3   Proposed Method

To establish the prospect of a visually salient block, numerous features must be exploited from within the HEVC codec. The main method behind our saliency estimation
comprises of 3 elements, namely: block structure, intra mode difference and residual
energy. These features are highly relatable to visual stimuli mechanisms such as intensity and orientation contrast. The block structure is complementary to salient features

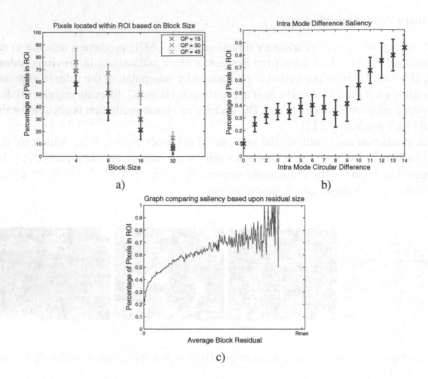

Fig. 4. Saliency correlation with a) block size b) intra mode difference c) average block residual

as the split flag in Fig. 1a) is signalled based upon the presence of a high colour or intensity contrast. Intra modes determine a suitable orientation approximation for the scene. By locating the presence of inconsistent variations within the determined intra modes, conspicuous regions can emerge. Large residuals arise when an accurate block prediction cannot be formulated, usually from the presence of peculiar block patterns. By pinpointing these block abnormalities, potential salient regions can be detected.

Firstly, we establish the relationship of these coding modes with the visual saliency by subjectively labeling the visual saliency of each of the pixels in the frames and analysing the corresponding coding modes. The experimental setup consists of 1000 different frames comprised from the publicly available MSRA-1000 database by Liu et al [14]. Subsequent ground truth ROI frames, governed by the outcome of subjective testing, have been manually created as part of the MSRA-1000 database. Section 3.1, Section 3.2 and Section 3.3 analyse the saliency relationship with each of the coding modes, whereas the new model is proposed in Section 3.4.

3.1 Block Size

Based on the PU block structure, a saliency estimation can be made. Fig. 3a) and Fig. 3b) show the original frame and corresponding PU block structure, respectively. The QP value partially determines the PU structure of the frame and must be

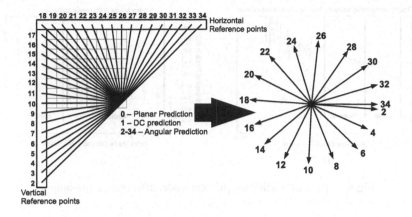

Fig. 5. Mapped circular difference modes

considered within our methodology. For lower QP values, an increased amount of block partitioning occurs. A relationship between block size and visual attention is derived from the results shown in Fig. 4a). Pixels located within the manually segmented ROI are compared with the partitioned block dimensions. A clear positive correlation can be established between the smaller block sizes and salient regions, from which block size saliency probability, S_{bl}, can be determined. The block size partitioning is determined by the level of homogeneous activity within each block, as greater partitioning occurs under the influence of an overall high block variance. Homogeneous regions within a frame portray low saliency characteristics so we expect larger block sizes to be less salient than finely partitioned areas.

3.2 Intra Mode Differences

By modelling the inconsistencies within angular intra prediction modes, an estimation to locate salient features can be performed. For example, a textured background, solely predicted from ADI modes 16 and 17, could contain prominent foreground objects depicted from a various combination of directional modes. By locating mode irregularities, salient conspicuous regions can be identified. Modes 2 and 34 are highly correlated so simple difference calculation between adjoining partitions is insufficient. In Fig. 5 we have re-mapped each prediction mode to a circular plain, where modes 34 and 2 are adjacent.

We define m_d as the absolute difference between consecutive modes $m1$ and $m2$, i.e $m_d = |m1 - m2|$. Let $\ominus$ represent the circular difference between 2 adjacent modes, which is calculated by:

$$m1 \ominus m2 = \begin{cases} m_d \text{ if } m_d \leq M/2 \\ M - m_d \text{ if } m_d > M/2, \end{cases} \qquad (3)$$

where M is the maximum range of possible modes, i.e 34-2. Modes 0 and 1 are omitted from our algorithm as they bare negligible resemblance towards angular prediction.

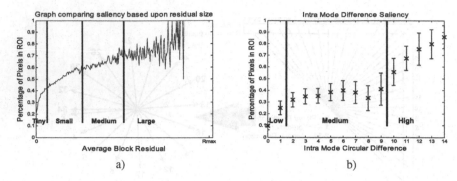

Fig. 6. a) Residual partitioning b) Intra mode difference partitioning

Fig. 3c) shows an intra mode map, derived from the ADI prediction mode choices, described in Fig. 5. Our method determines the circular difference between the horizontal and vertical consecutive modes, at block boundaries. If the intra mode map is given by $I_{i,j}$, the horizontal and vertical circular differences, $D_{i,j}^H$ and $D_{i,j}^V$, at the block boundaries can be determined by:

$$D_{i,j}^H = |I_{i,j} - I_{i+1,j}| + |I_{i,j} - I_{i-1,j}|,$$
$$D_{i,j}^V = |I_{i,j} - I_{i,j+1}| + |I_{i,j} - I_{i,j-1}|, \tag{4}$$

respectively. The block boundary differences are integrated over the entire block, in both horizontal and vertical directions, before linear combination into the absolute intra mode difference map, S_{int}, by Eq. (5):

$$S_{int} = (\int_0^N D_{i,j}^H \, dx + \int_0^N D_{i,j}^V \, dy)/4. \tag{5}$$

A correlation linking ADI prediction mode circular differences against ROI is shown in Fig. 4b). The graph shows a positive relationship between the location of high intra mode circular differences and salient data. Previous studies show neural stimuli are extremely sensitive towards orientation contrast creating visual saliency. [1] [2] The ADI prediction determines an overall suitable orientation which best defines each block, from which highly contrasting adjacent modes can decipher salient portions within the frame. Likewise, an abundance of neighbouring modes of similar orientation are highly probable to part of a common visually uninteresting region or background.

3.3 Residual Data

Residual data is the prediction error between the original frame and the block estimation, so low residual values arise when accurate block estimations are made. We exploit a relationship between the quantised residual energy and visual saliency to determine overall block saliency. Fig. 3d) shows the average quantised residual energy present within each block from the encoded frame, S_{res}. Fig. 4c) shows the positive correlation depicting the normalised residual energy of each block with the ROI, within the range $0 - R_{max}$, where R_{max} is the maximum possible residual value. From the graph, partitioned blocks containing a higher residual energy are more likely to be part of a visually salient region. High residual energy arises from large inaccuracies between the intra mode prediction and the original frame. These imprecisions can originate from obscure, salient regions, which are difficult to predict. Therefore it is highly probable that if a block contains large residual data it is also very likely to be visually salient.

3.4 Saliency Map Generation

Criteria for predicting visually salient blocks within the HEVC codec, can be determined. There is a high probability a block will be visually salient if it: is partitioned into a small size, accommodates a high average residual energy and contains large orientation inconsistencies between the surrounding prediction modes. The individual respective feature maps are combined into an overall saliency map using a refinement process based upon Bayesian probabilities [15]:

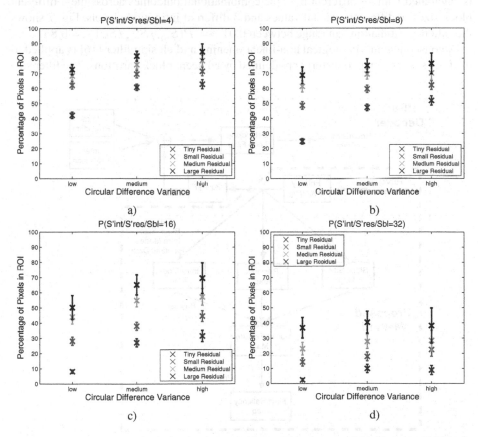

Fig. 7. Graphs showing $P(S'_{int}/S'_{res}/S_{bl})$ for all 48 possible outcomes a) block size = 4 b) block size = 8 c) block size = 16 d) block size = 32

Adjacent intra mode circular difference and residual data portraying similar saliency characteristics are classified into a smaller number of bins. 4 and 3 differing bin levels are used for the residual and intra mode difference maps, respectively. Thresholds are set for the residual and intra mode difference maps in Eq. (6) and Eq. (7), respectively, as follows:

$$S'_{res} = \begin{cases} Tiny & \text{if } S_{res} \leq R_{max} * 0.05, \\ Small & \text{if } R_{max} * 0.05 \geq S_{res} < R_{max} * 0.25, \\ Medium & \text{if } R_{max} * 0.25 \geq S_{res} < R_{max} * 0.50, \\ Large & \text{if } S_{res} \geq R_{max} * 0.50, \end{cases} \tag{6}$$

$$S'_{int} = \begin{cases} Low & \text{if } S_{int} \leq 1, \\ Medium & \text{if } 2 \geq S_{int} < 10, \\ High & \text{if } S_{int} \geq 10, \end{cases} \tag{7}$$

where S'_{res} and S'_{int} are the reclassified residual and intra mode difference data, respectively. Fig. 6a) and Fig. 6b) describe how the the data is partitioned. $P(S'_{int}/S'_{res}/S_{bl})$ is calculated with 48 different possible combinational outcomes across the 4 different block sizes, 4 different residual values and 3 different intra mode levels. Fig. 7 shows the saliency estimation can range between $0.01 >= P(S'_{int}/S'_{res}/S_{bl}) <= 0.85$.

A consequent morphological non-linear opening and closing filter [16] is applied to the final saliency map to distort gaps defined by adjacent block partitions. The filter size

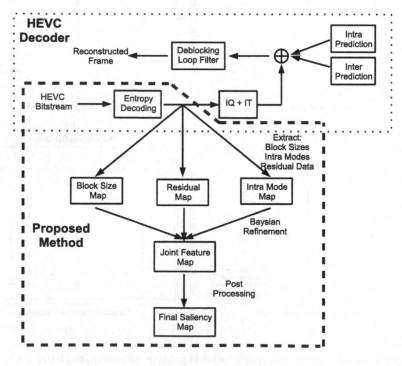

Fig. 8. Overall model diagram

needs to be substantial enough to connect gapping regions without completely distorting the saliency shape. An equation describing the final saliency map, S_T, is:

$$S_T^{(x,y)} = \psi.(P^{(x,y)}(S'_{int}/S'_{res}/S_{bl})), \tag{8}$$

where ψ is a morphological filtering operator. The overall system diagram is shown in Fig. 8.

4 Experimental Results

Fig. 9 compares our method to 3 existing state-of-the-art techniques, with a manually segmented ground truth frame. By subjective assessment, our model can accurately match the performance of the Itti, Ngau and RARE models, in Fig. 9b), Fig. 9c) and Fig. 9d), respectively, in terms of accurate saliency estimation. Fig. 9f) highlights the key salient areas computed by our model within the original frame.

Subjective assessment alone is not enough to justify the validity of results. Receiver operating characteristic (ROC) curves are shown in Fig. 10a) and Fig. 10b) to portray an objective evaluation of each saliency model. [17] Fig. 10a) displays the ROC curve for the corresponding frame in Fig. 9, whereas results across the entire MSRA-1000 database are shown in Fig. 10b). Table 1 shows the corresponding ROC Area Under the Curve (AUC) and average computational time per frame, for each model. The table values are computed as an average over all 1000 frames in the MSRA database and the average frame computational time, for each saliency model, does not include decoding the bitstream data. From the graph in Fig. 10b) and table our model is shown to have the second highest performance, accurately detecting salient regions, behind the RARE model. The major drawback of the RARE model is the exhaustive iterative procedure requires a large computation time to process each frame, in comparison to other existing techniques. The high complexity constraints deem the model highly unsuitable for any video saliency applications. The bottom row in Table 1 describes our proposed method having the second lowest computational complexity. The Ngau model, despite being capable of rapid scene analysis, performed the worst of all models when estimating salient regions. The simple algorithm searches for frame regions containing extreme coefficients and is limited to saliency estimation upon very basic scenes.

The main advantage of the proposed model is data required for saliency estimation is extracted directly from within the HEVC codec and re-mapped in a basic learning algorithm. As a result, only minimal added computational costs occur in excess of entropy decoding the HEVC bitstream, so our model can attain an accurate saliency estimation while maintaining a very low overall computational cost. This joint characteristic ensures our model is highly suitable for video saliency estimation. The other methods in comparison are pixel domain based so full HEVC bitstream decoding must be performed before the saliency algorithm is applied.

More results from our model are shown in Fig. 11 with the corresponding original images and segmented ground truth frames. The scenes are taken from the MSRA-1000 database, used throughout this work. The table portrays highly comparable ground truth frames and saliency estimation generated from our methodology.

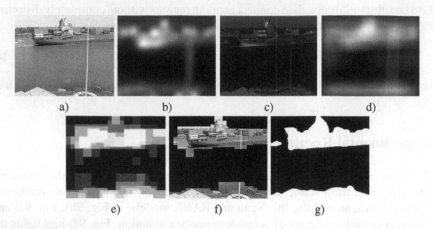

a)   b)   c)   d)

e)   f)   g)

Fig. 9. a) Original frame b) Itti model [2] c) Ngau model [4] d) RARE model [3] e) Proposed model f) Thresholded original frame using proposed model g) ground truth frame

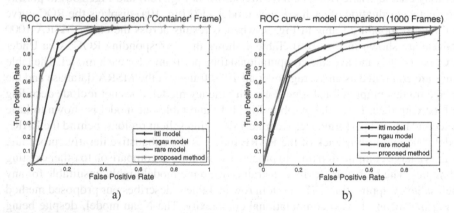

Fig. 10. ROC curves comparison to existing state of the art methodologies for - a) frame from 'Container' sequence b) entire MSRA-1000 database

Table 1. Saliency performance evaluation and computational time comparison of proposed and existing models

	itti model [2]	ngau model [4]	rare model [3]	proposed method
ROC AUC	0.807	0.766	0.874	0.827
Computational time (sec) [1]	0.311	0.059	6.572	0.129

---

[1] Computational time does not account for any HEVC bitstream decoding.

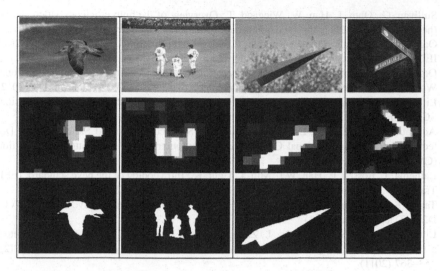

Fig. 11. Proposed method results taken from the MSRA-1000 database (Row 1: Original image, Row 2: Saliency regions using the proposed model and Row 3: Ground truth)

5 Conclusions

A novel HEVC domain visual saliency algorithm was presented within this work by uniquely exploiting features within the coding standard. Using partially decoded HEVC sequences from encoded intra frame data, the block partition size, residual energy and intra mode differences are combined using Bayesian probability refinement to constitute a definitive saliency model. Unlike existing methodologies, the HEVC bitstream does not need to be fully decoded to attain an accurate saliency estimation. The proposed method outperforms existing methodologies in terms of accurate saliency detection and computational time.

References

1. Treisman, A.M., Gelade, G.: A feature-integration theory of attention. Cognitive Psychology 12(1), 97–136 (1980)
2. Itti, L., Koch, C., Niebur, E.: A model of saliency-based visual attention for rapid scene analysis. IEEE Trans. on Pattern Analysis and Machine Intelligence 20(11), 1254–1259 (1998)
3. Riche, N., Mancas, M., Gosselin, B., Dutoit, T.: Rare: a new bottom-up saliency model. In: Proceedings of the International Conference on Image Processing (IEEE ICIP 2012), Orlando, USA, pp. 1–4 (2012)
4. Ngau, C., Ang, L., Seng, K.: Bottom-up visual saliency map using wavelet transform domain. In: 2010 3rd IEEE International Conference on Computer Science and Information Technology (ICCSIT), vol. 1, pp. 692–695 (July 2010)
5. Kim, J., Yi, C., Kim, T.: Roi-centered compression by adaptive quantization for sports video. IEEE Trans. on Consumer Electronics 56(2) (May 2010)
6. Wiegand, T., Sullivan, G.J.: The h.264/avc video coding standard. IEEE Signal Processing Magazine, 148–153 (March 2007)

7. Sullivan, G., Ohm, J., Han, W., Wiegand, T.: Overview of the high efficiency video coding (hevc) standard. IEEE Trans. on Circuits and Systems for Video Technology (99), 1 (2012)

8. Oakes, M., Bhowmik, D., Abhayaratne, C.: Visual attention-based watermarking. In: 2011 IEEE International Symposium on Circuits and Systems (ISCAS), pp. 2653–2656 (2011)

9. Oakes, M., Abhayaratne, C.: Visual saliency estimation for video. In: 2012 13th International Workshop on Image Analysis for Multimedia Interactive Services (WIAMIS), pp. 1–4 (2012)

10. Li, Z., Fang, T., Huo, H.: A saliency model based on wavelet transform and visual attention. SCIENCE CHINA Information Sciences 53, 738–751 (2010)

11. Achanta, R., Hemami, S., Estrada, F., Süsstrunk, S.: Frequency-tuned Salient Region Detection. In: IEEE International Conference on Computer Vision and Pattern Recognition (CVPR), pp. 1597–1604 (2009)

12. Lainema, J., Bossen, F., Han, W.J., Min, J., Ugur, K.: Intra coding of the hevc standard. IEEE Trans. on Circuits and Systems for Video Technology 22(12), 1792–1801 (2012)

13. Tan, T.K., Kanumuri, S., Bossen, F.: Enhancements to intra coding jctvc-d235. JCTVC-D235, Daegu, Korea (2011)

14. Liu, T., Yuan, Z., Sun, J., Wang, J., Zheng, N., Tang, X., Shum, H.Y.: Learning to detect a salient object. IEEE Transactions on Pattern Analysis and Machine Intelligence 33(2), 353–367 (2011)

15. MacKay, D.J.C.: Information Theory, Inference and Learning Algorithms. Cambridge University Press, New York (2002)

16. Mitra, S., Sicuranza, G.: Region-based filtering of images and video sequences: A morphological viewpoint. In: Nonlinear Image Processing, pp. 249–288. Academic Press (2001)

17. Gide, M.S., Karam, L.J.: Comparative evaluation of visual saliency models for quality assessment task. In: 6th International Workshop on Video Processing and Quality Metrics for Consumer Electronics, pp. 37–40 (2012)

Multiple Reference Frame Transcoding
from H.264/AVC to HEVC

Antonio Jesus Diaz-Honrubia, Jose Luis Martinez, and Pedro Cuenca

Albacete Research Institute of Informatics, University of Castilla-La Mancha, Spain
{Antonio.DHonrubia,JoseLuis.Martinez,Pedro.Cuenca}@uclm.es

Abstract. The emerging video coding standard, so called High Efficiency Video Coding (HEVC), has been recently developed by the ITU-T and JCT-VC groups to replace the current H.264/AVC standard. That standard has been very successful and it has been widely adopted in the last years. Hence, there will be a need for efficient conversion between the H.264/AVC standard to HEVC. In this paper, we present a fast motion estimation mechanism to speed up the transcoding process between H.264/AVC to HEVC. Because HEVC and H.264/AVC share a similar coding architecture, we try to exploit the information gathered in the H.264/AVC decoding algorithm by means of reducing the references frames checked. Experimental results show that the proposed transcoding mechanism can achieve a good tradeoff between coding efficiency and complexity in terms of motion estimation cost.

Keywords: H.264/AVC, HEVC, Transcoding, Reference frames.

1 Introduction

Recent advances in consumer electronics have resulted in a rapidly growing amount of user generated multimedia content, much of it in improved quality, including up to 4K definition video. Modern producing consumers wish to share their experiences and creative output with friends, relatives and like-minded people – nowadays mostly through social networks. Managing this media together with commercial content and the involved devices, protocols and services (for recording, storing, sharing and rendering) in a seamless fashion is a huge challenge. Moreover, how to make video be suitable for various device capabilities and dynamical bandwidths becomes very challenging as well.

Transcoding is one of the most promising technologies, which provides video adaptation in terms of bitrate reduction, resolution reduction or format conversion to meet various requirements. However, the emerging developments in video coding technology make transcoding much more complicated. A new developed video coding standard creates new requirements for transcoding existing formats to the new format for the interoperability of video contents.

Currently, video communication represents about half of the entire network traffic, with tendency for further increase. Therefore, techniques aiming at efficient

C. Gurrin et al. (Eds.): MMM 2014, Part I, LNCS 8325, pp. 593–604, 2014.

compression of video are of paramount importance; an example is how to potentially avoid the "spectrum crunch" which is foreseen due to increasing traffic in mobile networks. Substantial amount of effort have been made in this area during the recent past years, which resulted in the new generation of video compression standard, called High Efficiency Video Coding (HEVC) [1][2]. For similar quality, the current architecture of HEVC only consumes half of the transmission bandwidth of the previous H.264/AVC standard. The compression capability of HEVC establishes a new benchmark both in video and still image coding. Thus, HEVC can be expected as a successor to H.264/AVC. The wide use of the H.264/AVC standard today and the expected adoption of HEVC raises a new demand for AVC to HEVC transcoding. In practical, a video transcoder should make tradeoff between complexity and coding performance while making full use of the input bitstream to generate a new one.

In this paper, we present a novel frame-reference selection algorithm for inter-frame prediction to be used as part of a low complexity heterogeneous H.264/HEVC video transcoder. The proposed transcoding mechanism is based on the reuse of the reference frames gathered from the decoding stage in H.264/AVC. Moreover, a mapping between the HEVC partitions and the appropriate H.264/AVC area is needed. Experimental results show that the proposed transcoding mechanism can achieve a good tradeoff between coding efficiency and transcoding complexity.

The remainder of this paper is organized as follows: Section 2 includes a technical background of the new HEVC standard. Section 3 identifies the related work which is being developed about the topic. Section 4 introduces our proposed transcoding architecture. Experimental results are shown in Section 5. Section 6 concludes the paper and includes the future work.

2   Technical Background

HEVC introduces new coding tools respect to its predecessor H.264/AVC as well as it improves other already used [1][2]; all of them make it possible to notably increase the coding efficiency. One of the most important changes affects to the picture partitioning. HEVC dispense with the terms of Macro-Block (MB) and block for the Motion Estimation (ME) and the transform respectively and introduces three new concepts: Coding Unit (CU), Prediction Unit (PU) and Transform Unit (TU). This structure leads to a more flexible coding to suit the particularities of the frame. Each picture is partitioned into squared regions of variable size called CUs which replace the MB structure of previous standards. Each CU may contain one or several PUs and TUs and its size is limited from 8x8 to 64x64 pixels. To fix the size of each CU, first of all, a picture is divided into 64x64 pixels areas, each of which is called Coding Tree Unit (CTU), and then, each CTU can be partitioned into 4 smaller sub-areas of a quarter of the original area. This partitioning can be done with each sub-area recursively until it has a size of 8x8 pixels. Thus, a quad-tree structure is used, as it can be seen in Fig. 1.

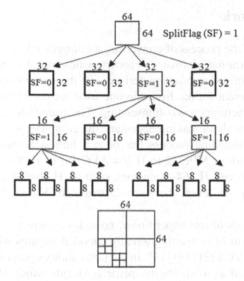

Fig. 1. CU splitting in HEVC

HEVC checks almost all these modes (Inter and Intra) to decide whether it splits a CU/PU/TU or not and it chooses the best Rate-Distortion (RD) case. Furthermore, in the case of inter prediction, for each of these partitions a ME algorithm is called; by default the HM reference software [3] defines the diamond search [4]. This wide range of possibilities makes HEVC to be much more computationally expensive than its predecessor H.264/AVC. As it can be seen in Table 1, the most expensive module in the new HEVC coder is the Inter-Prediction process, which accounts for the 77-81% of the total coding time [5].

HEVC introduces changes in other modules too, such as the Intra Prediction (where a whole of 35 different coding modes can be selected), the ME modes (it introduces Asymmetric Modes), new image filters or new transform sizes. However, since the ME is the most computationally expensive module, this paper is focused in it.

The HM 9.0 reference software introduces four configurations which are defined in [6]: "All Intra", "Low Delay" (with P or B frames) and "Random Access". This paper is focused on "Low Delay" and "Random Access" profiles, since "All Intra" does not carry out the ME.

Table 1. Consumed time by different HEVC modules [5]

Module	Consumed time (%)
Intra Prediction	1-2%
Inter Prediction	77-81%
Transform + Quant.	14-16%
Loop Filter	0,1-0,2%
Entropy coding	2-4%

3   Related Work

Video transcoding is the process of converting a compressed video stream previously encoded with a determinate format or characteristics into another video stream encoded with a different codec or characteristics. In the case of the video sequence is translated into a different format from the one used for encoding the input video sequence is known as heterogeneous transcoding. The transcoding process should perform the conversion without making necessary the complete process of decoding and re-encoding [7]. Transcoding process has been a hottest research topic in the last years in the framework of MPEG2 to H.264/AVC [8] or H.263 to H.264/AVC transcoders [9], also between H.264-extensions such as H.264/AVC to Scalable Video Coding (SVC) [10] or, even, between Distributed Video Coding (DVC) and H.264/AVC [11].

As far as the authors of this paper know, nowadays, there are few approaches that deals with the problem of converting already encoded streams with H.264/AVC into the new standard HEVC [12] [13] [14]. In [12], the authors propose a reuse of Motion Vectors (MVs) as well as to similarity metric to decide which HEVC CU partitions have to be tested. In a similar way, the approach presented in [13] is focusing on reducing the number of CU and PU partitions to be checked by means of an improved Rate-Distortion Optimization (RDO) metric.

In [14] a more complete proposal can be found. It combines parallelization at the coder side, taking advantage of the HEVC structures which were created with that propose, and reutilization of information fetched at the decoder side of H.264/AVC. In this second part of the proposal, it uses the frame resolution in order to restrict the quadtree splitting and it reuses the partitioning modes in order to select the available PU modes in each case.

However, a research work based on reusing the reference frames in order to accelerate the transcoder has not been found so far.

4   Proposed Transcoding Architecture

The proposed transcoding architecture is based on the reuse of the reference frames, which have already been chosen by H.264/AVC, in the HEVC encoder in order to reduce the coding time of the transcoder.

The motivation of reusing the reference frames is that the search in each one is very expensive computationally, since each frame has several CTUs and each CTU is recursively partitioned into many CUs and PUs, i.e. each CTU is partitioned in up to $4^0 + 4^1 + 4^2 + 4^3 = 85$ CUs and each one of these is divided in a whole of up to 17 different PUs, so the ME would be performed 1445 times per reference frame per CTU. This algorithm aims to reduce the number of frames where HEVC searches in the ME to just those where H.264/AVC searches in the area covered by the current partition of HEVC. Hence, a mapping between the HEVC partition and the appropriate H.264/AVC area is needed.

The given solution is to split the H.264/AVC MB in 4x4 pixels blocks, since it is the smallest subMB size in that standard and each one can reference a different frame, and assign the reference frame used to each block. Then, the corresponding area of HEVC is delimited by clustering the appropriate number of these blocks and storing each reference frame in a list without repetition of elements (i.e. the list {1, 2, 1} is the same than the list {1, 2}). In the example of the Fig. 2, the final list will contain the reference frames 1 and 2. Then, for each reference frame in HEVC, this list is checked and it will be only taken into account if it is contained in the list. Otherwise, the reference frame will be discarded for the current partition.

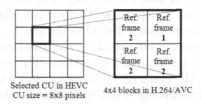

Selected CU in HEVC
CU size = 8x8 pixels   4x4 blocks in H.264/AVC

Fig. 2. H.264/AVC block fitting for a HEVC CU

This process could seem to elaborate a list which always contains all the reference frames due to the high amount of 4x4 blocks that fit in a CU, i.e. a CU of 64x64 pixels has $(64/4)^2 = 256$ blocks. However, a preliminary study has been carried out and it shows that this algorithm, on average and using 4 reference frames (the default amount in HM for "Low Delay" configurations), the 82.7% of the times that a PU encoding task is performed, it is only used 1 reference frame, while using 2, 3 and 4 frames the 14.3%, 2.5% and 0.5% of the times. This result could be explained by the fact that if a 4x4 block uses frame x as reference, the probability that the adjacent ones use the same frame x is very high due to the spatial correlation in both frames. Fig. 3 shows the architecture of the overall proposed transcoder.

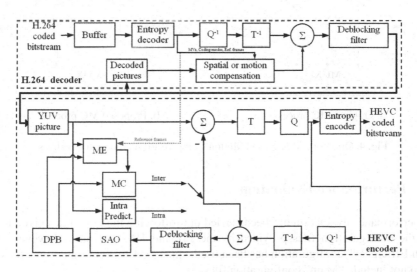

Fig. 3. Architecture of the proposed transcoder

This algorithm is easily extensible when two reference lists are present (i.e. for B slices); the solution is to follow the previously described algorithm twice: one for the List 0 (L0) and other for the List 1 (L1). This proposal will be referred in the remainder of this paper to as "Skip Reference Frames" (SRF).

However, the problem appears, specially, when using the "Random Access" configuration, since most frames only use two references in each list, which reduces the theoretical maximum time saving to a 50% in the accelerated loop. Changing default configuration in order to force each frame to have four reference frames might be a simple solution; nevertheless, it implies an excessive complexity increase. However, some frames, after being coded with H.264/AVC, discard the L1 to carry out the Motion Compensation (MC), so, this information can be used to reduce the complexity.

Fig. 4.a shows the original MC process for B slices. The proposed solution is not to calculate the Interpolated MC in the HEVC coder (as shown in Fig 4.b) if the L1 has been disabled for the corresponding H.264/AVC area. Disabling also the Backward MC could cause a greater loss in RD terms, so it has discarded to do it. This proposal will be referred to as "Skip Interpolated MC" (SIMC) in the remainder of the paper.

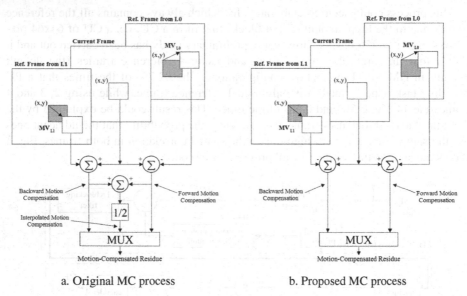

a. Original MC process   b. Proposed MC process

Fig. 4. Original and proposed Motion Compensation process for B slices

5   Performance Evaluation

The performance evaluation has been carried out following the guidelines given in [6] with the configuration files provided by the JM and HM software. However, only "High Efficiency" configuration has been used, since HM-like configurations files for JM do not include "Main" configuration files.

Firstly, all sequences have been coded with H.264/AVC using JM 18.4 [15]. Then, the resulting files have been transcoded using the JM 18.4 decoder and the HM 9.0 [3] encoder. This process has been carried out with the transcoder proposed in this paper and with the reference transcoder, which consists on decoding the H.264/AVC stream and recode it with the original HEVC coder.

For the reference transcoder and using "Low Delay P", "Low Delay B" and "Random Access" configurations, results shows that the ME module consumes the 80% of the HEVC coding time in average and the remaining modules only the 20%, as it can be seen in Fig. 5 (detailed for each configuration), which strengthens the motivation given in Section 2 about the need to speed-up this module.

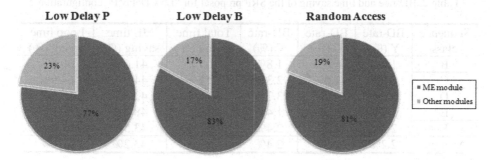

Fig. 5. Time consumed by the ME module in HEVC for each configuration

5.1   Results for Skip Reference Frames (SRF) Proposal

Results for the SRF proposal and for "Low Delay P", "Low Delay B" and "Random Access" configurations, in terms of BD-rate (the average bitrate increment) [16] and coding time saving for the *Total* coding time (including all HEVC modules), the *ME* module and the accelerated *Loop*, with respect to the reference transcoder, are shown in Table 2, Table 3 and Table 4. These results are the average values for all the sequences in a same class (a whole of 24 sequences). The accelerated *Loop* is the loop in the code where all the possible reference frames are checked.

It can be seen in Table 2 that the proposed transcoder for the "Low Delay P" configuration can reduce the *Total* coding time in a third and that it maintains the same quality than the reference transcoder while incrementing the bitrate only in a 2.2% for the Y component. The results for E class are especially significant, since this configuration is very suitable for videoconference sequences (like E class ones) and it achieves a 40.7% of the *Total* time reduction while incrementing only a 1.4% of the bitrate at the same quality. In addition, it can be seen that the *ME* time is reduced in an average of a 43% and that the target *Loop* is accelerated in a 62.6%, which is a pretty close value to 75%, the maximum theoretical value (as was expected because of the use of only 1 reference frame in the 83% of times that a CU is coded, as shown in Section 4).

Furthermore, it can be seen in Table 3 that for "Low Delay B" configuration, the BD-rate is similar to the previous one and the saved time is somewhat lower. This is due to the fact that, now, there are a whole of 8 possible reference frames for each 4x4 block, besides the Interpolated MC. These results remain good though, since the *Total* time saving is a 18.8%, while the BD-rate is only a 2.3% for the Y component and a 1.6% and a 1.5% for U and V components respectively.

However, Table 4 shows that, with a higher losing (it remains low though), the acceleration is much lower because of using only two reference frames for each list and the increase of candidates due to the use of two lists as described in the previous paragraph. As indicated before, a possible solution is to use the SIMC proposal, whose results are presented in the next sub-section.

Table 2. BD-rate and time saving of the SRF proposal for "Low Delay P" configuration

Sequence class	BD-rate Y (%)	BD-rate U (%)	BD-rate V (%)	Total time saving (%)	ME time saving (%)	Loop time saving (%)
B	1.9%	1.8%	1.8%	31.7%	41.6%	60.6%
C	2.0%	1.6%	2.3%	32.1%	44.0%	64.9%
D	3.6%	3.7%	3.7%	30.0%	42.6%	64.7%
E	1.4%	1.5%	1.4%	40.7%	48.4%	69.2%
F	2.2%	2.3%	2.4%	34.2%	43.5%	61.4%
Average	2.2%	2.3%	2.4%	33.3%	43.2%	62.6%

Table 3. BD-rate and time saving of the SRF proposal for "Low Delay B" configuration

Sequence class	BD-rate Y (%)	BD-rate U (%)	BD-rate V (%)	Total time saving (%)	ME time saving (%)	Loop time saving (%)
B	2.7%	2.1%	2.1%	20.3%	24.7%	31.5%
C	2.1%	1.3%	1.4%	23.5%	29.5%	37.3%
D	2.9%	1.8%	1.4%	25.9%	33.4%	45.2%
E	1.6%	1.2%	1.0%	11.3%	12.8%	17.2%
F	2.0%	1.3%	1.5%	15.6%	18.6%	25.5%
Average	2.3%	1.6%	1.5%	18.8%	22.7%	30.6%

Table 4. BD-rate and time saving of the first proposal for "Random Access" configuration

Sequence class	BD-rate Y (%)	BD-rate U (%)	BD-rate V (%)	Total time saving (%)	ME time saving (%)	Loop time saving (%)
A	3.6%	1.9%	2.4%	10.6%	13.8%	19.3%
B	3.0%	1.5%	1.6%	8.2%	10.2%	13.6%
C	3.4%	1.8%	1.8%	11.4%	14.5%	20.1%
D	3.8%	1.7%	1.9%	11.7%	15.3%	21.1%
F	1.9%	1.1%	1.3%	6.8%	8.3%	11.4%
Average	3.1%	1.6%	1.8%	9.3%	11.8%	16.1%

5.2   Results for Skip Interpolated Motion Compensation (SIMC) Proposal

In a similar way, Table 5 and Table 6 show the results for the SIMC proposal for "Low Delay B" and "Random Access" configurations. This proposal was originally developed for "Random Access" configuration; nevertheless, its use can be extended to any case where B frames are used, so, results for "Low Delay B" configuration have also been presented (not for "Low Delay P", since it does not include B frames).

Since, the Interpolated MC calculations are out of the loop of the reference frames, the *Loop* time saving present in the previous tables has been replaced in Table 5 and Table 6 for the Interpolated MC (*IMC*) time saving, being the Loop time saving the same than is present in the respective tables in the previous sub-section.

Hence, this proposal obtains a slightly higher BD-rate than the SRF proposal, while the overall time saving increases in an average of a 2.4%, achieving a 21.9% and a 11.0% of saving for "Low Delay B" and "Random Access" configurations. This improvement is due to a 41.7% and a 26.9% respectively of saving in the IMC.

Table 5. BD-rate and time saving of the SIMC proposal for "Low Delay B" configuration

Sequence class	BD-rate Y (%)	BD-rate U (%)	BD-rate V (%)	Total time saving (%)	ME time saving (%)	IMC time saving (%)
B	3.6%	2.7%	2.6%	23.8%	28.8%	48.0%
C	3.1%	2.3%	2.8%	27.2%	34.1%	57.0%
D	4.0%	2.5%	2.5%	30.4%	39.0%	67.7%
E	2.0%	1.5%	1.7%	12.9%	14.6%	21.0%
F	2.6%	1.8%	2.1%	17.6%	21.1%	30.6%
Average	3.1%	2.2%	2.3%	21.9%	30.8%	41.7%

Table 6. BD-rate and time saving of the SIMC proposal for "Random Access" configuration

Sequence class	BD-rate Y (%)	BD-rate U (%)	BD-rate V (%)	Total time saving (%)	ME time saving (%)	IMC time saving (%)
A	3.9%	2.4%	2.8%	11.7%	15.3%	29.1%
B	3.2%	1.6%	1.7%	10.3%	12.7%	25.3%
C	3.9%	2.2%	2.2%	13.4%	17.1%	36.0%
D	4.2%	2.1%	2.5%	14.2%	18.5%	37.1%
F	2.0%	1.2%	1.3%	9.0%	11.1%	21.8%
Average	3.4%	1.9%	2.1%	11.0%	14.0%	26.9%

As commented in Section 4, the Backward MC could have been skipped too. Nonetheless, as expected, skipping it causes a very severe lost in RD in "Random Access" configuration, not compensated with the acceleration obtained (it achieved an average total time saving of the 12.7% while losing a 9.9% of BD-rate for the Y component). Moreover, for the "Low Delay B" configuration, the results are exactly the same than for the SIMC proposal, since the L0 and the L1 contains the same reference frames, so the skipped frame in the L1 can also be found in the L0. This results strengthens the reasoning given of not propose a solution with this mechanism.

5.3   Overall Results

Finally, Fig. 6, Fig. 7 and Fig. 8 show the RD results for "Low Delay P", "Low Delay B" and "Random Access" configurations respectively. Two E class sequences are shown for both "Low Delay" configuration, while two A class sequences are shown for "Random Access", since these are the specially selected classes for each configuration in [14].

As it can be seen, the RD curves are quite similar for the reference and both proposed transcoders (the curves are even overlapped). The negligible drop in RD performance is justified by the reduction in computational complexity, still remaining much better RD results than H.264/AVC, whose curve is far below.

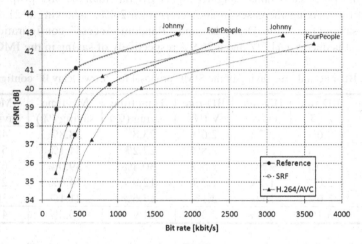

Fig. 6. RD performance for "Low Delay P" configuration for two E class sequences

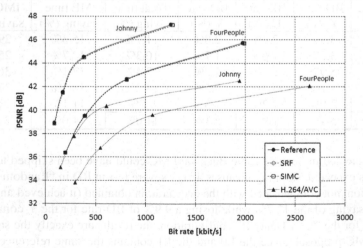

Fig. 7. RD performance for "Low Delay B" configuration for two E class sequences

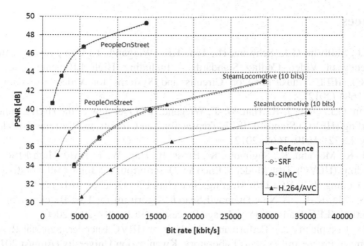

Fig. 8. RD performance for "Random Access" configuration for two A class sequences

6   Conclusions and Future Work

This paper presents a novel transcoding architecture based on fetching the coded reference frames in H.264/AVC to use them in the HEVC encoder. Results show, for "Low Delay P" configuration and SRF proposal, a total time saving of a 33.3%, which involves a 43.2% of time saving in the ME module and a 62.6% of time saving for the accelerated loop. This time saving can be obtained with a negligible RD cost.

However, for "Random Access" configuration, the most appropriate proposal, is the SIMC one, which gets a 11.0% of total time saving, which stands for a 14.0% and a 26.9% of time saving in the ME module and in the Interpolated MC respectively. In a similar way, results show that, for "Low Delay B" configuration, the SIMC proposal gets a total time saving of a 21.9%, which means a 30.8% and a 41.7% of time saving in the ME module and in the Interpolated MC respectively.

As the proposed transcoder obtains better results for "Low Delay P" configurations and for E class, it could be a good transcoder in a video conference scene.

As future work, more coded information in H.264, such as MB modes, could be used to improve this transcoding architecture, opening the possibility of accelerating the CUs partitioning or the decision modes for the TUs.

Reusing the MVs already coded by H.264/AVC could also be studied in order to obtain better predictors for the HEVC ME, but not to accelerate it, since the default algorithm is already very fast, improving the reference transcoder in RD terms.

Acknowledgements. This work has been jointly supported by the MINECO and European Commission (FEDER funds) under the project TIN2012-38341-C04-04.

Likewise, this work has also been supported by the Spanish Education, Culture and Sports Minister under grant FPU12/00994.

References

1. Ohm, J., Sullivan, G.J., Schwarz, H., Tan, T.K., Wiegand, T.: Comparison of the Coding Efficiency of Video Coding Standards – Including High Efficiency Video Coding (HEVC). IEEE Transactions on Circuits and Systems for Video Technology 22(22), 1669–1684 (2012)
2. Sullivan, G.J., Ohm, J.-R., Han, W.-J., Wiegand, T.: Overview of the High Efficiency Video Coding (HEVC) Standard. IEEE Transactions on Circuits and Systems for Video Technology 22(22), 1649–1668 (2012)
3. Kim, I.-K., McCann, K., Sugimoto, K., Bross, B., Han, W.-J.: HM9: High Efficiency Video Coding (HEVC) Test Model 9 Encoder Description. In: Joint Collaborative Team on Video Coding (JCT-VC) 11th Meeting, Shanghai, CN, October 10-19 (2011)
4. Zhu, S., Ma, K.-K.: A New Diamond Search Algorithm for Fast Blockmatching Motion Estimation. IEEE Transaction on Image Processing 9(2), 287–290 (2000)
5. Sim, D.: Complexity and Performance Analysis of HEVC Encoder. Seminar Research Report. Image Processing Systems Laboratory. Kwangwoon University (August 2012)
6. Bossen, F.: Common Test Conditions and Software Reference Configurations. In: Joint Collaborative Team on Video Coding (JCT-VC) of ITU-T SG16 WP3 and ISO/IEC JTC1/SC29/WG11 10th Meeting, Stockholm, SE, July 11-20 (2012)
7. Vetro, A., Christopoulos, C., Sun, H.: Video Transcoding Architectures and Techniques: an Overview. IEEE Signal Processing Magazine 20(2), 18–29 (2003)
8. Fernandez-Escribano, G., Kalva, H., Cuenca, P., Orozco-Barbosa, L., Garrido, A.: A Fast MB Mode Decision Algorithm for MPEG-2 to H.264 P-frame Transcoding. IEEE Transactions on Circuits and Systems for Video Technology 18(2), 172–185 (2008)
9. Fernandez-Escribano, G., Bialkowski, J., Gamez, J.A., Kalva, H., Cuenca, P., Orozco-Barbosa, L., Kaup, A.: Low-Complexity Heterogeneous Video Transcoding Using Data Mining. IEEE Transactions on Multimedia 10(2), 286–299 (2008)
10. Garrido-Cantos, R., De Cock, J., Martínez, J.L., Van Leuven, S., Cuenca, P.: Motion-Based Temporal Transcoding from H.264/AVC-to-SVC in Baseline Profile. IEEE Transactions on Consumer Electronics 57(1), 239–246 (2011)
11. Corrales-Garcia, A., Martinez, J.L., Fernandez-Escribano, G., Quiles, F.J.: Variable and Constant bitrate in a DVC to H.264/AVC transcoder. Signal Processing: Image Communication 26(6), 310–323 (2011)
12. Peixoto, E., Izquierdo, E.: A Complexity-Scalable Transcoder form H.264/AVC to the New HEVC Codec. In: International Conference on Image Processing (ICIP), Orlando, FL, USA (September 2012)
13. Zhang, D., Li, B., Xu, J., Li, H.: Fast Transcoding From H.264/AVC To High Efficiency Video Coding. In: IEEE International Conference on Multimedia and Expo (ICME), Melbourne, Australia, pp. 651–656 (July 2012)
14. Shen, T., Lu, Y., Wen, Z., Zou, L., Chen, Y., Wen, J.: Ultra Fast H.264/AVC to HEVC Transcoder. In: Data Compression Conference, Cliff Lodge, UT, USA, pp. 241–250 (March 2013)
15. Join Collaborative Team on Video Coding, Reference Software to Committee Draft, version 18.4 (2012)
16. Bjontegaard, G.: Improvements of the BD-PSNR model, ITU-T SG16/Q6, 35th VCEG Meeting, Berlin, Germany, Doc.VCEG-AI11, July 16-18 (2008)

Author Index